AN INTRODUCTION TO
TEXTILE BLEACHING

By

J. T. MARSH
M.Sc., F.R.I.C., F.T.I.

THIRD IMPRESSION

LONDON
CHAPMAN & HALL LTD.
37 ESSEX STREET W.C.2
1951

PREFACE

TEXTILE Science is a comparatively modern subject which arose and extended between the two World Wars of 1914 and 1939; this systematic and logical development is quite distinct from the work of the early pioneers to whom every praise is due.

Science originated in craftsmanship and technical pursuits and depended on the evidence of the senses; the practice of science still forms the necessary basis for its abstractions and speculations. It is essential to remember this fact and not to be misled into regarding Applied Science as a rather humble offspring of Pure Science. Science is indivisible, but the applied scientist concerns himself more with the objects and articles of our complex everyday life; the view that the applied scientist awaits the advances of his "pure" colleagues in the Universities, and then applies them, is a perversion.

Among the fundamental technical pursuits of early man were spinning and weaving, and it is rather remarkable that the objects in our most immediate and intimate environment should have escaped our scientific notice until quite recently.

From about 1500 B.C. there were few advances in applied chemistry until the French Revolution and the British Industrial Revolution; this sterile interval of more than 3,000 years, during which theory lagged far behind practice, was largely due to the division of society into the two great classes of workers and administrators. Practical applied science was a technique handed down by oral tradition among the under-privileged, who were unable to read or write and had not the leisure necessary for developing the theory from the practice. According to Lord Bacon, this division accounts for the paradox that the great fundamental discoveries are more ancient than philosophy, and that when contemplation began, the discovery of useful works ceased. Xenophon has stated that the mechanical arts carry a social stigma and are rightly dishonoured; they damage the body of those who work at them, and this physical degeneration results in a deterioration of the soul. Science was dishonoured and therefore weak.

The greater part of our modern industrial life originated with the Industrial Revolution and is based on applied science; without the science the industries could not exist. Our textile industries, however, our greatest single industry, remained fundamentally unaffected, and even to-day spinning and weaving are mechanised domestic pursuits grouped under a factory system in which mechanical power replaces manual labour. The well-known evils of the early textile factory system have left a tradition which is not easily eradicated; textile factories were not conducive to abstract generalisation from the multiplicity of particulars which science requires. The two great textile schools of Mulhouse and Manchester were responsible for most of the organised scientific work in the textile field up to 1914. Inside the factory . . . Theirs not to reason why, theirs but to bleach and dye.

The purification of textile materials, scouring and bleaching, followed a characteristic routine in which secret recipes were handed down from father to son surrounded by such an aura of mystery that it was impossible to approach the subject on a rational basis.

There are thus certain points in common between the scientific stagnation from 1500 B.C. to A.D. 1750, and the lack of scientific interest in textile materials and processes up to 1919 or thereabouts.

One of the greatest obstacles to an intelligent understanding of textile processes was the lack of precise knowledge of the constitution and structure of the textile fibres on which their behaviour depends; descriptive biology was not enough. The increasing use of manufactured chemical compounds, as against natural products, in the textile industries had the result that chemists were inevitably led to some consideration of the material which needed purifying and colouring by the new compounds, and finally the fibres aroused some interest in themselves as objects worthy of study. Through an unconscious type of combined operations, the foundations of Textile Science were laid by the laboratories of some universities, a very few enlightened industrial firms, and the trade research associations under the auspices of the Department of Scientific and Industrial Research of the British Government.

It is highly significant that the greater part of our knowledge of the constitution and structure of fibres has come from university laboratories with their complete freedom of choice of subjects to investigate and liberty to publish their findings; comparatively little, with one or two notable exceptions, has come from industrial laboratories. There seems to have been a tendency for the industry to neglect investigations of the fundamental properties of its own raw materials and to rely on the efforts of chemical manufacturers who are always in close contact with academic science and anxious to pursue the policy of service to the customer even if it necessitates research work on his behalf. Hence it is the former who synthesises fibres.

An important function of the research associations has been to collect and survey the available empirical developments in the textile industry, and to place them on a rational and scientific basis. In an old-established industry with its own traditions and terminology, the technical practice had to be translated into scientific phraseology and expressed with scientific exactitude; in this manner,

PREFACE

knowledge can be properly set down for the benefit of all. Measurements of quality have been devised, and the scientific control of some empirical processes has been responsible for very large annual savings.

A start has therefore been made to put our textile industries on a scientific foundation, but the injection of science into an old-established industry cannot be expected to yield the spectacular results associated with the formation of new industries in which the science and the scientist are welcome and essential features of recognised importance. It is only to be expected that the first bridgehead to be established was that associated with the work of chemists, namely bleaching, dyeing, printing and finishing. However, there are indications that the greatest penetration into the textile field on a scientific basis may come about by an outflanking movement with the regenerated and synthetic fibres, thus leaving the old areas of natural fibres as pockets of resistance which will be absorbed in time, perhaps voluntarily, or perish.

A generation of textile science has advanced on two broad fronts, first, a detailed examination of the characteristic properties of the textile fibres, and secondly, their behaviour in various circumstances.

Although the title of the present work is "An Introduction to Textile Bleaching," it is concerned with the occurrence and properties of the various textile fibres and the methods adopted for their purification; it deals with the first of the two sections of Textile Science outlined above. It is hoped to discuss the second in a subsequent volume on the physical and chemical finishes.

The chief aim of the present book is to help the younger technologists, particularly those just entering the textile industry. It is assumed that the reader will have a fair knowledge of chemistry; nevertheless, every effort has been made to present the subject-matter as simply as possible so that anyone connected with the bleaching, dyeing and finishing sections of the textile industry will find it helpful and informative.

The practice of bleaching, like that of other very old trades, is well known, but the principles on which it rests have not been clarified until comparatively recent times; hence it has been considered preferable to give more prominence to this aspect of the work in the present volume. This is all the more necessary in that there has been a tendency to belittle the value of scientific knowledge applied to bleaching because it has largely shown that the practice resulting from generations of empirical work is well-founded. It would indeed be remarkable if the contrary had been demonstrated, namely that bygone bleachers had failed in their task, for it is on their successful experimentation that modern scientific bleaching rests; this is a particular case of the generalisation that science originated in technical pursuits. Although its ultimate findings are of great value, the method of trial and error has been very costly; conservative bleachers have everything to gain by a close alliance with science. On the other hand, pure science alone may be insufficient for the solution of practical difficulties.

The modern industrial scientist, therefore, is fortunate in being able to draw on the resources and methods of science and of industry, as well as on the accumulated experience of both scientific and industrial practice where these are available to him. It is to record some of these principles and developments that the present work has been written; by understanding the present "state of the art" it is possible to improve our products and so render service to the community. This understanding is also the necessary first step in proceeding from the known to the unknown.

Thanks are due to Dr. Astbury, F.R.S., for a number of X-ray photographs, and also to Miss Alexander and Mr. N. F. Crowder of the Tootal Broadhurst Lee Company, for providing some photomicrographs of textile fibres.

Grateful acknowledgment of the loan of blocks is made to the Society of Dyers and Colourists, the Textile Institute, the *Textile Recorder*, the *Silk Journal and Rayon World*, Messrs. Mather and Platt, Sir James Farmer Norton, Ltd., Archibald Edmeston and Sons, Ltd., Hunt & Moscrop, Ltd., John Mitchell & Sons, Ltd., Samuel Pegg & Son, Petrie & McNaught, Ltd., Sellers & Company (Huddersfield), Ltd., Taylor, Wordsworth & Company, Ltd., and William Whiteley & Sons, Ltd. Courteous permission for the reproduction of illustrations has been received with thanks from the Institution of Mechanical Engineers, Messrs. Thomas Broadbent and Sons, Ltd., John Dalglish & Sons, Daniel Foxwell & Son, Ltd., Jackson and Brother, Ltd., Proctor & Gamble, Spooner Dryer and Engineering Company, Ltd., Tomlinsons (Rochdale) Ltd., and S. Walker, Ltd.

Finally, warmest thanks are given to numerous friends and colleagues with whom various sections of the work have been discussed, and in particular Dr. C. S. Whewell for undertaking the onerous task of reading typescript, galley-slips and page proofs.

J. T. M.

Manchester, 1945

CONTENTS

PART ONE
THE TEXTILE FIBRES

CHAPTER		PAGE
I.	GENERAL INTRODUCTION	1
II.	CELLULOSE FIBRES	5
III.	CELLULOSE CHEMISTRY	22
IV.	SILK	55
V.	WOOL	61
VI.	SYNTHETIC FIBRES	85
VII.	PHYSICAL PROPERTIES OF FIBRES	92

PART TWO
WETTING AND DETERGENCY

VIII.	WATER	102
IX.	ALKALIS	111
X.	SOAPS	119
XI.	WETTING AGENTS	133
XII.	DETERGENCY	144

PART THREE
SCOURING AND BLEACHING OF THE CELLULOSIC FIBRES

XIII.	INTRODUCTION TO COTTON BLEACHING	151
XIV.	SINGEING AND DESIZING	158
XV.	KIERS AND KIERING	167
XVI.	THE SCOURING PROCESS	185
XVII.	THE HYPOCHLORITE BLEACH	200
XVIII.	AROMATIC CHLORAMIDES AND CHLORITE	232
XIX.	GENERAL CONSIDERATIONS	237
XX.	THE PEROXIDE BLEACH	248
XXI.	THE SCOURING AND BLEACHING OF LINEN	262
XXII.	THE SCOURING AND BLEACHING OF RAYON	269
XXIII.	SPECIAL BLEACHING PROCESSES	279

PART FOUR
SCOURING AND BLEACHING OF ANIMAL FIBRES

CHAPTER		PAGE
XXIV.	SCOURING RAW WOOL	296
XXV.	SUBSIDIARY CLEANSING PROCESSES	317
XXVI.	FABRIC SCOURING	327
XXVII.	BLEACHING WOOL	354
XXVIII.	SILK DEGUMMING	371
XXIX.	SILK BLEACHING	401

PART FIVE
THE DRYING OF TEXTILES

XXX.	DRYING TEXTILES	415
XXXI.	DRYING MACHINES	426

PART SIX
TESTS FOR DAMAGE

XXXII.	DAMAGE IN THE CELLULOSIC FIBRES	449
XXXIII.	DAMAGE IN THE PROTEIN FIBRES	477

APPENDIX	493
BIBLIOGRAPHY OF BLEACHING	498
NAME INDEX	500
SUBJECT INDEX	506

PLATES

There are 154 illustrations numbered in one sequence. The 32 plates contain the following figures:

I.	FIGS. 1–4	⎫	
II.	FIGS. 5–6	⎬ *between pages*	8–9
III.	FIGS. 7–8a	⎥	
IV.	FIGS. 9–10a	⎭	
V.	FIGS. 21–22	*facing page*	54
VI.	FIGS. 23–24a		55
VII.	FIGS. 25–27		64
VIII.	FIGS. 28–29		65
IX.	FIGS. 35–36		88
X.	FIGS. 37–38		89
XI.	FIG. 46	⎫	
XII.	FIG. 47	⎬ *between pages*	168–169
XIII.	FIG. 48	⎥	
XIV.	FIG. 49	⎭	
XV.	FIG. 61	*facing page*	182
XVI.	FIGS. 62–63		183
XVII.	FIG. 65		200
XVIII.	FIG. 66		201
XIX.	FIG. 83	⎫	
XX.	FIGS. 84–85	⎬ *between pages*	312–313
XXI.	FIG. 86	⎥	
XXII.	FIG. 87	⎭	
XXIII.	FIGS. 90–91	⎫	
XXIV.	FIG. 92	⎥	
XXV.	FIG. 93	⎬ *between pages*	328–329
XXVI.	FIG. 94	⎥	
XXVII.	FIG. 95	⎥	
XXVIII.	FIG. 96	⎭	
XXIX.	FIGS. 112–114	*facing page*	416
XXX.	FIGS. 115–117		417
XXXI.	FIGS. 138–139		440
XXXII.	FIGS. 140–141		441

All other illustrations are in the text

PART ONE
THE TEXTILE FIBRES

CHAPTER I
GENERAL INTRODUCTION

FUNDAMENTALLY, a fibre is some form of matter which is several hundred times as long as it is wide or thick, but many materials of this type are unsuited to textile use. Other important properties are suppleness and strength.

The upper limit of length in textile fibres seems to be immaterial for threads may be formed by simply twisting the long filaments together; many of the artificial fibres may be made in any desired continuous length. The lower limit, however, is very important and some natural fibres have no textile value because they are too short for spinning into yarns. The limit for spinning appears to be about 5 mm. but on a commercial scale a minimum of about 1 cm. is required; it is interesting to note that this 5 mm. minimum fibre-length for textile yarns is the maximum for use in paper manufacture. Associated with length, is the diameter or fineness of the fibre; the coarser natural textile fibres are about 700 times as long as they are wide, but the finest may be 5,000 times as long as they are wide, and even this figure is greatly exceeded in real silk which may, perhaps, be termed a filament rather than a fibre. The coarse fibres, such as jute, can only be used for low-quality textile products such as sacks, etc.

Textile fibres must also possess sufficient strength, not only to withstand the mechanical operations of spinning and weaving or knitting, but also the various physical and chemical processes of bleaching, dyeing, and finishing; the strength in the wet state is obviously of importance, and some of the early artificial silks were a failure on this account. It appears that the limits of strength are about 1·2 grams per denier in the dry state and 0·7 gram per denier in the wet state. (The tenacity of rayons and some other fibres is estimated in grams per denier; the denier of a yarn is the weight in grams of 9,000 metres. It is based on an old weight, the denier, which is 0·05 gram, and the standard length for assessing the yarn number was 450 metres.)

In addition to good tensile strength, a satisfactory fibre for textile use must possess extensibility, for an inextensible fibre of high strength will be brittle and give poor service in actual wear. If the extensibility is largely elastic, then the fibre is of greater value, for plastic extension is often a disadvantage. Softness and suppleness are also important, for stiff fibres have little textile appeal, although they may be of value as bristles.

The crimp or waviness of the fibres, as distinct from long filaments, is of consequence in determining spinning value, as well as enabling the fibrous mass to retain a porous structure and hold entrapped air which is necessary for thermal insulation and the sensation of warmth; the elasticity of the crimp is an important feature of the most attractive fibre—wool.

The moisture relations of fibres intended for use as clothing require careful consideration, for the ability to absorb and desorb moisture is of great value from the hygienic standpoint; further, a fibre which does not absorb moisture to some extent would become wet by surface condensation and impart an unpleasant clammy sensation when worn next the skin. Most textile fibres are subject to great changes in physical properties as the absorbed moisture varies. The ability to wet with liquid water, as distinct from water vapour, is desirable in order that the material may be bleached and dyed in the processes of manufacture, and cleansed with soap and water in actual use as garments.

The density of a textile material should not be overlooked in assessing the importance of the various properties; if the density is very low, the material will not drape well and fall into graceful folds; on the other hand, a very high density is apt to be accompanied by a heavy, dull and characterless appearance.

Lastly, good textile fibres should be both abundant and cheap; they should also be free from attack by pests and disease.

CLASSIFICATION

There are many different methods of classifying the various textile fibres and one of the simplest is
   (1) Native fibres.
   (2) Manufactured fibres.

From the earliest times, until the beginning of the twentieth century, the whole textile field was contained within the four walls of linen, cotton, wool and silk; the possibility of manufacturing fibres, however, occurred both to Hooke in 1665 and to Reamur in 1734. Early attempts were made in 1884, but no commercial success was achieved until after the First World War (1914–1918) when regenerated cellulose came to the fore.

Natural fibres may be divided into three classes:
   (a) animal
   (b) vegetable
   (c) mineral.

Wool and silk are the chief animal fibres, but in addition to the animal hairs from sheep, useful fibres have been obtained from goats and camels, together with a certain quantity from rabbits. Cultivated silk is second in importance to wool among the animal fibres and may still be regarded as the aristocrat of textile materials.

The chief vegetable fibre is the cotton hair, and indeed, it may be regarded as the foremost textile material; other valuable vegetable fibres include flax, hemp, jute and ramie.

The mineral fibres are only of minor importance and have little textile value; asbestos is used on account of its resistance to heat.

Manufactured fibres may be divided into two chief categories:

(*a*) regenerated
(*b*) synthetic.

The regenerated fibres may again be classified according to the nature of the parent material, which is generally cellulose or protein. The commonest regenerated fibre is viscose rayon prepared from cellulose which has been treated with caustic soda and carbon disulphide, but substantial quantities of rayon are regenerated from solutions of cellulose in cuprammonia. It is also possible to regenerate cellulose from cellulose nitrate, but another derivative, cellulose acetate, forms the basis of acetate rayon; this is not strictly a regenerated cellulose but rather an extruded derivative of cellulose, however, it is possible to form filaments of regenerated cellulose from the acetate.

It has not been possible to regenerate satisfactory fibres from wool, but successful attempts have been made with silk. The greatest technical success has been achieved with casein from milk, although interesting fibres have also been regenerated from soya bean protein, globulins, fish-waste, leather, pea-nuts, and zein or the protein of maize. It must be remembered that regenerated cellulose is made from a cheap source and competes with an expensive native fibre—silk, whereas casein fibre requires a more expensive raw material and competes with a moderately cheap native fibre.

The synthetic fibres are sometimes confused with the regenerated fibres, and the term synthetic has been applied to all manufactured fibres; in actual fact, the regenerated fibres depend on some naturally occurring raw material for their origin and are often degenerated or degraded rather than synthetic. True synthetic fibres do not require raw material of high molecular weight as their starting material but may be built up or synthesised from the simple atoms. Pe Ce was the first synthetic fibre but the nylons are the most famous.

Classification of textile fibres may be arranged on different plans, as for example:

(*1*) Natural cellulose fibres;
(*2*) Natural protein fibres;

(3) Regenerated cellulose fibres, including cellulose acetate;
(4) Regenerated protein fibres;
(5) Synthetic fibres;
(6) Mineral fibres.

An alternative system of classification would be
(a) Cellulose fibres, native and regenerated;
(b) Protein fibres, native and regenerated;
(c) Synthetic fibres;
(d) Mineral fibres.

The relative importance of the various textile fibres may be seen from the list of world production figures for 1939.

PRODUCTION OF TEXTILE FIBRES

Fibre	Quantity
Cotton	13,800,000,000 lb.
Wool (scoured basis)	2,420,000,000 ,,
Flax	1,973,000,000 ,,
Rayon	1,147,280,000 ,,
Staple fibre	1,083,680,000 ,,
Silk	104,000,000 ,,

The general appearance of the chief textile fibres has often been described in various books; a detailed account is given in the well-known work of Matthews—"Textile Fibres," (Wiley, New York, 1947) so that a short description may now be considered adequate.

It is interesting to recall that the term "artificial silk" was replaced by "rayon" in 1924; since the advent of nylon, it has been suggested that the word "synthon" should apply to all synthetic fibres, and "prolon" to regenerated protein fibres.

Although the traditional textile fibres have a biological origin, coming from plants and animals, it is now possible to group them with the manufactured fibres under the one great comprehensive heading of long chain-molecules; the giant molecular-chains are built up from small repeating units in a fundamentally simple manner. As Astbury has remarked, the molecular chains lie along the fibre as the fibres themselves lie along the length of the yarn, so that a fibre may be regarded as a molecular yarn. The macro-structure is a reflexion of the micro-structure.

The nature, length, and grouping of the molecular chains are responsible for various typical physical properties of the fibres, which may be described by the word "toughness." The problem for the bleacher is to purify these textile materials, whether natural or manufactured, without adverse effect on their toughness.

CHAPTER II
CELLULOSE FIBRES

NATIVE CELLULOSE

THERE are two main types of native cellulose, the seed hairs such as cotton, and the bast fibres of which flax, jute, hemp and ramie are examples. The cotton hair is unicellular with a single solid apex, whereas bast fibres are multicellular and consist of completely enclosed tubes, pointed at each end. Another important difference is that the bast fibres actually form part of the plant structure whereas cotton is only a seed hair.

COTTON

The cotton plant grows about four feet in height and takes some five to nine months for growth and maturing. The boll or fruit generally cracks about 48 days after the appearance of the bud and is ready for picking two days later. It is necessary to separate the seeds from the hairs by a process called ginning, but after the removal of the hairs proper, there is still a residue of short fibres and fuzzy undergrowth known as linters which are used for paper, gun-cotton, wadding and rayon manufacture. These linters are removed from the seed by a delinter machine.

The individual cotton hair whilst in growth consists of a long single tubular cell, with one end attached to the seed. It is roughly cylindrical with a lumen or central canal running through it. When the enclosing pod has burst, and the hair is removed from the seed, the cell collapses and becomes a flat ribbon-like structure which, on exposure to light and air, becomes coiled into an irregular spiral band, with from 150–300 twists per inch. The twist is not continuous in any one direction and these convolutions are largely responsible for the spinning properties of cotton.

The hair consists of the central canal or lumen, the secondary thickening and the primary wall. The lumen contains the remains of the protoplasm and near the apex is the nucleus, responsible for the growth of the hair. The lumen also contains the endochrome, which gives the cotton its natural colour. The secondary thickening is composed of cellulose, which has been shown by Balls to be laid down in successive layers corresponding to the annual rings in tree growth. The primary wall or cuticle appears to be a protective layer, which shows spiral fibrils in both quick and slow spirals. The fibrils are more easily seen in old or damaged cotton, and are sometimes reversed in the same hair. It has been suggested that the fibrils act as springs and impart elasticity and flexibility to the cotton hair. With bleached cotton the external cuticle may be absent. Some idea of the anatomical structure of the cotton hair may be seen from its behaviour in swelling agents, such as cuprammonia. When the swollen hair is examined under the microscope, the swelling is not uniform, but appears as a distended tube tied at intervals after the fashion of a string of beads. The annular constrictions are parts of the cuticle which has elsewhere been ruptured by the swelling forces allowing the cellulose in the swollen condition to protrude in the shape of globules. The walls of the central canal can also be seen as the cellulose begins to dissolve in the swelling agent. (See Figs. 5 and 6.)

The dimensions of cotton hairs vary considerably with the different types of cotton, a general average length being one inch, whilst the diameter varies from 0·001 to 0·0005 inch. (This corresponds to 16–20μ; the micron, μ, is 10^{-4} cm.) The longest hairs—about 2·5 inches—have the least diameter. The hair is 3 to 4 times as broad as it is thick. Egyptian cotton is the most regular in length and diameter. The breaking load of the hair lies between the figures for silk and wool and varies from 3·5 to 10 g., but is considerably below either of them in showing only some 4% extensibility. Some varieties of cotton have a fair lustre but in general cotton cannot be considered as a lustrous material. The following data give a broad general view of typical cottons:

CHARACTERISTICS OF TYPICAL COTTONS

	Sea Island	Egyptian	American	Indian
Length	1·8–2·5 ins.	1·4–1·7 ins.	1·08–1·2 ins.	0·9–1·02 ins.
Fineness	1/1600 in.	1/1500 in.	1/1300 in.	1/1200 in.
Feel	soft	soft	fairly soft	harsh
Lustre	silky	fairly silky	fair	poor
Colour	cream	light brown	white	cream
Convolutions per hair	300	228	192	150
Breaking strain in g.	8	7·6	9	3·2

The Fine Structure of Cotton

The most complete account of the growth of cotton has been given by Balls, whose work was mainly confined to Egyptian cotton. The details of his investigations are contained in two books—"The Development and Properties of Raw Cotton" (A. and C. Black, Ltd., London, 1915) and "Studies of Quality in Cotton" (Macmillan, London, 1928), together with a series of papers in the "Proceedings of the Royal Society," 1919, *B90*, 542; 1922, *B93*, 426; 1923, *B95*, 72. Anderson and Kerr of the United States Department of Agriculture have reviewed the growth and structure of cotton, utilising the Mexican variety of American Upland cotton (Ind. Eng. Chem., 1938, *30*, 48).

The cotton hair exists as a single cell, about 1,200 times as long as it is wide, and is attached to the seed only at its base. Each hair originates from the seed coat, and the first evidence of formation is the appearance on the day of flowering of a slight swelling which rapidly elongates on the following day. The diameter of the hair is established soon after it originates, but the elongation continues for some 15–20 days and then suddenly ceases.

The fibre origin is not limited to the day of flowering, but the swellings which appear after the second or third day only produce the linters, whereas the cotton hairs of commercial value all commence their growth within the first two or three days.

Primary Wall

During the period of elongation, the protoplasm is only enclosed by a thin primary wall, which is about $0·5\mu$ in thickness and forms the cuticle in the mature fibre. The primary wall of the young hair possesses a coherent skeleton of cellulose from the first day of its appearance, but it does not respond very clearly to the usual tests for cellulose in the early stages of development, and this has led a number of investigators to assume that the primary wall is free from cellulose, and others that the material of which the wall is composed is not actually cellulose, but some closely related substance.

A third arrangement of the cellulose in the primary wall appeared when the stage of the polarising microscope was rotated so that the long axis of the fibre made an angle of about 45° with the planes of the nicols; a system of transverse strands of cellulose became visible and these too, anastomose. These three systems of strands which make the cellulose framework of the primary wall seem to be uniform over the entire surface and there is no evidence of the change of direction of the spiral which is so common in the secondary wall.

The primary wall does not respond to the usual staining tests, and gives no colour with zinc chloride-iodine. Staining with ruthenium red, however, demonstrates the presence of pectin, which may be extracted by treatment with boiling oxalic acid solution (0·5%) followed by hot ammonium oxalate solution (0·5%). Pectin appears to be confined to the primary wall as ruthenium red does not stain the inside of the fibre.

Secondary Layers

When viewed under the usual compound microscope, the second deposition of cellulose stands out in strong contrast to the structureless primary wall and even the first deposition of cellulose inside the primary wall is quite distinct when stained with zinc chloride-iodine.

The branching and rebranching strands wind in a steep spiral round the inner surface of the primary wall. The spiral makes an angle of 20–30° with the long axis of the fibre as compared with 70° for the spirals of the primary cellulose. The first secondary depositions appear quite suddenly.

One of the characteristic features of the first layer of secondary cellulose is the reversal of the spirals; there are areas in the wall where the cellulose threads change from right-hand to left-hand spirals, or vice versa. The number of such reversals varies considerably, but it is quite common to find fifty or more on a single hair; close double reversals are not uncommon. One type of reversal is that in which one set of spiral strands ends and a second system running in the opposite direction begins; the ends of the strands overlap at the reversal. The commonest method of reversal is a simple change in direction by bending in the form of an arc.

The second layer of secondary thickening does not necessarily follow the pattern produced on the previous day. The points of reversals are at different places and the direction of the spiral is often exactly the reverse of that in the first layer of the secondary wall. In many places, however, the pattern of the second layer conforms to that of the first layer.

It is difficult to follow the course of the subsequent deposition of cellulose, but in view of the behaviour of the hair on drying and also the result of swelling agents, some sort of pattern is established which is broadly followed by the later layers. The anastomosing character of the early depositions must be emphasised, for although the fibrils have a definite spiral path, they branch and rebranch with each other. The threads which form the spiral vary in length and also in diameter.

If the young fibres are freed from pectic material a coherent skeleton of cellulose remains, which can be identified by the usual tests. This skeleton is doubly refractive, it is soluble in cuprammonium hydrate, gives the characteristic cellulose reactions with zinc chloride-iodine and with potassium tri-iodide and 70% sulphuric acid; it also shows the typical X-ray pattern of cellulose.

The fibre wall, when examined under the usual compound microscope, shows no evidence of structure, but by means of a special staining technique Anderson and Kerr have demonstrated that the cellulose in the primary wall forms an open mesh of very fine thread-like structures with anastomosis (i.e. they have a branching habit, like arteries). It was observed that there were present two opposing systems of fine, spirally wound, threads of cellulose at an angle of about 70° to the long axis of the fibre. The cylindrical form of the young hair makes it possible to prove that the two systems of spiral threads exist in the same wall. Both right- and left-hand spirals were

PLATE I

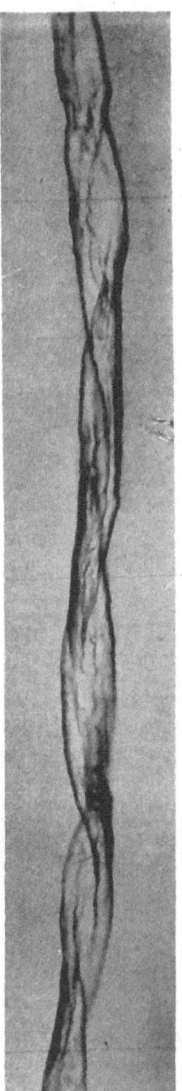

FIG. 1.

FIG. 3.

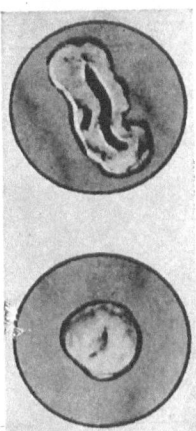

FIG. 4.

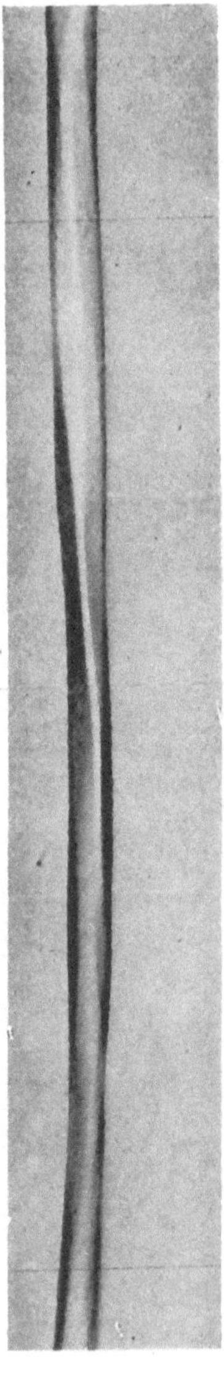

Figs. 1 and 2 are photographs of unmercerised and mercerised cotton hairs respectively, immersed in methylene iodide.

Magnification × 400 diameters approx.

Figs. 3 and 4 represent typical cross-sections.

Magnification × 700 diameters approx.

COTTON.

FIG. 2.

To face page 8.

PLATE II

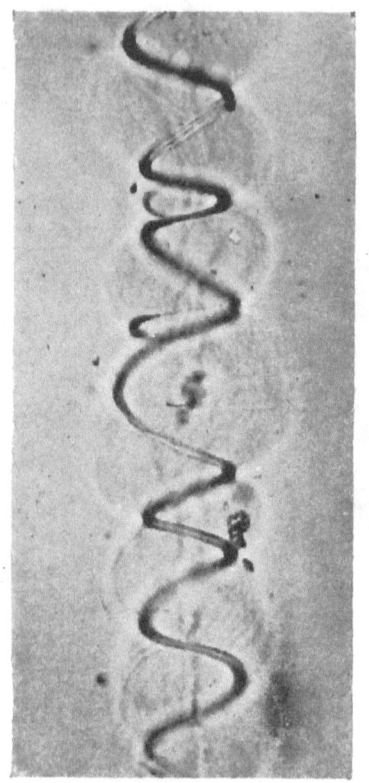 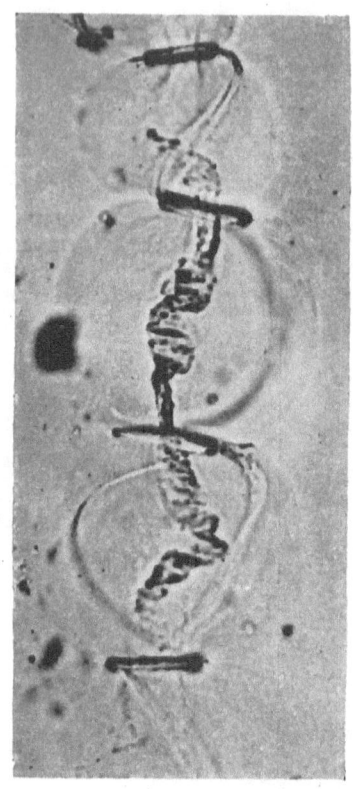

FIG. 5.   FIG. 6.

Cotton hair swollen with CS_2 and NaOH.
Bright (B.C.I.R.A.)

PLATE III

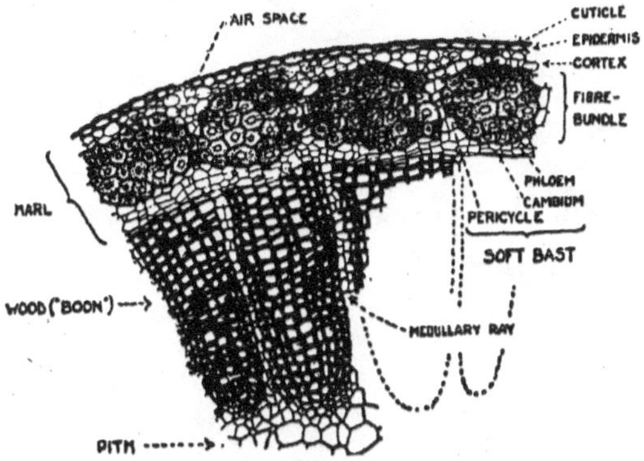

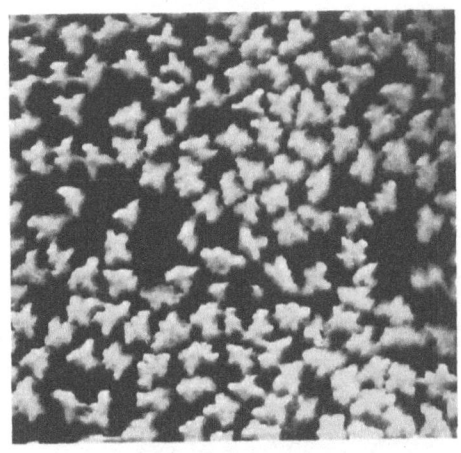

Fig. 7a. Diagram of cross-section of flax straw.

Fig. 8a. Photomicrograph of cross-section of viscose rayon.

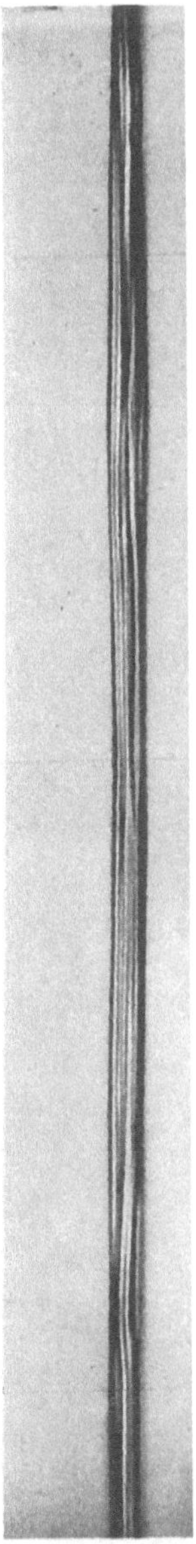

Fig. 7.
Flax.

Fig. 8.
Viscose rayon

PLATE IV

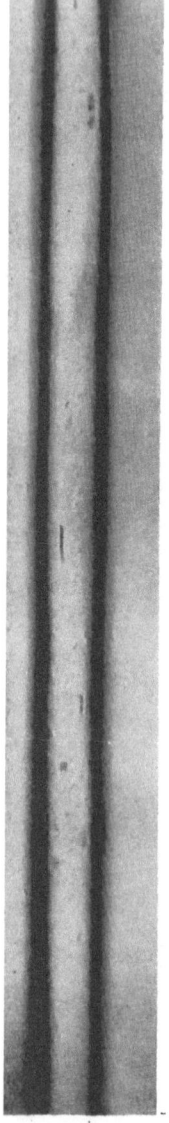

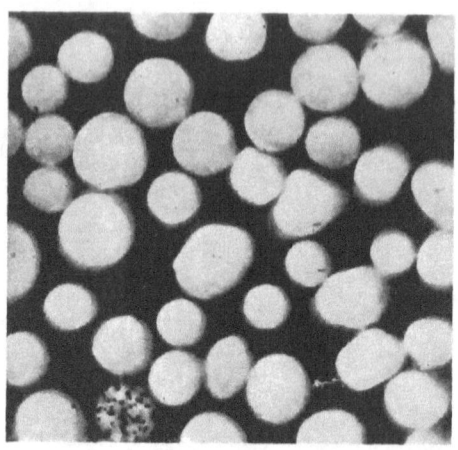

Fig. 9a.

Figs. 9 and 9a are from photomicrographs of cuprammonium rayon, and Fig. 10 from cellulose acetate (British Celanese)—R-O-X mounting medium.

Magnification × 400 diameters approx.

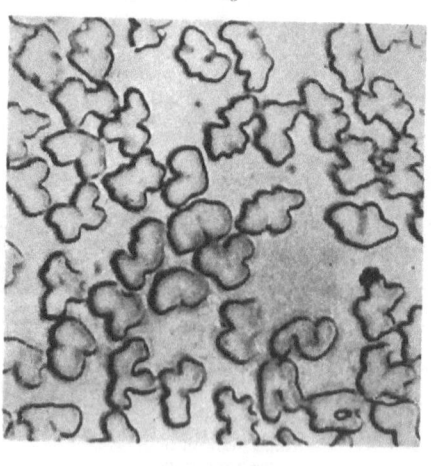

Fig. 10a.

Photomicrograph of cross-section of cellulose acetate yarn.

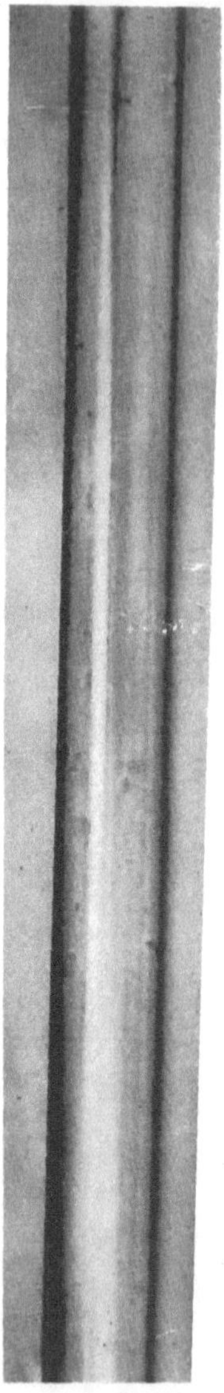

Fig. 9.

Fig. 10.

seen when the microscope was focused on the upper wall; they disappeared in the upper wall as the focal plane was lowered and appeared in the lower wall when the focus was established there.

The deposition of cellulose continues for at least twenty-five days until a few days before the boll opens. When this takes place, the fibres dry and the walls become twisted; the direction of these twists, or convolutions, conforms to that of the spiral in the greater part of the cell wall. Where a reversal takes places in the spirality, then a similar reversal or convolution occurs. There seems to be little doubt that the convolutions in the dry fibre are determined by the arrangement of the micellines the cell wall.

Growth Rings

One of the characteristic features of the section of the cotton hair is that when it is swollen, growth rings appear much in the same way as the annual growth rings in trees. The presence of these lamellæ has been known for many years, but Balls was the first to correlate them with the actual number of days during which the wall of the hair increased in thickness. He observed that the cotton plant ceased to grow during the hot afternoon period and assumed that the deposition of cellulose also stopped at this time and resumed during the night and morning. The lamellæ were, therefore, regarded as a sign of discontinuous periodic growth.

These growth rings may be used to distinguish between cotton hairs which are produced early in the season and those which are produced later. In the earlier fibres, the growth rings are relatively broad and may occasionally reach a diameter of $0\cdot3\mu$, but the rings formed in fibres which develop later in the season are much thinner and rarely exceed $0\cdot14\mu$ in diameter.

Kerr has confirmed the observation regarding correlation between the number of lamellæ and the number of days during which secondary thickening takes place, but, in contrast to Balls's hypothesis of intermittent growth, Kerr has suggested that the deposition of cellulose is a continuous process which varies in its rate according to temperature.

When cotton is grown under constant illumination and at a constant temperature of about $30°C$., no growth rings appear. When the temperature is varied under conditions of constant lighting, then growth rings appear. Similarly, when the temperature is maintained constant and the artificial lighting is turned on and off at twelve-hour intervals, then indistinct growth rings appear.

One important feature of this investigation is that although it is possible to control the growth ring pattern by the above means, yet it has not been found possible to control either the pattern of the spirals, or to influence the reversals.

Further work (*ibid.*, 1939, *9*, 325) showed that fibres produced under constant illumination are finer and of lower breaking load than normal cotton hairs; the intrinsic strength, however, is the same.

Summary

The fine structure of the cotton fibre may be summarised as follows:

1. A primary wall contains cellulose and pectic substances. The cellulose micelles in this wall are grouped into delicate anastomosing threads which have at least two systems of orientation: (*a*) a flat right-hand spiral, (*b*) a flat left-hand spiral, and probably also (*c*) a transverse position. All three systems seem uniform over the entire surface of the fibre cell.

2. A secondary wall is composed of many lamellæ of cellulose. The lamellæ are not separated from one another by non-cellulosic substances but represent dense and less dense areas of cellulose. The layers are formed of systems of spirally wound branching threads, and the direction of the spiral is reversed at frequent intervals.

3. Frequently the pattern of spirals first appearing in the secondary wall is not similar to that in subsequent layers of the wall. Most of the layers of the wall, however, follow a pattern that is established soon after secondary thickening has begun.

General

A number of important microscopical observations on the structure of plant fibres have been made during the last hundred years; the existence of fibrils in cell walls was reported by Meyen in 1838, whilst in 1852 Agardh was able to separate the spiral lines and demonstrate the fibrillar nature of the structure. Striations on the surface of the fibre and stratifications in the cross-sections were observed by Nageli in 1877, whilst Wiesner in 1886, by means of a treatment with acids at fairly high temperatures, was able to obtain from fibres fine particles which he termed dermatosomes.

During recent years investigations have been directed more towards the isolation of the structural unit, as only a superficial observation is possible on the intact cell wall. Ritter (J. Forestry, 1930, *28*, 533) isolated a number of structural units from wood fibre, first, the fibrils and bundles of fibrils, and secondly, smaller spherical units.

Farr and Eckerson (Boyce Thompson Inst. Contrib., 1934, 6, 189 and 309), in their study of the development of the cotton hair demonstrated the presence of small particles of uniform size which were present in bead-like strands and considerably larger in size than Ritter's spherical units. These particles in chain formation appear to form a single fibril in the cell wall and they may be separated even from the mature cotton fibre. The question of these "particles" being the structural unit of cellulose is still under consideration as they are considerably larger than the micelles. Farr and Eckerson believe them to be the fundamental, biological, structural unit, but the question is one which must await further investigation. The chemical dissection of the cellulose wall of the cotton hair into fibrils, and the further dissection into dermatosomes, fusiform bodies, etc., is only accomplished as the result of severe chemical treatment so that some caution must attend the reconstruction of the hair from data of various investigators.

FLAX

Flax appears to be the earliest vegetable fibre to be used industrially and the plant was grown in almost every country, but in modern times its production is confined chiefly to France, Ireland, Belgium, Holland, Russia, and North America. The plant is an annual and grows from three to four feet in height. When it has reached its full growth the plant is pulled up or cut down and subjected to a process termed rippling whereby the leaves and seeds are removed by a series of upright forks. The flax stalk or straw is then tied in bundles for the purpose of retting, which involves the decomposition of the woody matter enclosing the cellulose fibres. The process is one of fermentation and may be carried out by stagnant water in pools, in slow-moving streams, by exposing to sun and dew for about two weeks, or by various chemical methods. The older processes are still the most popular. The pectic matter which holds the fibres together must not be completely removed by retting as its presence in controlled amounts maintains the strength and elasticity of the flax—over-retted flax is brittle and weak.

The retted flax is dried and subjected to a process called scutching which involves passage through several pairs of fluted rollers, which break up the woody matter. The stalks are then placed in a machine with a rotating cylinder, on which are a number of wooden blades to remove the loosened woody matter. This procedure is known as heckling or hackling. The flax is then sorted according to quality, and is termed line whilst the waste is known as tow and used for the manufacture of twine and thin ropes.

The flax straw contains about 27% of flax which is the inner bark or bast fibre. The flax fibre is multicellular with a tapering pointed end and a narrow lumen. It shows occasional longitudinal striations and peculiar cross-like dislocations termed nodes. The cell wall is uniform in thickness and the cross section is polygonal.

Flax varies in length from a few inches to three feet, but the average length of good flax is twenty inches. The individual cell, however, is from 0·25 to 2·5 inches in length and has a diameter of 0·005 to 0·001 inch. There are about 32 cells in the fibre.

The flax fibre is much stronger than the cotton hair, and the lustre is good except in the case of Egyptian flax which has a dull appearance. The colour is a pale yellowish white but some varieties are grey. The amount of moisture in the fibre varies, but the accepted regain is 12% compared with 8·5% for cotton. Flax is a better conductor of heat than cotton, and it is this property which gives the much appreciated feel of coolness to linen and also makes it valuable for surgical bandages. Flax is more easily disintegrated by chemical means than cotton and the fibre is considerably weakened during the bleaching processes.

VISCOSE

The viscose process is utilised to manufacture the greater part of the world's output of rayon. The reaction which produces an alkali-soluble xanthate of cellulose, was discovered by Cross, Bevan and Beadle in 1892 (B.P. 8700).

The first stage in the process is to treat wood pulp with 18% NaOH solution at about 20°C. for 1 to 2 hours until thorough and uniform penetration takes place. The sheets of alkali-cellulose are then pressed to remove excess of alkaline liquor, and disintegrated or shredded in a machine with revolving arms until the "crumbs" of soda-cellulose are formed.

The alkali-cellulose is transferred to vessels, under thermostatic control, where it is allowed to "age" for 2 to 3 days. The "ageing" step involves oxidation of the alkali-cellulose during which there is a depolymerisation or reduction in chain-length of the cellulose molecules. The oxidation of the alkali-cellulose is very susceptible to temperature change.

The crumbs of alkali-cellulose are then "xanthated" by rotating in a churn with carbon disulphide, the usual amount being 32%, calculated on the cellulose. During the xanthation, the alkali-cellulose changes in colour from white to yellow and finally orange. The time of reaction is usually 3 to 4 hours, during which the temperature is controlled to between 20 and 25°C.

$$\text{Cell.OH} \longrightarrow \text{Cell.ONa} \longrightarrow \text{Cell.CSSNa}$$

When xanthation is complete, the solid is dissolved in dilute caustic soda to give a solution containing 7 to 8·5% cellulose and about 6·5% sodium hydroxide. The solution is carefully filtered.

The viscose solution is unstable and gradually decomposes by a combination of hydrolysis and saponification, but refrigeration will prevent this. Normally, however, it is arranged that the whole process depends on the day of spinning or regeneration determining

CELLULOSE FIBRES 13

the timing of the previous steps. The "ripening" of the viscose is accompanied by an increase in viscosity; ripening is essential to satisfactory spinning which depends on the ease of coagulation. The "ripening number" may be determined by estimating the number of c.c. of 10% ammonium chloride solution necessary to coagulate 20 g. of viscose. Most viscose solutions are ripened until the ripeness number exceeds 10; unripe viscose gives "milky" filaments and is apt to contract unduly on drying. During the later stages of the ripening process, the vessels containing the viscose solution are placed under vacuum in order to remove any air bubbles and dissolved gases which would interfere with the smooth flow of the viscose and produce broken filaments.

When ready for spinning, the viscose is subjected to air pressure and forced into the pumps which deliver an even flow of viscose to the spinnerets, passing through a candle filter on the way. Modern spinnerets are usually made of tantalum, and the number of holes they contain decide the number of filaments which comprise the yarn. For continuous filament viscose, the number of apertures varies from 18 to 120, but for staple fibre 500 to 2,000 holes may be utilised. The finer the single filament required, the smaller the orifice, but the filament denier is not decided by the size of the aperture alone; the composition of the spinning bath and the rate of spinning also play their part in determining the filament denier.

As the viscose is extruded into the spinning bath it commences to coagulate and the filaments are drawn out or attenuated by reason of a higher speed of collecting the filaments relative to their delivery. A high degree of elongation during spinning increases the strength of the filaments but decreases the extension at break.

Decomposition with salt baths alone is too rapid and tends to produce weak filaments; on the other hand, if the filaments are coagulated by sulphuric acid alone, they are insufficiently hardened when they come into mutual contact and do not make a satisfactory yarn. For some years now, the spinning baths for viscose have included both sulphuric acid and a sulphate; many factors in the spinning of viscose are so inter-related that one cannot be dogmatic about spinning baths. In general, however, spinning baths include 8 to 10% H_2SO_4, 14 to 20% Na_2SO_4, 1% $ZnSO_4$ and 4 to 10% of sucrose; in some cases the sugar is omitted, and 4 to 5% $MgSO_4$ added to the bath. The temperature of the spinning bath is usually 40 to 45°C. The characteristic serrated cross-section of modern viscose filaments is due to the presence of the metallic salts and is probably caused by osmotic pressure on the "skin" of the filaments.

A certain amount of time is necessary for the decomposition of the viscose and in the commonest form of spinning (the pot method), the filaments which emerge from the spinnerets are led horizontally for

12 to 25 inches through the bath to a hook, after which they are drawn vertically from the bath to a rotating glass "Godet wheel," followed by a descent to the spinning pot or Topham box. The pot is about 7 inches in diameter and rotates at 6,000 to 10,000 r.p.m. about a vertical axis. The yarn falls into a funnel immediately over the axis of the pot and is pulled through the tip of this funnel by centrifugal force which throws the yarn against the inner wall of the

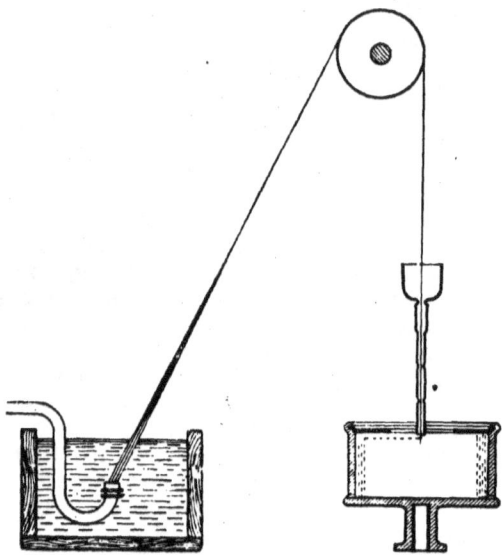

FIG. 11.—Diagram of the spinning of viscose rayon.

pot. The rotation causes the filaments to be twisted into a thread with about 2·5 turns per inch; a traverse motion distributes the yarn from top to bottom of the pot and builds up a cake at a rate of 60 to 70 yards per minute. As the walls of the pot are perforated, a great deal of the adhering liquor from the bath is removed by the centrifugal force. After two hours or so the pot is fully charged, and the cake of yarn removed.

At one time it was customary to wind the yarn into hanks for the subsequent processes, but there is a growing tendency to treat the material in cake form to avoid mechanical damage. The cakes are stacked and compressed by end plates through which the various liquids are forced, passing down the hollow centres to percolate through the yarn to the outside. The first treatment is a thorough

CELLULOSE FIBRES

washing with water to remove the sulphuric acid and other constituents of the spinning bath. This is followed by removal of the sulphur in the filaments with sodium sulphide solution of 1 to 2% concentration, after which the yarn is again washed. A bleaching process is then carried out with either hypochlorite or peroxide, followed by further washing; in some cases, the bleaching step is omitted as it may be effected more conveniently in the knitted or woven fabric. Where the bleach is omitted from the sequence of steps, the desulphurised cakes are washed with acid to remove the alkalinity, washed again with water, and in all cases, a softening agent is added.

Hydro-extraction removes as much water as possible, but the final drying is a somewhat delicate process as it is important that the outside of the cake should not dry before the inside and set up strains in the yarn; efficient drying may take a week or more.

Properties

The moisture absorption of viscose and other regenerated cellulose rayons at all relative humidities is higher than that of cotton; Urquhart and Eckersall (J.T.I., 1932, *23*, 163) established that the absorption ratios of commercial rayons, relative to cotton, lay between 2·12 and 1·75. The moisture content of 11% is generally accepted as standard for viscose rayon.

Regenerated cellulose exhibits considerable swelling when immersed in water, the increase in volume amounting to almost 100%, or 40% in diameter. Whereas the wet strength of cotton is 110 to 120% of the dry strength, regenerated cellulose shows a diminution in strength when wet, amounting to 50% approximately. The extension at break is 20% in the dry state and 22% when wet.

The strength of rayons is usually expressed as tenacity or grams per denier. Ordinary viscose and cuprammonium rayons have a tenacity of 2, but the "strong" viscose rayons have tenacities of 4·5; their extension at break is 12% in the dry state and 18% when wet.

The density of rayon is 1·53, the same as that of native cellulose.

Rayon filaments are generally made in three deniers, 1·5, 3·0 and 4·5; the corresponding diameters of the filaments are 14 to 15μ, 19 to 21μ and 24 to 26μ.

Staple Rayon

Rayon staple fibres originated in Germany during 1914–1918; until 1930, however, production diminished but improved methods of manufacture have caused a greatly increased output.

The fundamental chemical processes are identical with those of the ordinary viscose production, but continuous methods are employed as far as possible. Whereas in the case of continuous filaments, the number of holes in the spinneret varies from 20 to 120, for staple fibre they range from 500 to 2,000.

According to one method the "tow" is cut while it is in the wet state and is then finished in bulk; an alternative method is to keep the continuity of the filaments until the material has been dried and then cut the "rope" into the required staple. The shrinkage and crimping of the fibres which results from the wet method is claimed to be an advantage.

English staple rayon, as produced by Courtaulds, is available in various lengths and deniers.

FIBRO

Bright		Matt	
1·5 denier	$1\frac{7}{16}$ in.	1·5 denier	$1\frac{7}{16}$ in.
3·0 ,,	$1\frac{7}{16}$ in.	3·0 ,,	$1\frac{7}{16}$ in.
4·5 ,,	4 in.	3·0 ,,	4 in.
4·5 ,,	6 in.	4·5 ,,	2 in.
8·0 ,,	8 in.		

The 6-inch and 8-inch "Fibro" as well as the 4-inch variety are intended for worsted manufacture, the 2-inch Fibro is used in the woollen trade, and the $1\frac{7}{16}$ Fibro for the cotton trade. All of these qualities are also used for 100% Fibro fabrics.

NITRATE RAYON

The cellulose nitrate process for the production of artificial silk is the oldest of such processes, and originated with Swan. The general scheme simply involves the treatment of cellulose with nitric acid, in presence of sulphuric acid, to form the nitrate which is soluble in a mixture of alcohol and ether. The viscous solution is then forced through fine orifices, either into air (dry spinning) or into a liquid (wet spinning), when solidification of the extruded filament occurs.

In the actual spinning process, which is the extrusion of the filaments through the fine orifices, the individual filaments are very weak, and it is usual to combine some 10 to 25 together to form a suitable thread, which hardens on account of the evaporation of the solvent, and is improved further by extension.

The threads are wound into hanks and "denitrated" by means of ammonium hydrosulphide, or calcium hydrosulphide, in 3% solution at 40°C. for about three hours, as a result of which the cellulose is regenerated, and the nitrogen content falls from about 11% to a minimum of 0·05%. The loss in weight is about 30% and the tensile strength has fallen about 25%, but the rayon is now as soft as natural silk and has lost its high inflammability. Owing, however, to the reduced tenacity in the wet state, the subsequent operations of washing and bleaching need great care.

CELLULOSE FIBRES 17

The two main manufacturing advantages of nitrate rayon were the relative stability of the spinning solution, i.e. the cellulose nitrate solution, and the small amount of waste during manufacture.

The manufacture of rayon from cellulose nitrate seems to have ceased about 1935.

CUPRAMMONIUM RAYON

The solubility of cellulose in cuprammonium hydrate was utilised by the Vereinigte Glanzstoff-Fabriken A.G. of Elberfeld, for the production of rayon. The raw material is usually bleached cotton linters, the preliminary treatment of which is carried out not only with a view to purification but also to increase its solubility. If the cotton is overbleached, then watery solutions will be produced which are unsuitable for "spinning" on account of the low viscosity. The cotton material is not dried but hydro-extracted, and stored in a damp condition when it is more readily soluble. The cuprammonium hydrate or Schweizer's reagent was originally prepared, in the Glanzstoff process, by the action of ammonia, water, and air on electrolytic copper. As the solubility of the cellulose is dependent on the copper content, it is necessary to cool the cuprammonium solution, both during its preparation and storage, in order to obtain solutions as rich in copper content as possible. Temperatures of below 5°C. are employed, as otherwise decomposition occurs with precipitation of cupric hydrate. The solvent and storage power of the cuprammonium hydrate may be increased by the addition of caustic soda solution or by additions of beet sugar, which has been denatured by treatment with copper sulphate.

The moist bleached linters are dissolved at a low temperature with powerful agitation, and a concentration of 6–8% of cellulose is obtained as the greater the concentration of cellulose, the stronger is the rayon produced from the solution.

The cotton dissolves in about six to eight hours, but an entirely homogeneous solution is not obtained so that it is necessary to filter the spinning solution through iron filter presses, the last being made of 200 mesh steel or nickel filter gauze. De-æration next takes place to remove air bubbles from the solution by exposing it to reduced pressure in storage tanks. A certain amount of ammonia is also removed and the viscosity of the solution consequently is increased. The solution is then forced through spinnerets under 1·5 to 2 atmospheres pressure, into the coagulating bath.

The cuprammonium solution of cellulose may be caused to decompose by means of either sulphuric acid or alkali, with consequent regeneration of the cellulose. The original coagulating bath was composed of sulphuric acid containing about 500 g. per litre in order to produce a modified parchmentising action with beneficial effects on the tensile strength. In the Glanzstoff process, however, about 30% caustic soda solution was used as the coagulating bath with the addition of 6% of sugar or sodium lactate. The temperature was usually maintained at 50°C. The effect of the sugar in the bath was to produce filaments containing less precipitated copper, but in all cases it was necessary to remove the copper by rinsing in 2% H_2SO_4, either directly after the coagulating and washing or after the later twisting and reeling processes.

The Glanzstoff process yielded fairly coarse filaments and its manufacture was discontinued in 1914. The production of very fine filaments was made possible by Thiele and elaborated by the firm of J. P. Bemberg, A.G., of Barmen-Rittershausen, filaments finer than those of natural silk being made. In this modern method of manufacture the preparation of the spinning solution takes place in an entirely different manner. The beaten cellulose is intimately mixed with freshly precipitated cupric hydroxide to form cupric hydroxide cellulose which is soluble in concentrated ammonia. The cupric hydrate is prepared by running concentrated caustic soda solution at 0°C. into a concentrated copper sulphate solution at 15°C. with constant stirring until all the copper is precipitated as the hydroxide. The thick liquid is then mixed with the well-beaten cotton linters and mixing continued for 30 minutes. The mixture is then pressed and washed until the wash-water is free from sodium sulphate. The pressed mass or cakes containing about 40% of water are minced and squeezed through sieves to form fine threads, which are subjected to a vacuum in order to remove enclosed air. The necessary ammonia is added in a vertical dissolving vessel which is fitted with an agitator. After about 24 hours a concentrated solution of cellulose is obtained and this is diluted to 7–8% by the addition of the necessary aqueous ammonia. The customary additions to the spinning solution, the de-æration and filtering are carried out as in the older process.

The essential feature of the stretch spinning process is that the spinning solution emerges under slight pressure from the spinnerets into a feebly coagulating liquid such as water which moves with increasing speed in the direction of the forming filaments, which solidify incompletely at first, and are stretched by the rapidly flowing coagulating bath and a mechanical device. Any air must be removed from the coagulating liquor in order to avoid the formation of bubbles which would cause the filaments to break. When the filaments have been drawn to the required fineness, they are coagulated completely by means of a second bath consisting either of dilute sulphuric acid or a solution of a metallic salt, such as 15% ferrous sulphate. The water used in the first coagulating bath should be as pure as possible as its slight coagulating action is due to the removal of some ammonia and copper from the spinning solution as it is forced through the fine orifices of the spinnerets.

Although cotton linters is the main starting material for the manufacture of cuprammonium rayon, wood pulp has been used with some success in England. It has been found, however, that the rayon produced, which nevertheless bleaches readily, is of a pale cream colour whereas that from cotton linters is almost completely white. It is also interesting to note that according to the data of Ridge, Parsons and Corner (p. 467) the fluidity range of cuprammonium rayons is considerably lower than those from viscose. There is also some evidence to show that the fluidity of cuprammonium rayon from cotton linters is slightly lower than that from wood pulp. This is in agreement with similar observations on the viscosity of cellulose acetates and nitrates, i.e. the viscosity is a measure of the quality of the starting material as distinct from indicating the effects of the chemical actions of solution and regeneration.

ACETATE RAYON

Acetate rayon is not formed from regenerated cellulose but from a compound of cellulose and acetic acid. It is, therefore, a cellulose derivative, but is included here for convenience.

Cellulose acetate was first discovered by Schutzenberger (Compt. rend., 1865, *61*, 485), but the product was considerably degraded. The early processes were also all concerned with the fully acetylated product which is only soluble in chloroform, but partial hydrolysis, discovered by Miles in 1903, produced an acetate which was soluble in acetone. Recent work, however, indicates that hydrolysis is not the only factor. Cellulose triacetate contains 62·5% of combined acetic acid, but the acetone-soluble acetates range from 51·5 to 60% combined acetic acid. Taniguchi and Sakurada (J.S.C.I., Japan, 1938, *41*, 72) showed that cellulose triacetate becomes soluble in acetone if mixed with the acetone-soluble acetate. Deripasko and Drujan (J. Phys. Chem. Russ., 1937, *10*, 798) fractionated a cellulose acetate containing 54·7% acetic acid and found the highest amount of combined acetic acid to be 58·5%.

The usual process of manufacture is to form the primary acetate (triacetate), and then the secondary acetate in which five-sixths of the available hydroxyl groups are substituted.

The starting product is bleached cotton linters which are treated with an acetylating mixture consisting of approximately equal parts of acetic acid and acetic anhydride, together with a small amount of sulphuric acid. The temperature should not be allowed to rise above 35°C. Acetylation under these conditions generally requires 5 to 8 hours and the progress of the reaction is followed by following the solubility of the product.

The second stage is the formation of the secondary acetate by adding water and maintaining the mass at 20°C. for 3 days. Samples are withdrawn towards the end of this period, and tested; the secondary acetate should swell but not dissolve in chloroform, and it should be soluble in a hot mixture of equal parts of benzene and alcohol containing a little water.

Any free sulphuric acid is neutralised with sodium acetate, and the cellulose acetate precipitated by the addition of water. The flakes are allowed to settle and then thoroughly washed with water. As the acetate contains small amounts of sulphuric acid esters, these are removed by boiling with 0·02% H_2SO_4 for 1 to 2 hours, after which the flakes are washed, hydro-extracted and dried.

Secondary cellulose acetate is the only cellulose derivative in the field of commercial rayon production.

The purified secondary acetate is dissolved in a suitable solvent for spinning, filtered and de-ærated. Where the wet spinning method is employed, the solution of cellulose acetate in acetone can be spun, i.e. coagulated as a number of continuous filaments, into baths of hydrocarbons, oils or aqueous solutions of salts. It has also been possible to "spin" from the acetylating mixture containing the secondary acetate without going through the processes of isolation, purification and re-solution. In the wet spinning processes, the thread is collected on bobbins, washed, twisted and reeled, as in the case of the other methods of rayon manufacture.

Most cellulose acetate, however, is spun by the dry system, which consists in projecting the acetone or acetone-alcohol solution of the secondary acetate (5 to 20% concentration) into a closed compartment whereupon the solvent is evaporated by means of hot air. The uniformity of the filament is obtained by "spinning" vertically downwards, while a current of warm, dry air flows upwards. In these circumstances the material is not coagulated too rapidly, but remains plastic, so that the filaments are improved by the drawing action of the spinning spool below the enclosed vessel into which the solution is projected through the spinnerets.

The spinning shaft is 6 to 10 feet in height and the rate of delivery may be 100 to 200 yards per minute.

Regulation of the temperature is essential, and the speed of the current of air must be uniform.

It is possible to draw off and recover the greater part of the solvent vapours, some of which is mixed with warm air and returned to the enclosed vessel or cell, as this contributes to the formation of uniform filaments. It has been found that in order to obtain filaments with a round cross-section, the spinning atmosphere must contain more solvent vapour than when filaments with a flatter cross-section are required.

The technical success of these processes for textile purposes is largely due to the work of the Dreyfus brothers.

Recent work has been directed to the improved tensile strength of acetate rayon produced by stretching; this phenomenon is in accordance with similar observations on micellar orientation in regenerated cellulose production. The cellulose acetate filaments may be stretched in wet steam or water at 100 to 120°C. according to the methods of B.P. 438,588–91; 438,584–7; 438,655–6; 438,786; 443,707 and 443,773. Stretching acetate rayon filaments in a plastic state may also be accomplished by swelling agents which are organic liquids (or vapours with solvent properties) according to B.P. 454,580 or by balanced mixtures of solvents and non-solvents (B.P. 453,155).

Simultaneous hydrolysis and stretching of acetate rayon has been responsible for great advances in tensile strength of the regenerated cellulose. The subject is covered in a large number of patent specifications from which it appears that the use of inorganic bases is accompanied by definite disadvantages such as the production of a core of unchanged ester surrounded by an outer layer of pure cellulose. Liquid organic bases, particularly aliphatic bases, give even results. For instance, according to B.P. 417,220, acetate rayon which has been softened and stretched, is passed through 20% methylamine solution at 60°C. for 3 minutes. A regenerated cellulose may be obtained which has been stretched 200% and possesses a tenacity of 2·5 to 3 g. per denier, and an extension of 12%. B.P.429,103 utilises ethylene diamine which is particularly valuable in this connection. The products contain less than 1% acetic acid after hydrolysis, and if necessary their affinity for colouring matters which dye cellulose, may be improved by treatment with NaOH or KOH (B.P. 501,768). These processes are believed to form the basis of "*Fortisan.*" It may be that the advantages of these processes over stretch spinning viscose into strong H_2SO_4 lies in the comparative mildness of the processes, thus allowing more control and producing a uniform result.

The moisture content of cellulose acetate is lower than that of cellulose on account of the blocking of the hydroxyl groups. The normal moisture content of acetate rayon is 6·5%. For the same reason acetate rayon swells to very slight extent in water.

The density of acetate rayon is 1·33. The melting-point is 250°C.

The ordinary acetate yarn has a tenacity of 1·5 and an extension at break of 20%, but the specially strong yarns from acetate have tenacities of 6·8 with an extension at break of only 7%.

The ordinary acetate yarns are susceptible to the action of hot water and should not be treated wet above 85°C. or there is a loss in lustre. Saponification, of course, may be produced with alkaline liquors. Hence it is of great importance to take suitable precautions during the scouring and bleaching of acetate rayon.

CHAPTER III
CELLULOSE CHEMISTRY

CONSTITUTION

ANALYSIS of cellulose gives the following composition: C 44·4%; H 6·2% and O 49·4%, corresponding to the empirical formula $C_6H_{10}O_5$. Consideration of the structure of native cellulose, its colloid properties and inertness in the chemical sense led to its being regarded as $(C_6H_{10}O_5)_n$; the formation of derivatives such as the tri-nitrate and tri-acetate point to the presence of three hydroxyl groups per glucose residue.

The relationship between cotton cellulose and *glucose* has been known for very many years and yields of some 95 to 98% were obtained by Fleschig in 1883, by Ost and Wilkening in 1910, and by Willstätter and Zechmeister in 1913; the glucose was estimated by reducing power and optical rotation and obtained by hydrolysis of cotton with dilute H_2SO_4 or dilute HCl.

The actual isolation of glucose from cotton was accomplished by Monier-Williams (J.C.S., 1921, *119*, 803) and confirmatory evidence was produced by Irvine and Hirst (J.C.S., 1922, *121*, 1,585). The position of the three hydroxyl groups was established in a series of papers by Denham and Woodhouse (J.C.S., 1913, *103*, 1,735; 1914, *105*, 2,357 and 1917, *111*, 244) on methylation. Similar investigations by Irvine and Hirst (J.C.S., 1922, *121*, 1,585) confirmed the presence of hydroxyl groups in the *2 : 3 : 6* positions.

Combined hydrolysis and acetylation of cotton gives considerable yields of the octa-acetate of a disaccharide known as *cellobiose*; this was isolated by Franchimont in 1879 and examined by Maquenne and Goodwin in 1904. Haworth and Hirst (J.C.S., 1921, *119*, 193) showed that this sugar consisted of two glucose residues united by a *1 : 5* or *1 : 4* β-glucosidic linkage and suggested that a similar linkage was represented in cellulose. Hydrolysis of the methylated product gave equal amounts of tetramethyl and trimethyl glucose and the latter was identical with the *2 : 3 : 6* trimethyl glucose of Denham and Woodhouse.

A very important advance in our understanding of the chemistry of the sugars was made by Haworth (Nature, 1925, *116*, 430; J.C.S., 1926, p. 89) when he formulated the structure of glucose as a six-membered ring, instead as the previously accepted five-membered structure. The use of atomic models is really essential to the full understanding of the position, for the problems are mainly stereochemical. Representation of the formulæ on flat paper is difficult,

CELLULOSE CHEMISTRY

but differences of configuration may be shown by distribution of hydroxyl groups above and below the plane of the paper. For example, the α and β forms of glucose differ only in stereo-chemical arrangement.

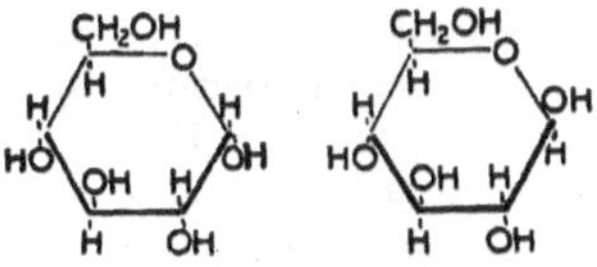

α-glucose   β-glucose

The relation between glucose and pyran is shown by

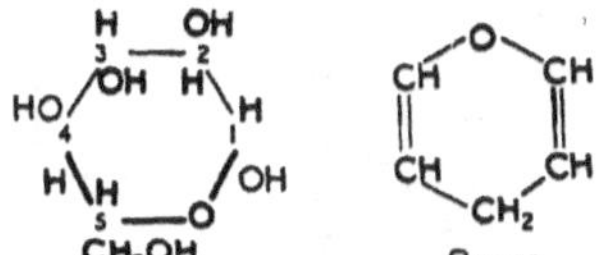

Pyran

The most symmetrical of all methods of linking the glucose is between the carbon atoms in the *1* and *4* positions; this linkage also occurs in maltose and hence in starch, but in this case α-glucose is involved. Comparison of atomic models shows that in maltose the hydroxyl group in the *4* position is below the plane of the ring and in a position to combine with the hydroxyl of the other glucose residue and the two rings are then in alignment. This condition can only be realised in the case of cellobiose by turning one of the β-glucose units through 180° so as to bring the hydroxyl in the *1* position of one unit below the plane in order to unite with the hydroxyl in the *4* position of the other glucose unit. This new conception of cellobiose was confirmed by Haworth, Long and Plant (J.C.S., 1927, p. 2809).

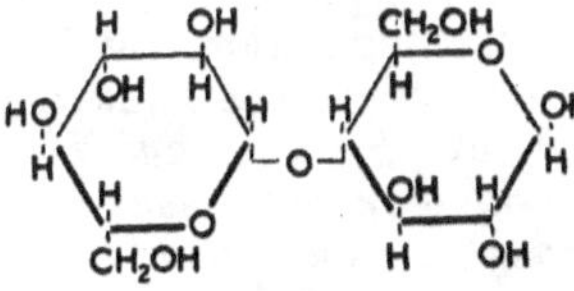

Cellobiose

Prior to this time there had been tentative suggestions by Tollens and by Freudenberg that cellulose was a long chain structure, but the general opinion of a small structural unit was strengthened by the early X-ray evidence. Cellulose was, of course, recognised as a highly polymerised material and it was assumed that some of its characteristic properties were due to the formation of large aggregates produced by forces of association.

The general advance in the technique of X-ray analysis due to the work of W. H. and W. L. Bragg produced data which enabled models to be constructed of compounds whose properties were obscure unless a three-dimensional picture was available. Sponsler and Dore had been collecting data on the structure of cellulose for some years and were of the opinion that the cellulose units were long

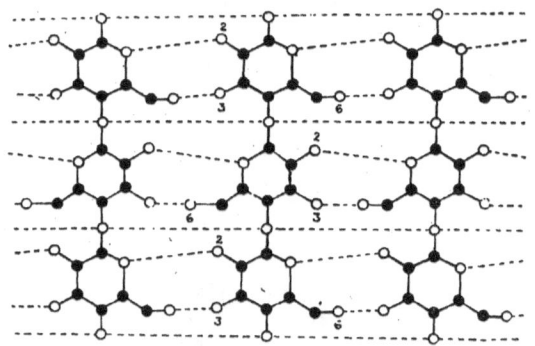

FIG. 12.—Corrected version of Sponsler and Dore's diagram of the arrangement of glucose residues in cellulose.

chains of glucose residues lying along the length of the fibre. When models were constructed and compared with the X-ray diagrams for ramie fibre it was found that Haworth's ring-structure for β-glucose agreed with the lattice spacings, but instead of the *1 : 4* linkage throughout Sponsler and Dore (Colloid Symposium Monograph, 1926, *4*, 174) adopted alternate *1 : 1* and *4 : 4* linkages.

The idea of alternate glucosidic and ether linkages presented difficulties to chemists and the general scheme of interpretation was re-examined, at Haworth's suggestion, by Meyer and Mark, who found that the structural arrangement would allow for the cellobiose or *1 : 4* linkage throughout. *Sponsler and Dore recognised that the constituent units are arranged in continuous chains which lie parallel to the fibre axis and through the unit cell of the X-ray determinations.* The glucose units are bound longitudinally by primary valencies and laterally by secondary valencies. The primary valencies account for the chemical and mechanical stability of the fibre; the secondary valencies are easier to rupture, which explains the relatively low transverse strength. Further, this idea covers the swelling of cellulose by liquids, which do not attack it chemically, but may enter between the chains and widen the space between them. Esterification and etherification are possible without loss of fibre form on the assumption that the new groups will accommodate themselves in the spaces between the chains.

If the hypothesis of the molecular chain is correct, then in each terminal glucose residue there should be one more group open to methylation than in the other residues forming the chain. At one end of the chain it is the reducing group which should be open to methylation and the methoxyl group would be split off again on hydrolysis, but the non-reducing terminal hydroxyl group when methylated should withstand hydrolysis. This means that hydrolysis should yield a certain proportion of tetramethyl glucose which had not been found previously; indeed, its absence has been regarded as evidence for some of the early alternative structures of cellulose. Haworth and Machemer (J.C.S., 1932, *134*, 2372), however, were able to supply the final piece of chemical evidence for the molecular chain theory by the isolation of tetramethyl glucose from the hydrolysis of trimethyl glucose. This is illustrated from the following formula:

The modern view of the constitution of cellulose is represented by

Cellulose

Cellulose cannot be regarded as a homogeneous substance; the work of Schmidt (Ber., 1934, *67*, 2037) indicates that even carefully purified cellulose contains some carboxyl groups which means that glucuronic acid residues form part of the chain-molecule. It may be assumed that native cellulose contains approximately one glucuronic acid residue for every hundred glucose residues. Apparently the plant is able to oxidise cellulose to glucuronic acid, from which xylan may be produced by decarboxylation, as shown by the appropriate formulæ.

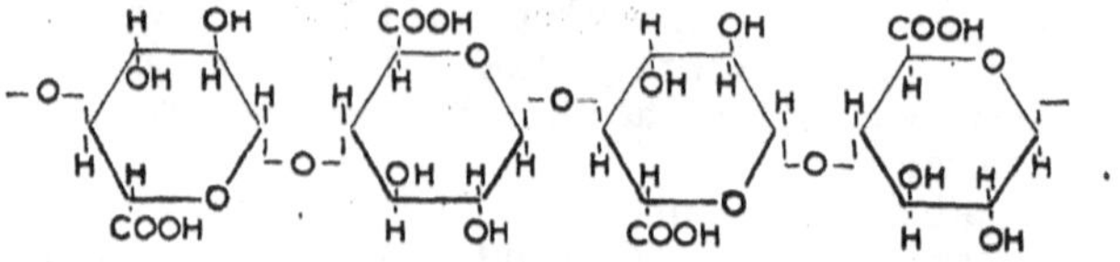

Polyglucuronic acid

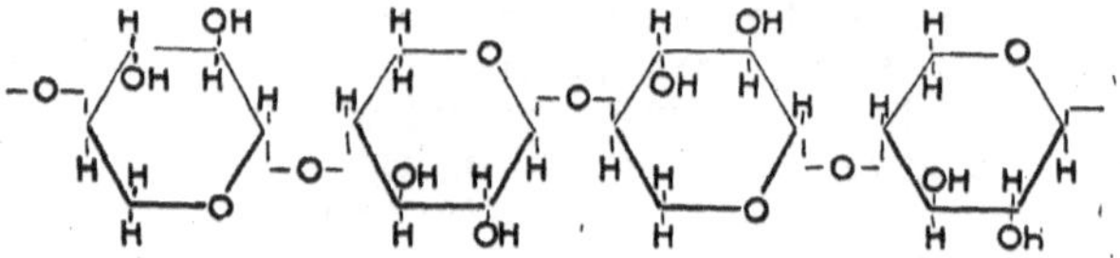

Xylan

The work of Bone (J.S.D.C., 1934, **50**, 307) is interesting in this connection, for it shows that when water is evaporated from cotton, the boundary region develops an increased affinity for Methylene Blue.

The amount of xylan in cotton is negligible, but jute contains about 12% of xylan, whereas flax and hemp only contain 2%.

MOLECULAR WEIGHT

The following data, taken from a discussion at the Faraday Society Symposium (Trans. Farad. Soc., 1933, **29**) show a wide range of values.

COMPARISON OF MOLECULAR WEIGHTS

Material	Mol. wt.	Method
Cellulose from linters	20,000–40,000 (minimum)	Chemical — tetramethyl glucose; Haworth and Machemer.
Cellulose acetate	120,000	Viscosity; Staudinger.
Cellulose acetate	35,000 (average)	Osmotic pressure; Buchner and Samvel.
Cellulose linters	40,000	Ultra-centrifuge; Stamm.
Cellulose	16,000	Chemical; Schmidt.
Ramie	24,000–32,000	X-ray; Mark.

Various methods have been applied to different products so that no comparison is really satisfactory. Staudinger's recent work, however, gives higher results than those previously obtained, and his contribution to the problem is based on a property of substances of high molecular weight, the high viscosity of their solutions in relatively low concentrations. Most of his publications refer to the "degree of polymerisation" from which the molecular weight may be obtained by multiplying by 162 in the case of cellulose and a corresponding figure for the derivatives. The degree of polymerisation is the number of glucose residues in the molecule, and for native cellulose this is about 2,000 (Ann., 1936, *526*, 72). An interesting table which follows the degree of polymerisation through the manufacturing processes for artificial silk shows the following figures:

DEGREE OF POLYMERISATION

	Cuprammonium rayon.	Viscose rayon.	Nitrate rayon.	Cellulose acetate.
Raw linters	1,400	—	—	1,400–700
Bleached linters	700	—	—	700
Wood pulp	—	700–900	—	—
Spinning solution	400–500	—	$\sim$500	250–350
Rayon	400–500	300–450	$\sim$200	250–350

Staudinger (Papier Fabrikant, 1938, *36*, 381) has given a list of the constants to be used in viscosity calculations for the various solvents for cellulose and its derivatives. Good agreement has also been shown for the degree of polymerisation according to the osmotic pressure, viscosity and ultra-centrifuge methods of determining molecular weight.

DEGREE OF POLYMERISATION

	Viscosity Staudinger	Ultracentrifuge Kraemer
Native cellulose	2,000 to 3,000	3,500
Purified cotton linters	—	1,000 to 3,000
Wood pulp	900 to 1,500	600 to 1,000
Commercial rayons	250 to 500	200 to 600
Cellulose nitrate lacquers	300 to 600	500 to 600
Commercial cellulose acetate	200 to 350	175 to 360

These data are taken from the work of Staudinger (Papier Fabrikant, 1938, *36*, 474) and Kraemer (Ind. Eng. Chem., 1938, *30*, 1200).

MOLECULAR ARRANGEMENT

The anisotropic character of cellulose, the directional nature of its physical properties, was known from its optical behaviour for some time before the definite theory of structure was proposed by Nageli (Die Starkekorner; Schultub, Zurich, 1858), who postulated that all cellulose fibres were built up of submicroscopic *crystalline* particles termed "micelles." Many properties were explained by Nageli on this basis, but the theory was not generally accepted at the time. The investigation of the double refraction of textile fibres by Ambronn (Ber. Saechs. Ges. Wiss., Leipzig, 1911, *63*, 249) revived interest in the micelle theory.

The work of von Laue (Interferenz-Erscheinungen bei Röntgenstrahlen; Ber. de l'Acad. des Sc. Bavaroise—June, 1912) established recognition of the fact that a crystal behaves as a three-dimensional grating to X-radiation, and this indicated a regularity of interatomic distances and hence of internal structure. The X-ray technique was utilised by Herzog and Jancke (Z. Phys., 1920, *3*, 196) to show that cellulose has a definite crystalline structure which is the same whatever the source of material. Polanyi (Naturwiss., 1921, *9*, 288) made the first observations on the size of the unit cell—the smallest unit which still possesses the geometrical properties of the whole crystal lattice—assuming the arrangement to be of the rhombic-quadratic type. The dimensions were calculated as follows:

The Unit Cell of Cellulose

a	8·65 to 8·75 Å	Horizontal.
b	10·25 to 10·35 Å	Vertical period along the fibre axes.
c	7·8 to 7·9 Å	Forming an angle with "a".

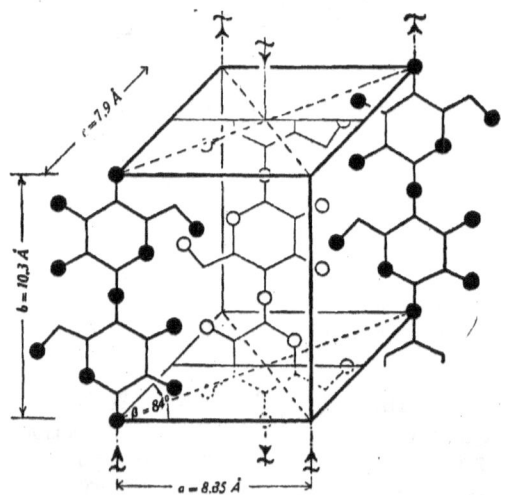

Fig. 13.—The crystal unit of cellulose, according to Meyer.

It must be remembered that at this time it was thought that cellulose was an association of glucose or cellobiose anhydrides, so that when calculations showed that the unit cell could accommodate four glucose residues, it was assumed that the cellulose molecule was identical with the unit cell of the X-ray work.

CELLULOSE CHEMISTRY

The work of Sponsler and Dore (see page 24) in 1926 was an important departure from the prevalent idea, even if their chemical considerations had to be revised. This was done by Meyer and Mark (Ber., 1928, *61B*, 593), who showed that a model could be produced which agreed with the chemical requirements of Haworth and the X-ray data of Polanyi.

The crystal unit is the pattern which repeats itself in all directions and the long chains run through it along the "b" axis; it must not be confused with the cellulose molecule or with the micelle.

New X-ray data by Meyer and Misch (Ber., 1937, *70*, 266) entailed some revision of the old model, mainly on the lines of alternating orientation in each network of chains, and the calculated intensities required by the new model showed good agreement with the observations.

Sponsler (Trans. Farad. Soc., 1930, *26*, 813) has expressed the view, based on the probable structure of the cellulose in the cell wall of the native material, that the plane of the glucose residues lies at an angle of about 45° to the wall surface, which is, therefore, studded with hydroxyl groups; this results in a highly reactive surface.

MICELLE AND CHAIN MOLECULE

Among the views expressed by Mark and Meyer (Ber., 1938, *61B*, 593) was the theory that cellulose is not composed of molecules in the usual sense but of micelles (*cf.* Nageli, page 27) made up of cellobiose chains in parallel arrangement, held together by intermicellar forces which are similar to Van der Waals forces. The chains are not necessarily all of the same length, but from the breadth of the X-ray interference lines and diffusion co-efficients it was estimated that there must be about 1,500–2,000 glucose residues per micelle.

Hengstenberg and Mark (Z. Kryst., 1928, *69*, 271) showed that in the ramie fibre the micelle is present as a rhombus measuring 500 to 600 Å along the fibre axis and 50 by 55 Å across the axis. The estimate of micelle size was in good agreement with that previously advanced by Herzog (J. Phys. Chem., 1926, *30*, 457) and was verified by Clark (Ind. Eng. Chem., 1930, *22*, 474) by the use of relatively long X-ray wave lengths.

A comparison between the micellar size of native cellulose and viscose rayon by Hengstenberg and Mark (Z. Krist., 1928, *69*, 271) gave the following results:

SIZE OF MICELLE

	Perpendicular to fibre axis	Parallel to fibre axis
Ramie	55 Å	600 Å
Viscose rayon	41 Å	305 Å

This decrease in micellar size supports the view that the cellulose chains have been broken in the processes of manufacturing rayon. It is also in agreement with the lower tensile strength of rayon, particularly in the wet state.

The active hydroxyl groups with which the surface of the micelles is studded provide the forces which hold the micelles together; if the micelles are reduced in size, then the tensile strength must fall, particularly in the case of random orientation. With smaller micelles there is also a better and more rapid penetration by water, which is adsorbed on the hydroxyl-studded surface and so covers those groups which normally hold the structure together. The cellulose swells, and as the forces are operative to a less extent and on a smaller area

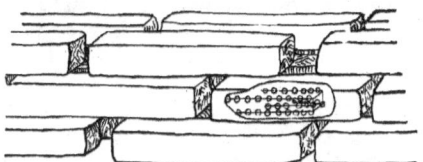

Fig. 14.—Simple diagram of micelles and chain molecules of cellulose.

than for native cellulose, the tensile strength in the wet state is very poor.

In the simplest hypothesis the micelles may be likened to bricks in a wall, overlapping each other in much the same way. These discrete brick-like submicroscopic crystalline particles are orientated with respect to the axis of the fibre and account for the estimate of 75% of cellulose existing in the crystalline state, the remaining 25% being amorphous matter separating the micelles and allowing them to move as units during intermicellar swelling and to be dispersed during dissolution. In this structure the mechanical properties were supposedly due to the cohesive forces between the individual crystallites, in which the length of the cellulose chain molecule was some 600 Å; this may be a minimum value, for more recently Meyer (Ber., 1937, *70*, 266) has estimated 1,000 to 1,500 Å for the length of the micelle.

The theory of continuous structure, according to the school of Staudinger, envisages much longer chain-molecules, 10,000 to 15,000 Å, which have come together in such a manner that the crystalline regularity is interrupted by regions which behave as amorphous matter to X-rays and also to swelling and dispersing agents.

The simple micellar hypothesis is no longer rigidly upheld, for it is now clear that the molecular order only approximates to crystalline perfection, and further, it is difficult to account for the high tensile strength and other mechanical properties of native cellulose by postulating a simple brick-like structure. More recent views developed by Astbury, Gerngross, Hermann, Frey-Wyssling, Kratky, Mark and others embrace the possibility of some of the chain-molecules persisting through the micelles, or the distribution of chain-endings giving the effect of regularly arranged units. Some of these speculations have been represented diagrammatically by Frey-Wyssling (Die Stoffausscheidungen der hoheren Pflanzen; Springer; Berlin, 1935). Fig. 15. The possibility of individual

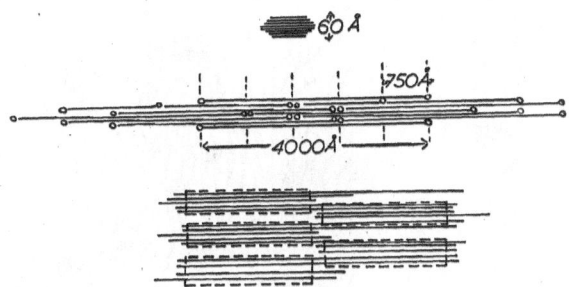

Fig. 15.—Micelle and chain molecule.

chain-molecules persisting through the micelles reconciles the conflicting deductions derived from X-ray and viscosity data, and is in agreement with the mechanical properties of cellulose fibres.

Further developments are illustrated in Fig. 16, the chain molecules being shown as lines which are not always parallel but occasionally come together in orderly manner (as indicated by the darker lines) to give the effect of discrete crystalline particles. Nevertheless these aggregates or chain bundles are linked together by chains which extend from one region of regular arrangement to another, protruding from the ends of the crystallites as amorphous fringes. In the diagram, chain endings are shown by A and A', which come within the crystalline region, which extends from B to B'.

Where chain molecules are free, their ends may be readily separated, but if the ends lie within the crystallite it is necessary to overcome not only the weak forces between the ends of the chains but also the Van der Waals forces along the entire length of the chains within the crystallite; hence the strength is increased by the adjacent molecular chains. This conception helps to explain the fact that orientated regenerated cellulose shows only a slight fall in tensile strength as the degree of polymerisation falls from 560 to 300, but exhibits a greater decrease in strength when the degree of polymerisation falls below 200; similarly with native cellulose, where there is little change as the degree of polymerisation falls to 800 or so, but a great change at 600 and below. When the molecular chain is longer than the micelle its ends are distributed at random within the micelle and the weakness of chain endings is reinforced by the crystal forces holding the chains together; with shorter chains, however, the end of the chain is apt to come near the end of the crystallite so that the strength is lower.

More recent developments in the structure of cellulose have come from consideration of regenerated cellulose. The chain molecules of

FIGS. 16 and 17.—Micelle and chain molecule.

cellulose are not necessarily straight; indeed, since free rotation is possible at every glucosidic linkage, there is a tendency to agglomeration where the chains are sufficiently free. On the other hand, the secondary valency forces of the hydroxyl groups tend to bring about an alignment of adjacent chain molecules, producing an orderly arrangement; the secondary valencies cause a crystalline arrangement.

The agglomerated state predominates when the cellulose exists in dilute solution, and the crystalline state predominates in the solid form, but even here there are some agglomerated molecules whose presence becomes more prominent on swelling. Individual micelles are no longer held to exist as fundamental structural elements but, as in other aspects of cellulose chemistry, the old term is still employed with a different meaning. Micelles are regions where the crystalline tendency predominates but they have fringes of chain molecules in which the disorderly array accounts for the presence of "amorphous" material (see Fig. 17). The persistence of some of the chain molecules through the micelles explains the high tensile strength of a system of small units and reconciles the requirements of X-ray and viscosity measurements.

One of the difficulties of the older conceptions of the structure of cellulose was to explain the cohesion of highly swollen regenerated cellulose on the basis of non-swelling crystallites, for the X-ray pictures of dry and wet cellulose show no difference. The modern view assumes that water penetrates between the micelles and into the fringes or clusters of molecular chains which open and close, umbrella fashion, on wetting and drying.

Boulton, Delph, Fothergill and Morton (J.T.I., 1933, *24*, 113P) pictured cellulose as a three-dimensional network of more or less

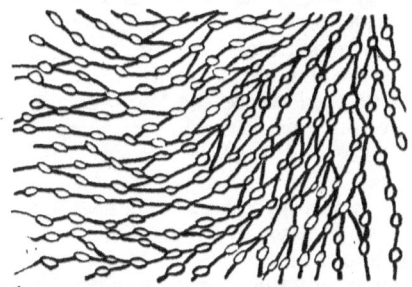

Fig. 18.—Diagram of the network basis of the micellar system of cellulose.

parallel chain molecules which occasionally produce regions of crystalline regularity. Between the crystallites, the molecular chains form, in the swollen condition, an open network through which small molecules easily pass. The network basis has been extended further in recent years, and the micellar system of cellulose is now pictured as a porous structure in which regions of crystalline arrangement are suspended by fringes or regions predominated by the tendency to agglomeration.

Further developments are being made by Hermans and Kratky (Kolloid Z., 1939, *86*, 245), who are attempting to bring the mechanical properties of viscose filaments with different amounts of orientation into some sort of scientific system. Kratky's publications (ibid., 1938, *84*, 149; 1939, *88*, 78) envisage a network of long chains of micelles held together by cross-linkages of micelles. This is represented in Fig. 18, where micelles are shown as lines and the amorphous intermediate areas as ellipses. Two types of structure were postulated (*a*) the ideal loose network which on deformation behaves as separate chains, and (*b*) the ideal close network; a number of equations were developed on this basis. In the loose network, the micelles are suspended in the swelling medium without any mutual interaction, but in the close network, the swelling is such that owing to the expansion of the pores a considerable extensibility is established, but there is always some interlocking of the whole porous system which prevents a slipping of the chain bundles over one another.

Deformation of the close network produces a volume contraction due to a tightening up of the net.

Highly extensible, swollen, isotropic filaments of regenerated cellulose have been examined by Hermans and Kratky, and during extension it was found that increase in orientation accompanied extension up to 80%, but in the further extension from 80 to 120% no further orientation took place. The cluster of chain molecules protruding from the micelles and acting as hinges must, therefore, be regarded as elastic and not as inextensible. During extension, when the point is reached where the transverse connections of the micelles offer appreciable resistance to stretching, then an extension of the molecular fringe takes place without any improvement in micellar orientation.

The *intermicellar system* of cellulose has been examined by Frey-Wyssling (Protoplasma, 1937, *27*, 372). Cellulose was impregnated with solutions of gold and silver salts and crystals of the metals precipitated by reduction; these produce both dichroic and X-ray effects. It was estimated that the particles of gold and silver deposited in the cellulose fibres have diameters of about 100 Å, which is about twice the width of the micelles. Hence the fibres must contain coarse capillaries, which accommodate the colloidal dyestuffs, but fine capillaries of the order of 10 Å are also postulated, water and molecular solutions being presumed to penetrate into these. The two capillary systems possess branching habits (anastomosis) and merge into one another forming a hetero-capillary system.

The data of Frey-Wyssling are perhaps open to criticism on account of the swelling process during impregnation with the salt solutions; 100 Å must be regarded as a maximum value. The theory of dyeing has also contributed to our knowledge of pore size and the present position has been reviewed by Boulton and Morton (J.S.D.C., 1940, *56*, 145). Dry viscose sheet is impermeable to alcohol, benzene and picric acid, which pass easily through the swollen material. Hence the capillary canals may be estimated to be about 5 Å or less in the dry material and 20 to 30 Å in the water-swollen state. In the swollen state all fibres take up dyestuffs, but in the unswollen condition viscose rayon and cotton mercerised under tension are almost unstained, whereas cuprammonium rayon and the native fibres absorb dyes almost as easily as in the swollen state. Presumably the capillary system of the latter class contains comparatively large canals.

It is frequently stated that the amount of pore space in cotton is some 30 to 40% of the structure, but a recent publication by Peirce and Lord (J.T.I., 1939, *30*, 173) indicates that the volume of the minute crevices in dry cotton is less than 1% and increases to about 3·5% at a moisture content of 6%.

The micro-fibrils of Frey-Wyssling, mentioned above, must not be confused with the fibrils seen in native cellulose under the microscope, for the micro-fibrils are submicroscopic and consist of chains of micelles. Fibrils, on the other hand, may actually be obtained from native cellulose by a combination of chemical and mechanical treatment and may even be dissected further to smaller particles known as fusiform bodies, dermatasomes, etc. (See page 10.)

The dimensions of some of the components which go to make up cotton cellulose have been tabulated by Frey-Wyssling, as shown on page 36.

Orientation

In discussing the orientation of the micelles in cellulose, Herzog (Naturwiss., 1921, *9*, 320) pointed out that in ramie and other bast fibres the micelles are orientated parallel to the fibre axis, in cotton they are turned spirally round the axis, whereas in rayon this orientation may be missing, but it can be produced by stretching.

When X-ray photographs of cellulose are examined, interference spots seem to be more sharply defined where the fibres have a high degree of orientation, but where the orientation is less perfect, the spots tend to spread into arcs of circles. These features may be noticed in Figs. 16 and 17.

The high degree of orientation in the ramie fibre was the reason for its almost exclusive use in most of the X-ray examinations of cellulose, particularly in the early days.

Morey (Text. Res., 1934, *4*, 491) made important contributions to the knowledge of micellar orientation and spiral structure of fibres, by measurements of the polarisation of the fluorescence from fibres dyed with strongly fluorescent dyes; quantitatively, the method is perhaps open to criticism. He found that the highest orientation occurred in flax, whereas it was hitherto thought that ramie was the best orientated cellulose fibre. The method also showed the degree of spiral arrangement in the native fibres to be higher in flax than in ramie. The following results were obtained when a 5% dyeing was used from Thioflavine S (Du Pont).

FLAX was the most highly orientated of the fibres examined, giving an orientation of 73%, with 5 to 10% of unorientated structure. The highest value observed was 85%. In view of the

TEXTILE BLEACHING

From Frey-Wyssling (Die Stoffausscheidung der höheren Pflanzen; Springer; Berlin.)

The figures refer to the number of times the smaller structure is contained in the larger, the value for the chain molecule is arbitrary. The Ångström unit, Å is 10^{-8} cm.

	MACROSCOPIC	MICROSCOPIC			SUBMICROSCOPIC		AMICROSCOPIC	
	Section. $(0\cdot01)^2 \mu \times 50$ mm.	Lamella. $0\cdot4 \times 10\pi \times 5 \times 10^4 \mu$	Fibril. $0\cdot4 \times 0\cdot4 \times 100\ \mu$	Dermata-some. $0\cdot4 \times 0\cdot4 \times 0\cdot5\ \mu$	Micelle. $60 \times 60 \times 750$ Å	Chain molecule. $7\cdot5 \times 750$ Å	Cellobiose residue $7\cdot5 \times 10\cdot3$ Å	Glucose residue. $7\cdot5 \times 5\cdot2$ Å
Cellobiose residue								2
Chain molecule							about 75	about 150
Micelle						about 100	$7\cdot5 \times 10^3$	$1\cdot5 \times 10^4$
Dermatasome					3×10^4	3×10^6	$2\cdot2 \times 10^8$	$9\cdot5 \times 10^8$
Fibril				200	6×10^6	6×10^8	$4\cdot4 \times 10^{10}$	9×10^{10}
Lamella			4×10^4	8×10^6	$2\cdot4 \times 10^{11}$	$2\cdot4 \times 10^{13}$	$1\cdot8 \times 10^{15}$	$3\cdot6 \times 10^{15}$
Section.		25	1×10^4	2×10^8	6×10^{12}	6×10^{14}	$4\cdot5 \times 10^{16}$	9×10^{16}
Hair. $(0\cdot01)^2 \pi \times 50$ mm.	1	25	about 1 million	about 1/10 milliard	order of 1 billion	order of 1 billiard	order of 1/20 trillion	order of 1/10 trillion

preference for ramie in X-ray work, this high value is surprising, but the angle of spiral is greater, 5·5°, which means a greater spreading of each X-ray spot into an arc is obtained than with ramie.

RAMIE fibre gave a mean orientation of 69%, but the spiral which runs in the same direction as in flax was 3·5° only and showed much less variation. (Average deviation of $\pm 1°$, compared with $\pm 3°$.)

HEMP had a mean orientation of 59% with an average spiral of 0°.

COTTON was chiefly characterised by the absence of well-defined values for orientation or spiral. The mean orientation was only 39% with a deviation of $\pm 15\%$, compared with $\pm 7\%$ for flax and $\pm 6\%$ for ramie, but in spite of this great variation it is possible to assign an average orientation value. For the angle of spiral, however, the values ranged from 45° in one direction to 50° in the other, though it is usually accepted as lying between about 25° and 30°. Where some sort of average value is desired, the X-ray method is to be preferred, for this gives an integrated value over a considerable volume of fibres at once.

It is well known that cellulose fibres are very strong in the longitudinal direction; this is another example of their anisotropic character. The fact that well-orientated viscose fibres have a tensile strength of 80 Kg./sq. mm. as compared with 100 Kg./sq. mm. for flax shows the importance of the position of the micelles. Orientated viscose has a strength not much below that of flax, whilst the strength of cotton varies from a quarter to three-quarters of this strength although cotton cellulose has a longer molecular chain-length than viscose.

The extensibility of cellulose fibres also depends on the arrangement of the component micelles. When perfectly orientated, the micelles of a fibre under tension will slip on each other with a consequent diminution in cross-sectional area until the fibre breaks with an extension of about 2%. Owing to the parallel arrangement of the micelles, considerable force is required to overcome the friction, which may be assumed to be proportional to the area of contact, so that the tensile strength is relatively high. If, on the other hand, the degree of orientation is low, the effect of tension is to bring about extension and orientation—the extension at break is relatively high and the breaking load low.

The relative proportions of crystalline and amorphous cellulose are matters of some consequence, and until recently their determination was based almost entirely on measurements of the intensity of the diffuse background radiation in the X-ray diffraction pattern. More recently, however, it has been considered that a chemical approach is possible; if solid cellulose is submitted to a suitable chemical change, the percentage of the amorphous and accessible region may be calculated from the extent of the initial rapid reaction which is confined to the amorphous area. For instance, if cellulose is brought into solution by H_2SO_4 or HCl, the hydrolysis to glucose proceeds rapidly in the homogeneous system; in the heterogeneous system, however, the reaction proceeds rapidly at first and then almost reaches a standstill. This is assumed to be due to the difference in accessibility of the amorphous regions and the crystallites.

Results from various sources have been collected and tabulated on page 39.

THE MOISTURE RELATIONS OF CELLULOSE

It is a well-known fact that cellulose contains a certain amount of moisture even when it feels "dry." When, however, the cotton is dried by heating and then exposed to ordinary cold air, it quickly absorbs most of the moisture it has lost on drying. This property is common to all substances which have large surfaces and exist in the colloid state. The amount of moisture in cellulose varies with the relative humidity of the surrounding atmosphere and with the temperature. It is obvious that the amount of moisture in the cotton is an important factor when buying and selling are considered, as raw cotton and yarns are sold by weight. The accepted regain is therefore fixed at 8·5 %, the moisture being expressed as a percentage of the dry weight. The moisture contained in cotton is also referred to as the moisture of "condition," and the term moisture content strictly refers to the amount of moisture contained in conditioned cotton expressed as a percentage of the weight of the material in its conditioned state.

Whilst both relative humidity and temperature affect the moisture content, the former is the more important, for if it is kept constant the amount of water does not vary over a small range in temperature. On the other hand, there is considerable variation with a change in humidity. This variation has been studied by many observers, but there can be no doubt that our present knowledge of the absorption of moisture by cellulose in largely due to the work of Urquhart and Williams, published in a series of papers in the *Journal of the Textile Institute* since 1924.

The results showed that there are two possible values of moisture regain, a lower value if the cotton was initially drier than the atmosphere and a higher value if it had been wetter. The difference between the two sets of observations had been noted previously, but Urquhart and Williams showed that a real hysteresis existed in the case of cotton. The taking up of water was termed "Absorption" and the giving up of water "Desorption."

The moisture regain of cotton definitely depends on its previous heat treatment, heating to a high temperature reducing the capacity to absorb water. The hysteresis shown by soda-boiled cotton is

Amorphous Component of Cellulose

Material	Method	Amount	References
Linters	TlOEt	0.4%	1. Purves; J.A.C.S., 1944, *66*, 59
Ramie	TlOEt	0.25%	2. Purves; Paper Trade J., 1940, *110*, 29
Ramie	Infra-red	Very small	3. Ellis & Bath; J.A.C.S., 1940, *62*, 2859
Linters	X-ray	Almost none	4. Kratky; Silk & Rayon, 1939, *13*, 480
Cellulose	—	10%	5. Meyer; Ber., 1937, *70*, 266
Rayon unstretched	—	60%	6. Mark; Ind. Eng. Chem., 1942, *34*, 449
Rayon stretched	—	30%	7. Mark; Ind. Eng. Chem., 1942, *34*, 449
Viscose rayon	HCl/FeCl$_3$	21%	8. Nickerson; ibid., 1941, *33*, 1022; 1942, *34*, 85, 1480
Linters	—	5%	9. Nickerson; ibid., 1941, *33*, 1022; 1942, *34*, 85, 1480
Linters mercerised	—	11%	10. Nickerson; ibid., 1941, *33*, 1022; 1942, *34*, 85, 1480
Cellulose	Diazomethane	0.4%	11. Reeves & Thompson; Contrib. Boyce Thompson Inst. 1939, *11*, 55
Ramie	Esterify/X-ray	30–50%	12. Hess & Trogus, Kolloid Z., 1934, *57*, 168
Cotton fibres	Periodic Acid	1–2%	13. Mark et. al; Ind. Eng. Chem.; 1943, *35*, 1083
Cotton linters	,, ,,	6%	14. Mark et. al; Ind. Eng. Chem., 1943, *35*, 1083
Bemberg rayon	,, ,,	7.4%	15. Mark et. al; Ind. Eng. Chem., 1943, *35*, 1083

greater than that for raw cotton. It was found that the moisture content of soda-boiled cotton is 22·6% at 100% relative humidity; the measurements were made at 25°C.

Urquhart and Williams (J.T.I., 1925, *16*, 155) examined the absorption of moisture by cotton mercerised without tension, in order to obtain accurate knowledge of the moisture content at all humidities, and also the effect of varying the concentration of the mercerising solution.

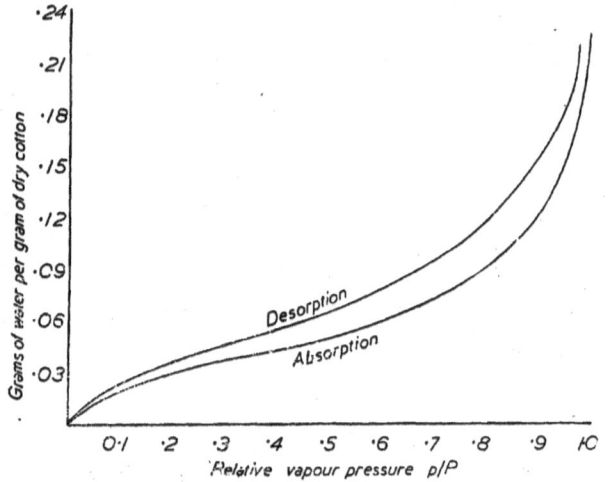

FIG. 19.—The absorption and desorption of water vapour by soda-boiled cotton at 25°C.

They noticed that the ratio of the moisture content of mercerised cotton to that of the original soda-boiled cotton at the same relative humidity is appreciably independent of that humidity. The mean values from the nineteen readings were 1·57 in absorbing and 1·46 in the desorption of moisture. These ratios are therefore indicative of the extent to which the absorptive capacity of the material has been altered by mercerising without tension in 15% sodium hydroxide solution.

The shape of the curves in Figs. 19 and 39 may be explained by a two-phase theory of water absorption. A relatively small amount of water appears to enter into combination with the dry cellulose, heat being evolved; this water contracts and does not give the X-ray diagram of liquid water. It seems likely that this water is absorbed on the hydroxyl groups of the cellulose, whereas the remaining 20 to 25% of water which can be taken up is bound by capillary forces. The later form of the curves, therefore, is determined by capillary condensation.

Collins (J.T.I., 1930, *21*, 313) measured the swelling of cotton hairs under various conditions of humidity, and found a fairly close parallelism between the extent of change in dimensions and the amount of moisture absorbed, between the temperature limits of 20° to 100°C.

In passing from dry air to water, the increase in area of cross-section is about 44% and the increase in length about 1%, when the hair is fully wetted.

The breaking load of cotton increases with increasing moisture content; the "wet-strength" is 110 to 120% of the dry strength.

THE CHEMICAL RELATIONS OF CELLULOSE

From the structure of cellulose, the effect of chemical reagents may be considered under three sections: first, those reagents which are capable of reacting with the hydroxyl groups of the glucose residues; secondly, reagents which cause a diminution in the length of the molecular chain; and, thirdly, reagents which alter the relative positions of the chain-molecules.

DERIVATIVES

In the organic chemistry of cellulose the chief feature is the presence of three functional groups which are alcoholic in nature; those hydroxyls in the *2* and *3* positions belong to secondary alcohols and that in the *6* position to a primary alcohol.

The general formation of cellulose derivatives is a matter of the substitution of these groups, and it is now becoming common to show this more clearly by using the formula $C_6H_7O_2(OH)_3$ in place of $C_6H_{10}O_5$. Thus cellulose trinitrate may be written $C_6H_7O_2(ONO_2)_3$; cellulose triacetate $C_6H_7O_2(OCOCH_3)_3$ and the trimethyl ether of cellulose as $C_6H_7O_2(OCH_3)_3$.

The chemical reactions of cellulose are complicated by several factors, one of which is the degree of orientation of the material, and this also exerts a great influence on the physical properties, as shown on page 94. The extent of orientation controls the rate of penetration of the reactant into the interior of the fibre and so influences not only the simpler process of dyeing, but also the mechanism of chemical reaction; for instance, Elöd and Schmid-Bielenberg (Z. phys. Chem., 1934, *B25*, 27) found that the velocity of acetylation of cellulose increased in the order of decreasing orientation of the structural units in flax, ramie and cotton.

The early views on the nature of the lower derivatives of cellulose supposed these to be mixtures of cellulose and the tri-substitution product, but this idea has given way to the more complex conception of heterogeneous micellar reactions.

With purified fibres, it is assumed that the reagent penetrates the fibre, and first attacks the surface of the micelle, which is transformed into the tri-derivative. The reagent then penetrates into the micelle and produces a partially reacted area in which only some of the hydroxyl groups are substituted and which lies between the fully-substituted surface and the untreated interior. It would appear that in many cases the heterogeneous nature of reactions with cellulose is due to both micellar structure and fibre structure. The less organised parts of the structure would tend to react before the crystalline regions, and this may produce a random arrangement of substituted groups along the cellulose chains. The outer sections of the fibre and the external chains of the crystallites, probably tend to be more highly substituted than the interior. Recent work on the cellulose acetates by Sakurada (J.S.C.I., Japan, 1938, *41*, 381) indicates that chemical reaction first takes place in the non-crystalline fringes which unite the micelles and then spreads along the chain molecules to the micelles.

The two great classes of derivatives of cellulose are the esters and ethers. Cellulose nitrate is the best-known example of an inorganic ester and cellulose acetate of the organic esters. Cellulose formate may be made by the action of formic acid itself and is unique in this respect, for other organic esters necessitate the use of acid and acid anhydride; acid chlorides are used for the preparation of the higher esters.

Cellulose ethers have been investigated to a great extent, and the best known are the methyl, ethyl, benzyl and hydroxyethyl celluloses. Methyl cellulose is generally prepared by the action of methyl chloride or sulphate on soda cellulose and ethyl cellulose may be made in a corresponding manner. The solubility of the ethers varies according to the degree of substitution, ethers of the highest degree of substitution being soluble in organic solvents; water-soluble ethers have a lower degree of substitution, and the alkali-soluble ethers the lowest. Hydroxyethyl ethers may be made by treating soda cellulose with ethylene chlorhydrin but an alternative method is to use ethylene oxide.

MODIFIED CELLULOSE

The chemical modifications of cellulose do not include the typical organic derivatives such as esters and ethers, but refer to the early products of degradation—the water-insoluble modifications produced by the action of acid or oxidising agents. The former are still referred to as "hydrocelluloses" and the latter as "oxycelluloses," probably on account of the belief of early investigators that these were chemical compounds which could be recovered from the portion of the material which, unlike native cellulose, was soluble in cold alkali of moderate concentration.

The early literature on hydrocellulose has been well reviewed by Clifford (J.T.I., 1923, *14*, 169); Clifford and Fargher (J.T.I., 1922, *13*, 189) have given a good summary of the work on oxycellulose.

At one time there was a little confusion between hydrocellulose, the result of acid hydrolysis of cellulose—and hydrate cellulose, the result of swelling with sodium hydroxide solutions. No such confusion exists to-day, but it is perhaps interesting to point out that the modified celluloses have new chemical properties, whereas the mercerised cellulose has the old chemical properties to an enhanced degree. This has been shown very clearly by Neale (Trans. Farad. Soc., 1933, *29*, 228) in tabular form, which is reproduced in the following pages.

The chief methods used for the evaluation of the modified celluloses are copper number, solubility number and fluidity. These tests are measures of the chain length of cellulose.

Hydrocellulose and oxycellulose are of importance in bleaching vegetable fibres, for the former may be formed by careless acidification or "souring" of the goods, and the latter may be produced by the uncontrolled action of bleaching agents.

Hydrocellulose

Hydrocellulose formation is a relatively simple matter, merely involving scission of the molecular chain of cellulose. The term hydrocellulose was originally applied to the powdery material which resulted from the action of acids on cellulose but the modern definition is confined to products which still retain their fibrous form. In technical practice, damage by the action of acids is generally confined to a partial loss of tensile strength and the powdery hydrocelluloses have received little attention for the past twenty years.

The sensitivity to acid is a weakness of cellulose and appears in strong contrast to the behaviour of the animal fibres.

Hydrocellulose may be produced in two ways: treatment with hot dilute acids (the acid-steeping method), and impregnation followed by drying (the acid-drying method).

One of the best and most systematic surveys of the preparation and properties of hydrocellulose was made by Birtwell, Clibbens and Geake (J.T.I., 1926, *17*, 145). The fall in tensile strength of cotton caused by the action of acids is accompanied by a fall in the viscosity of the material when dissolved in cuprammonium hydroxide solution, and by a rise in the reducing value or copper number. Both

Properties which Change with "Degradation" of Cellulose

		Tensile strength dynes/cm²	Fluidity in 0·5% Solution	Copper No. (Braidy) Gms. Cu. reduced by 100 g. of Cellulose	Absorption of Methylene Blue (Basic dye).
Natural cotton cellulose scoured with 1-2% NaOH under 10-40 lb./sq. in. pressure		$4·9 \times 10^9$	1·5	0·02	0·3 to 0·8
"Degraded" cotton cellulose	Treated HCl 200 g./l. for 24 hrs. at 20°C.	$1·5 \times 10^9$	34·5	2·44	do.
	Oxidised by alkaline hypobromite 0·32 g. O_2 consumed by 100 g. cellulose.	$2·4 \times 10^9$	31	0·5	3·0
	Oxidised by hypochlorous acid 0·32% O_2 consumed	—	29	3·4	1·1
"Activated" cotton cellulose	Treated 25% NaOH without restraint on swelling or shrinkage, washed, air dried	These properties are unaffected by swelling or activation of cellulose.			
	As above, but without allowing shrinkage	—	—	—	—
	Treated without shrinkage, but dried at 110°C.	—	—	—	—
	Swollen in 65-70% H_2SO_4 without restraint (Time of action— 5 mins.)	See	refs. (20, 6,	12, 7, 8, 3)	
Cellulose both "degraded" and "activated"	Viscose rayon	About 4×10^9	40	1·1	—
	Cupra rayon (Bemberg)		28	0·5	—
REF. Nos.		19, 12, 6	9, 4, 12, 21	4, 6, 21	3, 4

References.
  2. Birtwell, Clibbens, Geake and Ridge . . J.T.I., 1920, *1*, 285.
  3. Birtwell, Clibbens and Ridge . . . J.T.I., 1923, *14*, 297.
  4. Birtwell, Clibbens and Ridge . . . J.T.I., 1925, *16*, 13.
  6. Clibbens . . . . J.T.I., 1923, *14*, 217.
  9. Clibbens and Geake . . . . . J.T.I., 1928, *19*, 77.
 12. Farrow and Neale . . . . . J.T.I., 1924, *15*, 157.
 17. Neale . . . . J.T.I., 1931, *22*, 320.
 18. Neale . . . . J.T.I., 1931, *22*, 349.

CELLULOSE CHEMISTRY

Properties which Increase with Swelling or "Activation" of Cellulose

Absn. of water gms./gm. at 50% Relative Humidity.	Absn. of NaOH from N/2 soln.	Absn. Ba(OH)$_2$ from N/5 soln.	Rise in Cu. No. (Schwalbe) after 15 mins. in boiling 5% H$_2$SO$_4$	Cu. No. (Braidy) after 2 hrs. in N/10 KOH, N/10 KBrO at 18° C.	Absn. Sky Blue FF (Direct dye) at 100° C.
	Milli-Equivalents per Glucose Unit				
0·055 Ratios	42·5	71	2·2	1·5	0·15%
0·97	Unaffected by hydrolysis or oxidation of the cellulose. (See refs. 17 and 18)		Tests obviously invalidated by great degradation of cellulose, but unaffected by mild degradation. (See ref. 2)		
0·97	—	—	—	—	—
	Expressed as Ratios Relative to Scoured Cotton.				
1·50	2·55	2·70	1·7	1·6	1·8
1·35	1·96	2·05	—	1·45	1·6
1·2	1·89	1·99	—	1·45	—
1·83 (max.)	3·20	3·20	—	2·54 (max.)	—
2·0	3·6	4·0	5	—	0·9
1·84	3·4	3·8	5	—	2·9
27, 25, 2, 26, 28	17	18	22	2	29

19. Peirce . . . . J.T.I., 1923, *14*, 170.
21. Ridge, Parsons and Corner . . . . J.T.I., 1931, *22*, 118.
22. Schwalbe . . Z. angew. Chemie., 1908, *21*, 1931; 1909, *22*, 197.
25. Urquhart and Williams . . . . J.T.I., 1924, *15*, 138, 433.
          1926, *17*, 38.
26. Urquhart and Williams . . . . J.T.I., 1927, *18*, 55.
27. Urquhart, Bostock and Eckersall . . J.T.I., 1932, *23*, 135.
28. Urquhart and Eckersall . . . . J.T.I., 1932, *23*, 163.
29. Neale.

of these properties reflect the average chain length of the cellulose material, so that one might reasonably expect to find some intercorrelation between these properties and the tensile strength.

There is also a definite relation between the strength and the copper number of the hydrocelluloses, a diminution of 10% in breaking load corresponding to a rise of 0·25 in copper number, and a decrease of 80% in tensile strength to a rise of 3·5. The copper number is, however, of relatively little general value as a measure of the extent of acid tendering, owing to the fact that it is greatly reduced by alkali boiling under conditions which have little effect on the viscosity.

Birtwell, Clibbens and Geake (loc. cit.) also made a quantitative examination of the increased affinity for basic dyes as the result of the action of sulphuric and phosphoric acid of high concentration on cotton, compared with the decreased affinity in the general case of acid attack. The initial fall in absorption is followed by a rapid rise when the concentration passes a certain value. This is due to retention of the acid.

Oxycelluloses

Systematic investigation of the *oxycelluloses* by Birtwell, Clibbens and Ridge (J.T.I., 1925, *16*, 13) showed that the products fall between two extreme types, one characterised by great affinity for Methylene Blue and abnormal retentive power for alkali, the other by high reducing power as shown by copper number, and excessive loss in weight on boiling in alkali. Both types show a decreased tensile strength.

The highly acidic type of oxycellulose is formed by alkaline oxidising agents, and appears to be due to the action starting at the more exposed carbon atoms in the *6* position of the glucose residue; cleavage of the molecular chain probably takes place at the same time. The alcoholic group is converted to aldehyde and then to carboxyl as shown by Kalb and von Falkenhausen (Ber., 1927, *60*, 2514).

$$
\begin{array}{cc}
\underset{\text{Acidic}}{\begin{array}{c} \mathrm{CH} \\ \diagup \quad \diagdown \\ \mathrm{O} \quad\quad \mathrm{HOCH} \\ | \quad\quad\quad | \\ \mathrm{HOOCCH} \quad \mathrm{HCOH} \\ \diagdown \quad \diagup \\ \mathrm{CH} \end{array}}
&
\underset{\text{Reducing}}{\begin{array}{c} \mathrm{CH} \\ \diagup \quad \diagdown \\ \mathrm{O} \quad\quad \mathrm{CHO} \\ | \\ \mathrm{HOCH_2CH} \quad \mathrm{CHO} \\ \diagdown \quad \diagup \\ \mathrm{CH} \end{array}}
\end{array}
$$

The high-reducing type of oxycellulose, which is also alkali-sensitive, is probably formed by a cleavage of the glucose ring between the carbon atoms in the *2* and *3* positions as discussed by Davidson (J.T.I., 1938, *29*, 195).

When oxycelluloses of the reducing type are boiled with dilute alkali under suitable conditions they become indistinguishable by chemical means from normal unmodified cotton. The viscosity measurement of such material reveals the change, no matter what the subsequent treatment may have been. Oxycelluloses characterised by increased affinity for basic dyes are not altered by boiling in alkali.

The type of oxycellulose which is produced by hypochlorite depends chiefly on the alkalinity or acidity of the hypochlorite solution, as is shown in the curves of Fig. 68.

Clibbens and Ridge (J.T.I., 1927, *18*, 135) pointed out that the rate of oxycellulose formation is a term which cannot possess a precise significance, as the word "oxycellulose" denotes the product of a group of reactions rather than a unique chemical individual. Examinations were made of the rate of consumption of oxygen by cotton cellulose and also the rate of change of properties which vary with oxidation, i.e. copper number, methylene blue absorption and viscosity. No general correlation can be expected between the various properties, as a rapid loss in strength may be accompanied by either a slow or rapid change in solubility in alkali.

In the case of cotton treated with hypochlorite solution of varying acidity or alkalinity the results are shown in Figs. 70-72. The rate of increase of copper number is slow with alkaline liquors and at pH 13 the cotton is scarcely affected. With neutral hypochlorite, the rate of increase of copper number is very high, but falls again with hypochlorous acid. The maximum effect of the neutral solution is again seen in relation to the rate of rise in methylene blue absorption and the rate of fall in viscosity.

Hypobromite oxidation, when examined under conditions of relatively high alkalinity, showed a much greater rate of oxidation than hypochlorite. Alkaline hypobromite, however, under certain conditions acts as a sort of automatic copper number regulator—a fact which is utilised in the measurement of the reactivity of mercerised cotton.

Solubility of Modified Cellulose

Oxycelluloses suffer a considerable fall in viscosity as the result of an alkaline boil, and differ in this respect from the hydrocelluloses. The extent of the viscosity change varies according to the manner in which the oxycellulose is formed, but is greater for neutral hypochlorite oxycelluloses.

The percentage loss in weight on boiling for four hours with 1% NaOH solution at ordinary pressure is about six times the copper number when the latter does not exceed 2·5; for modification which results in a copper number greater than 2·5, hydrocelluloses experience a greater loss, and oxycelluloses a smaller loss of weight than corresponds with this relation, and the divergence becomes wider with increasing modification of the cellulose.

The fluidity of acid-tendered material is not altered by kier-boiling, nor is there any alteration in the fluidity of cotton oxidised with alkaline hypochlorite. With other oxidising agents, both tendering and fluidity are increased by the alkaline boil, but the most marked rise in fluidity on boiling is shown by the neutral hypochlorite oxycelluloses, and the rise is comparatively slight with the materials oxidised by dichromate, which latter, however, show the greatest additional loss of strength on boiling with alkali.

Solubility at Room Temperature

It has been known almost since the chemically-modified cotton celluloses were first described, that they possess the property of dissolving to some extent in sodium hydroxide solutions at room temperature and could be precipitated without much change. It was probably this property which led the early investigators to attempt to isolate an individual product to which the name "oxycellulose" or "hydrocellulose" might properly apply.

Birtwell, Clibbens and Geake (loc. cit.) established that with the hydrocelluloses formed from bleached cotton, the solubility is completely defined if the copper number is known, irrespective of the mode of acid attack. In the case of the oxycelluloses, however, different relations obtain, for although in any one series of oxidised cottons formed by the action of the same oxidising agent under the same conditions for varying times the copper number defines the solubility, yet for different oxidising agents or the same agent under different conditions, there is no general relationship between solubility and copper number. There appears to be a general correlation between the solubility of modified cotton in $10\ N-2\ N$ NaOH and its fluidity in cuprammonium solution, which is independent of the manner of modification.

Solubility at Low Temperatures

Davidson (J.T.I., 1934, 25, 174), in his examination of the solubility of modified cellulose in alkali, established that the solubility is at a maximum at a certain alkali concentration for any temperature, and that as the temperature is lowered from the normal, the maximum solubility is greatly increased; it also occurs at a lower alkali concentration.

Oxycelluloses of the hypochlorite series are more soluble than hydrocelluloses of the same fluidity, but after an alkali-boil these oxycelluloses show a fluidity-solubility relation nearly the same as that found for hydrocelluloses from unmercerised cotton. On the other hand, oxycelluloses from swollen cotton have a relatively low solubility for a given fluidity.

Davidson also established that when modified cottons are boiled under pressure with dilute alkali, the residual material is more soluble in NaOH solution at $-5°$C. than the original modified cotton, in spite of the loss in weight. The increase in solubility is slight for hydrocelluloses from unmercerised cotton, but is considerable with hydrocellulose from mercerised cotton and also with oxycelluloses of the hypochlorite series.

The solubility of any modified cotton, measured under optimum conditions at $<5°$C. was found to increase with increasing fluidity —samples with very high fluidities being completely soluble. Hydrocelluloses from unmercerised cotton, whether boiled after modification or not, have approximately the same fluidity-solubility relation, but hydrocelluloses from mercerised cotton are much less soluble than material of equal fluidity similarly prepared from unmercerised cotton.

General

Although the properties of the various oxycelluloses vary according to the method of formation, there is always a fall in tensile strength and a rise in fluidity; the relation between these two properties has shown that the loss in strength for a given rise in fluidity is greatest for the hydrocelluloses and least for the oxycelluloses prepared by the action of dichromate in oxalic acid solution. If the modified celluloses are subjected to an alkali boil, the tensile strength is diminished and the fluidity increased to an extent which also depends on the manner of modification—the dichromate oxycelluloses suffering the greatest additional loss in strength. The changes of fluidity and strength produced by boiling are such that the fluidity-strength relation for the boiled material is approximately the same whatever the method of modification. The behaviour of the dichromate oxycelluloses and those prepared by oxidation in non-alkaline solutions is difficult to interpret on the theory of cellulose as a chain molecule, but Davidson (J.T.I., 1934, 25, 174; 1938, 29, 195) suggested that oxidation does not necessarily result in the scission of the chain molecule, but that the alcohol groups in the glucose residue are first oxidised to an aldehyde or that the ring structure may be broken, causing a weakening of the glucosidic linkage. This weak linkage may then be broken by boiling the oxycellulose with weak alkali or even with water and also by treatment

50   TEXTILE BLEACHING

with an alkaline solvent such as cuprammonium hydrate—processes which are without much action on hydrocellulose and have comparatively little effect on the oxycelluloses prepared in alkaline solutions.

As a result of the fluidity measurements on cellulose nitrates derived from oxycelluloses, Davidson was able to divide these into two classes:

(a) Those prepared from alkaline oxidising solutions which give almost the same relation between cellulose nitrate fluidity and cellulose fluidity as hydrocelluloses; the cellulose nitrate fluidity of these is little affected by alkaline treatment of the original oxycellulose.

(b) Those prepared with neutral or acid oxidising solutions, which give considerably lower cellulose nitrate fluidities than hydrocelluloses of equal cellulose fluidities; the cellulose nitrate fluidity of these may be greatly increased even by cold treatments of the original oxycellulose with feebly alkaline solutions.

There is very strong evidence that the oxycelluloses prepared by the action of acid or neutral oxidising agents contain alkali-sensitive linkages in the molecular chain.

DISPERSED CELLULOSE

The dispersion of cellulose may be accomplished by strong bases, strong acids, solutions of inorganic salts, and by certain specific agents such as cuprammonia.

Bases

Dispersion by strong bases takes place at room temperatures and in solutions of moderate concentration. The best-known example is the use of sodium hydroxide of 55 to 60°Tw. concentration for mercerising cotton. Strong swelling takes place and if the alkali is removed with the yarn under tension, then the cross-sections of the cotton hairs assume a circular shape which imparts a high degree of lustre to the mercerised material. Maximum swelling takes place at 35°Tw. NaOH but there is a preferential absorption of alkali during the process and this necessitates the use of higher concentrations. Cotton actually behaves as a weak monobasic acid and its iso-electric point has been determined as lying between pH 2·4 and 2·8 by Gavoret (Compt. rend., 1938, *206*, 1299) and confirmed at pH 2·5 by Sookne and Harris (Text. Res., 1941, *11*, 307).

Other alkaline hydroxides also exert a swelling action on cotton, and Heuser and Bartunek (Cellulosechem., 1925, *6*, 19) have shown increases in the diameter of cotton hairs of 97% with 9·5% LiOH, 78% with 18% NaOH, 64% with 32% KOH, 53% with RbOH and 47% with 40% CsOH. It will be noted that the order is that of the supposed hydration of the ions.

As the temperature of the solution is lowered, the amount of swelling increases but cotton does not dissolve in alkaline hydroxides unless it has been degraded by oxidation or acid hydrolysis. Regenerated cellulose, on the other hand, dissolves readily because of its shorter chain-length; 8% NaOH will dissolve rayon at room temperatures.

The maximum swelling of cotton by alkaline hydroxides takes place in solutions of moderate concentration. Strong organic bases such as the quaternary ammonium hydroxides, also exert a dispersing action on native cellulose and a number of them are sufficiently powerful to act as solvents, particularly the benzyl substituted ammonium hydroxides (B.P., 439, 806). Solvent action usually occurs at concentrations of 25% and over, with strong swelling at lower concentrations. Dimethyldibenzyl ammonium hydroxide appears to be the most powerful solvent. Brownsett and Clibbens (J.T.I., 1941, *32*, 32) have examined the action of various bases on cellulose and compared the results with their viscosities in equivalent aqueous solutions; the more viscous bases form the better solvents.

The dispersion of cellulose has been described in some detail by Marsh (*Mercerising*; Chapman and Hall; London, 1941).

Acids

Strong acids are also capable of dispersing native cellulose. Although concentrated solutions are generally required, swelling with sulphuric acid starts as the concentration rises above 50%, and passes through various stages until solution occurs. For "mercerising" with sulphuric acid the usual strength is 62% and parchmentising starts at 68 to 69% concentration. Above 78%, sulphuric acid does not parchmentise cotton but there is a rapid swelling followed by carbonisation and solution. Ordinary concentrated hydrochloric acid only causes slight swelling of cotton, but more concentrated solutions are capable of swelling and dissolving cotton; 36·3 to 37·2% HCl causes "mercerising", a slightly higher concentration (37·2 to 38·5%) produces high swelling and transparency, whilst 38·9% and over, causes gelatinisation and solution. Concentrated phosphoric acid also causes swelling followed by solution. A mercerising action is also produced by nitric acid of 67·5 to 74·5% concentration; stronger swelling, solution and nitration takes place in solutions of 77 to 80% nitric acid. The higher concentrations also produce parchmentising effects.

Salts

The swelling of cellulose in sodium hydroxide solution, and in sulphuric acid solutions was discovered by Mercer (B.P., 13,296 of 1850); he also observed the action of concentrated solutions of inorganic salts. In this case, however, high temperatures are necessary. This phenomenon was later examined by von Weimarn (Kolloid Z., 1912, *11*, 41), who stated that any salt will dissolve cellulose if applied under suitable conditions of temperature and pressure. The most efficient salts are those which easily become hydrated and are very soluble in water, e.g. thiocyanates of barium, calcium and strontium, and the iodides of sodium, calcium, strontium and lithium.

Herzog and Beck (Z. physiol. Chem., 1920, *111*, 287) stated that only certain salts possess solvent powers, which depend on the hydration of the ions and increase with it. The solvent powers of salts follows from the series $NH_4 < K < Na < Li$; $Ba < Sr < Ca$; $\frac{1}{2}SO_4 < Cl < Br < I < CNS$.

H. E. Williams (J.S.C.I., 1921, *40*, 221) studied the action of thiocyanates and concluded that salt solutions which have solvent powers of cellulose must have a positive heat of dilution.

Calcium thiocyanate solutions are very efficient solvents; a solution whose concentration produces a boiling point of 135 to 150° C. will dissolve cotton when heated to 80 to 100°C. Solutions of higher and lower boiling-points do not dissolve cellulose but exert a swelling and parchmentising effect. Cold solutions of 60% concentration produce a mercerising action and hot solutions a parchmentising effect.

Strong swelling is produced by cold saturated solutions of potassium iodide and the effect is increased by warming; barium iodide exerts a similar effect.

Zinc chloride in 40 to 60% concentration, produces swelling on gentle warming, and solution on further heating.

Specific Reagents

Cuprammonium hydrate is the best-known example of a specific reagent which disperses cellulose in virtue of chemical reaction; its action was discovered independently by Mercer and by Schweizer (J. pr. Chem., 1857, *22*, 109).

Nickel ammonium hydroxide also dissolves cellulose and solutions of other hydroxides in ammonia have also been suggested. Cuprammonia, however, remains the most popular solvent and is used for the production of regenerated cellulose and also as a medium for viscosity determinations as a measure of damage.

General

It has been suggested by Meyer (*Natural and Synthetic High Polymers IV*, Interscience Publishers, New York, 1942) that the hydroxyl groups of cellulose are amphoteric in character; in certain circumstances, they may form oxonium cations, as in inorganic

CELLULOSE CHEMISTRY 53

acids, or metallic complexes in solutions of inorganic salts. On the other hand, the hydroxyl groups may dissociate, liberating hydrogen ions, and forming anions to which water may be bound to form a hydrated anion; this may occur with alkalis and strong organic bases. Hence the hydroxyl groups of cellulose may be transformed into anions or cations according to the reagent used.

The swelling and the great imbibition of water, may be due to electrostatic forces tending to produce a uniform distribution of the charges in the whole liquid so that the cellulose salt expands; this effect probably predominates in presence of strong solutions of inorganic acids and salts.

With moderately concentrated solutions of alkali, the swelling of the cellulose has been explained by Neale (J.T.I., 1929, *20*, 373) in terms of the Donnan membrane equilibrium.

The dispersion of cellulose by cuprammonium hydroxide solution is not easy to understand; it has been suggested that the strong base forms a soluble basic salt with the cellulose, but it is also possible that the copper enters into a hydrophilic complex formation with the hydroxyl groups in the *2* and *3* positions.

Dispersed cellulose does not exhibit any chemical properties which are not possessed by native cellulose; it does, however, show the old properties to an enhanced degree.

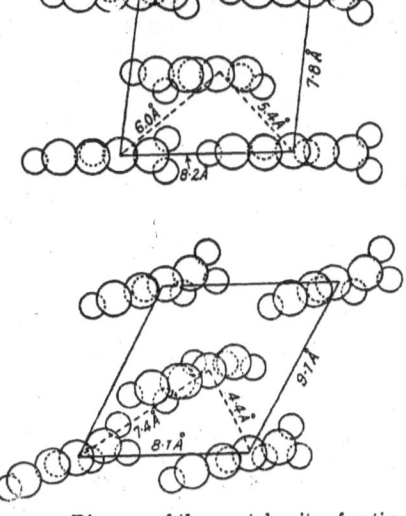

Fig. 20.—Diagram of the crystal units of native and dispersed cellulose.

The increase in moisture sorption and the affinity for dyestuffs are well known, and the increased absorption of alkali from dilute solution has been utilised as a measure of dispersion.

The activation of the old properties is displayed by an increased rate of hydrolysis with acids and this, too, has been utilised as a measure of swelling.

The effect of dispersing agents as revealed by X-rays shows that the planes of the molecular chains have moved apart as seen in Fig. 20. This is in agreement with the increased sorptive capacity, and the accessibility to chemical reagents produces greater reactivity.

Tensile strength, fluidity and copper numbers are not adversely affected by the mercerising process in which the cotton is merely swollen. With the regenerated celluloses, however, there is degradation as well as dispersion. Although the series cotton, mercerised cotton, and rayon represent three stages of increasing dispersion of cellulose, the attack on the cellulose during the manufacture of rayon gives to it some of the properties of modified cellulose. The low tensile strength, the high swelling and the increased activity necessitate careful handling in all aqueous liquids. Hence all wet finishing processes are a potential source of mechanical damage on account of the lower tensile strength, not only in the dry state, but also in the wet and highly swollen condition. Although cellulose is resistant to hot dilute alkaline solutions, yet it is sensitive to dilute acid and to oxidising agents. Cellulosic fabrics are of great value on account of their laundering capabilities, but the drastic conditions of scouring cotton and even linen cannot be applied to rayon. Furthermore, in the bleaching processes, the use of dilute acid and of oxidising agents entail greater risk with rayon than with cotton on account of the activated state of the former.

PLATE V

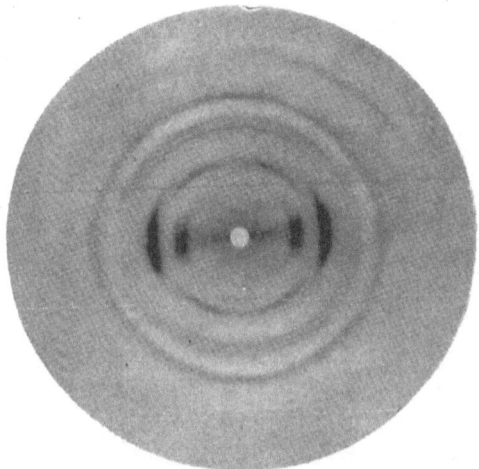

FIG. 21. X-ray photograph of native cellulose.

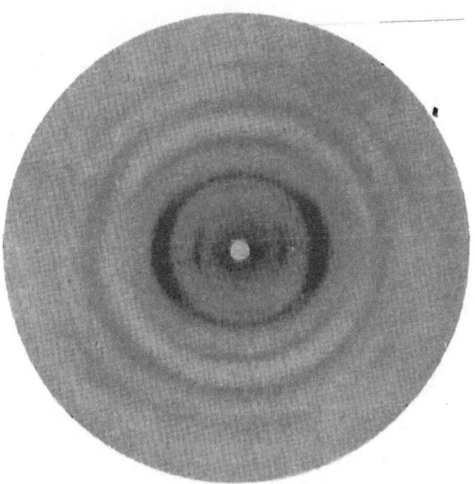

FIG. 22. X-ray photograph of dispersed cellulose.

Courtesy of Dr. Astbury, F.R.S.

To face page 54.

PLATE VI

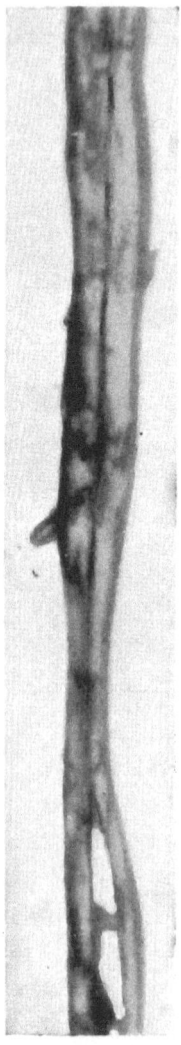

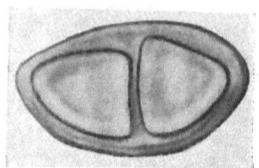

FIG. 23a. Photomicrograph of cross-section of raw silk. (× 1000 diameters)

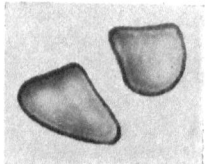

FIG. 24a. Photomicrograph of cross-section of degummed silk. (× 1000 diameters)

(Figs. 23 and 24 are × 400 diameters.)

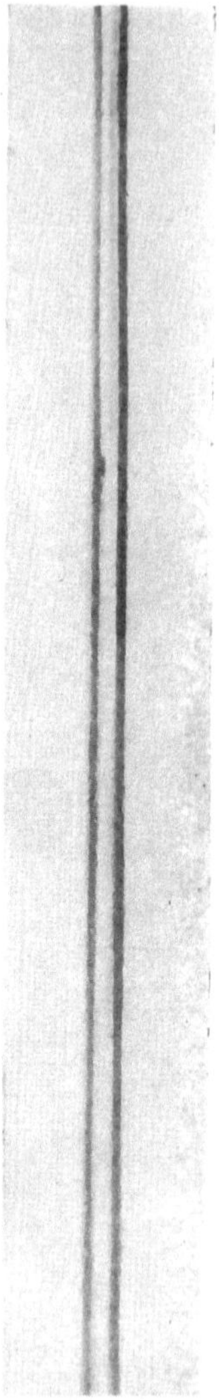

FIG. 23. Raw Silk

FIG. 24. degummed silk.

Courtesy of N. F. Crowder

To face page 55.

CHAPTER IV
SILK

THE larvæ of numerous species of moths are capable of extruding a solidified secretion, but the most important producer of silk is the mulberry worm or *Bombyx mori*. The great majority of the Bombyx moths are bred under controlled conditions with a view to the production of the best type of silk. Nevertheless, within recent times the improvements which have been made in the processing of wild silk, although a coarser and stiffer product, have increased its value. Wild and cultivated silks are both utilised in commerce.

The silkworm is hatched from the eggs laid by the moth, and after a period of 32 to 38 days it passes to the pupal stage. As a measure of self-protection, the insect spins a cocoon around its body by extruding the contents of two silk glands through spinnerets on each side of the head. During the formation of the cocoon, the insect moves its head in a figure-of-eight manner.

The twin filaments or brins are composed of fibroin and cemented together with sericin; the composite thread or bave, solidifies on meeting the air. There has been considerable doubt and controversy over the exact method by which the threads are produced, and the whole question has been reviewed by Bergmann (Text. Res., 1939, *9*, 329). The silk fluid appears to be an intimate mixture of fibroin and sericin at first, but later separates into a two-layer mixture. Shortly after entering the common duct which terminates the excretory tubes, the silk fluid passes between two glands, known as the glands of Fillippi, and through a silk press or narrow passage formed by muscular construction. The muscles may restrict the flow of fluid and therefore control the diameter of the final filaments.

The decrease in diameter which takes place towards the inside of the cocoon seems to indicate muscular fatigue.

The silk is in a viscous fluid state as it passes through the press, the restriction of which tends to orientate the micelles within the filament. Several theories have been put forward to account for the coagulation of the filament, and include suggestions of oxidation action of enzymes and the influence of salts. Foa (Koll. Zeit., 1912, *10*, 12) has examined the solidification properties of the silk fluid, and found that although it readily solidifies on standing, it does not solidify on boiling. The conclusion was reached that solidification is actually brought about by the mechanical actions of stretching and pulling which are brought to bear on the fluid immediately it leaves the gland. Rubbing, stirring, pressure and shearing action all contribute to solidification, the last factor having been demonstrated by Ramsden (Nature, 1938, *142*, 1120).

Were the insects in the cocoons allowed to live, they would eject an alkaline fluid to soften the cocoon and then emerge as moths; such a course would reduce the amount of available silk. The general procedure, therefore, is to heat the cocoons in an oven at 60 to 70°C. for 3 hours, or to steam them for 10 minutes; a certain proportion is kept and allowed to develop naturally for purposes of reproduction.

The cocoon weighs from 1 to 3 g. and may contain about 3,000 to 4,000 yards of silk of which 600 to 1,200 yards may be reeled. The remainder is worked up in a manner similar to that used for making worsted yarn; wild silk is not reeled or thrown.

In the reeling process, four to eight cocoons are placed together in hot water, to soften the gum, and the threads are drawn through guides and wound into skeins on the reel. Considerable skill is required to maintain uniformity of strength and fineness. Very little twist is imparted during the reeling operation which generally takes place near the centre of silk production. Two or three threads are combined and twisted during the throwing process which produces a strong and uniform yarn; this process is similar to the production of doubled yarns and is performed in the silk weaving and knitting districts.

The raw silk or grege from first-class cocoons is termed organzine and is mainly used in the warps of woven cloths, sixteen threads generally being used together in the yarn; the silk from the inferior cocoons is termed tram, and is generally used as weft, twelve threads forming the yarn.

The waste silk which comes from the exterior of the cocoon, and is made of short and tangled fibres, together with any damaged fibres, is termed floss and is used for the manufacture of spun silk, or schappe. The waste silk is degummed before processing.

The raw silk fibre is a double filament as produced by the silk-worm, and it has been found advantageous to weave or knit before degumming, the silk gum or sericin acting as a natural size and protecting the fibres from abrasion during manufacture of the fabric.

Under the microscope, the twin filaments or brins of raw silk are readily visible and often appear separate from one another for considerable distances; the surface is apt to be irregular on account of the torn or broken areas of sericin. Transverse fissures are sometimes visible, being caused by rupture of the sericin through twisting or bending.

The fibre or twin filament has an average diameter of 0·018 mm.; the single filaments vary from 9 to 11 μ in diameter. Degummed silk is roughly triangular in cross-section and presents a very uniform appearance. The smooth transparent filament rarely exhibits striations but treatment with dilute chromic acid causes fine longitudinal markings to appear. Wild silks, on the other hand, are broad and thick, the cross-section tends to be flat and the fibre shows very distinct striations.

Chemical Structure

Raw silk threads consist of sericin and fibroin, together with small amounts of mineral salts, wax, and colouring matter. Although there is variation between the various types, average figures are 76% fibroin, 22% sericin, 1·5% fat and wax, and 0·5% mineral salts. Cocoon wax is unevenly distributed and its composition has not been defined.

The main constituents, sericin and fibroin, are proteins consisting of carbon, hydrogen, nitrogen and oxygen; the absence of sulphur differentiates silk from animal hairs. Sericin is soluble in hot water, hot soap solution and dilute caustic alkali; its removal separates the bave into two brins which are composed of practically pure fibroin, and are much softer and more lustrous than the raw silk. Sericin may be precipitated from solution by alcohol, tannin, lead acetate, stannous chloride or iodine. Formaldehyde reacts with sericin and renders it insoluble in hot water and soap solutions; chromium salts produce a similar effect.

Fibroin is insoluble in water and in organic solvents but it dissolves in mineral acids, alkaline hydroxides, ammoniacal solutions of copper or nickel, and in solutions of certain salts such as zinc chloride, sodium and lithium thiocyanate and calcium chloride. It is possible to regenerate the silk but this process has not been developed commercially. Hydrolysis of fibroin by strong acids produces a number of amino acids of which glycine, alanine and tyrosine are the most important being present to the extent of 43·8, 26·4 and 13·2% respectively.

$$\begin{array}{ccc} NH_2 & NH_2 & NH_2 \\ | & | & | \\ CH_2 & CH.CH_3 & CH.CH_2.C_6H_4OH \\ | & | & | \\ COOH & COOH & COOH \\ \text{Glycine} & \text{Alanine} & \text{Tyrosine.} \end{array}$$

Chemical Structure

According to the classical work of Fischer, polypeptides may be built up in chain-molecules by the condensation of the simple amino-acids

TEXTILE BLEACHING

$$R.CH\underset{COOH}{\overset{NH_2}{\diagup}}$$

Amino acid

$$NH_2CHCOOH + NH_2CHCOOH \longrightarrow NH_2CHCONHCHCOOH$$
$$\quad\;\;\; R \qquad\qquad\quad R' \qquad\qquad\qquad\qquad R \quad\;\;\; R'$$

Condensation of amino acids

It is therefore concluded that the chain-molecule of fibroin is of the following type

$$\cdots NH-CH_2-NH-CO-CH-CH_2-NH-CO-CH-CH_2-NH \cdots$$
$$\qquad\qquad\qquad\qquad\;\; R \qquad\qquad\qquad\qquad\;\; R$$

$$\cdots NH-CH_2-NH-CO-CH-CH_2-NH-CO-CH-CH_2-NH \cdots$$
$$\qquad\qquad\qquad\qquad CH_3 \qquad\qquad\qquad\;\; CH_3$$

Silk Fibroin

Molecular Structure

The molecular weight of silk-fibroin has not yet been settled, and part of the difficulty is due to the fact that there is not complete agreement among the results obtained by various workers. Some of the most recent results have been provided by Bergmann and Niemann (J. Biol. Chem., 1938, *122*, 577), who also utilised some results of Vickery and Bloch (ibid., 1931, *93*, 105) to calculate the number of amino-acid residues per molecule. If the ratio of histidine is taken as unity, then there are 4 lysine, 12 arginine, 162 tyrosine, 648 alanine and 1,296 glycine residues making in all 2,592 ($2^n \times 3^m$). For an average molecular weight of 84 for the amino-acids, the probable minimum molecular weight of fibroin is 217,700.

As fibroin is a typical protein, it combines with acids and alkalis to an extent determined by the pH of the medium; the question of the isoelectric point is somewhat confusing on account of conflicting data, but recent work of Harris (Text. Res., 1939, *9*, 374) indicates pH 3·6. Other investigators record pH 4·2.

X-ray examination of silk fibroin by Meyer and Mark (Ber., 1928, *61*, 1932) established the presence of an amorphous mass containing crystallites of fully extended polypeptide chains which lie side by side to make up the micelles which are orientated parallel to the length of the fibre. It was determined that the length of an amino-acid residue was 3·5 Å, and it was found that the repeat period of the

SILK 59

chain in the direction of the fibre axis was 6·95 Å, corresponding to two amino-acid residues, mainly glycine and alanine; the distance between the polypeptide chains is between 4·4 and 6·1 Å.

According to Meyer and Mark, the dimensions of the unit cell of silk fibroin are 7·0 Å along the b period which runs along the length of the fibre, the other two dimensions being 9·68 Å along the a axis, and 8·80 Å along the c axis; the angle between the a and c axes is 75°50'.

Properties

Like all native fibres, silk has an affinity for water which it absorbs at all humidities; the official regain is 11%. At the same relative humidity, the regain under conditions of desorption is always greater than with adsorption. The swelling of dry degummed silk when exposed to the atmosphere varies with the relative humidity; percentage increase in diameter has been measured by Denham and Dickinson (Trans. Farad. Soc., 1933, *29*, 300), the swelling reaching 16·3% at 100% relative humidity.

The tensile strength of silk, expressed as tenacity (grams per denier) is 2·8 to 4·9; the wet strength is 75 to 85% of the dry strength. The extension at break varies from 18 to 25%.

The elastic properties of silk enhance its value as a textile fibre; the recovery from extension takes the usual form of an initial rapid contraction followed by a slow creep—the so-called elastic aftereffect, but even in saturated atmospheres the original length is not regained. Some interesting data have been given by Denham and Lonsdale (Trans. Farad. Soc., 1933, *29*, 305).

The density of raw silk lies between 1·30 and 1·37, but lower values have been recorded for the boiled-off material, Matthews (Textile Fibres, 1947, p. 712) giving a figure of 1·25.

The electrical insulation powers of silk are well known, but from the textile standpoint its chief value lies in its softness and lustre. Silk is also plastic and this property has been utilised to improve the lustre of silk fabrics by particular methods of embossing which produce water-mark effects such as the moire finish.

Silk fibroin is not readily attacked by warm dilute acids, but it dissolves with decomposition in concentrated acids. Moderate concentrations of mineral acids bring about considerable shrinkage which may be utilised in the production of crepe effects. Organic acids are less destructive and may be employed for increasing the scroop or characteristic crunching effect of silk; tannic acid is readily absorbed by silk, to the extent of 25%, and is not readily removed by washing.

Dilute alkali has little action on the fibroin, but at higher concentrations, there is some hydrolysis accompanied by a loss of strength and lustre. Low concentrations and high temperatures produce a similar effect.

Degummed silk is sensitive to oxidising agents, and even prolonged exposure to light will cause damage, in presence of oxygen. When oxidising agents are used in connection with the bleaching of silk, considerable care must be exercised to avoid damage; peroxides are preferred to permanganates as they produce less degradation.

Chlorine is readily absorbed by silk with the production of chloramines.

The action of certain metallic salts has already been mentioned (p. 57) in connection with solubility, and this property has been utilised to measure the viscosity of the material as a test for damage by chemical means. Another interesting property of silk is its affinity for salts of certain metals, and this has been employed to increase the weight of silk goods; silk weighting is commonly produced by tin salts.

CHAPTER V
WOOL

NATURE has provided sheep with a protective covering of wool whose main function appears to be that of keeping the body-temperature normal. The ancestor of the domestic sheep moulted, or shed this covering, every spring, whereas the wool of the domestic sheep will continue to grow if not shorn, and may reach a final length several times as long as the annual growth. The covering of the wild sheep differs in another respect, in that it consists of two distinct coatings; the inner covering of short woolly fibres differed from the outer coat of long coarse hairs.

Evolution and careful breeding, however, have developed the under-coat so that it is now the main covering of sheep, but nevertheless some wild sheep, as well as goats and camels, mainly yield hairs, (mohair, cashmere, camel-hair, alpaca and vicuna) which are shed annually. With the exception of the Merino, a certain proportion of the fleece of sheep still consists of the long, coarse and somewhat brittle fibres or hairs, and it is difficult to draw a sharp line of demarcation between wool and hair.

Wool, hair and fur, all have their origin in the dermis or underskin and grow from a follicle or pit-like depression in the form of a sheath, which is surrounded by suint glands and sebaceous glands. The former secrete potassium salts of various fatty acids which preserve the wool from the action of sunlight, and the latter secrete wool fat which lubricates the fibre and preserves it from mechanical damage during growth, prevents matting of the fibres, and acts as a waterproofing medium.

The growth of the fibre has been described in some detail by Duerden (J.T.I., 1926, *17*, 268). A full-grown hair consists of two main parts, the root and the stem; if a hair is pulled from the skin, part of the follicle comes with it so that the root presents a slightly bulbous appearance. The stem is roughly cylindrical and tapers to a point, but once cut, the wool continues to grow with a blunt end so that a tapering end may be regarded as evidence of lambs' wool. In addition, however, lambs' wool is of high quality.

There is considerable variation in the length, fineness and number of crimps in wool, not only from one type to another, but also within the same staple. Fine wools vary in length from 1·5 to 5 inches, medium wools from 2·5 to 6 inches and the long wools from 5 to 15 inches. Although the fibre length plays an important part in assessing quality, there is no simple relation between length and quality; the measurement of length is affected by the number of crimps and curls so that the stretched length may vary from 1·2 to 1·9 times the natural length. In general, however, the greatest length is seen in the coarsest wools. From the data of Duerden (J.T.I., 1929, *20*, 93), it appears that the number of crimps and the fineness of the fibre both increase with increase of wool quality; best merino (above 120s) had 27 to 30 crimps per inch and is 14 to 15µ wide, whereas the lowest quality merino (56s) has 5 to 7 crimps per inch and is 25 to 29µ wide. Many types of wool, however, are anomalous in this respect. Although the reason for the waviness of wool is not yet clear, it has been pointed out by Lefroy, according to Barker and Norris (J.T.I., 1930, *21*, 1) that the number of crimps produced in four-monthly periods tends to be constant, but that the length of wool produced in these periods varies, and with it, the size of the crimps. For instance, a lock of wool produced in twelve months showed 30 crimps along its length, but the size of the waves divided it into three regions, very large waves being found at the root end, large waves at the tip and fine waves in the middle. The length of wool grown in each period of four months was 1·16 inches (10 crimps) at the tip, 0·81 inches (10 crimps) in the middle, and 1·31 inches (10 crimps) at the root end. The number of crimps appears to be a periodic function of time, and has been confirmed by Norris and Rensburg (ibid., 1930, *21*, 481) for Merino, New Zealand Romney and Crossbred wools.

According to the *American Wool Handbook*, the average width of the various wools is 17µ for the fine wools (Super Merino), 24 to 32µ for the medium wools, and 40µ for the long wools.

The tensile strength and extension of wool is related to its fineness, but it is difficult to give average figures on account of the great variation between the individual fibres, as shown in the following table according to Taenzer (Arch. Naturgesch., 1925, *91*, A No. 9).

RELATION BETWEEN STRENGTH, EXTENSION AND FINENESS

Fineness	Strength	Extension
µ	g.	%
18	1·5–6·0	9–51
25	10–12	13–47
34	17–23	2–38
50	39–49	10–79

The tensile strength of wool decreases with increase in moisture content, and its extension increases.

The outer surface of wool and hair is covered with flat, irregular, overlapping scales which seem to vary in size and shape. The work of Nathusius in 1864, however, showed that the size of these scales is the same for various types of wool but that the extent of overlapping creates an apparent difference, which is due to variations in the rate of growth relative to the number of scales. The shape of the scales is determined by the diameter of the fibre so that in the finest wool, the scale will pass completely around the fibre giving it the appearance of inset flower-pots. With the coarser-wools, the scales overlap in two directions.

The full size of the scales is approximately 28μ long and 36μ wide, but the complete area is not exposed on account of the overlapping which, in the fine wools, may be such that only 8 to 10μ of the scale length is visible. With the coarser fibres, the amount of scale exposed is greater and may even amount to 18μ in length. The amount of visible scale is an important means of differentiating wool from related fibres and also for distinguishing between different types of wool. Minimum overlapping produces a smooth appearance as the free edges tend to fit more closely so that the wool presents a hair-like and lustrous effect.

The number of scales per given unit of length, such as 100μ, may be of value for purposes of identification.

In all cases, the overlapping of the scales is such that the free ends project outwards and towards the tip of the fibre, giving the edge its characteristic serrated appearance. On this account, the fibre can only move amongst its neighbours in the direction of its root end —a fact which is of importance in felting.

The cortical layer forms the major proportion of wool and is found immediately below the epithelial scales. The cortex is composed of elongated cells which give a striated appearance to wool, when viewed under the microscope, as these spindle-shaped cells are partly visible through the scales. Although they cannot be distinguished very clearly in this manner, yet it is possible to isolate them in different ways, the commonest of which is to steep the wool in ammonia for several days at room temperatures when the fibre may readily be disintegrated. Various bacteria are also capable of disintegrating wool as shown by Gabriel (J.T.I., 1932, *23*, 171), who isolated the cortical cells from Lincoln, Romney and Merino wools; the cortical cells varied in length from 70 to 150μ but the average length (110μ) appears to be independent of the quality of the wool. The cells are about 2 to 5μ wide and 1·2 to 1·6μ thick. Wool may also be disintegrated by 0·25% trypsin solution at pH 8·6 in two days at 35 to 40°C., according to Burgess (ibid., 1934, *25*, 289). The epithelial scales may also be isolated in the same manner.

In the case of hairs and the coarse wools, dark patches or continuous bands are often seen under the microscope, but these rarely appear in the finer wools. These dark portions reveal the presence of the medulla, or marrow, of the fibre as it consists of hollow cells which scatter the light and therefore appear dark. The medulla consists of superimposed cells of various shapes, and the diameter of these cells varies from 1 to 7μ. The size of the medulla differs in various fibres and may even occupy 25 to 33% of the diameter. The medulla of some animal hairs, from rabbit, squirrel and cat, is characteristic of the particular fibre, but with the true wools if the medulla is present, then it occurs in isolated and discontinuous fashion.

The cross-section of wool varies in shape but a large proportion of the sections presents a circular appearance, particularly in those wools of the highest spinning quality. The oval or circular shape may be measured by the ratio of the axes, and the spinning qualities classified as very good (below 1·2), medium (1·2 to 1·22), and fair (above 1·22).

The density of wool is 1·305; the work of King (J.T.I., 1926, *17*, 53; 1927, *18*, 274) appears to be the most important in this connection.

CHEMICAL CONSTITUTION

Analysis of wool shows that it is composed of 50% carbon, 22 to 25% oxygen, 16 to 17% nitrogen, 7% hydrogen and 3 to 4% sulphur, but these data are only approximate as the fibre itself is not homogeneous chemically. The nitrogen content has been shown to vary by Barritt (J.S.C.I., 1928, *47*, 69), whilst Barritt and King (J.T.I., 1929, *20*, 151) found that the sulphur content of different wools varied by as much as 25%. Further, wool fibres vary in composition along their length as shown by Bonsma (J.T.I., 1931, *22*, 305), and this variation was shown to be due to the nutritional value of the pasture during the period of growth. Additional factors in causing variation are exposure to light and air, for McMahon and Speakman (Trans. Farad. Soc., 1937, *33*, 844) showed that 14% of the total sulphur in the tip of wool may be lost on this account.

Although the scales and the cortex have been shown to have the same sulphur content, yet they differ in composition as shown by the Pauly test for tyrosine and histidine which gives positive evidence for the cortex and negative for the cuticle.

Simple analysis, therefore, presents no satisfactory solution to the problem of the constitution of wool, but it has been established that if wool is boiled for several hours in 20% HCl, it may be decomposed or hydrolysed into a number of amino-acids of general formula $NH_2.CHR.COOH$ where the radical R depends on the particular amino-acid. These different amino-acids have been isolated and estimated, with the following result:

PLATE VII

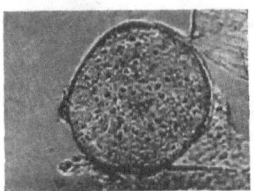

Fig. 27. Photomicrograph of cross-section of wool. (× 600 diameters)

Fig. 25 refers to merino wool and Fig. 26 to cross-bred wool. (× 400 diameters)

WOOL.

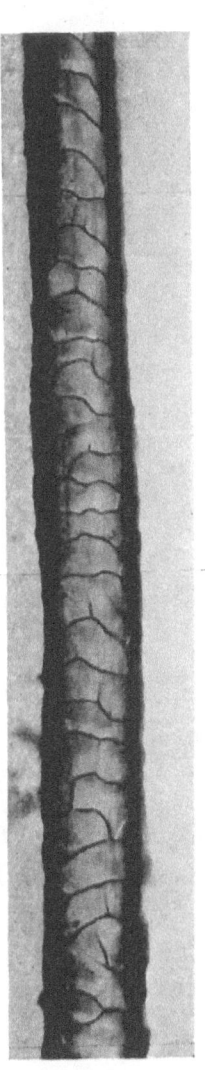

Fig. 25.

Fig. 26.

To face page 64.

PLATE VIII

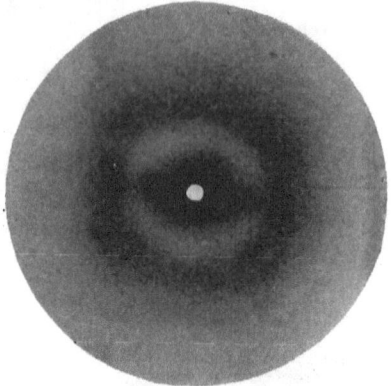

Fig. 28. X-ray photograph of stretched Cotswold wool (β-keratin).

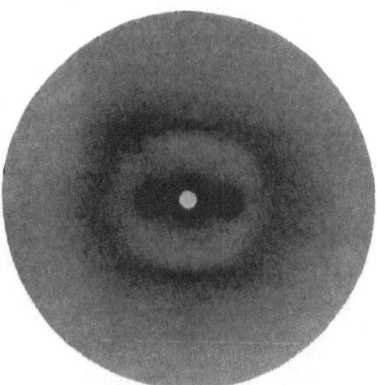

Fig. 29. X-ray photograph of unstretched Cotswold wool (α-keratin).

Courtesy of Dr. Astbury, F.R.S.

HYDROLYSIS PRODUCTS FROM WOOL

Glycine	0·6%
Alanine	4·4%
Valine	2·8%
Leucine	11·5%
Serine	2·9%
Proline	4·4%
Aspartic acid	2·3%
Glutamic acid	12·9%
Cystine	13·1%
Tyrosine	4·8%
Tryptophane	1·8%
Arginine	10·2%
Histidine	6·9%
Lysine	2·8%

Much of this quantitative work must be accepted with reserve on account of the difficulties of accurate analysis, and the fact that wool cannot be regarded as a discrete chemical compound. Nevertheless, when the common factors are taken into consideration, it is possible to build up a satisfactory chemical structure.

Milder conditions of hydrolysis of the wool keratin are capable of producing a number of peptides of higher molecular weight, so that it may be inferred that in the intact protein, the original amino-acids were linked together to form polypeptides in accordance with the views of Fischer.

$$\cdots CO-NH-\underset{R_2}{CH}-CO-NH-\underset{R_1}{CH}-CO-NH-\underset{R_4}{CH}-CO-NH-\underset{R_3}{CH}\cdots$$

The conception of the molecular chain is capable of explaining many of the physical and chemical properties of wool, but does not present a complete picture.

Fischer's generalisation that proteins arise from the condensation of α-amino-acids, by means of the peptide linkage, —CO—NH—, was of great value in elucidating the structure of silk which consists mainly of two simple amino-acids, glycine and alanine. Meyer and Mark (Ber., 1928, *61*, 1932) established that the repeat period for amino-acid residues was 3·5 Å and suggested that silk fibroin was built up of alternating residues of alanine and glycine. With the larger number of different amino-acid residues in keratin, the possible permutations of the chain structure would be enormous, but Astbury found that there was a certain dimensional regularity of 3·4 Å in wool keratin (J.S.C.I., 1930, *49*, 441).

The case of cellulose is obviously much simpler, as there is only the simple residue of β-glucose to consider.

X-ray work on cellulose and silk gave the results which might be expected from theory, but similar work on wool was not particularly promising until Astbury made use of its elastic properties In just the same way as Katz (Chem. Zeit., 1925, *49*, 353) discovered that stretched rubber gave a good crystal photograph, so Astbury (loc. cit.) found that the photograph of wool in the stretched state closely resembles that of natural silk. The remarkable elastic properties of wool are therefore accompanied by changes in X-ray photographs, for when wool is stretched 100%, then the distance of the molecular repeat period in the X-ray photograph (10·2 Å) was found to be twice that shown by unstretched wool (5·1 Å). The fully extended form was termed β-keratin and the folded normal configuration α-keratin; this intramolecular transformation is also shown by feather-keratin and myosin.

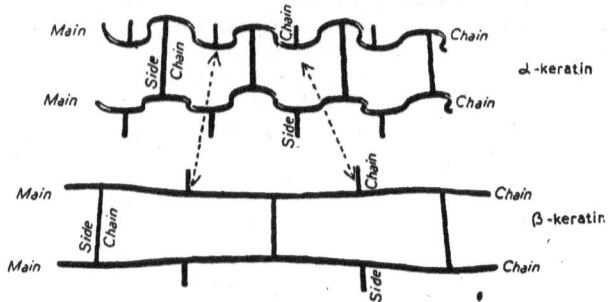

FIG. 30.—Diagram of the transformation on stretching α-keratin.

The unit cell of α-keratin has the following dimensions: (*a*) 27 Å, (*b*) 5·15 Å, and (*c*) 9·8 Å; β-keratin (the extended form of keratin) gave (*a*) 9·3 Å, (*b*) 3·4 Å, and (*c*) 9·8 Å.

The folded or coiled structure has recently been revised by Astbury (Chem. and Ind., 1941, *60*, 491) and is shown below:

WOOL

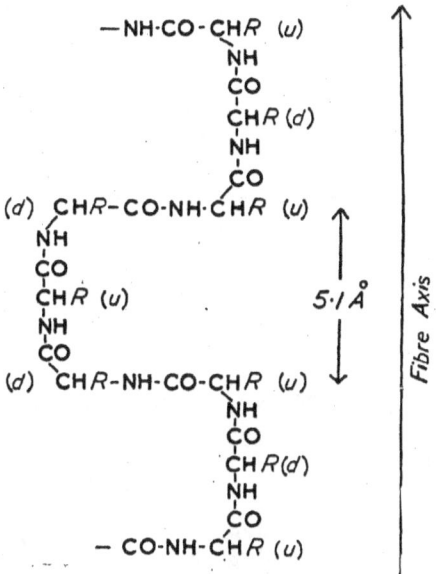

The side chains R lie normal to the plane of the paper, (u) denoting UP and (d) DOWN.

Now the great strength of native fibres is mainly due to the cohesive forces along the sides of the chain molecules, but in the case of wool there are additional forces due to specific amino-acids. Cystine, for instance, has two amino groups and two carboxyl groups and is arranged so that an amino group and a carboxyl group can be incorporated into each of the peptide chains, in a similar manner to the rung of a ladder, as suggested by Astbury.

$$\begin{array}{cc}
-\mathrm{CH} & -\mathrm{CH} \\
\mathrm{NH} & \mathrm{NH} \\
\mathrm{CO} & \mathrm{CO} \\
\mathrm{CH} \cdot \mathrm{CH_2 \cdot S-S \cdot CH_2} \cdot \mathrm{CH} \\
\mathrm{NH} & \mathrm{NH} \\
\mathrm{CO} & \mathrm{CO} \\
-\mathrm{CH} & -\mathrm{CH}
\end{array}$$

Other acids containing two carboxyl groups, when incorporated into the chain, leave one carboxyl group free; similarly, those containing two amino groups will leave one free, so that salt linkages are also possible:

```
-CH                        -CH
  \                          \
   NH                         NH
   /                          /
  CO                         CO
   \                          \
    CH CH₂CH₂COOH              CH CH₂CH₂CH₂ CH₂ NH₂
    /                          /
   NH                         NH
   \                          \
    CO                         CO
   /                          /
 -CH                        -CH


   /CO                                            \CO
    \                                              /
     CH CH₂ CH₂COO⁻  NH₃⁺ CH₂CH₂ CH₂CH₂ CH
    /                                              \
   NH            Salt linkage                       NH
    \                                              /
     CO                                            CO
```

Two types of linkage are therefore postulated: (a) the sulphur linkage or disulphide bond, and (b) salt linkages.

Speakman (Trans. Farad. Soc., 1933, *29*, 148) has shown that the salt linkages play an important part in the reactivity of the fibre, and demonstrated that the acid and basic side chains are combined to form salt linkages (ibid., 1934, *30*, 539). Similarly, Speakman's work (J.T.I., 1936, *27*, 231P) has established the presence of the disulphide groups in wool and proved the existence of cystine bridges between the peptide chains.

The modern view of the skeleton structure of wool keratin is shown in Fig. 31.

The equivalence of the acid and basic side chains was deduced by examination of the elastic properties of wool in solutions of different pH values. The resistance to extension in unbuffered solutions is independent of pH between the values of 4 and 8; in buffered solutions the values were pH 5 to pH 7. It was also found that the resistance to extension showed a linear relation to the amount of acid combined with the wool; at pH 1 the whole of the acid can be accounted for in terms of the lysine, hystidine and arginine content of the wool. From the standpoint of salt linkages therefore, the attraction between the positive and negative ions hinders the extension of the fibre in water by impeding the unfolding of the main polypeptide chains. The salt linkages are broken in acid solution and facilitate extension of the fibre; alkaline solutions exert a similar effect which is complicated by hydrolysis of the cystine

WOOL

link. A study of the elastic properties of de-aminated fibres in buffer solutions of varying pH, by Speakman and Stott (Nature, 1938, *141*, 414), revealed that the resistance to extension is independent of pH between the values of 5 and 1, which would be expected if all

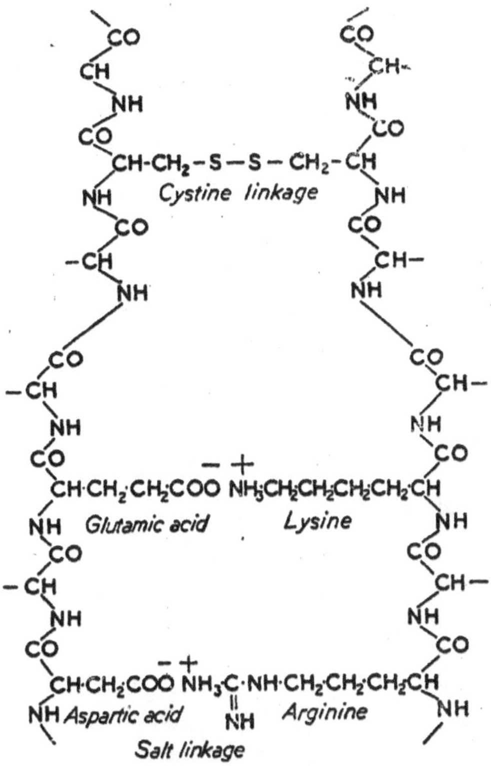

```
    \CO                           \CO
    CH                            CH-
    \NH                           \NH
    CO                            CO
    /                             /
    CH-CH₂-S — S — CH₂-CH
    \NH    Cystine linkage        \NH
    CO                            CO
    /                             /
   -CH                            CH—
    \NH                           \NH
    /                             /
    CO                            CO
   -CH                            CH—
    \NH                           \NH
    CO                            CO
    /            − +              /
    CH·CH₂·CH₂·COO  NH₃CH₂CH₂CH₂CH₂CH
    \NH  Glutamic acid    Lysine  \NH
    CO                            CO
    /                             /
   -CH                            CH—
    \NH                           \NH
    CO          − +               CO
    /                             /
    CH·CH₂·COO  NH₃·C·NH·CH₂·CH₂·CH₂·CH
    \NH Aspartic acid  ‖          \NH
              Salt linkage NH  Arginine
```

FIG. 31.—Diagram of the grid structure of wool.

the salt linkages were ruptured by complete de-amination of the basic side-chains. The liberation of carboxyl groups was demonstrated by the sharp rise in the curve between pH 5 and 7.

The titration curve for wool, due to Speakman and his co-workers (Trans. Farad. Soc., 1934, *30*, 539; 1935, *31*, 1425; 1937, *33*, 844), is closely related to that showing the reduction in resistance to extension in liquids of various pH values; determinations of the heat of reaction between wool and acids by Speakman and Stott (ibid., 1938,

34, 1203) gave results in agreement with the hypothesis of salt linkages.

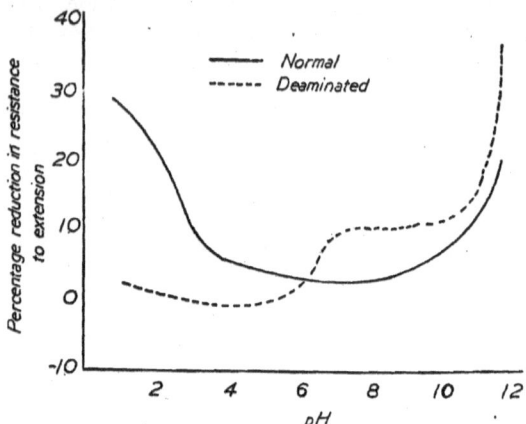

FIG. 32.—Reduction in resistance to extension of wool in solutions of various pH values.

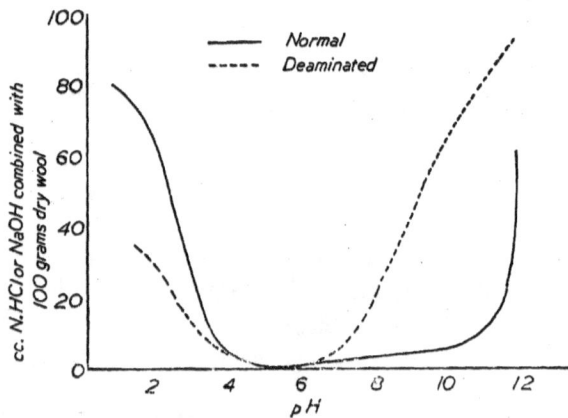

Fig. 33.—Combination of acid and alkali with wool.

As previously mentioned, the fact that cystine is a diaminodicarboxylic acid implies a cystine link between the peptide chains; variations in sulphur content between different wools and different parts of the same wool may be interpreted as variations in the number of disulphide bonds. The fact that von Weimarn (Koll.

Zeit., 1926, *46*, 120) found keratin to be the most insoluble of the proteins is in accordance with the existence of covalent bonds between the polypeptide chains, as such bonds would cause a reduction in solubility.

Cleavage of the disulphide bond should therefore bring about increased solubility, and it seems probable that agents such as sodium sulphide, which dissolve wool readily, act in virtue of the reduction of cystine to cysteine. Unfortunately, there is a strong possibility of side reactions with these alkaline reagents but Speakman and Nilssen (J.T.I., 1941, *32*, 83) were able to utilise chlorine peroxide, which attacks tyrosine, cystine, histidine, and tryptophane, although other amino acids and simple peptides are unaffected according to Schmidt (Ber., 1922, *55*, 1529). Wool treated with chlorine peroxide, although more resistant than some related protein fibres, was found to dissolve completely in lithium thiocyanate solution from which the greater part (85%) could be precipitated by alcohol.

Additional evidence in favour of the cystine link has been provided by a study of super-contraction which occurs on treatment with reagents such as sodium sulphide, sodium bisulphite, potassium cyanide and silver sulphate, as discussed by Speakman (J.S.D.C., 1936, *52*, 335). The folds of the main polypeptide chains are stabilised by the cross-linkages, so that the fact that the above reagents, which are known to break the disulphide bond, cause a contraction to less than the original length is additional evidence for the reality of the cystine link. Again, those fibres which had been treated with chlorine peroxide were found to contract 58% when boiled in decinormal hydrochloric acid for an hour.

The fact that certain reagents, such as sodium sulphite, which are known to attack cystine, do not bring about super-contraction is explained by Speakman (J.T.I., 1941, *32*, 83) on the basis of the formation of new linkages between the basic side chains and the hydrolysis products of the cystine link. If the basic side chains are removed by de-amination, then super-contraction may be realised with sodium sulphite and also with borax.

The molecular structure of wool, as generally accepted, consists of folded polypeptide chains joined by salt linkages and cystine bridges.

Molecular Weight

Wool has not been synthesised as yet, but Aberhalden and Fodor (Ber., 1916, *49*, 561) have produced compounds containing nineteen united amino-acids. The main peptide chain of wool, however, is believed to contain about 576 amino-acid residues. On account of the insolubilising effect of the cross-linkages in wool keratin, it will be understood that the viscosity methods of determining molecular weight, which have been so fruitful with cellulose, cannot be applied to keratin. However, it has now been suggested that the amino-acids in proteins are put together according to some definite plan, so that the side chains occur in a regular order. Bergmann and Niemann (J. Biol. Chem., 1936, *115*, 77; 1937, *118*, 301) have been able to show that the frequency of occurrence of the different amino-acid residues in the peptide chains of egg albumin and cattle hæmoglobin could be expressed in the form $2^n \times 3^m$, where n and m are integers. Calculation of the least common multiple of the frequencies led to the determination of the minimum number of amino-acid residues and hence to the molecular weight. With egg albumin, for instance, the minimum number of residues (*cf.* degree of polymerisation) was found to be 288, giving a molecular weight of 35,712 which is in good agreement with Svedberg's value of 34,500.

Astbury and Bell (Nature, 1941, *147*, 696) have applied these observations to wool and obtained a figure of 68,000 for the molecular weight.

Molecular Arrangement

The physico-chemical methods of Speakman (Trans. Farad. Soc., 1933, *29*, 148; J.S.D.C., Jubilee Number, 1934, p. 43) and the X-ray determinations of Astbury and Sisson (Proc. Roy. Soc., 1935, A, *150*, 533) agree in suggesting that the folded polypeptide chains are linked together in *one plane* by the salt and cystine linkages, and that several of these planes are superimposed to form the crystallite of the fibre.

The thickness of the crystallites is about 200 Å as a maximum, and their length may be approximately ten times this value.

Measurements of the work necessary to effect 30% extension of the wool fibre in various media showed that this reached a maximum with n-butyl alcohol, indicating that water and the lower alcohols were able to enter the pores of the wool substance and exert a swelling action. The pores in the dry fibre will, therefore, have a diameter of the order of the size of propyl alcohol, corresponding to 6 Å. The wool fibre increases 17·5% in diameter when immersed in water and it has been calculated that the pore size in the swollen state corresponds to 35 to 40 Å. (Speakman, Trans. Farad. Soc., 1932, *26*, 61.)

The work of Frey-Wyssling (Protoplasma, 1937, *27*, 372), on the intermicellary spaces in fibres by the deposition of gold and silver particles, has also been applied to wool; estimates of 60 Å are recorded. Presumably, this must be regarded as a maximum for swollen wool, as the technique employed would involve swelling the wool with aqueous solutions.

MOISTURE RELATIONS

Dry wool rapidly absorbs water from the atmosphere at first, but more slowly as equilibrium is approached. As with other fibres, the amount of water absorbed depends on the relative humidity and temperature, and whether the moisture content comes from absorption or desorption—a real hysteresis exists. The absorption of moisture is considerably affected by temperature, and is greatest at low temperatures; there is a linear relation between moisture and temperature for low relative humidities, but at high humidities there is a definite minimum around 40°C., according to Speakman and Cooper (J.T.I., 1936, *27*, 183). When wool is dried from regains below saturation, the absorptive power decreases with increasing temperature of drying. This loss in absorptive power may be restored by allowing the wool to reach saturation with water vapour, except in those cases where wool has been heated over water at a high temperature. Wool is much more sensitive to attack by hot water than cotton, and at temperatures above 55°C. the disulphide bond is attacked and hydrolysed as shown by Speakman and Cooper (loc. cit.).

The accepted moisture regain of scoured wool is 18%; it has the greatest absorptive capacity of any native textile fibre, amounting to 33·95% at 100% relative humidity, under which conditions there is an increase of 31·8% in the area of the cross-section.

The wet strength of wool is 80 to 90% of the dry strength; the extension at break increases, however, being 28 to 48% in the dry state and 35 to 60% in the wet state. The elastic properties of wool as affected by humidity have been examined in some detail by Speakman (J.T.I., 1927, *18*, 431; J.S.C.I., 1930, *49*, 209; Trans. Farad. Soc., 1929, *25*, 92).

As previously explained, Astbury has shown that the elastic properties of wool are mainly due to the crumpled nature of the main polypeptide chains, and the reversible transformation of α- and β-keratin.

The elastic properties of wool are strongly influenced by moisture. It is well known that if wool is stretched in water or a saturated atmosphere to not more than 30% of its original length, and then released in the same medium, the return to the original length will be complete. Dry fibres, however, do not behave in a similar manner but exhibit an initial rapid recovery followed by a slow creep known as the elastic after-effect. The path of elastic recovery is not identical with that of extension but shows elastic hysteresis. It is also well known that if extended fibres are exposed to steam before immersion in water takes place, there is no contraction in length; this is the phenomenon of permanent set, which plays a large part in crabbing and blowing. The set is permanent to temperatures lower than that originally employed.

Speakman (loc. cit.) has shown that the torsional rigidity of wool is also sensitive to changes in relative humidity and that Young's modulus decreases with increasing relative humidity. At a constant relative humidity, the resistance to extension shown by wool decreases with rise in temperature; the elastic properties also vary with the rate of loading. If the extension in water is rapid, wool may be stretched about 34% without significant change in properties and this observation has been of value in the use of calibrated fibres for determining the effect of various reagents. When the fibre is stretched slowly, or maintained in the extended state for some time there is a permanent weakening particularly at high humidities. This was demonstrated by Speakman (Proc. Roy. Soc., 1928, *103*, 377) who determined the load-extension curves by rapid extension in water up to 30% and then observed the rate of decay of tension. Fibres which had been maintained in the extended state for given periods of time were released, and the load-extension curve re-determined in water; the reduction in the work required for 30% extension was greater as the time of stretching increased, thus demonstrating the decay of tension in strained fibres. This property is associated with hydrolysis of the disulphide bond, which is stable to acid solutions in which the decay of tension is small, but is attacked by alkaline solutions in which the fibres readily become plastic.

The high degree of elasticity in water is shown by the fact that at room temperatures, with slow loading, it is possible to stretch wool 70%, and the fibres recover completely in water, returning to their original length.

Astbury and Woods (Phil. Trans., 1933, *232*, 333), however, have demonstrated that stretched fibres fail to return to their original length in humidities below saturation; the extent and rate of recovery increase with the relative humidity, a fibre which does not return to its original length in an atmosphere of low humidity recovering completely in water.

At high temperatures, as previously mentioned, wool acquires peculiar elastic properties associated with permanent set. Fibres which have been steamed or immersed in boiling water whilst extended do not contract when placed in cold water; they do, however, show contraction in steam. This effect has been examined in some detail by Astbury and Woods (loc. cit.) who steamed fibres for various times at 50% extension, and then released them in steam until no further contraction took place. The lengths were measured and the extension determined. Steaming at 50% for 30 minutes followed by release in steam shows a permanent extension of 20%,

15 minutes only produces 10% permanent extension in steam, but steaming for only 2 minutes, although producing a permanent set at 50% extension to cold water, shows a contraction to 30% *less* than the original length when released in steam. This new effect of super-contraction was examined by X-ray methods, and it was shown that the cross-linkages were attacked with further folding of the poly-peptide chains. It was subsequently shown by Speakman (J.S.D.C., 1936, *52*, 335) that reagents, such as sodium bisulphite and potassium cyanide, which are known to attack the disulphide bond, also cause super-contraction.

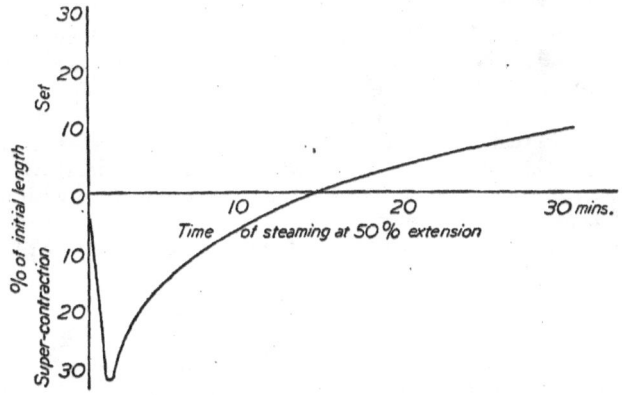

FIG. 34.—Super-contraction and set of wool on steaming.

Returning to the work of Astbury and Woods (loc. cit), and the observation that when fibres are stretched to 50% extension for 15 minutes or more in steam, they fail to return to their original length when released in steam, it would appear that the system becomes stabilised by some form of linkage rebuilding. This was examined by Speakman (loc. cit.), who determined that de-aminated fibres were incapable of acquiring permanent set even after 6 hours steaming, and also that the setting power of ordinary fibres decreases as their sulphur is removed by baryta. New linkages must apparently be formed by reaction between the basic side chains and the breakdown products of the cystine linkage. The action of steam is to produce a sulphinic acid from the cystine linkage as follows:

$$R.CH_2.S.S.CH_2R' + H_2O \rightarrow R.CH_2.S.OH + R'.CH_2.SH.$$

and a possible system of rebuilding may be represented by

$$R.CH_2.S.OH + H_2N.R' \rightarrow R.CH_2.S.NH.R + H_2O.$$

The suggestion that disulphide bond breakdown is a preliminary to permanent set is supported by the fact that stretched fibres take on a permanent set more readily as the pH rises to 9·2; above this value there is a decrease in setting power, probably on account of the sulphinic acid being unstable in alkaline solution. Further, fibres with permanent set were found to contain free sulphydryl groups and were also shown to be stable to boiling sodium bisulphite solutions.

On the assumption of the formation of the new linkage, —S.NH—, it is possible to explain the action of sodium sulphite as follows:

$$R.CH_2.S.S.CH_2.R + Na_2SO_3 \rightarrow R.CH_2.S.Na + R.CH_2.S.SO_3.Na$$
$$R.CH_2.S.SO_3.Na + H_2N.R' \rightarrow R.CH_2.S.NH.R' + NaHSO_3$$

and the action of sodium bisulphite in the following manner.

$$R.CH_2.S.S.CH_2.R + NaHSO_3 \rightarrow R.CH_2.S.Na + R.CH_2.S.SO_3.H$$
$$R.CH_2.S.SO_3.H + H_2N.R' \rightarrow R.CH_2.S.NH.R' + H_2SO_3.$$

Before leaving the discussion of the moisture relations of wool, it may be of interest to draw attention to a recent suggestion of Cassie, King and Atkins (Nature, 1939, *143*, 163) and amplified by Cassie (J.T.I., 1940, *31*, 17) that textile fibres adsorb moisture with evolution of heat, and as wool fibres absorb more moisture than other fibres, they are supreme in avoiding temperature changes at the skin.

CHEMICAL REACTIONS
Water

Wool is not immune to the action of WATER. Although there is no reaction at room temperatures, the fibre is damaged by water at 50°C. probably by cleavage of the disulphide bond in the following manner:

$$R.CH_2.S.S.CH_2.R + H_2O = R.CH_2.SH + HO.S.CH_2.R$$

The initial reaction is followed by the elimination of sulphur in accordance with the equation

$$R.CH_2.S.OH = R.CHO + H_2S$$

as shown by the work of Speakman and McMahon (Trans. Farad Soc., 1937, *33*, 844). Harris (J. Res. Nat. Bur. Stand., 1938, *20*, 563) has found that the sulphur lost by wool on exposure to light and air was liberated as H_2S, whilst Meunier and Rey (Compt. rend., 1926, *183*, 596) showed that the hydrogen sulphide is rapidly oxidised to sulphur dioxide and sulphuric acid.

Breakdown of the disulphide bond reduces the number of cross-linkages in the fibre, which therefore swells to a greater extent in aqueous solutions than does untreated wool as demonstrated by von Bergen (Textilber., 1923, *4*, 23; 1925, *6*, 745; 1926, *7*, 451). The treated wool has an increased affinity for dyestuffs and a decreased resistance to extension.

Acid

The action of ACIDS has been considered to some extent in discussing the constitution of wool; aqueous solutions of mineral acids can attack both the peptide chains and the salt linkages. At higher temperatures wool dissolves completely yielding simple peptides and finally amino-acids. The cystine linkage, however, is not attacked. The action of dilute solutions of mineral acids does not result in permanent damage to wool, and may therefore be used to remove cellulosic impurities.

It has been shown by Speakman and Hirst (Trans. Farad. Soc., 1933, *29*, 148) and confirmed by Steinhardt and Harris (Text. Res., 1940, *10*, 181) that wool will adsorb considerable amounts of acids at room temperatures, the exact amount being determined mainly by the pH of the solution. Combination with acid starts at pH 4·8 and maximum acid-combining capacity occurs at pH 1, where 80 cc. of acid combine with 100 g. of dry wool. The chief change produced by the acid is to reduce the work necessary to stretch the fibres 30%, as previously mentioned. There is close agreement between the amount of combined acid and the reduction in the work necessary for extension. This may be explained on the basis that the salt linkages are the means of union between wool and acids. When the acid is removed by washing, the salt linkages are restored and the original properties of the wool return. This conclusion is supported by work on de-aminated fibres, as a result of which it was shown by Speakman and Stott (Nature, 1938, *141*, 414) that the reduction in work necessary to effect 30% extension in water was the same as that required to stretch an untreated fibre in acid at pH 1.

$$R.CH_2.CH_2.CH_2.CH_2.NH_2 + HNO_2 \rightarrow R.CH_2.CH_2.CH_2.CH_2.OH$$

De-aminated fibres show equal resistance to extension in solutions of pH 1 and pH 5.

The fission of salt linkages by acid is also demonstrated by the work of Speakman and Stott (Trans. Farad. Soc., 1934, *30*, 539), who found that the swelling of wool in solutions of strong acids increased as the pH values diminished.

The behaviour of weak acids is different from that of strong acids at the same pH values; concentrated solutions of weak acids produce intense swelling, the amount of combined acid is much greater than with the stronger acids, and the fibres are more easily stretched. The peculiar action of acids such as oxalic and chloracetic may be explained by the Donnan membrane hypothesis, and it has been suggested that the swelling, the heat of reaction, and the large extent of combination are due to a separation of layers of cross-linked polypeptide chains which comprise the micelles. The high swelling-pressure offers points of resemblance to the dispersion of native cellulose by sodium hydroxide solutions of moderate concentration.

Alkali

Wool does not combine with ALKALI below pH 7; the resistance to extension of wool in $N/5$ salt solution containing various amounts of caustic soda, or hydrochloric acid, is independent of pH between the values of 5 and 7. On this account, it was suggested by Speakman and Hirst (Trans. Farad. Soc., 1933, *29*, 148) that although wool is an amphoteric colloid, and combination with acid starts at pH 4·8, this value must not be regarded as the isoelectric point, but rather as the beginning of an isoelectric region which extends from pH 5 to pH 7.

The extent of combination increases rapidly with pH and is accompanied by decreased resistance to extension. Above pH 10, however, the wool cannot be washed free from alkali which causes severe damage on account of its attack on the disulphide bond. Below pH 10, the absorption of alkali can be explained in terms of the salt linkages.

As the elastic properties of wool are independent of pH between the values of 4 and 8, it has been suggested that this is the safest region for processing, but with increase in salt concentration the limits narrow to pH 5 to 7, and according to Whewell (Text. Rec., 1941, November, p. 29) at 80°C. it is only a point.

Except for very dilute solutions of alkali, there is very rapid degradation of wool, and at high temperatures complete dissolution occurs. The amount of degradation may be measured by estimating the loss in work necessary to stretch the fibres, or by the actual amount of wool dissolved. As the pH of the solution in which the wool is immersed increases from pH 5, the effect of temperature becomes critical. For instance at pH 13 there is a sudden increase in the destructive action at 30°C., and the wool dissolves at 90°C. At low temperatures there is little damage at pH 13, but at 50°C. there is a sudden increase in damage at pH 10. The safe limits of textile processing for wool are decided by pH and temperature.

The action of alkali on the disulphide bond is shown in the following manner:

$$R.CH_2.S.S.CH_2.R \rightarrow R.CH_2.SH + HO.S.CH_2.R$$
$$R.CH_2.S.OH \rightarrow R.CHO$$

The action of calcium and barium hydroxides may differ from that of sodium hydroxide but sulphur is removed in all cases.

It has been found that alkali-treated wools do not exhibit super-contraction when boiled with solutions of sodium bisulphite, and this may be attributed to the formation of new linkages, such as $R.CH_2.S.CH_2.R$, according to Speakman and Whewell (J.S.D.C., 1936, *52*, 380) or $R.CH:N.(CH_2)_4.R$, by interaction of the aldehyde groups with the amino-groups of the salt linkage, according to

Phillips (Nature, 1936, *138*, 121). The number of these new linkages, however, is far less than those broken by the alkali.

Returning to the destructive action of alkali on wool, the effect of 0·1% NaOH solution for 20 minutes at 60°C. is to cause 70% reduction in breaking load and 12% loss in weight; 0·2% NaOH, under similar conditions, showed 85% reduction in strength and 18% loss in weight. As little as 0·01% NaOH brings about a loss of 43% in strength, and of 2% in weight. Even ammonium hydroxide has a harmful effect under similar conditions, 0·1% NH_4OH causing a loss in strength of 44% and a loss in weight of 3·7%.

More concentrated solutions of sodium hydroxide, however, have been shown by Buntrock (Farben Zeit., 1898, *9*, 69) to act in a different manner. At room temperatures, an immersion of 10 minutes brings about a decrease in strength up to 15% NaOH, after which there is an increase in strength, which amounts to 30% in the case of 38% NaOH solution. The increased strength is accompanied by an increase in lustre, and a "mercerised" effect is produced. At higher temperatures, however, the wool is attacked.

Freney and Lipson (Council for Scientific and Industrial Research, Australia, Pamphlet No. 4, 1940) found that treatment of wool with 58% NaOH for 10 minutes at 45°C. gave non-felting properties to the material. The effect of alkali in alcohols was also examined and it was shown that non-felting results could be obtained by treatment with 4 to 6% KOH in methylated spirits at 23 to 28°C. for 1·5 to 2·5 minutes, or in 0·5 to 1·25% KOH for 3 minutes, at 15 to 25°C. (Journal of the Council for Scientific and Industrial Research, 1941, *14*, 25). Hall and Wood, of the Tootal Broadhurst Lee Co., Ltd., have produced good non-felting wool by using certain mixtures containing a preponderating amount of cheap hydrocarbon in the solvent mixture, as described in B.P. 538,428 and 538,396. A mixture of solvents is utilised, one liquid (butyl alcohol) being a solvent for the alkali, and the other a non-solvent; it is believed that the non-solvent diluent forces the alkali on to the wool. A typical example is to treat wool with 0·6% NaOH in a mixture of 1 part of butyl alcohol and 9 parts of white spirit for one hour at 18 to 20°C.

The FELTING of wool is one of its most characteristic properties and varies according to the type of wool, the long lustre wools showing relatively little tendency to felt when milled, in acid or alkaline liquors.

Attempts to derive a strict relationship between scaliness and milling properties, however, have not been successful.

The old theory of the interlocking of the serrations of the scales causing felting has now been abandoned as it cannot be reconciled with actual observations. In order for felting to occur, the fibre must possess scale structure, ease of deformation and recovery, together with some crimpiness which may be a minor factor. The "earth-worm" theory of Arnold (Leip. Monat. Text. Ind., 1929, *44*, 463) postulates that frictional movement extends the hair in the direction of its root, as the scale structure prevents movement towards the tip; the moist fibre tends to counteract this elongation in virtue of its elasticity and when pressure is released the hair contracts in the direction of its root, drawing other fibres with it. Shorter (J.S.D.C., 1923, *39*, 270) has suggested that fibre travel, due to elasticity, is closely associated with the milling process. These two ideas must be combined with Speakman's observations that acid and alkali facilitate the extension of the wool fibre in a remarkable manner. Speakman, Stott and Chang (J.T.I., 1933, *24*, 273) have shown that the effect of pH on extension is similar to the effect on milling shrinkage which is at a minimum between pH 4 and 8; hence wool will not felt in neutral liquors. Similarly, the optimum milling temperature of 45°C. has been shown to be due to decreasing powers of recovery from extension above that temperature.

Reducing Agents

Alkaline REDUCING AGENTS can exert a powerful destructive action on wool, the disulphide bonds being attacked at pH 11 or above. Sodium sulphide solution is well known for its rapid powers of gelatinising and dissolving wool; its use as a depilatory in the unhairing of hides is equally well known. The sulphides of calcium and barium also attack wool by fission of the cystine link.

Speakman (J.S.C.I., 1931, *50*, 1) has examined the action of sodium sulphide in some detail. There is an increase in diameter of 400% in 0·15 N Na_2S solution in 16 minutes but at about 200% increase in diameter local blisters indicate that the cuticle is more resistant than the cortex. With more dilute solutions (0·0325 N), it was found that sodium sulphide brings about a contraction in length followed by an extension; after 54 minutes, there was a longitudinal contraction of 6% followed by an extension of 100% in 236 minutes. The maximum swelling in diameter was 133%, reached in 86 minutes; but the original diameter returned after washing, although the length was reduced to 20%, less than the original length.

Sodium sulphide also possesses the property of reversing the permanent set which can be induced by steaming in a state of strain. Whereas sodium hydroxide removes permanent set and causes 10% shrinkage, sodium sulphide induces shrinkage and then extension.

The reaction between wool and sodium sulphide is accelerated by increase of time or concentration.

The action of reducing agents such as sodium sulphite or bisulphite has already been considered in discussing the constitution of wool (see page 71).

The well-known super-contraction produced by hot solutions of sodium bisulphite was described by Elsasser in G.P. 233,210.

Speakman has shown that part, at least, of the damage caused by the action of acid or alkaline reducing agents on the disulphide bonds may be repaired through treatment with oxidising agents or salts of polyvalent metals (J.S.D.C., 1936, *52*, 423). Methods of this type form the basis of permanent set at low temperatures as described in B.P. 453,700 and 453,701. Speakman, Stoves and Bradbury (J.S.D.C., 1941, *57*, 73) have shown that after the fission of the disulphide bond by sodium sulphite-bisulphite at pH 6, the bonds may be re-formed by treatment with the acetates of mercury, copper or iron. After more severe methods of breakdown, rebuilding may be accomplished by treatment with ammonia followed by oxidising agents, or sodium tungstate or aluminate, together with a swelling depressor.

The rejoining of reduced disulphide bonds by organic non-polar links has been studied by Patterson, Geiger, Mizell and Harris (Text. Res., 1941, *11*, 379). Strongly alkaline solutions of thioglycollic acid ($HS.CH_2.COOH$) will dissolve wool, but at pH 4·5 a large excess of 0·2 M thioglycollate will bring about reduction. The treated wool was washed, and brought into contact with a molar phosphate buffer solution at pH 8, in which the alkylating agent was dissolved or suspended. Suitable alkylating agents are methyl iodide, ethyl bromide, benzyl chloride, ethylene dibromide and trimethylene dibromide. The bifunctional reagents forms cross-linkages of the bis-thiopolymethylene type which are remarkably stable to alkali.

Oxidising Agents

OXIDISING AGENTS are capable of producing drastic changes in the wool molecule, and here again the main attack is on the disulphide bonds. As hydrogen peroxide is used in commercial bleaching processes, its action must be controlled.

Smith and Harris (Am. Dyes. Rep., 1936, *25*, 180, 183, 383 and 542) have examined the oxidation of wool. When immersed in hydrogen peroxide solutions of varying concentrations at 50°C. for 3 hours, there was a sharp fall in the cystine content, the nitrogen content and an increased loss of weight in $N/10$ NaOH solution (1 hour at 65°C.) when the concentration reached a value between 4 and 6 vol. With 2-vol. H_2O_2 for 3 hours at various temperatures, it was found that the damage increases rapidly above 50°C.; hence a temperature of 50°C. and a concentration of about 5 vol. appear critical. The pH of the solution has no effect below pH 7, but as the pH rises the alkali solubility increases and the cystine content decreases.

Various oxidation products are possible, such as R.SO.S.R, $R.SO_2.S.R$, R.SO.SO.R, $R.SO_2.SO.R$ and $R.SO_2.SO_2.R$; the two last compounds do not give a dark coloration (lead sulphide) with lead acetate, but the first three products give a positive reaction.

The photochemical oxidation of wool is well known and the formation of acid has often been noted; Smith and Harris (loc. cit.) determined that the oxidation was accelerated in presence of acid but decelerated by alkali.

According to Elöd and his co-workers (Textilber., 1942, *23*, 313), the oxidation of wool by hydrogen peroxide is accompanied by de-amination.

Halogens

CHLORINE and bromine, in presence of water, are capable of powerful attack on wool; under controlled conditions, however, the use of chlorine has commercial interest. It may be used for improving the affinity for dyestuffs, as discovered by Mercer, and also for imparting a lustre to the "Oriental rugs." Its chief application is for the production of non-felting wool, and this process has been examined by many investigators, namely, Trotman (J.S.C.I., 1922, *41*, 219; 1926, *45*, 20, 111), Speakman and Goodings (J.T.I., 1926, *17*, 607) and Edwards (J.S.C.I., 1932, *51*, 243). It is generally considered that the non-felting result is due to the characteristic scale structure becoming inoperative, in so far as shrinkage is concerned, by the formation of a gelatinous layer. Most aqueous processes attack the wool and impair its wearing ability, so that a compromise must be reached between unshrinkability and wear. Attack may be restricted to the surface of the fibres by the use of gaseous chlorine (B.P. 417,719) or by treatment with sulphuryl chloride in white spirit (B.P. 464,503).

An interesting action of concentrated chlorine water may be seen in the Allworden effect; a series of small bubbles is formed along the hair and the scales gradually disappear. This effect does not take place with damaged fibres (Z. angew. Chem., 1916, *29*, 77).

Metallic Salts

The action of METALLIC SALTS on wool varies with the nature of the metallic radical. The neutral salts of the alkali metals have little effect and are scarcely absorbed even from boiling solutions. The salts of the heavy metals, such as iron, chromium, copper and tin, and also salts of the alkaline earth metals, are capable of reacting with wool to form linkages of the type $-S-M-S-$, where M represents an atom of a metal. Extremely stable linkages may be formed and care should be taken to avoid metal stains in processing.

When wool is treated with chromates or bichromates, in neutral or acid solution, oxides of chromium are produced within the fibre and function as mordants in the dyeing process. The first action is the absorption of chromic acid from the solution, and as much as 10% may combine with the basic side-chains of the keratin; the acid is then reduced, either by the reducing agents in the bath, or by disulphide bond attack.

The creping action of salts such as calcium thiocyanate, may be attributed to the production of hydrosulphides when steamed under slightly alkaline conditions; the hydrosulphides attack the disulphide bond and cause super-contraction. The action of thiocyanates was first described by Siefert (Z. angew. Chem., 1899, *69*, 86).

Organic Compounds

ORGANIC COMPOUNDS may react with many of the active groups in keratin, and a large number of derivatives is theoretically possible. In general, however, reaction is limited to compounds of low molecular weight except in the presence of swelling solvents. Wool has been acetylated, with a consequent diminution in the affinity for acid dyes as described by Munz and Haynn (Chem. Zeit., 1922, *46*, 895). Phosgene has been combined with the basic side-chains of the keratin to give —N:CO side-chains. Quinone also combines with the basic side-chains, as shown by Meunier (J.S.D.C., 1911, *27*, 214), and there is a considerable increase in strength which seems to point to new cross-linkages being formed. The strengthening effect diminishes with de-aminated wool, according to the extent of de-amination, investigated by Speakman and Stoves (see J.T.I., 1941, *32*, 83).

Formaldehyde at pH 6 to 7 may combine with wool to give cross-linkages of the type $R.NH.CH_2.NH.R$, but the strengthening effect is less than that of quinone, according to Speakman (J.T.I., 1941, *32*, 83). The effect of formaldehyde has been examined by many workers, and it is recognised that some degree of resistance to alkali is conferred by this treatment. The work of Trotman, Trotman and Brown (J.S.D.C., 1928, *44*, 49) showed that the formaldehyde compound is hydrolysed by warm alkali, so that the effect is not permanent. Formaldehyde treatment affords no protection against damage by chlorination.

REGENERATED PROTEINS

It has not been possible to regenerate satisfactory fibres from wool which has been dissolved in an appropriate solvent, but silk, on the other hand, has been regenerated in filament form.

Artificial fibres from other proteins have attracted considerable attention during recent years, and if they have not had the same commercial success as rayons, it must be remembered that this may not be entirely due to technical considerations. Rayons compete with real silk which is expensive, and are made from a cheap material —wood.

The commonest artificial protein fibre is prepared from casein which is not particularly cheap and must compete with wool which is relatively cheap.

Casein-wool has been known for some time, the first suggested use of casein for textile fibres having been disclosed by Todtenhaupt in B.P. 25,296 of 1904. Within more recent times, however, the production of Lanital, according to the processes of Ferretti, has attracted great attention; the chief specifications are B.P. 483,731; 483,807; 483,808; 483,809; 483,810.

The casein is produced by the acid treatment of milk under certain definite conditions, following which the casein is dissolved in sodium hydroxide solution and allowed to mature, when it becomes more viscous, given certain definite temperature conditions. When the required volume and viscosity have been obtained, the temperature is lowered slightly and the viscous solution is spun into filaments by coagulation in an acid bath containing aluminium salts and formaldehyde.

The treatment with formaldehyde and with aluminium salts is of great importance from the standpoints of tensile strength and resistance to swelling. It appears that the formaldehyde will form links with the amino-groups in adjacent chains; similarly the aluminium is capable of bridging the chains through the carboxyl groups.

Lanital is the best-known casein fibre, but there is an American fibre termed Aralac, and also an English product.

Casein fibres are not particularly strong, figures of the order of 0·9 g. per denier being given; the wet strength is 50% of the dry strength. The extension at break in the dry state may be as high as 50%, and in the wet state may even reach 100% on account of the great swelling in water. The casein fibres are mainly used in admixture with the native fibres.

Another interesting method of making artificial protein fibres is given in B.P. 467,704 by Astbury, Bailey and Chibnall. Globular proteins, such as ground-nut globulin or soya bean globulin, may be dissolved in an aqueous solution of urea and allowed to degenerate or de-nature, during which process the solution gradually becomes more viscous. Filaments may be produced from this viscous solution by spinning into a suitable bath which usually contains inorganic salts, together with acid and sometimes formaldehyde solution.

These processes are of particular interest because the globular proteins are now known to consist of large round molecules which are formed of coiled polypeptide chains. During the degeneration or de-naturation, the globular configuration breaks down and the polypeptide chains unfold and then agglomerate into parallel bundles similar to those found in β-keratin. The above method is believed to be used in the manufacture of Ardil.

CHAPTER VI
SYNTHETIC FIBRES

VINYON

It has been known for some years that vinyl chloride, $CH_2:CHCl$, was capable of being polymerised to form a plastic material and considerable work was done on drawing the polymer into fibrous form, particularly in Germany, where what is believed to be the earliest synthetic fibre, PeCe, was produced. Unfortunately, the polymeric vinyl chloride suffers from certain defects which restrict its commercial exploitation.

There is, of course, a large number of substances which exist in the monomeric state and which are capable of polymerisation, and some of these are capable of admixture in various proportions. In the work on polymerisation in general, as distinct from its application to synthetic fibres, it was found that a mixture of two polymers gave a result, which might be predicted from consideration of each of the component parts. On the other hand, in certain circumstances, it was possible to mix two different monomeric substances and polymerise them together; this is called copolymerisation. These copolymers often have properties which are quite different from those of a mere mixture of polymers. A very interesting application of this principle has been used in textile production in order to form a truly synthetic fibre which is called Vinyon. This material has been developed to a certain commercial extent in the U.S.A. because the two monomers vinyl chloride $CH_2:CHCl$ and vinyl acetate $CH_2:CH.O.CO.CH_3$ are fairly cheap products arising from modern developments in the American petroleum industry.

Vinyon is a copolymer of 89% of vinyl chloride and 11% of vinyl acetate and the mixture is polymerised up to a molecular weight of about 20,000. The copolymer may be dissolved in acetone and spun into fibres by the "dry spinning system," similar to that used for cellulose acetate filaments. Moderately coarse filaments of the copolymer are spun and these are later stretched several hundred per cent, when they acquire very high tensile strength and can also be drawn out finer than real silk.

The following table gives comparative data for Vinyon, two rayons and real silk:

Vinyon suffers from certain disadvantages as a textile fibre; it is thermoplastic and softens below the boiling-point of water which makes certain hot finishing processes difficult to execute. The softening on heating is accompanied by shrinkage in length. Vinyon

TEXTILE BLEACHING

VINYON AND OTHER FIBRES

Fibre	Tensile strength wet (g. denier)	Extensibility wet (per cent.)
Vinyon, not stretched	1·0	120·0
Vinyon, medium stretched	2·3	25·0
Vinyon, highly stretched	4·0	18·0
Acetate rayon	0·85	36·0
Viscose rayon	1·0	25·0
Degummed silk	3·4	26·3

will not burn but only melts. It is also soluble in certain chlorinated hydrocarbons and certain organic esters. It is very highly water-repellent and absorbs practically no moisture from the atmosphere. One of its most important properties is the resistance to chemical reagents, both concentrated acid and concentrated alkali.

The copolymerisation of vinyl acetate and vinyl chloride presumably takes place in the following manner:

$$CH_2:CH \atop O.COCH_3 \quad +CH_2:CH \atop Cl \quad \longrightarrow \quad -CH_2-CH-CH_2-CH- \atop O.COCH_3 \quad Cl$$

NYLON

The word "nylon" is a generic term for a group of substances known as polyamides; it does not refer to a single substance. The polyamides may be made in two ways; first, by the self-condensation of amino-acids, or, secondly, by the condensation of diamines with dibasic acids. Naturally, many polyamides have been investigated and two of them have reached commercial importance, both prepared by the second method outlined above. They are referred to as nylon 66 and nylon 610; the simple system of nomenclature is based on the number of carbon atoms in the diamine and in the dibasic acid respectively. Hence, nylon 66 is made from hexamethylene diamine and adipic acid, whereas nylon 610 comes from hexamethylene diamine and sebacic acid. Most yarn consists of nylon 66, and monofil bristles are made of nylon 610.

The method of preparing the nylon 66 is to heat equivalent amounts of adipic acid and hexamethylene diamine together to form a salt which is comparatively stable and can be stored until required. The salt may also be purified by recrystallisation from alcohol and water if necessary. The super-condensation of the salt is generally effected by heating in an autoclave, stabilisers being present to control the viscosity and molecular weight; when the required condensation product has been obtained, the molten nylon is extruded in ribbon-form and cut into small portions on cooling. For forming fibres, the molten product is extruded through spinnerets and wound on to bobbins; at this stage, the tensile strength is low, but when cold-drawn to four times their original length, a high degree of strength and elasticity is obtained.

Many of the early products have been described by Carothers and Hill (J.A.C.S., 1932, *54*, 1579). They prepared linear condensation superpolymers by means of self-esterification of substances such as ω-hydroxy decanoic acid. The poly-esters are not highly viscous in solution and show no signs of colloidal behaviour. The molecular weights vary from 800 to 5,000. The term "superpolymer" refers to linear polymers having molecular weights above 10,000, and these exhibit colloidal behaviour and show some of the properties associated with naturally occurring high polymers. The designation α-esters is also used for polyesters having molecular weights from 800 to 5,000 and ω-ester for the superpolymers. Owing to the type of condensation product, the molecular weight can be determined by direct chemical means, as the terminal groups are still present. Whereas the α-polyesters dissolve readily in cold chloroform, the ω-esters dissolve much more slowly, first imbibing the solvent and swelling to form a highly viscous solution.

Carothers and Hill have shown that these linear condensation superpolymers may be drawn into strong pliable transparent fibres. Filaments of the superpolyesters from hexadecamethylene dicarboxylic acid and trimethylene glycol have been made by "spinning" a chloroform solution by extruding the viscous solution through an ordinary rayon spinneret into a warm chamber to evaporate the chloroform, and if sufficient tension is applied the filaments formed without tension exhibit the phenomenon of "cold drawing," i.e. when stress is gently applied a transparent, lustrous, orientated, fibre is formed with a sharp boundary at the junction of the transparent and opaque sections of the filament. The breaking strength of transparent filaments is about six times that of the opaque. The tensile strength of 16–24 kg./sq. mm. compares favourably with that of cotton (about 28 kg.) and natural silk (35 kg.). The wet tenacity is fully equal to the dry. The filaments also showed remarkable springiness and elastic properties.

Filaments have also been made from superpolyesters of ω-hydroxy decanoic acid, ω-hydroxypentadecanoic acid, ethylene glycol and sebacic acid, trimethylene glycol and adipic acid and ethylene glycol and succinic acid.

Van Natta, in the same laboratories (Du Pont de Nemours & Co.), has shown that it is not possible to spin continuous filaments from the polyester of hydroxydecanoic acid until the molecular weight reaches about 7,000 and that the phenomenon of cold drawing does not appear until the molecular weight reaches 9,000. Carothers and

Hill conclude that a useful degree of strength and pliability in a fibre requires a molecular weight of 12,000 and a molecular length of 1,000 Å.

In addition to the polyesters described above, it has also been found possible to produce giant molecules by condensing amino acids with themselves. During the course of this research it was found that the molten material could be drawn out in the form of a long fibre, somewhat resembling silk; even after the fibre was cool, it could be drawn out still further to several times its original length. Such a phenomenon had not been previously observed in the case of any crystalline organic compound.

One of the simplest methods of nylon production is the reaction of the amino acid with itself, as previously stated, and this has been described in B.P. 461,236 where one or more amino monocarboxylic acids are subjected to a fairly extensive heat treatment. It is necessary to choose the amino acid so that the NH_2 group is as far removed as possible from the COOH group. In the case of the general formula for amino acids, $NH_2(CH_2)_n$ COOH, the value for n must not be less than 5, otherwise cyclisation of the amino acid occurs, producing imides.

These superpolymers are formed by the tail of one molecule reacting with the head of its neighbour.

$$NH_2(CH_2)_9COOH + NH_2(CH_2)_9COOH + NH_2(CH_2)_9COOH$$
$$-NH_2(CH_2)_9CO.NH(CH_2)_9CO.NH(CH_2)_9CO.NH(CH_2)_9CO-$$

Filaments produced by these methods are cold drawn to give some 300% elongation and have been found to be exceptionally strong, tough and pliable; the tenacity is from 3–4 grams per denier, which is higher than that of any rayon. The superpolyamide fibres are not sensitive to temperature or humidity and possess remarkable recovery from elastic stretching.

Consideration of the polyamides from amino acids shows that the reacting groups are two amino groups and two carboxyl groups, but it has been shown in B.P. 461,237 that the same type of condensation may be brought about if the two amino groups are on one molecule and the two carboxyl groups are on the other, i.e. superpolyamides are produced from diamines and dicarboxylic acids. It is believed that this method is utilised in the production of nylon on a commercial scale.

The general scheme is to condense together compounds of the following types. $NH_2(CH_2)_x NH_2$, where x is at least 4 and $COOH(CH_2)_y COOH$ where y is at least 3. The radical of dibasic carboxylic acid is that fragment remaining after the two acidic hydroxyls have been removed from its formula. Thus the radical of carbonic acid is $-CO-$; that of adipic acid is $-CO-CH_2-CH_2-CH_2-CH_2-CO-$. The radical length is the number of atoms in

PLATE IX

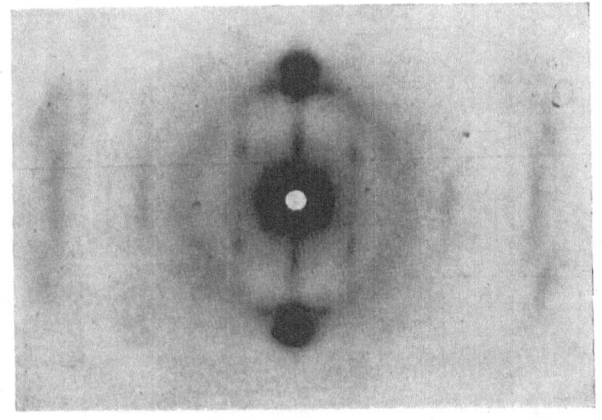

Fig. 36. X-ray photograph of Nylon.

Courtesy of Dr. Astbury, F.R.S.

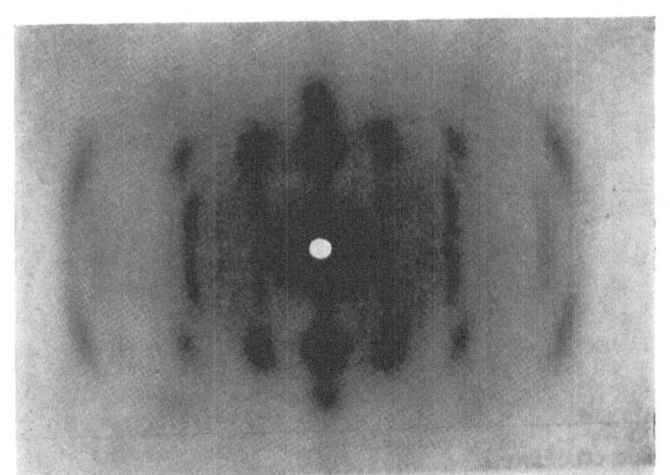

Fig. 35. X-ray photograph of natural silk.

To face page 88.

PLATE X

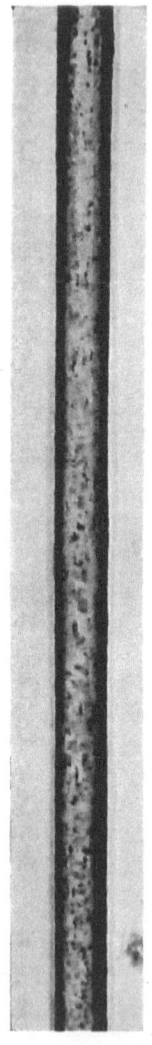

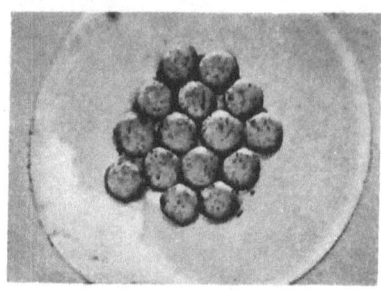

FIG. 38. Photomicrograph of cross-section of nylon. (× 300 diameters)

Figs. 1, 2, 7, 8, 9, 9a, 10, 10a, 23-26, and 37 are × 400 diameters.

Fig. 10a is by courtesy of Miss Alexander. Fig. 7a is by courtesy of the Linen Industry Research Association. Figs. 1-4, 7, 8, 8a, 9, 9a, 10, 25-27, 37, and 38 are by courtesy of Dr. W. Kling of Bohme-Fettchemie Gesellschaft, m.b.H., Chemnitz.

FIG. 37.
Nylon.

To face page 89.

SYNTHETIC FIBRES

the chain of the radical. Thus the radical length of carbonic acid is 1 and that of adipic is 6.

Similarly the radical of a diamine is the fragment remaining after one hydrogen has been removed from each amino group and the radical length is the number of atoms in the chain of the radical. Thus the radical of ethylene diamine is —NH—CH_2—CH_2—NH— and its length is 4, while similarly the radical length of pentamethylene diamine is 7.

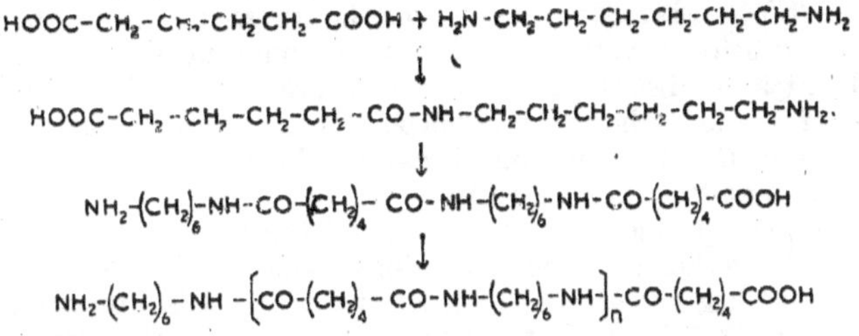

HOOC-CH_2-CH_2-CH_2-CH_2-COOH + H_2N-CH_2-CH_2-CH_2-CH_2-CH_2-CH_2-NH_2

↓

HOOC-CH_2-CH_2-CH_2-CH_2-CO-NH-CH_2-CH_2-CH_2-CH_2-CH_2-CH_2-NH_2.

↓

NH_2-$(CH_2)_6$-NH-CO-$(CH_2)_4$-CO-NH-$(CH_2)_6$-NH-CO-$(CH_2)_4$-COOH

↓

NH_2-$(CH_2)_6$-NH-[CO-$(CH_2)_4$-CO-NH-$(CH_2)_6$-NH-]$_n$-CO-$(CH_2)_4$-COOH

Obviously the unit length of a polyamide is the sum of the radical lengths of the diamine and acid used in its synthesis. In order to obtain a suitable superpolymer the total length of reacting radicals must be at least 9. From this type of product, which is formed from two different molecular species, it is possible to produce exceedingly fine filaments with a denier of 0·2. The density of the fibres or filaments is a little greater than unity, 1·14, and the melting-points range from 167°–278°C., the highest being obtained from tetramethylenediamine, $NH_2(CH_2)_4NH_2$ and adipic acid, COOH $(CH_2)_4$COOH. Hexamethylenediamine, and adipic acid may also be used, similarly pentamethylenediamine and sebacic acid.

HOOC.$(CH_2)_4$COOH + H_2N$(CH_2)_6$$NH_2$
adipic acid   hexamethylene diamine

↓

—OC$(CH_2)_4$CO $\Big[$ NH$(CH_2)_6$ NH.CO$(CH_2)_4$CO $\Big]_x$ NH$(CH_2)_6$NH—
polyamide.

This gives a superpolyamide whose unit length is 14.

It is interesting to re-write this formula for comparison with that of silk fibroin (see p. 58).

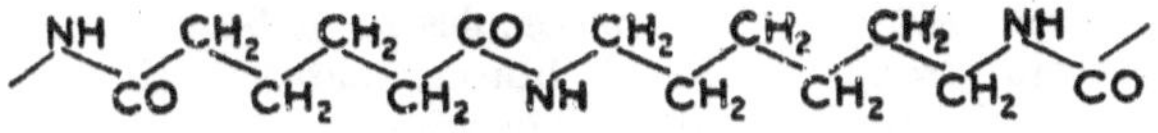

NH CH_2 CH_2 CO CH_2 CH_2 CH_2 NH
  CO CH_2 CH_2 NH CH_2 CH_2 CH_2 CO

Nylon

An improvement in the method of manufacture was described in B.P. 474,999, it being found beneficial first to isolate the salts from the base and the acid as a separate operation and then to heat them. The salts are much more stable than the diamines and can be stored until ready for the polymerisation process. Further, it is possible to purify the salts by recrystallisation and thus facilitate the standardisation of the end product.

The salts are subjected to heat treatment, e.g. hexamethylenediamine adipic acid salt is heated with an equal weight of mixed xylenols (b.p. 218–220°C.) for three hours in an atmosphere of nitrogen, utilising a vessel provided with some means of returning the solvent as it distils. The warm mixture is poured into a large volume of alcohol where the superpolyamide separates as a white powder (m.p. 248°C.) and the fibres may be spun either from the molten mass or from the solvent.

In a further specification (B.P. 487,734) the polyamides produced by heating together diamines and dibasic acids with a total radical length of not less than 7 and at least one of which contains oxygen or sulphur in the chain of atoms separating the reactive groups are condensed with decamethylenediamine; for instance, salicyl acetic acid could be utilised.

Properties

As previously stated, one of the most interesting properties of nylon is that it can be cold drawn to three or four times its original length, depending upon the particular polyamide.

The molecules are originally arranged in random fashion, but during the drawing they become orientated, that is, they are parallelised and brought much closer together.

This orientation is not only an interesting scientific point but is largely responsible for the great industrial value of nylon, because after this drawing process, it becomes exceedingly strong, probably owing to the fact that decreased inter-molecular distances produce increased molecular attraction.

Not only has nylon high tensile strength, but it also shows almost 100% recovery from elastic extension. Its water-absorption is quite low, amounting to 3·5% at 60% R.H., under which conditions viscose rayon takes up about 12% moisture.

It is interesting to note that the research took approximately ten years and cost more than a million dollars.

The most important synthesis of fibres demonstrates the correctness of the idea that long-chain molecules must be put together in order to obtain strong, organised, fibrous material; the molecular weight of nylon is approximately 20,000–40,000.

The similarity between silk and the nylons with regard to molecular structure is quite interesting. In the case of silk, the groups of NH and CO occur very frequently and are only separated by a group such as CH_2 or $CH.CH_3$. This is in agreement with the analysis of silk which gives simple amino acids by hydrolysis. With the polyamides produced according to B.P. 461,236 and 7, the groups of CO-NH occur much less frequently along the molecular chain, and the neighbouring CO—NH arrangement is separated by five CH_2 groups. The repeat pattern of the fibre consists of at least seven atoms excluding hydrogen.

In the case of polyamides produced according to B.P. 474,999 the CO—NH grouping is reversed in alternate repeats, and appears as NH—CO, so that the actual repeat pattern here is about twice as long as in the first example.

The American specifications dealing with nylon production and the preliminary work are U.S.P. 2,071,250 to 2,071,253 inclusive, 2,130,947 and 8.

Nylon filaments do not swell in water and are not affected by boiling in 5% NaOH solution. The general resistance to swelling of the chemical reagents, with the exception of phenols, may be due to the large number of CH_2 groups in the structure causing the NH—CO groupings to be widely separated.

The resistance to chemical attack makes nylon inert to reagents such as the common organic solvents, dilute alkalis, and dilute acetic acid; on the other hand, 5% hydrochloric acid at the boil renders nylon brittle and finally destroys it. As previously mentioned, nylon dissolves in phenol, *m*-cresol and xylenol at room temperatures; strong acids such as sulphuric, hydrochloric and even formic cause the fibres to swell and eventually disintegrate. Concentrated acetic acid will also exert a solvent effect when hot.

Nylon is susceptible to oxidising agents.

An interesting property of nylon is that of acquiring some permanent set in hot water or steam; this is utilised in setting stockings by "pre-boarding" after knitting, so that they will retain their original shape during wear and laundering.

CHAPTER VII
PHYSICAL PROPERTIES OF FIBRES

Some interesting comparisons of the various properties of textile fibres have been compiled from various sources. For instance, the lengths and diameters of the chief fibres are shown in the following table:

Dimensions of Textile Fibres

Fibre	Diameter in μ Min.	Max.	General	Length of fibres
Silk (raw)	12	26	13	300 to 1,000 yds.
Cotton	15	30	18	0·5 to 2 ins.
Viscose rayon	11	34	18	Continuous or cut
Acetate rayon	14	52	20	,, ,,
Flax	15	25	20	0·75 to 1·5 ins.
Flax bundle	120	250	200	12 to 20 ins.
Wool	15	80	21	4 to 16 ins.
Camel hair	20	28	24	2 to 2·5 ins.
Human hair	50	100	80	12 to 36 ins.

The diameters of some of the common textile fibres have also been given in the following table:

Diameters of Textile Fibres

Fibre	Diameter in microns (10^{-4} cm.)
Coarse wool	30 to 40μ
Fine wool	15 to 25
Cotton	15 to 20
Silk (degummed)	9 to 12
Rayon 1·5 denier	14 to 15
,, 3·0 ,,	19 to 21
,, 5·0 ,,	24 to 26
,, 8·0 ,,	30 to 33
,, 10·0 ,,	35 to 40

It must not be assumed that it is possible to exchange any rayon filaments of equal denier and obtain the same effect in the fabric; the cover of the filaments is also determined by the density. For example, the densities of the chief fibres are seen in the following table.

The densities of nylon, acetate and viscose are 1·14, 1·33 and 1·52, so that the diameters of a 10-denier filament would be nylon 35·2μ,

PHYSICAL PROPERTIES OF FIBRES

acetate 32·6μ and viscose 30·5μ. A glass filament of 10 denier would be 23·8μ wide, as its density is 2·5.

Densities of Textile Fibres

Fibre	Density
Cotton	1·52
Acetate	1·33
Wool	1·30
Nylon	1·14

Comparison of Viscose and Acetate Rayons

Denier	Viscose	Acetate
1	9·6 μ	10·3 μ
5	21·6 μ	23·0 μ
10	30·5 μ	32·6 μ
15	37·4 μ	39·7 μ
20	43·2 μ	46·1 μ
25	48·3 μ	51·5 μ

An alternative method of comparing some of the common textile fibres is on the denier basis, as shown in the following table:

Fibre Deniers

Silk	1·0
American cotton	1·8
Indian cotton	2·25
Merino wool (70s)	4·3
Crossbred wool (58s)	7·5
Flax	5·6
Jute	14·0

Reference has already been made to the fact that rayon may be spun in a number of different deniers as required, and it may also be cut into various lengths. Popular qualities for spun rayon are discussed on p. 16.

The methods of expressing the strength of fibres are sometimes apt to be misleading. Measurements of breaking load are unrelated to the cross-sectional area of the fibre and hence have little real value, because the breaking load will vary according to the area of the cross-section of the fibre.

The tensile strength may be expressed in terms of the unit of cross-sectional area either as lb. per sq. in., or as Kg. per sq. mm.; with the development of continuous filaments of regenerated and synthetic fibres, there is a tendency to express the strength in terms of the yarn numbering system or denier (the denier is the weight in grams of 9,000 metres of the yarn). The tenacity is expressed in grams per denier.

Tenacity is not directly interchangeable with tensile strength and some confusion may be caused on this account; values of tensile strength are directly comparable, but values of tenacity are not, because the density of the fibre is included in its tenacity. Hence two fibres may have the same tenacity but different tensile strengths, because their densities differ and therefore their cross-sectional areas.

The Strength of Fibres

Fibre	Tenacity (g./den.)	Tensile strength (lbs./sq. in.)
Saponified stretched acetate	7·0	136,000
Glass	6·5	213,000
Ramie	6·5	125,000
Flax	6·3	121,000
Cotton	6·0 to 4·0	120,000—80,000
Nylon	5·0 to 4·5	73,000—65,000
Silk (degummed)	4·9 to 2·8	78,000—45,000
Steel	4·6	460,000
Viscose rayon (strong)	4·6 to 3·4	89,000—66,000
Viscose rayon	2·4 to 1·8	46,000—25,000
Acetate rayon	1·8 to 1·2	30,000—20,000
Wool	1·7 to 1·0	28,000—17,000
Casein fibre	1·0 to 0·6	17,000—10,000

The effect of orientation, particularly in the regenerated fibres, is well known, and the following data illustrate this point:

Effect of Orientation

	Normal fibre	Stretch-spun
Viscose	25 Kg./mm.^{2}	80 Kg./mm.^{2}
Acetate	20 Kg./mm.^{2}	100 Kg./mm.^{2}

The effect of orientation on tenacity is also shown in the following table:

Orientation and Tenacity

Fibre	Tenacity
Normal viscose	1·85
Orientated viscose	5·88
Cuprammonium rayon	1·75
Orientated cuprammonium rayon	4·95
Acetate rayon	1·38
Orientated acetate rayon	4·95
Fine filament acetate with extreme orientation	8·52

PHYSICAL PROPERTIES OF FIBRES

Reference has already been made to the importance of the moisture relations of textile fibres. With regard to the amount of moisture held by a fibre at a certain humidity, there is sometimes a little confusion between moisture content and moisture regain; the former is expressed as a percentage of the total weight, and the latter as a percentage of the dry weight. Many measurements are related to a standard atmosphere, which in Great Britain and the U.S.A. is 65% R.H. and 70°F. (21°C.), but in continental Europe is 65% R.H. and 20°C. Standard regains are commonly, but not universally accepted.

MOISTURE IN TEXTILE MATERIALS

Fibre	Regain	Content
Cotton	8·5%	7·8%
Silk	11·0%	9·9%
Flax and hemp	12·0%	10·7%
Jute	13·75%	12·0%
Wool	16·0%	13·8%
Wool tops (oil combed)	19·0%	16·0%
Wool tops without oil	18·25%	15·5%
Worsted yarn	18·25%	15·5%
Wool yarn (carded)	17·0%	14·5%
Wool noils	14·0%	12·3%
Wool noils, scoured and carbonised	16·0%	13·8%
Wool and worsted fabric	16·0%	13·8%
Viscose and cuprammonium rayon	11·0%	10·0%
Acetate rayon	6·0%	5·6%
Nylon	4·0%	3·8%
Vinyon	0·0%	0·0%

The amount of moisture absorbed at 100% R.H. is also of interest in connection with some of the common fibres.

WATER ABSORPTION AT 100% R.H.

Silk (scoured)	35%
Wool (scoured)	33%
Cotton (scoured)	25%
Viscose	45%
Acetate	18%
Nylon	8%

The breaking load of the various textile fibres is affected to greater or less degree by the moisture they contain. An important consideration in all wet treatments of textile materials is the ratio of the strength when wet to that when dry.

Wet Strength of Textile Fibres
(as percentage of air-dry)

Cotton	110 to 120%
Wool	80 to 90%
Silk	75 to 85%
Viscose, ordinary	45 to 55%
Viscose, strong	55 to 65%
Acetate	65 to 70%
Nylon	88 to 90%
Lanital	25 to 30%

It will be noted that the regenerated fibres generally suffer a substantial decrease in tensile strength on wetting with water.

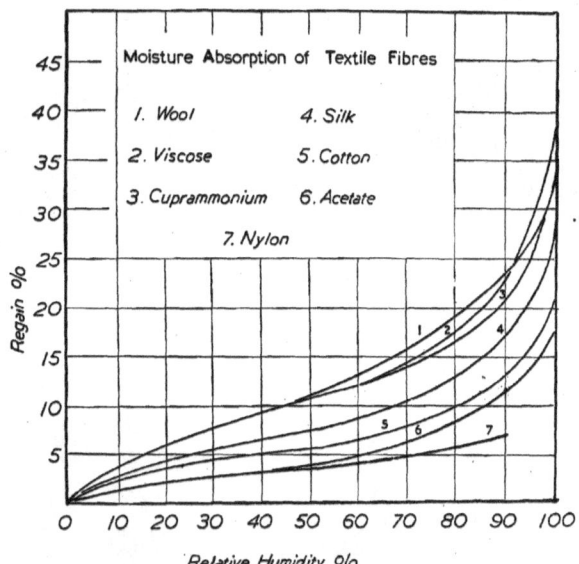

Fig. 39.—Comparative moisture absorptions of the main textile fibres.

The amount of swelling which the textile fibres undergo when immersed in water is also of interest; some tables of this value are apt to be misleading, for it is not always clear whether the results are calculated on the dimensions in the completely dry state or in the conditioned state, i.e. at about 65% R.H.

PHYSICAL PROPERTIES OF FIBRES

SWELLING OF TEXTILE FIBRES ON WETTING
(from 65% R.H.)

Fibre	Increase in area of cross-section
Silk	46%
Wool	25%
Cotton	30%
Viscose	35–60%
Acetate	6– 9%
Nylon	1– 3%

The extension at break of the various fibres is also affected by the amount of moisture they contain.

COMPARISON OF DRY AND WET EXTENSIONS

Fibre	Dry (%)			Wet (%)		
Cotton	5·7	to	12·5	6·1	to	13·2
Viscose rayon	7·8	,,	25·8	13·0	,,	42·8
Cuprammonium rayon	16·9	,,	19·6	17·0	,,	29·0
Acetate rayon	20·9	,,	30·0	28·9	,,	30·0
Wool	28·0	,,	48·3	34·8	,,	60·5
Silk	13·5	,,	17·2	30·0	,,	30·1
Lanital	6·1	,,	50·0	83·0	,,	110·5
Flax	1·8	,,	—	2·2	,,	—

The above data refer to a comparison of the physical properties of the fibres in their conditioned state and when wet with liquid water; the properties also change as the fibres pass from the completely dry state through atmospheres of increasing relative humidity.

ALTERATION OF STRENGTH AND EXTENSION WITH HUMIDITY

	Tensile strength in 10^3 Kg./cm.^{2}			
	0% R.H.	50% R.H.	70% R.H.	100% R.H.
Ramie	2·84	3·98	3·88	0·93
Cotton	2·89	4·12	3·78	7·81
Wool	1·45	1·47	1·09	0·66
Silk	3·56	4·23	3·78	5·19
Viscose rayon	1·56	0·82	0·94	0·50

	Extension at break (%)			
Ramie	1·7	2·3	1·8	4·0
Cotton	1·6	7·6	7·0	8·2
Wool	13·2	33·5	38·8	46·3
Silk	13·5	24·0	27·0	32·7
Viscose rayon	12·7	12·1	14·7	18·3

TEXTILE BLEACHING

MECHANICAL PROPERTIES

(B. H. Wilsdon, "The Times Trade and

Property	Unit	Condition	Synthetic fibre Nylon	
			Min.	Max.
Tensile strength	Kg. per sq. mm.	Dry		51
	Kg. per sq. mm.	Wet		48
	Percentage	Wet/Dry		94
Maximum stretch	Percentage	Dry	26	60*
		Wet	30	57
Modulus of elasticity	Kg. per sq. mm.	Dry		505
Hooke's law range	Stretch (%)	Dry		4
Extension and	% of initial length	Dry	16	100
Recovery	% of extension	Dry	91	24
Regain	Gms. of water per 100 gms. wool	30% R.H.		2·0
		80% R.H.		5·5
	Difference	—		3·5

* Variable.

The molecular structure of most fibres exhibits a certain amount of orientation which indicates that although the fibre is essentially composed of substances of high molecular weight, yet it is not amorphous, but here and there the molecular chains are in parallel array which is sufficiently pronounced to show crystallinity.

The extent of the crystalline nature of the various fibres cannot be measured with precision, but the following table indicates the general order:

CRYSTALLINE NATURE OF FIBRES

Fibre	Crystallinity
Flax	Very high
Nylon	High
Cotton	Medium
Silk	Medium
Wool	Low
Rayons	Low to high

of Fibres
Engineering Supplement," October 1939)

Natural Fibres			Artificial Fibres		
Wool	Silk	Cotton	Viscose	Acetate	Casein
Min. Max.	Min. Max.	Min. Max.	Min. Max.	Min. Max.	Min. Max.
13.5 21.6 11.5 16.6 79.0 97	46 70 47 52 86 95	21 80 24 83 99 113	18 40 8.6 20 42 65	16 21 10 12 59 70	10 12 5.2 5.6 43 54
28 74 35 80	7 25 24 36	5.7 12.5 6.1 13.2	7.8 26 9 43	21 30 26 32	Variable 83 110
260–400 5	700 3	600 at 65%R. 3	700 1	500 1	Plastic
28 74 100 80	10 14 53 45	6 10 40 30	Practically none	Maximum 5%	No Recovery
8.8 19.0 10.2	6.5 17.4 10.9	4.0 10.2 6.2	8.0 18.0 10.0	3.0 9.0 6.0	16.3 20.4 4.1

With rayons the degree of crystallinity is largely determined by the amount of stretch applied during their manufacture.

The susceptibility to various physical influences may be deduced from the data previously collected, but the textile fibres also differ in their resistance to common chemical reagents.

Cotton is prized on account of its resistance to hot alkaline solutions, and this property makes it a valuable material for fabrics which will withstand repeated washings; cotton, however, is readily degraded by acids and also by oxidising agents.

Wool, on the other hand, although resistant to acids, is readily affected by alkalis and must therefore be washed carefully in the chemical sense; the elasticity of wool and the presence of scales necessitate careful manipulation during laundering if felting is to be avoided.

The regenerated cellulose rayons are more susceptible to physical and chemical damage than the native cellulose fibres and need handling with care, particularly in the wet state; they are also more

COMPARATIVE EXTENSIBILITY OF YARNS
(Loasby, J.T.I., 1943, *34*, 45P)

Load g./den.	Nylon 45 den.	Silk 4/14	Wool	Acetate	Cel. Fort 60	Viscose 300	Durofil 60	Tenasco 150	Vinyon 160
0·4	1·0	0·2	—	0·5	0·3	0·5	0·3	0·3	0·3
0·8	3·3	0·6	1·6	1·6	0·6	2·5	0·6	1·0	1·5
1·2	5·7	1·0	2·7	5·0	1·0	9·0	1·0	4·6	5·0
1·6	7·6	1·5	3·7	—	1·5	13·7	1·6	8·5	9·7
2·0	9·3	4·0	4·8	—	2·2	—	2·6	11·3	12·3
2·4	11·0	8·5	6·0	—	2·8	—	3·3	13·7	14·5
2·8	12·0	11·3	7·6	—	3·3	—	4·0	—	—
3·2	13·5	17·2	10·0	—	3·8	—	4·4	—	—
3·6	14·5	—	—	—	4·2	—	4·8	—	—
4·0	—	—	—	—	4·6	—	—	—	—

readily degraded by acids and oxidising agents. Acetate rayons, although they do not swell in water to any appreciable extent, are readily hydrolysed by hot alkaline solutions and need careful treatment on that account.

Lastly, the vegetable fibres may be bleached by solutions of alkaline hypochlorites, but these have a peculiar effect on wool; hydrogen peroxide may be used for the bleaching of all the textile fibres except nylon.

PART TWO
WETTING AND DETERGENCY

CHAPTER VIII
WATER

WATER is the universal cleansing agent, and is often sufficient in itself for many textile washing operations. In dealing with oils and fats, however, it is usually necessary to employ an assistant to help in their removal by the aqueous phase.

One of the commonest cleansing agents which is added to water is soap, but it is common knowledge that soaps behave differently in different waters; it is difficult to form a lather in hard water without using considerable soap, for to a certain extent the hard water softens itself at the expense of the soap.

The total hardness of water consists of temporary hardness or permanent hardness or both. Temporary hardness is due to calcium or magnesium bicarbonate and disappears after the water is boiled; permanent hardness is due to other calcium or magnesium salts, and is so-called because it is not removed by boiling. In order to soften water, it is necessary either completely to remove the calcium and magnesium compounds or to replace the calcium or magnesium by other metals whose salts do not produce hardness and are harmless in other respects.

The British official degree of hardness is the Clark degree which corresponds to the presence of calcium carbonate or its equivalent to the extent of 1 grain per gallon or 1 part per 70,000 parts of water; other impurities in water are generally expressed as parts per 100,000, and the hardness of water may also be expressed in this manner. French degrees of hardness are measured as parts of calcium carbonate or its equivalent per 100,000, and German degrees are expressed as parts of lime, or calcium oxide, per 100,000 parts. 1° Clark is equal to 1·43 French degrees or 0·8 German degrees or 0·83 United States degrees.

Although the hardness of water is generally expressed as calcium carbonate, this does not mean that the water contains calcium carbonate (which is almost insoluble), but that the actual amounts of calcium and magnesium salts are merely expressed as the equivalent amounts of calcium carbonate for convenience as a basis for analysis and comparison.

WATER

Waters have been classified into four main groups as follows: 0° to 4°, soft water; 4° to 7°, moderately hard; 7° to 20°, hard water; and above 20°, very hard water.

The total hardness of water may be measured by titration with a soap solution. For example, with Wanklyn's method, the soap is prepared by dissolving 10 g. of pure white Castile soap in 600 c.c. of 90% alcohol and diluting to 1 litre with distilled water; the hardness is measured by the amount of solution necessary to form a permanent lather with 70 c.c. of water. A control test is first made with 70 c.c. of distilled water, and the correction figure obtained is subtracted from the reading for the water under test. The hardness of the water, according to the Wanklyn method, is equal to the grains of calcium carbonate per gallon, plus one.

Clark's solution is first prepared by dissolving 80 g. of pure oleic acid in alcohol and neutralising with alcoholic potassium hydroxide. It is then standardised against a standard hard water and diluted with 50% alcohol until 1 c.c. of soap solution is equivalent to 0·001 g. $CaCO_3$. The standard hard water is prepared by dissolving 1 g. of pure Iceland spar in dilute HCl and evaporating the solution to dryness; the residue is dissolved in water, and again evaporated, the process being repeated until the residue is neutral. The solution is then diluted to 1 litre with distilled water. In using Clark's solution, 50 c.c. of the water under test is placed in a large flask and the standard soap solution is added in small amounts, 1 or 0·5 c.c., until a permanent lather begins to form. With magnesia, a false lather may be seen, but this disappears on the addition of a further 0·5 to 1 c.c.; if the lather is stable, however, then the additional amount is not included in the final result. Tables are available giving the hardness in degrees from the amount of soap used. The number of c.c. of soap solution required for 50 c.c. of the water, multiplied by 2, gives the parts of $CaCO_3$ per 100,000; if 70 c.c. of water are used, the c.c. of soap solution gives the parts per 70,000 i.e. grains per gallon.

Possibly the most convenient method is the estimation of hardness with acid. The temporary hardness may be measured by titration with $N/10$ nitric acid. 100 c.c. of water are coloured yellow by the addition of a few drops of methyl orange, and the amount of acid required to produce a pink colour is noted; this number of c.c. is multiplied by 5 to give temporary hardness in parts of $CaCO_3$ per 100,000. If temporary hardness is required according to the English system, then 70 c.c. of water should be taken for the estimation. From another sample of the original water, the sulphates, and the soluble salts of calcium and magnesium are removed, together with carbonates, by adding a known excess of sodium carbonate solution. The water is then boiled, and titrated with standard acid. This gives the permanent hardness.

Magnesium salts, however, may come under two heads, for magnesium bicarbonate causes temporary hardness and other magnesium salts produce permanent hardness. Magnesium hardness may be estimated by neutralising 100 c.c. of water with $N/10$ H_2SO_4 in presence of methyl orange and boiling until all the carbon dioxide is expelled. The solution is then mixed with 50 c.c. of filtered lime water of known equivalence in terms of $N/10$ acid, and the liquid raised to the boil on a water-bath, corked and allowed to cool. The solution is finally made up to 200 c.c. with distilled water and allowed to settle, after which it is filtered rapidly whilst covered to prevent absorption of atmospheric carbon dioxide. 100 c.c. of the water is then titrated with $N/10$ H_2SO_4 utilising phenol phthalein as indicator. If 50 c.c. of the lime water required a c.c. of $N/10$ acid, and 100 c.c. of the liquid required b c.c., then $(a-b)$ c.c. of acid corresponds to the magnesium hardness in 50 c.c. of the original water. Each c.c. of acid corresponds to magnesium hardness equivalent to 5 parts of calcium carbonate per 100,000 parts of water.

Our supplies of water fall into three classes: (*a*) moorland water, (*b*) other surface water, and (*c*) well or spring water.

Moorland waters are commonly used in Scotland and are generally of a low degree of hardness; they are often coloured and are slightly acid, which renders them corrosive. The colour may be removed by small amounts of sodium aluminate or aluminium sulphate (filter alum). The hardness of these waters is generally of the permanent type.

The other surface waters are often slightly alkaline, but their nature depends to a large extent on the mineral character of the region over which they flow. Colour and suspended matter may be removed by coagulants.

Some well waters are similar to surface waters but others contain considerable amounts of carbon dioxide either dissolved in the free form, or as bicarbonates. The hardness may be removed as described later, but it must be remembered that the presence of carbon dioxide may cause corrosion difficulties in the boiler plant and turbines.

For use in textile processes, it is generally necessary to have clear water; turbidity may be due to large or small particles suspended in the water, and with the former, the normal settling process in the "lodge" or reservoir is adequate. Small and colloidal suspensions, however, may need different treatment to clarify the water; they may be removed by a filter of sand or by special methods of coagulation and adsorption.

Aluminium hydroxide has found extensive use in this field but is often derived from aluminium sulphate which is an acid salt, without coagulating powers. When an alkali is added, however, or in presence of temporary hardness, aluminium hydroxide is precipitated and the suspended matter coagulated. As aluminium hydroxide is soluble in both acid and alkali, complete precipitation is only obtained over a very narrow pH range, so that careful control is necessary to obtain proper coagulation. To remove colour or matter suspended in the colloidal state, the hydroxide should be precipitated on that side of the neutral point which produces a precipitate of opposite charge to that of the suspended impurities. When this is done, colloidal aggregation takes place and the solid matter may be removed by sedimentation or filtration.

Aluminium sulphate may be used alone in waters of temporary hardness, but clarification is sometimes imperfect, probably on account of the optimum pH value being disturbed by the liberation of carbon dioxide. The addition of a little alkali generally assists the process.

Sodium aluminate has also been used as the alkaline precipitant and should be added before the alum as it appears to "seed" the growth of the aluminium hydroxide. The required amount of sodium aluminate is small, usually 0·1 to 0·25 grain per gallon, and after the preliminary trials, it is generally kept constant but changes in the composition of the water are dealt with by altering the amount of alum. This method is stated by Hendry (J.S.D.C., 1942, *58*, 153) to have certain advantages; it is more efficient in clarification and

WATER 105

colour removal, and control is simpler as the pH range for efficient treatment is wider—the amount of residual alumina is also less.

SOFTENING

During the past twenty years, investigations into the methods of softening water have been intensified with a view to increasing the efficiency of fuel and the generation of power. It has been found that many waters which are suitable for the production of steam at 150 lbs. per sq. in. are not satisfactory for pressures of 600 to 700 lbs. per sq. in.; when the pressure required is as high as 1,500 lbs. per sq. in. the water should contain no significant amount of dissolved solids and less dissolved oxygen and carbon dioxide than distilled water.

The methods of water-softening for boiler feeds have naturally a wider scope than for power alone; industry in general and the textile industry in particular has benefited from the recent advances.

The simplest method for softening water consists in the removal of any calcium or magnesium which may be present in the form of the soluble bicarbonates or sulphates; these metals are the chief source of trouble in the bleaching, dyeing and finishing sections of the textile industry owing to their well-known tendency to form insoluble soaps. For boiler-feed purposes, however, it is often necessary to remove all dissolved substances which may produce scale.

There are three main divisions into which the methods of softening water may be classified: (*a*) precipitation of the calcium and magnesium compounds, (*b*) replacement of the cation which is not desired, by a cation which forms soluble salts, e.g. sodium, and (*c*) removal of both cations and anions to give a water containing little or no dissolved solids. It is possible to combine certain of these processes.

The general methods of softening water were discussed by the Society of Chemical Industry in 1941 and a short report of the proceedings has been given by Parker (Chem. and Ind., 1941, *60*, 795).

LIME may be used to remove temporary hardness when added in sufficient quantity to convert the soluble calcium bicarbonate into the insoluble carbonate; the lime also removes dissolved carbon dioxide. At the same time, the lime converts any magnesium bicarbonate in the water into the relatively insoluble magnesium hydroxide.

The lime process, however, does not remove permanent hardness due to the presence of dissolved sulphates and chlorides of calcium and magnesium. The water is not completely softened and may have a residual hardness of 2 to 4 parts per 100,000 parts expressed as calcium carbonate. Some difficulty is apt to occur in bringing about rapid and complete precipitation and sedimentation. Coagulating agents such as aluminium sulphate and sodium aluminate have been used to assist precipitation, and improvements have been effected by returning some of the precipitated calcium carbonate to the water as it is being treated with lime in such a manner that the water flows through the carbonate and increases the size of the particles so that they are more readily removed by sedimentation. Another difficulty lies in the disposal of the sludge, for water of temporary hardness amounting to 10 parts per 100,000 will produce about 1·5 tons of air-dry sludge per million gallons.

The LIME-SODA process will remove both temporary and permanent hardness; it consists in the addition of lime and sodium carbonate in sufficient amounts to precipitate the calcium salts as carbonate and the magnesium salts as hydroxide. The sulphates of calcium and magnesium originally present in the water are replaced by the equivalent amount of sodium sulphate, so that if the permanent hardness was high, then the softened water will contain considerable quantities of soluble sodium salts. With this process, some of the difficulties of slow precipitation and sedimentation may be obviated by heating the water beforehand. The last traces of precipitate may be removed by filtration through sand, but as this may result in increasing the quantity of silica in the water, anthracite of graded particle size may be used as an alternative.

SODA AND CAUSTIC SODA may be utilised in place of the lime-soda treatment but is a more expensive method. The concentration of dissolved solids in the treated water is greater than with the previous process and the alkalinity of the water is also increased. The quantity of sludge is less, as one-half of the carbon dioxide in the bicarbonates and all the excess carbon dioxide are converted into the soluble sodium carbonate.

LIME AND BARIUM CARBONATE OR HYDROXIDE have been used in the U.S.A., but the barium compounds are more expensive than sodium carbonate or hydroxide. Special precautions are necessary to remove the precipitated calcium carbonate, barium carbonate and sulphate and the methods appear to be confined to the treatment of water containing large amounts of sulphates, under conditions where the production of dissolved sodium sulphate is a disadvantage.

BASE-EXCHANGE ZEOLITES have been employed for the softening of water for many years; the process depends on the property of base-exchange which is possessed by certain hydrated alumino-silicates. The compound is graded into particles of suitable size and when hard water is passed through a column of the sodium alumino-silicate, the calcium and magnesium are removed from the water and replaced by sodium.

When the sodium in the aluminosilicate has been replaced by

WATER

calcium and magnesium, the material may be regenerated or reactivated by treatment with sodium chloride solution.

There are three types of zeolite in commercial use: (a) natural glauconites, (b) treated clays and (c) synthetic products from sodium silicate and aluminate. Their base-exchange values vary and the volume of water which can be treated by 1 cubic foot of base-exchange material to reduce the hardness by 10 parts per 100,000 parts is 700 gallons with the glauconites, 1,100 gallons with the clays and 1,800 gallons with the synthetic zeolites.

The amount of salt required for reactivation is generally from 2 to 4 times the weight of the calcium carbonate equivalent of the hardness removed; in addition, the regenerated material must be washed with water and the quantity required is usually from 1 to 4% of the volume of water softened. There is some loss of material due to disintegration and reduction in efficiency, the treated clays being weakest in this respect.

The base-exchange process is very suitable for the treatment of water for household purposes and for institutions; it also meets the requirements of laundries where the temporary hardness of the water does not exceed 20 parts per 100,000 parts. Compared with the previous processes, the base-exchange method gives a softer water with a hardness of not more than 1 part per 100,000 parts, and there are no difficulties of precipitation and sedimentation; it also adjusts itself to any variation in the hardness of the water being treated. On the other hand, the total concentration of dissolved solids is higher and the alkalinity is increased; the process is not suitable for water above 40 or 50°C. and its efficiency may be impaired by the presence of iron compounds or suspended solids and colloidal matter.

Carbonaceous base-exchange materials may be prepared by treating products such as coal and starch with strong sulphuric acid. Patented processes are responsible for the production of large quantities of base-exchange material from coal on a commercial scale. The exchange value is almost as great as that of the synthetic aluminosilicates and the products are not destroyed by dilute solutions of mineral acids. In addition to possessing the power of becoming reactivated with sodium chloride, the carbonaceous products may be treated with acid to give the hydrogen compound, after which the calcium and magnesium in the water may be replaced by hydrogen. In this manner, bicarbonates and sulphates are replaced by carbon dioxide and sulphuric acid so that the treated water will contain very little dissolved solid, where the original hardness was due to bicarbonates of calcium and magnesium.

EXCHANGE-RESINS were developed in 1934, when it was found that resins from phenols and tannins had base-exchange properties, and resins from aromatic bases possess acid-exchange properties (B.P. 450,308; 450,309; 474,361). An interesting paper in this connection has been contributed by Adams and Holmes (J.S.C.I., 1935, *54*, 1).

The resins are not destroyed by dilute solutions of acids or alkalis, and the value of some of the base-exchange resins is slightly higher than that for the synthetic aluminosilicates. They may be reactivated with sodium chloride to replace calcium and magnesium with sodium, or they may be regenerated with acid and used to replace the calcium and magnesium with hydrogen as with the carbonaceous base-exchange materials described above.

The acid-exchange resins are unique; not only are they the only substances with acid-exchange properties, but their exchange value is twice as great as that of the base-exchange resins.

The use of a base-exchange material followed by an acid-exchange resin will give a water containing little or no dissolved solid.

According to Harwood (Chem. and Ind., 1941, *60*, 760), the lime-soda plant is economical to operate from the standpoint of the cost of precipitants but requires considerable space and strict control. Where the water contains considerable amounts of magnesium salts, the successful operation of the lime-soda system becomes more difficult and necessitates the use of a coagulant such as sodium aluminate; the disposal of sludge sometimes presents a difficulty. The lime-soda softener produces water of 2 to 3° hardness by the ordinary methods and the water is usually of low total alkalinity. The base exchange plant costs more to operate and with water of 20° hardness is not economical in comparison with the lime-soda method. Where the hardness exceeds 15 to 16° hardness, the use of a zeolite plant tends to give water with too much sodium bicarbonate for many textile purposes.

The economic side of hard water may be illustrated by the example of water of 10° total hardness, every 100,000 gallons of which will destroy 9·75 cwt. of soap, costing about £20 (in 1941).

Prior to 1920, it was customary for most laundries not to soften the water, but to use soap and soda or silicated alkali as detergents; to-day, however, nearly all laundries use soft water. In the old method the fabrics became loaded with insoluble compounds, consisting mainly of the sticky calcium and magnesium soaps, which impaired the handle of the material and caused suspended soiling matter to adhere to the cloth giving it a grey colour; with wool, the deposition of these insoluble soaps accentuated the tendency to felt.

PHOSPHATES have been employed in connection with the difficulties occasioned by hard water; sodium phosphate, Na_3PO_4, is sometimes used as a water-softener. Simple treatments with the theoretical amounts of trisodium phosphate in the cold under favourable conditions are not sufficient to produce very soft water, but a preliminary treatment with lime or with lime and sodium carbonate enables the residual hardness to be removed by trisodium phosphate.

The reaction depends on the formation of the insoluble calcium and magnesium phosphates.

Considerable quantities of alkaline salts are used as cleansing agents particularly for glass-ware, and sodium phosphate, meta-silicate and carbonate seem to be preferred on the grounds of efficiency and economy; one failing is the insolubility of their alkaline earth salts, so that precipitation takes place when more than a trace of calcium or magnesium is present. Investigation of other phosphates has led to the discovery of the extraordinary properties of sodium hexametaphosphate $(NaPO_3)_6$, commonly known as Calgon. This compound forms a soluble complex ion with calcium or magnesium, and although not a detergent, yet it prevents the formation of insoluble salts and soaps of the alkaline earths.

It has been suggested that this takes place in the following manner:

$$Na_2[Na_4(PO_3)_6] + Ca^{..} \longrightarrow Na_2\left[Na_2Ca(PO_3)_6\right] + 2Na^{.}$$

$$Na_2[Na_4(PO_3)_6] + 2Ca^{..} \longrightarrow Na_2\left[Ca_2(PO_3)_6\right] + 4Na^{.}$$

Calcium soaps are kept in solution according to the following scheme:

$$Ca(RCOO)_2 + Na_2[Na_4(PO_3)_6] \longrightarrow 2RCOONa + Na_2[Na_2Ca(PO_3)_6]$$

Calcium carbonate is dissolved in a similar manner:

$$CaCO_3 + Na_2[Na_4(PO_3)_6] \longrightarrow Na_2CO_3 + Na_2[Na_2Ca(PO_3)_6]$$

For many purposes, therefore, the addition of hexametaphosphate to wash waters or rinsing waters is a useful alternative to a water-softening plant or to the use of the modern sulphated alcohols. The latter have a dispersing action on lime soaps, but the hexametaphosphate reacts stoichiometrically with the calcium compound. When the hexametaphosphate is added to hard water in the absence of soap, additions of about one-third of the amount required for complete softening produce a turbidity which clears on the addition of further hexametaphosphate; the clear solutions become cloudy on heating probably on account of decomposition and transformation of the metaphosphate to orthophosphate. Dispersing agents of the sulphated alcohol or sulphonated alcohol class are not sensitive to temperature in the same manner. These organic reagents produce continuous decrease in turbidity as their concentration is increased in presence of calcium or magnesium soaps, but with the hexametaphosphate there is only a small effect at first, followed by a rapid decrease in turbidity as the concentration necessary for softening is reached. The organic reagents have a dispersing action on freshly precipitated calcium soaps but not on the dried soaps, according to Lindner (Textilber., 1936, *17*, 861), whereas the hexametaphosphate reacts chemically with the dried lime soap.

Sodium pyrophosphate, $Na_4P_2O_7$, has been introduced for a similar purpose, but Gilmore (Ind. Eng. Chem., 1937, *29*, 584) has found that the hexametaphosphate is much more efficient. In presence of soap, the amounts of hexametaphosphate or pyrophosphate required to produce the same reduction in Ca ion concentration at pH 10 bear the ratio of 1 : 9 when estimated by Clark's method. The amounts of hexametaphosphate and pyrophosphate which dissolve the same quantity of calcium sulphate bear the ratio of 1 : 5, and moreover, the hexametaphosphate is more effective in preventing calcium deposits.

The *Trilons A* and *B* may be regarded as organic analogues of sodium hexametaphosphate in the action with calcium and magnesium.

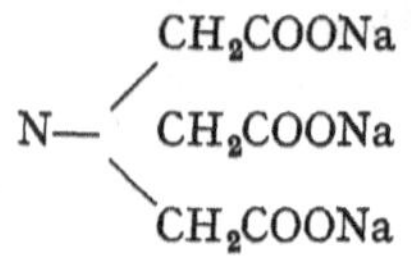

$$N\begin{cases} CH_2COONa \\ CH_2COONa \\ CH_2COONa \end{cases}$$

Trilon *A*

$$\begin{matrix} NaOOCCH_2 \\ \end{matrix} N-C_2H_4-N \begin{matrix} CH_2COONa \\ \end{matrix}$$
$$NaOOCCH_2 \qquad\qquad CH_2COONa$$

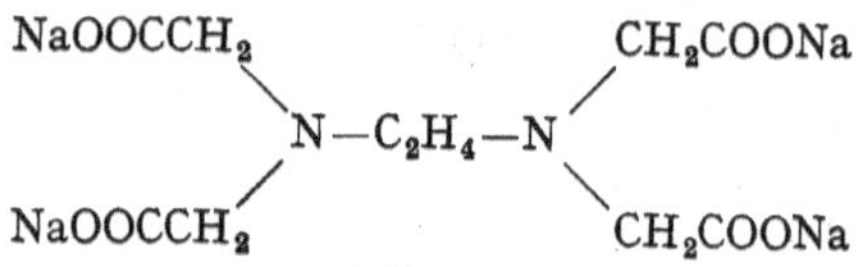

Trilon *B*

CHAPTER IX
ALKALIS

THE two chief alkalis used in the textile industries are sodium hydroxide and sodium carbonate; approximately one-sixth of the world's output of each of these compounds is absorbed by our textile industries.

The mercerising of cotton and the manufacture of viscose rayon require considerable quantities of caustic soda but neither of these processes is normally regarded as a cleansing operation. The scouring of cotton, however, is fundamentally a cleansing operation in which the goods are boiled for periods up to twelve hours in solutions of caustic soda whose average concentration may vary between 1·5 and 2% according to circumstances. Flax, hemp and jute are very susceptible to the action of hot caustic alkali which disintegrates the fibre bundles into the individual fibres of about one-tenth of the original length; sodium carbonate is therefore preferred for scouring processes.

The animal fibres are very sensitive to hot dilute sodium hydroxide solutions in which wool dissolves quite readily. Under less drastic conditions, the wool becomes impoverished, assumes a yellow colour, and is brittle and weak with a harsh handle. Sodium carbonate is therefore preferred. When wool contains less than 10% of grease, it will not withstand a pH much above 9·5, so that caustic soda, if used at all, must be limited to the preliminary scour, and the residual grease removed at pH 9. Wool suint is a useful agent for removing the grease by emulsification and may be maintained at pH 9 by the addition of sodium carbonate; the latter stages of grease removal are effected by mixtures of soap and sodium carbonate, by soap and finally by rinsing with water.

The removal of silk gum from real silk is effected by soap, by soap and borax, or by soap and ammonia.

There is a twofold reason for the use of alkalis in the textile purification processes, first, the saponification of some of the fatty impurities which occur naturally in the native fibres produces "soaps" which can emulsify the non-saponifiable fats and waxes, and secondly, to maintain the optimum alkalinity of the soap solutions where these are utilised. Alkalis also play a part in water-softening as they precipitate calcium and magnesium salts from hard water. Pyrophosphates delay the formation of these precipitates, whereas the correct use of sodium hexametaphosphate prevents their precipitation, owing to the formation of a soluble complex.

The commercial form of hexametaphosphate sold as Calgon is actually slightly alkaline owing to the presence of some 10% of tetrasodium pyrophosphate and a trace of sodium sesquicarbonate.

Active and inactive alkali are terms sometimes encountered in discussions on cleansing and detergency with respect to alkalis. Active alkali is that portion of the alkali content, expressed as Na_2O, which maintains a pH above 9·5 approximating to the phenolphthalein colour change.

Liddiard (Chem. and Ind., 1941, *60*, 480, 684 and 713) has given a useful review of the use of alkalis in cleansing processes; the following table shows the activities of various commercial alkalis.

ACTIVITIES OF VARIOUS COMMERCIAL ALKALIS

Alkali	Total Na_2O content	Active Na_2O (pH 9·5)	Active Na_2O %	pH of soln. 1%
Caustic soda	76·4	75·0	97	13·5
Soda ash	56·2	28·1	50	11·4
$Na_2CO_3.H_2O$	48·0	24·0	50	—
Washing soda	21·7	10·8	50	—
$NaHCO_3$	37·0	0	0	8·4
Sodium sesquicarbonate	41·2	11·1	27	9·9
Sodium metasilicate (5 H_2O)	29·6	24·9	84	12·5
Sodium sesquisilicate (11 H_2O)	36·8	31·0	86	—
Sodium orthosilicate (3 H_2O)	52·1	46·2	89	—
Trisodium phosphate (11 H_2O)	27·3	8·7	32	12·0
Disodium hydrogen phosphate (12 H_2O)	18·1	0	0	—
Tetrasodium pyrophosphate	46·6	1·9	0·4	10·1
Calgon	30·4	0	0	7·2

Sodium hydroxide is usually supplied as 97 to 98% NaOH and is available in flake, block or powder form as well as a saturated solution. Caustic soda exists entirely in the active form. It is a poor wetting and emulsifying agent, and with oils and grease which are difficult to emulsify it cannot be used alone; it does not assist soap greatly as it is apt to break the emulsion. A further point which requires attention is its power of being absorbed by the textile fibres. Potassium hydroxide is rarely used in industrial practice for economic reasons. Sodium peroxide is seldom used as a cleansing agent, but the available oxygen imparts bleaching powers.

Sodium carbonate is commercially available in three forms, anhydrous (soda ash or 58° alkali), monohydrate (the most stable form) and washing soda (the decahydrate). Soda ash is supplied as a dry powder and has a wide commercial use, but the other two varieties are mainly intended for domestic consumption. Sodium carbonate contains equivalent amounts of active and inactive alkali, but the bicarbonate is inactive unless boiled to form some carbonate. Sodium sesquicarbonate is $Na_2CO_3.2\ Na\ HCO_3.2\ H_2O$. The wetting and emulsification powers of all the carbonates is poor when used alone, but they contain some hydrolysed molecules in solution which assist the surface action of soap solutions. Soda ash is also useful in reducing permanent hardness but gives rise to precipitates with salts responsible for temporary hardness.

The soluble alkaline silicates have varied uses. The most common form is the liquid silicate of soda known as water-glass; this is sold in two forms, $Na_2O : SiO_2$ 1 : 2 and 1 : 3, corresponding to 140°Tw. (54%) and 78°Tw. (38%) respectively. Sodium metasilicate is a crystalline solid usually marketed with 5 H_2O; 84% of the Na_2O content is in the active form but higher values are shown by the orthosilicate (89%) and the sesquisilicate (86%). The silicates are the best wetting and emulsifying agents amongst the alkalis; with hard water they give rise to calcium and magnesium silicates which are semi-colloidal in character and finally separate as a non-adherent sludge.

The orthophosphates, particularly trisodium orthophosphate, have applications in washing processes, but tetrasodium pyrophosphate is more valuable in detergent solutions on account of its powers of preventing the deposition of calcium precipitates. Calgon, which is mainly sodium hexametaphosphate is still more valuable in all fields of detergency as it prevents the precipitation of salts from hard waters. One-third of the Na_2O content of trisodium phosphate is available as active alkali, and a small amount of active alkali is available in tetrasodium pyrophosphate; all the other phosphates contain no active alkali. Trisodium phosphate exhibits fair wetting and emulsifying powers and offers definite assistance to soap, but the other phosphates have no surface activity. One of the most important features of the phosphates is their use with hard water; the orthophosphates are good water-softeners particularly for most of permanent hardness, but with water of temporary hardness they may precipitate tricalcium phosphate. Calcium salts of pyrophosphoric acid are soluble in water, as also is the complex calcium derivative of sodium hexametaphosphate.

Borax is generally considered to be a mild alkali although part of its alkalinity is in the active form; the comparatively low alkalinity is utilised in the washing of silk and delicate fabrics of all types. In certain circumstances, it is a useful addition to soap as it actually reduces the alkalinity of the solution; it finds an outlet in borax soaps for certain purposes such as the well-known borax-stearin softener for cotton. Borax itself is a poor emulsifying agent but assists soap in this work. It is sometimes used in the degumming of real silk.

Ammonia is a relatively weak alkali in solution but also finds a use with soap in the degumming of real silk.

As previously stated, although the alkalis themselves do not generally improve wetting or emulsification, they may assist soap in this respect by increasing the pH to the optimum. With excessive quantities of grease, alkalis may cause breakdown of the emulsion, for although soap and alkali together have a high capacity for removing dirt, impurities, or soiling matter, the capacity for suspending the foreign matter is diminished by large amounts of alkali; hence the amounts of soap and alkali need careful adjustment according to the type of impurity it is intended to remove and also according to the temperature at which the solution will be employed. The work of Holden and Vowler (*Technology of Washing*, British Launderers' Research Association, London, 1937, p. 11) is of great interest in this connection.

The measurement of the concentration of alkali is often carried out by determining the specific gravity of the aqueous solution and tables are available for comparing the specific gravity, which may be measured by Twaddell's hydrometer, with the percentage composition. It should be emphasised that this method only applies to pure solutions, and with used or "dirty" liquors the density is a false guide to the concentration of alkali, which is best measured by titration with acid solutions according to the ordinary methods of volumetric analysis.

pH

Where any one alkali is under consideration, its "alkalinity" may be expressed by the concentration of the aqueous solution, that is, by the amount of the alkali in the liquor. Different alkalis, however, as is well known, differ in alkalinity even when present in equal concentrations; the terms strong alkali and mild alkali are commonly employed. Hence it becomes necessary to define the alkalinity by its intensity as well as by its concentration, and this is done by the pH system of measurement, which enables comparisons to be made between the effectiveness or potentiality of alkaline solutions of caustic soda, sodium carbonate or soap.

Similar considerations apply to acids, for if different solutions of acids are being compared, then their acidity may be defined on the pH scale, which takes account of the fact that a concentrated solution of a weak acid may be less acidic or capable of less intensive activity than a dilute solution of a strong acid.

The pH value is a number which is used to express the concentration of hydrogen ions in an aqueous fluid, and thus indicates the reaction of the liquid, that is, its acidity, alkalinity or neutrality. Now, according to the theory of electrolytic dissociation, any liquid of which water is a constituent contains free, positively charged,

ALKALIS

hydrogen ions (H+) and also negatively charged hydroxyl ions (OH⁻); when the numbers of these two ions are exactly balanced the liquid is neutral.

All acids have a common action in furnishing hydrogen ions in solution, and the strength of the acid is governed by the quantity of hydrogen ions in the solution; the greater the degree of dissociation, the stronger the acid. Similarly, bases and alkalis furnish hydroxyl ions, and the strength of the alkali is determined by its dissociation and the quantity of hydroxyl ions it provides. The term pH stands for the potential of hydrogen.

According to modern theories, a substance must be an electrovalent compound if it is to ionise; many pure acids do not conduct electricity and must be regarded as co-valent compounds. The conductivity of their solutions is due to the formation of the H_3O^+ ion and not the H^+ ion, water being capable of co-ordination. The solution and ionisation of a strong acid in water may be represented in the following manner:

$$HA + H_2O \leftrightharpoons H_3O^+ + A^-$$

The term hydrogen ion concentration refers to the H_3O^+ or hydroxonium ion, although the term is rarely used.

The conception of pH values depends on this theory of electrolytic dissociation as developed by Arrhenius who postulated that when a salt is dissolved in water, a portion splits up into ions which are carriers of equal and opposite charges of electricity. Similarly with acids, which dissociate to give an excess of hydrogen ions, and with bases which dissociate to give an excess of hydroxyl ions; the greater the dissociation, the "stronger" the acid or base.

The actual concentration of these ions is very small, but the product of their concentrations is always constant:

$$OH^- \times H^+ = K_\omega = 10^{-14}.$$

Hence if one of the concentrations is known, the other can be calculated.

With water, the concentrations of H^+ and OH^- are equal and the hydrogen ion concentration is 10^{-7} (0·0000001 g./l).

The exponent 14 has been made the basis of the pH scale by Sørensen, and although the subject is now familiar to most textile chemists, a further and comprehensive account may be seen in the work of Britton (*Hydrogen Ions*, Chapman and Hall, London, 1942). A useful pamphlet on pH values, written by Cocking, has been published by British Drug Houses, Ltd., of London.

As the above figures are somewhat unwieldy, it is general to use the logarithm of the reciprocal of the hydrogen ion concentration which is termed the pH value. On this basis, the pH of water is 7 which represents neutrality; it may seem odd that both acidity and alkalinity may be represented on the same scale, but this is due to the fact that acids and alkalis have *opposite* properties rather than different qualities.

The range of the pH scale is approximately equidistant of either side of neutrality, 6 N.HCl having a pH value of —0·3, and 7 N.KOH having a pH value of 14·5. The value pH 1 corresponds to a hydrogen ion concentration of 0·1 g./l, and the value pH 14 to 0·00000000000001 g. of hydrogen ions per litre.

Hence the intensity of the acid solution ranges from pH 1 to the neutral point of pH 7, whereas the intensity of alkalinity ranges from pH 14 to the neutral point of pH 7; this means that acid intensity increases as the figure approaches unity (falling numerical value) whereas the alkali intensity increases as the figure approaches 14 (rising numerical value).

Buffers

Although the description of the acidity or alkalinity of a solution by its pH value has a very great practical significance, it must be realised that for most textile purposes the relation between the pH and the concentration is almost as important as the pH itself; in other words, there is an important relation between the amount of the alkali and its intensity. For example, two solutions may be prepared from sodium hydroxide and from sodium carbonate to give the same pH value; the concentration of NaOH will be much less than that of the Na_2CO_3. However, these alkaline solutions have to do some work in textile cleansing operations and in the course of the work alkali is consumed, as for example, in the cleansing of wool. An alkaline solution of NaOH at pH 10 contains so little alkali that its pH would soon fall, but on the other hand, a solution of Na_2CO_3 at pH 10 contains so much alkali that it could do more work before its pH is substantially affected. The ability to maintain a certain pH value and yet perform useful work is called buffer capacity, and solutions with good buffer capacity are known as buffer solutions; they damp out the effects that might otherwise be exerted by acid and alkali on the solution.

This controlled pH is of great importance in textile cleansing processes, for the solutions must be maintained between certain values to be effective on the one hand and to be safe on the other; the degumming of silk and the scouring of wool are two examples of this aspect of pH, for the solutions must be capable of purifying the raw material but must not damage it.

Some strongly alkaline solutions rapidly lose their power, whereas other less active solutions tend to maintain their alkalinity by progressive hydrolysis; this reserve alkalinity which comes into action between certain pH limits constitutes the buffer effect. Buffer

ALKALIS

substances are usually the salts of a weak acid or base, and prevent a sharp change in pH on the addition of acid or alkali. For example, sodium acetate dissociates into a certain number of Na^+ ions and an equal number of CH_3COO^- ions; if HCl is added to this solution, the following system is established:

$$H^+ + Cl^- + CH_3COO^- + Na^+ \rightleftharpoons Na^+ + Cl^- + CH_3COOH$$

from which it is seen that free acetic acid has been produced, but this is a weak acid and only slightly dissociated. Hence the added hydrogen ions from the HCl do not become effective in the solution.

The addition of HCl to water causes the pH to fall because of the large release of hydrogen ions, but in presence of sodium acetate there is a resistance to the rise in acidity, i.e. a buffer effect.

In the scouring processes, the buffered system should be capable of absorbing large quantities of acid or alkaline dirt without losing its activity; alternatively, they must be capable of replenishing the alkali removed during the scouring operation. The best buffer systems consist of mixtures; for instance, a mixture of sodium carbonate and bicarbonate is effective because the addition of acid only converts a little carbonate into bicarbonate and conversely, the addition of alkali only converts bicarbonate to carbonate. With both reagents in equal amounts there is little change in the effective alkaline power; mixtures of carbonate and bicarbonate are very useful around pH 10·2.

Very many textile bleaching and finishing processes depend to greater or less degree on the use of solutions buffered at a favourable pH, and although this was reached by empirical methods, the reasons for certain operations are now understood and point the way to a better and fuller control of the old processes as well as to improved processes and new processes.

The importance of pH control may be exemplified by considering wool and silk which are susceptible to alkaline attack and yet must be scoured in solutions in the region of pH 9 for effective cleansing. With both fibres, there is a decrease in the strength of the liquor as scouring or degumming proceeds owing to the consumption of alkali and soap in the process; additional alkali has to be provided, for if the pH falls below certain limits, the cleansing is ineffective. Cotton is less sensitive to alkali and scouring may proceed at pH values between 11 and 13, but in the bleaching process with hypochlorite solutions, the cellulose is very rapidly attacked at pH 7 as will appear later; some hypochlorite solutions are buffered around pH 9 to 10 to avoid a drift into the dangerous neutral region during bleaching. Again, the peroxide bleach for all fibres is generally carried out under slightly alkaline conditions and care must be exercised with the animal fibres.

It should not be forgotten that wool and silk are both amphoteric, behaving as bases to strong acid solutions, and as acids to alkaline solutions; the cellulose fibres, too, possess slightly acid properties and are capable of absorbing sodium ions from solutions of sodium salts.

CHAPTER X
SOAPS

As previously mentioned, although water is the universal cleansing agent it is usually necessary to employ an assistant which may be soap; within recent times there has been considerable work on new detergents to replace soaps. The use of organic solvents as cleansing materials is limited in scope, and Fuller's earth is only rarely used in modern times.

The theory of detergent action is most complex and many of the phenomena are imperfectly understood. Adam (J.S.D.C., 1937, *53*, 121) states that it is possible to trace the relationship between at least five different actions: (*a*) the lowering of the surface tension of water, (*b*) the wetting of a continuous solid surface by water, (*c*) the penetration of water into the porous solid, (*d*) dispersion or emulsification of grease and fatty matter, and (*e*) detergent action or the detachment and removal of grease from its adhesion to a solid. The work of Adam (*supra*) and also Palmer (J.S.C.I., 1941, *60*, 56) shows that when oiled fibres are placed in detergent solutions, the oil first forms flat globules which finally become detached and float away as emulsion particles. Hence, in scouring, two very important factors must be the interfacial tension between the oil (or grease) and the solution, and the adhesion between the fibre and the oil or fat. The first factor depends on the nature of the emulsifying agent, and the second factor depends on the fibre and on the oil or wax.

The reduction of the surface tension of water and the wetting of the solid are essential preliminaries to the removal of oils, fats and waxes from the various textile fibres to which they are attached.

WETTING

In the wet processing of textile materials, it is essential that there should be thorough and even wetting by water or the aqueous solution, but water itself has only moderate wetting ability. Liquids behave as if they were covered or surrounded by a skin, which acts along the surface and tends to keep it intact; this force or surface tension, is shown in the tendency of the liquid to adopt the minimum possible surface and form a sphere, and therefore be prevented from spreading and wetting another surface. For a drop to spread, it is necessary for this surface tension to be overcome.

With the aqueous solutions of inorganic salts, there is little difference between the surface tension of the solution and the solvent, but the aqueous solutions of some organic compounds, such as the fatty acids and amines, have lower surface tensions that water alone, and this property is often exhibited in solutions of low concentration. The best-known examples are the soaps, such as sodium oleate, but one of the most striking developments of recent years has been the multiplicity and variety of wetting agents.

To reduce the surface tension, the molecules of the dissolved substance must become adsorbed at the air-liquid surface, and they will do this when the field of attractive forces around their molecules is less than that around the water molecules. With soaps and other paraffin-chain salts, the long hydrocarbon-chains move to the surface and partly cover it, so that the field of force of the free surface approaches that of a hydrocarbon.

FIG. 40.—Contact angles.

The wetting of a surface involves the reduction of the interfacial tension so that the water spreads as a continuous film instead of remaining as drops; for perfect wetting, the contact angle between the liquid surface and the solid surface should be zero, a state which is reached when the liquid is attracted by the surface more than it attracts itself.

Wetting may therefore be brought about in two ways: (*a*) by reducing the surface tension or cohesion of the liquid, and (*b*) by increasing the adhesion between the liquid and the solid; both methods may operate together.

Textile materials are often coated with wax, grease or oil which possess a hydrophobic or water-repellent character; the wetting agent contains hydrocarbon groups together with one or more solubilising groups in the molecule. The hydrocarbon-chains will lie on the solid-liquid surface and bring with them the hydrophilic groups which are responsible for the solubility of the wetting agent; these groups at the surface will increase the adhesion between the solid and the water. The hydrocarbon head and the polar tail of the wetting agent act as a type of bridge between the solid and the water, for the polar tails project into the water, and the hydrocarbon head attaches itself to the hydrocarbon on the surface of the fibre. A similar picture accounts for the preliminary mechanism of emulsification or dispersion which diminishes the interfacial tension between two liquids, or between the solid and the liquid, and so produces a large increase of interface which facilitates dispersion.

Liddiard (Chem. and Ind., 1941, *60*, 480) points out that oils and similar materials have a relatively low surface tension, and water, with a surface tension of about 73 dynes per cm., does not readily wet these compounds until its surface tension is reduced; alkalis alone are poor wetting agents.

Wetting Agents

Material	Concn.	θ_a	θ_r	Surface tension
Caustic soda	1%	104°	84°	68·55 dynes/cm.
	5%	95°	80°	73·19 ,,
Sodium metasilicate	1%	99°	87°	70·66 ,,
	5%	64°	54°	69·81 ,,
Sodium carbonate	1%	102°	90°	69·28 ,,
	5%	98°	80°	64·44 ,,
Trisodium phosphate	1%	94°	82°	63·49 ,,
	5%	70°	60°	73·19 ,,
Sodium silicate (180°Tw.)	1%	101°	88°	70·87 ,,
	5%	92°	79°	68·76 ,,
Calgon	1%	104°	83°	68·97 ,,
London tap water	—	105°	101°	70·87 ,,
Pure water	—	109°	100°	73·01 ,,
Sodium oleate	1%	20°	—	29·11 ,,
	5%	10°	—	28·26 ,,

θ_a is the contact angle against paraffin wax when advancing into the solution, and θ_r when receding from the solution.

Wetting and emulsification play important parts in the purification of textile materials and are associated with detergency; for this reason the emulsifying powers of various agents may be tested by their ability to decrease the interfacial tension of petroleum ether and the water in which the agent is dissolved. The drop number is a convenient method.

The Donnan pipette is capable of measuring the number of drops per given volume delivered from a tube and tip of definite dimensions; the dynamic interfacial tension may be estimated in this manner and affords an indication of the emulsifying powers of the emulsifying agent. The addition of certain compounds to an oil may reduce the oil-water interfacial tension to a considerable extent, as shown by the greatly increased number of drops from the pipette. For instance, the addition of caustic soda to saponifiable oils will increase the number of drops, but has little action with highly refined mineral oils; if stearic acid is added there is also little effect. With both stearic acid and caustic soda, however, there is considerable reduction of the interfacial tension due to soap formation.

The reduction of interfacial tension by soaps depends to a large extent on the molecular size of the fatty acid, although there are other factors which play a part. The lower fatty acids, up to butyric acid, exert little effect, but with lauric acid which contains 12 carbon atoms per molecule, the reduction of the interfacial tension is considerable, and with stearic acid the effect is still greater. When the number of carbon atoms in the fatty acid exceeds 18, the reduction of the interfacial tension decreases with increase of molecular weight.

Soaps or salts of fatty acids with less than 12 carbon atoms behave much like electrolytes even in fairly concentrated solutions, but those with a higher number of carbon atoms do not behave like electrolytes, even in dilute solutions, and exhibit abnormal elevation of boiling-points, depression of freezing-points, etc.; the solutions are colloidal, not molecular.

McBain and his co-workers discovered that soap solutions possess high electrical conductivity, which indicates that they must contain something in the nature of electrolytes, in addition to the colloid. It was established that there is a gradual transition from colloid to crystalloid as the soap solutions become more dilute, and a micelle theory was advanced to account for the various observations. An ordinary electrolyte such as sodium chloride dissociates in solution into the simple ions Na^+ and Cl^-, and the electrical conductance of the solution is determined by the number of ions present. Soaps also dissociate into ions (e.g. Na^+ and $C_{17}H_{35}COO^-$), but the solutions exhibit colloid properties as shown by viscosity, osmotic pressure, and protective-colloid action. Sodium stearate is the salt of a weak acid and the dissociated anion gives rise to a certain proportion of undissociated stearic acid, so that a corresponding proportion of hydroxyl ions or free alkali is formed.

$$C_{17}H_{35}COO^- + Na^+ + H_2O \longrightarrow C_{17}H_{35}COOH + Na^+ + OH^-.$$

For ordinary electrolytes, the equivalent conductivity of the solutions is determined by the concentration of Na and OH ions which can be estimated by titration with acid. With soaps, however, the equivalent conductivity is too high to be accounted for solely by this concentration, and McBain concluded that the colloidal particles must possess an equivalent conductivity comparable with that of ordinary ions; as the colloid particles were highly charged micelles which functioned as ions, they were termed ionic micelles.

As soaps are able to form colloidal particles in a reversible manner out of simple ions, they are classed among the colloidal electrolytes which are an intermediate class between ordinary colloids and ordinary electrolytes.

Hartley (*Aqueous Solutions of Paraffin Chain Salts*, 1936) has pointed out that when a warm clear aqueous solution of soap is cooled, an opaque silky curd is formed consisting of a suspension of the aggregated soap molecules. With gentle heat, it is possible to cause the soap to pass through a viscous gel state to the mobile solution again. The mobile solution is the important condition for detergency and contains no visible particles but colloidal aggregates are undoubtedly present, consisting of 10 to 100 long-chain ions, each carrying a negative charge. The micelle is the nucleus of a localised high negative charge, as a consequence of which a number

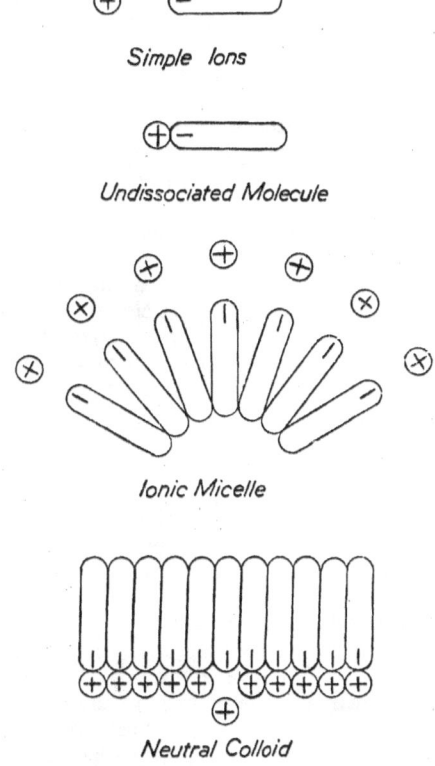

FIG. 41.—Ions, ionic micelles, and neutral colloids.

of positively charged sodium ions (gegenions) are attracted to the micelle surface.

The comparatively high conductivity in very dilute solution is assumed to indicate a state of complete ionisation into simple ions such as $C_{17}H_{35}COO^-$ and Na^+, but as the concentration increases, there is a fall in conductivity which indicates micelle formation. The concentration at which micelle formation starts is approximately the same for the different soap-like colloidal electrolytes, but in each case the concentration depends on the temperature of the solution, the length of the long-chain ion, and, to some extent, on the solubilising group and the gegenion.

Reference has already been made to the fact that these colloidal electrolytes have a long-chain hydrocarbon head and hydrophilic or solubilising tails; the ions therefore will have dual properties for the ionic tail (COO^-) will have a strong affinity for water, but the hydrocarbon head has no such affinity and seeks a more compatible phase. In the interior of a soap solution, the ions so arrange themselves to provide their own second phase, aggregating into micelles with the hydrocarbon heads in the interior and the ionic tails forming the surface. The paraffin chains are presumed to be in a liquid state of aggregation and the micelle may be regarded as a drop of paraffin surrounded by ionic tails—a view which is supported by the solvent action of soap solutions for solids and liquids which are insoluble or very slightly soluble in water.

The possibility of other aggregates is a matter of controversy. McBain suggests a neutral colloid particle in more concentrated solutions, and Lawrence (Trans. Farad. Soc., 1935, *30*, 109) describes a neutral or secondary micelle made from primary micelles. Chwala and Martina (Textilber., 1937, *18*, 725) consider that four phases may exist in solution: (*a*) simple ions, (*b*) pre-micelles, (*c*) the ionic micelle, and (*d*) the charged neutral particle; the pre-micelle is considered to play an important part in the washing process on account of its smaller size and greater mobility, and the charged neutral particle consisting of undissociated soap molecules with a number of attached ions, acts as a powerful protective colloid. McBain believes that the colloidal properties of soap are largely due to this particle.

It has already been stated that wetting agents contain hydrocarbon and solubilising groups in their molecules, that is, they are composed of both polar and non-polar groups. Langmuir (J.A.C.S., 1917, *39*, 1848) has shown that the polar portion directs the molecule into water, and the non-polar or oil-soluble portion is directed against the air or will be attracted by any surface with which the aqueous solution comes into contact. Wetting agents and detergents must include a correct balance between their polar and non-polar constituents in order to be satisfactory.

With the lower fatty acids, there is complete solubility in water and no detergent effect, but as the chain length increases, the sodium salts of the fatty acids display detergent and wetting properties reaching an optimum from lauric to myristic acid. Above stearic acid, however, the solubility of the salt is considerably diminished and the balance is weighed down on the hydrophobic side. The introduction of a double bond, which is hydrophilic in effect, restores the balance, making sodium oleate soluble in water with excellent detergent and wetting properties. Too many double bonds, however, although increasing the solubility, overshadow the hydrophobic character so that linolenic acid, with three double bonds, has little wetting ability in spite of the chain length, $C_{17}H_{29}COOH$.

This question of balance is important in considering the old soaps as well as the newer wetting agents and detergents.

SOAPS

SOAPS have been used for about 2,000 years, but it is not certain whether they were first utilised as pomades or detergents; their manufacture on a commercial scale appears to date from 1791. The term "soap" may be applied in general to the metallic salts of the higher fatty acids, but, in particular, it is confined to the sodium or potassium salts. Although soaps can be prepared commercially by neutralising the fatty acids with the hydroxides or carbonates of sodium or potassium, the common method is to decompose fats and oils with sodium or potassium hydroxide. The triglyceride is saponified to form soap and glycerol, as follows:

$$(C_{17}H_{33}COO)_3C_3H_5 + NaOH \longrightarrow C_{17}H_{33}COONa + C_3H_5(OH)_3.$$

The direct formation of soap by neutralising oleic acid is shown by

$$C_{17}H_{33}COOH + NaOH \longrightarrow C_{17}H_{33}COONa + H_2O.$$

Fats are the glycerides of fatty acids, and may give three different fatty acids, in the form of their sodium salts or soaps, on saponification with alkali.

$$\begin{array}{l} CH_2O.CO.R^1 \\ | \\ CH\ O.CO.R^2 + 3NaOH \longrightarrow \\ | \\ CH_2O.CO.R^3 \end{array} \begin{array}{l} CH_2OH \\ | \\ CHOH + R^1COONa + R^2COONa + R^3COONa. \\ | \\ CH_2OH \end{array}$$

It has been found possible within recent years to produce fatty acids on a large scale by treating fats with acids, steam, lime or enzymes and recover the glycerin at the same time; the fatty acids can be used for making soap on a small scale as required in many textile works. Not only are the difficulties of glycerin recovery obviated, but the fatty acids may be neutralised with sodium carbonate instead of the more expensive caustic soda. Many of the commercial fatty acids contain small amounts of neutral fats, but these may be saponified with sodium hydroxide solution of 32·5°Tw.

The somewhat crude fatty acids are generally termed "oleines" of "stearines," according to whether they are liquid or solid at room temperatures; for textile purposes, the oleines should not solidify much over 10°C. According to the method of preparation, it is possible to have saponification oleines or distillation oleines.

As might be expected from the terminology, the oleines consist chiefly of oleic acid, and the stearines of stearic acid. Attention must be drawn, however, to another meaning which may be attached to olein—the sodium or ammonium derivative of sulphonated ricinoleic acid, known as Turkey Red Oil (see p. 133), is often termed olein oil. In general, it may be assumed that the term "oleine" when used in the woollen and worsted trade refers to crude oleic acid, but the cotton industry understands that "olein oil" refers to a derivative of sulphonated castor oil. The crude liquid oleic acid is sometimes known as red oil. The position is further complicated by the fact that in pure chemistry, "olein" means the glyceride of oleic acid.

Coconut and palm-kernel oils may be used as a source of butyric (C_3H_7COOH), caproic ($C_5H_{11}COOH$), caprylic ($C_7H_{15}COOH$), capric ($C_9H_{19}COOH$), lauric ($C_{11}H_{23}COOH$) and myristic acids ($C_{13}H_{27}COOH$). Most animal and vegetable fats contain palmitic ($C_{15}H_{31}COOH$), and stearic acids ($C_{17}H_{35}COOH$). Oleic acid ($C_{17}H_{33}COOH$) with a double bond is found in olive and neatsfoot oils and also in tallow, while linoleic acid ($C_{17}H_{31}COOH$) with two double bonds is the main constituent of linseed oil. The unsaturated hydroxy acid, ricinoleic acid ($C_{17}H_{32}.(OH).COOH$) comes from castor oil.

The common hard soaps are the sodium salts of the fatty acids, the corresponding soft soaps being the potassium salts. The saponification products of fats are all soluble in the hot mixture and the sodium soaps are separated by virtue of their insolubility in sodium chloride solution which is added to the mass until the sodium soap rises to the surface as a curd, when it is removed, washed, re-melted with water and cast into cakes. The glycerol is removed by evaporation under reduced pressure and the salt recovered at the same time.

Bar soap generally contains 25 to 30% of moisture, and flake soap contains 5 to 10% of moisture.

Potassium soaps cannot be salted-out as they are decomposed by sodium chloride; generally the mixture of glycerol and potassium soap is sold without removing the latter, but sometimes potassium soaps are made without glycerol, by neutralising the fatty acid.

It must be realised that the term "soft" as applied to soap does not necessarily imply the potassium compound, for a hard soap may be the result of the saponification of a hard fat such as tallow, but with the same fatty acid the sodium salt is harder than the corresponding potassium salt.

Whereas sodium soaps are precipitated unchanged from their aqueous solutions by the addition of common salt and other soluble sodium salts, soaps from different fatty acids require different concentrations of salt solution, and on this account marine soaps which lather with sea water are made from coconut and palm oils.

Rosin soap is often utilised in the kier-boiling of cotton goods; rosin is made from the crude turpentine obtained from pine-wood after removal of the turpentine by distillation. The chief constituent of rosin is abietic acid, $C_{18}H_{27}.COOH$, which dissolves to form a soap when boiled with sodium carbonate or hydroxide. Rosin soap is often added to commercial soaps on account of its strong detergent powers.

Unless soaps are made from a pure fatty acid such as oleic or stearic acids, they consist of the sodium salts of the mixture of fatty acids present in the original fats and oils, and the nature of the soap depends on the character of the glycerides in the original oil or fat. Most commercial soaps contain some free alkali, which, estimated as Na_2CO_3, may vary from a trace to as much as 1·5%.

As previously stated, soaps are slightly soluble in cold water, but dissolve readily on warming; cooling the solution forms a gel which liquefies again on warming. Solutions of soap are decomposed by mineral acids, or by any acid which is stronger than the fatty acid of the soap, and the whole of the fatty acid liberated forms a white mass of curds. The action of many soluble metallic salts on soap solutions is to form an insoluble metallic soap, as with the calcium and magnesium salts in hard water. Zinc and aluminium soaps, however, are sometimes produced intentionally on fabrics for waterproofing effects.

With the unsaturated soaps, the ease of oxidation of the fat, the fatty acid, and the soap, increases with the extent of unsaturation and affects the tendency towards rancidity and stickiness so that it is preferable to avoid the highly unsaturated compounds. In soap mixtures, however, it is sometimes advisable to have a certain amount of unsaturated soap as the presence of the unsaturated radicals produces a more liquid soap.

The extent of unsaturation may be measured quantitatively by the well-known iodine value, which is the number of grams of iodine absorbed by 100 g. of the compound under standard conditions. Another characteristic of fats is their saponification value, which is the number of grams of potassium hydroxide required to saponify 1 g. of fat; the lower the molecular weight of the fat, the higher will be its saponification value, for example, the saponification value of coconut oil (lauryl and myristyl) is 246 as compared with 190 for olive oil in which the fatty radical is oleyl.

As previously mentioned, salts of the fatty acids start to exhibit soapy properties when their molecular weight is sufficiently high, but as the molecular weight increases there is a diminution in solubility. The lathering powers are also connected with the fatty acid, lauric acid being outstanding.

An important feature of the making of soap, therefore, is the selection of fats, or blends of fats, to give a soap suitable for the particular purpose, with the required consistency, solubility and lathering power. The following data from Webb (*Modern Soap and Glycerine Manufacture*; Davis Bros., London, 1927) show that a wide range of soaps may be produced from a variety of fats.

PROPERTIES OF SOAPS

Soap from	Colour	Consistency	Detergency	Lather
Coconut oil	Pale-yellow to white	Extremely hard	Excellent	Rapid, foamy, impermanent
Cottonseed oil	Buff to yellow	Medium to soft	Good	Rapid, thick, medium persistence
Olive oil	Green	Soft	Very fair	Abundant, close, persistent
Tallow	Buff to white	Very hard	Good	Slow, thick, permanent
Linseed oil	Yellow	Soft	Good	Greasy, thick, permanent

With a more scientific approach to the question of stock blending for soap manufacture, new standards have been developed for estimating the soap-making potentialities of oils. For example, the I.N.S. factor is the difference between the saponification value and the iodine value of the oil (Saponification Value—Iodine Number= I.N.S. Factor); with increase of I.N.S. factor, the soaps from the oils are harder, less soluble, less prone to rancidity, better colour, but poorer in respect of detergent and lathering powers. Estimation of the mean I.N.S. factor of a blend of oils enables one to determine in advance the type of soap produced on saponification. The I.N.S. values range from 250 for coconut oil to 15 for linseed oil.

The soap solubility ratio (S.S.R.) is the ratio of the I.N.S. factors of the mixture of oils to the sum of the I.N.S. factors of those oils in the mixture with a factor greater than 130 (excluding palm kernel and coconut oil); high S.S.R. values indicate good lathering and solubility. The use of the I.N.S. and S.S.R. values has proved very successful and of greater consequence than the titer of the oil which is related to the melting-point of the fatty acid and the solubility of the soap.

The calcium and magnesium soaps are sticky masses which clog the fibres and prevent penetration of the liquor hindering detergent action, as well as wasting soap. The deposited calcium or magnesium soaps cannot be removed by rinsing, and in most cases bring about a discoloration of the fabric on ageing. The weakness of soap to hard water is the chief reason for the development of the newer wetting agents and detergents, but there is the additional disadvantage of the limited pH range for stable emulsions with dirt and the fatty matter of raw native fibres, outside which coagulation or re-precipitation on the fibre may occur.

A peculiar property of soaps is associated with the concentration of the solution; detergent power and wetting efficiency reach a maximum at low concentrations, estimated at 0·2 to 0·6%, and diminish as the concentration increases. Sodium laurate has the greatest wetting power of the saturated aliphatic soaps at room temperatures; the soaps from coconut and palm-kernel oils possess high detergent powers at relatively low temperatures of 30 to 40°C. The higher members of the series are more efficient at higher temperatures, the stearate around 65°C. and the palmitate at 80°C.

TITER OF FATTY ACID

Fatty Acid		Titer	Appearance of turbidity in soap solution
Lauric	$C_{11}H_{23}$—	44°C.	40–45°C.
Myristic	$C_{13}H_{27}$—	54°C.	50–55°C.
Palmitic	$C_{15}H_{31}$—	62°C.	60–65°C.
Stearic	$C_{17}H_{35}$—	71°C.	65–70°C.
Arachidic	$C_{19}H_{39}$—	77°C.	75–80°C.
Oleic	$C_{17}H_{33}$—	14°C.	20–30°C.

The titer of the fatty acid is the temperature at which the molten material solidifies again.

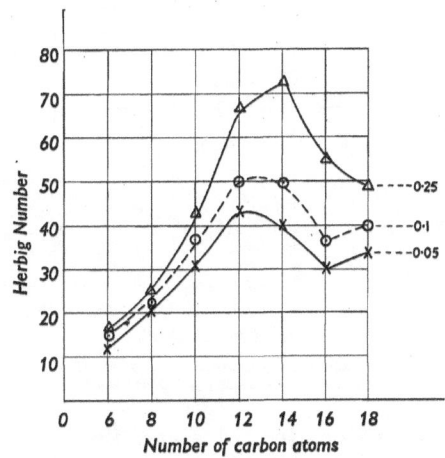

FIG. 42.—Relation between Herbig number and length of alkyl chain in sodium salts of saturated fatty acids at 60°C.

Venkataraman (J.S.D.C., 1940, 56, 503) has examined the wetting power, calcium soap dispersing powers, interfacial tension, drop number and protective colloid action of a number of soaps from sodium caproate (C_6) to sodium stearate (C_{18}). The Herbig number (see p. 144), which measures wetting power, increases up to C_{14} and then falls, but the drop number increases up to C_{16} in 0·25% solutions; in 0·1% solution the Herbig number is at a maximum with

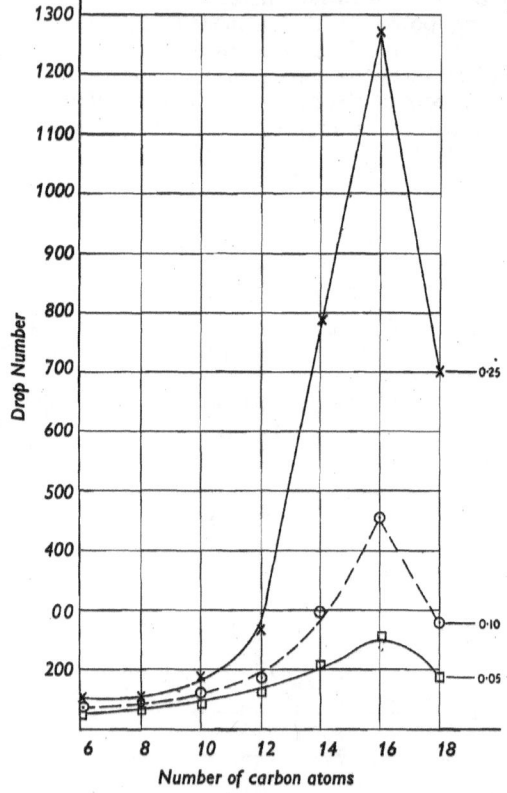

Fig. 43.—Relation between drop number and length of alkyl chain in sodium salts of saturated fatty acids at 60° C.

C_{12}. In the range of unsaturated soaps, the oleate is the best wetting agent.

The resistance to hard water increases up to C_{10} for the saturated soaps; sodium stearate is better than any of the unsaturated products. For protective colloid·action, sodium caprate (C_{10}) is the best of the soaps. Sodium ricinoleate is superior to the oleate as an agent for dispersing calcium soaps, but is inferior in its action as a protective colloid.

The potassium soaps are generally more soluble than the corresponding sodium soaps, and exert their optimum detergent powers at a slightly lower temperature, approximately 5 to 10°C.

Both sodium and potassium soaps hydrolyse in water, and the extent of hydrolysis depends on the concentration, the nature of the soap and the temperature. The subject has been discussed by McBain (Third Report on Colloid Chemistry, 1920, p. 2).

HYDROLYSIS OF SOAP SOLUTIONS

Concentration	Na palmitate % hydrolysis	K palmitate % hydrolysis
$1·0\ N$	0·20	0·08
$0·75\ N$	0·30	0·31
$0·5\ N$	0·37	0·65
$0·1\ N$	1·28	1·25
$0·05\ N$	2·22	2·02
$0·02\ N$	—	5·6
$0·01\ N$	6·6	6·8

The hydrolysis of soap solutions above decinormal (3%) is only a fraction of a per cent, and is not very different for sodium or potassium soaps. In more dilute solutions, however, soap solutions are distinctly alkaline and the alkalinity persists even in presence of a large excess of palmitic acid. For instance, McBain found that a 10% excess of palmitic acid only reduced the alkalinity by about one-half and that even in presence of 100% excess of acid the OH ion was $0·00004\ N$. This slight amount of alkalinity prevents the existence of more than minute traces of free fatty acid in any soap solution, so that any solid product of hydrolysis cannot be fatty acid but an acid soap. It is equally important to note that when excess of alkali is added, it is not taken up by the soap present, but remains almost entirely in the free condition.

The effect of temperature is important as the hydrolysis-alkalinity decreases with the lowering of the temperature.

HYDROLYSIS OF SOAP SOLUTIONS AT 90°C.
($0·1\ N$ solution or 20 to 30 g./l.)

Soap		Hydrolysis % Potassium	Sodium
Laurate	$C_{11}H_{23}COONa(K)$	2·9	2·5
Myristate	$C_{13}H_{27}COONa(K)$	3·3	4·5
Palmitate	$C_{15}H_{31}COONa(K)$	7·4	7·5
Stearate	$C_{17}H_{35}COONa(K)$	15·0	13·0
Oleate	$C_{17}H_{33}COONa(K)$	8·1	8·6

When the various fatty acids are compared it is found that the degree of hydrolysis increases rapidly as the homologous series is ascended.

The action of soap is determined to some extent by the fibre with which it is utilised. When soap splits up in water into alkali and fatty acid, the wool absorbs more alkali than acid and so becomes alkaline even from a neutral soap. Removal of the alkali leaves free fatty acid, so that the bath becomes charged with acid soap which is deposited on the wool; this effect is greater in dilute solutions as there is less soap available for the removal of the fatty deposit. This absorption of alkali from soap solutions is such that wool may "kill" a soap solution, and for this reason it is customary to employ small amounts of alkali with soap for certain purposes. Excessive amounts of alkali, however, will precipitate soap.

The alkalinity of soap solutions is not the chief defect, however, the weakness to acid conditions, and above all to hard water having led to the search for improved products which comprise the thousands of wetting agents and textile auxiliaries which appeared between 1920 and 1940.

Within recent times, soaps have been made for special purposes from triethanolamine, for example, triethanolamine oleate; it is used as a soap for dry-cleaning and as an emulsifying agent for wool creams. In 5% solution, triethanolamine oleate gives a pH value of 7·8.

Although the ethanolamine soaps give solutions of much lower pH values than the ordinary soaps, yet they still have an alkaline reaction. Neutral or slightly acid products have been made by the esterification of the glycols with fatty acids; some of these compounds have been discussed by Bennet (Ind. Chemist, 1932, *8*, 223). The highest detergent and dispersing qualities seem to come from mixtures of the mono- and di-esters, for example, the diglycol stearate is a mixture of mono- and distearate and not only gives very stable emulsions, but when added to the extent of 5% to ordinary soaps it is capable of extending the life of the lather with a consequent saving of cost. Diglycol oleate, laurate and palmitate are also known; they are insoluble in water but are used as dry-cleaning soaps. They may, however, be used in conjunction with ordinary soaps as stated above.

CHAPTER XI
WETTING AGENTS

Sulphonated Products

THE so-called SULPHONATED OILS represent the first attempt to overcome the weakness of soaps towards hard water. Turkey red oil was the forerunner of these compounds and owed its origin to the work of Mercer in 1846, when "sulphated oils" were prepared by treating olive oil with sulphuric acid. These oils were utilised in the dyeing and printing of Turkey Red.

The Turkey red oils are also known as olein oils, and must not be confused with the oleines which are the crude liquid mixtures of fatty acids obtained by the splitting of fats (see p. 125).

Modern Turkey red oil is made by treating castor oil with sulphuric acid, removing the excess of acid, and neutralising the product with sodium, potassium or ammonium hydroxide. The reaction is somewhat complicated, but consists mainly in sulphuric acid ester formation

$$C_6H_{13}.CH(OH)CH_2.CH:CH.C_7H_{14}.COOH \rightarrow C_6H_{13}.CH(O.SO_3H)CH_2.CH:CH.C_7H_{14}.COOH.$$

With olive oil, which is the triglyceride of oleic acid, the sulphuric acid reacts at the double bond

$$C_8H_{17}.CH:CH.C_7H_{14}.COOH \rightarrow C_8H_{17}.CH(O.SO_3H)CH_2.C_7H_{14}.COOH.$$

The introduction of the sulphuric acid residue renders the fatty acid soluble in water and imparts a certain amount of resistance to acid as compared with soap. The SO_3H group is capable of forming soluble calcium and magnesium salts in hard water, which is an advance on soap, but the Turkey red oils have poor detergent properties; nor are they capable of dispersing existing lime soaps.

The original oils contained some 3 to 7% of combined sulphuric acid, and about 1900 this was increased to 10% in Monopol soap; more recently the Prestabit V Oils of the I.G. have been prepared with 20 to 30% of combined H_2SO_4. The more highly "sulphonated" products have greatly increased wetting powers and improved resistance to lime and acids; unfortunately, this is at the expense of the fatty character.

When true sulphonates have been formed, such as sulphonated palmitic acid $C_8H_{17}.CH(SO_3H).COOH$, it was found that they were unable to disperse lime soaps, as the original colloid nature of the fatty chain was more than counterbalanced by the extreme hydrophilic nature of the COOH and SO_3H groups.

Intrasol (Stockhausen) is an example of a truly sulphonated oil, which may be used to prevent the precipitation of lime soap.

It has also been possible to prepare products which are allied to the sulphated oils, by amidation or esterification of the carboxyl group in ricinoleic acid followed by treatment with sulphuric acid, on the assumption that the carboxyl group is responsible for poor resistance to lime. Avirol AH extra is believed to be the sulphuric ester of butyl ricinoleate, and Humectol C the sulphuric ester of the amide of ricinoleic acid. These agents have remarkable wetting powers, but little or no detergent or emulsifying action.

NAPHTHALENE SULPHONIC ACIDS were used in Germany immediately prior and subsequent to 1918, in the form of their sodium salts, and represent an important departure from the fatty basis of previous work; these compounds contain no carboxyl groups. Although the resistance to acids and to lime was excellent, the detergent effect was much below that of soap. The excellent wetting powers, however, led to the manufacture of propyl, isopropyl and isobutyl derivatives of naphthalene sulphonic acids such as Nekal S, Leonil SS and Neopermin H; sodium salts of the sulphonated hydrocarbons include Nekal A, Nekal BX and Leonil S. The Invadines are sodium salts of alkyl phenylene sulphonates. Compounds of this type have also been mixed with organic solvents to form such trade products as Neopermin (cyclohexanol), Floranit M (amyl acetate), Laventin BL (terpenes), and Flerhenol PF (trichlorethane). The addition of glue to the sulphonated hydrocarbons gives emulsifying agents such as Nekal AEM and Leonil LE.

It will be remembered that molecules with a dual hydrophilic and hydrophobic nature will form wetting agents provided the groups are properly balanced, and that the position of the hydrophilic group permits orientation at the interface. In soaps, the polar group is at the end of the molecule, but in most of the so-called sulphonated oils, the polar group is in the centre of the molecule. With the polar end group, the molecules can lie close together and a large number will be required to cover the surface because although the molecule is long, the diameter is small; with the central polar group, however, relatively few molecules will be required to form a surface because the orientated molecule is relatively short but much greater in area.

Calsolene Oil HS of I.C.I. makes use of this property and so possesses very good wetting powers; its formula is as follows:

$$CH_3.(CH_2)_3.CH(C_2H_5).CH_2\diagdown$$
$$CH.O.SO_3Na$$
$$CH_3.(CH_2)_3.CH(C_2H_5).CH_2\diagup$$

Another compound based on the same molecular architecture has the formula

WETTING AGENTS

$$CH_3.(CH_2)_7.O.CO.CH_2$$
$$CH_3.(CH_2)_7.\overset{|}{C}H.SO_3Na$$

Caryl and Ericks (Ind. Eng. Chem., 1939, *31*, 44) have described the extraordinarily high wetting powers of Aerosol OT, which is the dioctylester of sodium sulphosuccinate (U.S.P. 2,028,091). The formula of this compound is as follows:

$$C_8H_{17}.OOC.CH_2$$
$$C_8H_{17}.OOC.\overset{|}{C}H.SO_3Na.$$

Of the many possible esters of sodium sulphosuccinate, Caryl (Ind. Eng. Chem., 1941, *33*, 731) concludes that for good wetting the number of carbon atoms in the two non-polar chains must lie between 14 and 22.

Solvents

There are also available many solvent emulsions for improving the detergent properties of soaps and allied compounds; the oldest of these preparations is Tetrapol which consisted of a mixture of 84% Monopol soap and 15% carbon tetrachloride.

In mixtures of the type under consideration, the solvents are mainly hydrocarbons and chlorinated hydrocarbons; with recent years there has been a tendency towards the use of sulphonated naphthalene derivatives.

Of the simpler types, Hexoran also contains carbon tetrachloride, while Trioan and Astol A contain trichlorethylene.

Soaps, together with solvents, form the basis of such preparations as Terpurile, Cycloran and Solventol.

Solvents may be mixed with sulphonated oils, such as Turkey Red oil, Monopol soap and Avirol, to give products of which Avivan, Koloran and Flerhenol may be regarded as typical, in addition to the above-mentioned Tetrapol.

Detergent properties may also be improved by the addition of hydrogenated phenols of which cyclohexanol may be regarded as the prototype

$$\begin{array}{c} CH_2 \\ CH_2 \quad\quad CHOH \\ | \quad\quad\quad\quad | \\ CH_2 \quad\quad CH_2 \\ CH_2 \end{array}$$

This is sold under the name of Hexalin; Methylhexalin is a mixture of the three isomeric methylcyclohexanols. Similar trade products are Savonade, Texapon, Texalin, Sextol and Hydralin.

TEXTILE BLEACHING

Other solvent preparations are based on substances such as tetrahydronaphthalene and decahydronaphthalene, sold under the names of Tetralin and Decalin.

[Structural formulas of Tetralin (left) and Decalin (right)]

Hydranaphthal is supposed to be a mixture of Tetralin and a soap.

As previously mentioned, the modern tendency is towards the use of sulphonated naphthalene derivatives, and some alkyl derivatives of the naphthalene sulphonic acids are sold under names such as Leonil S, Nekal A, Oranit, Neopermin N; Neopermin probably contains a solvent in addition.

Condensation Products

Two parallel developments followed the work on sulphonation: (a) the blocking of the carboxyl group as in the Igepons, but still maintaining a fatty chain; and (b) avoiding the use of carboxyl groups and utilising fatty alcohols which would react with sulphuric acid; the carboxyl group is altered by reduction, in this case.

FATTY ACID CONDENSATION PRODUCTS have also been utilised in the production of scouring agents; the best-known examples are the Igepons. Igepon A is formed by the esterification of oleic acid with an aliphatic hydroxy-sulphonic acid and its probable formula is $C_{17}H_{33}COOCH_2CH_2SO_3Na$, whereas Igepon T comes from the condensation of oleic acid with an amino-sulphonic acid and has a formula corresponding to $C_{17}H_{33}CONHCH_2CH_2SO_3Na$, or $C_{17}H_{33}CON(CH_3)CH_2CH_2SO_3Na$. Both of these products are constructed from a fatty acid residue which is attached by an aliphatic bridge to the SO_3Na or solubilising group, and the balance between the hydrophobic and hydrophilic groups, coupled with their molecular dispersion in solution, renders these agents particularly valuable in wool scouring. It is claimed that for a given fat content they are from three to five times as efficient as soap.

High claims are made for these agents, and it has been stated that they possess good wetting powers, emulsifying capacity and detergent action. The Igepons also disperse lime soaps to a very high degree and possess the added advantage of being chemically neutral, an important point in wool scouring. The ester form of Igepon A renders it susceptible to alkali hydrolysis in pressure kiering, but

WETTING AGENTS

the acid-amide linking of Igepon T is more stable. The Igepons possess good detergent powers at low temperatures as well as high temperatures, whereas the optimum temperature for soap is 40 to 45°C.

The Lamepons are somewhat similar to Igepon T, being derived from oleic acid and the hydrolysis products of certain proteins.

The probable formula is $C_{17}H_{33}CONHR_1(CONHR_2)_xCOONa$, the preponderance of CONH groups imparting powers of dispersing lime soaps in spite of the COOH group.

The Meliorans are mainly compounds of fatty aromatic ketones such as $C_{15}H_{31}.CO.C_6H_4.SO_3Na$; they possess good detergent properties and are capable of dispersing lime soaps.

The Ultravons are derived from o-phenylenediamine and are considered to have the following type of formula:

$$C_{17}H_{35}\diagup\overset{NH}{\underset{N}{\diagdown}}\diagdown C_6H_4SO_3Na \qquad C_{17}H_{35}\diagup\overset{NH}{\underset{N}{\diagdown}}\diagdown C_6H_3(SO_3Na)_2$$

Ultravon K   Ultravon W

The monosulphonate is very effective in dispersing lime soaps and the disulphonate has good detergent powers.

It will be noted that in these compounds, outlined above, the solubilising or hydrophile group is attached to end of the molecule, and the products are anion-active.

Between the extremes of soap and the Igepons, as representing colloidal and molecular solutions, it has been found possible to produce compounds of interesting character by adjusting the balance of hydrophobic and hydrophilic groups. With soap, it had been assumed that the carboxyl group was responsible for most of the disadvantages, and interest was focused on the SO_3H group as a superior alternative solubilising group. This strongly acid residue is apt to overpower the fatty residue in certain circumstances, so that in the scouring of cotton, for instance, the Igepons are slightly inferior to soap. Instead of adjusting the hydrophilic group, however, it is possible to maintain the carboxylic group and adjust the hydrophobic group or fatty residue as in Medialan A, which is $C_{17}H_{33}.CO.N(CH_3).CH_2.COONa$, an oleyl sarcosine compound, and somewhat similar to Igepon T, which is the condensation product of oleic acid and N-methyl taurine.

Although the carboxyl group is retained, the drawbacks of soap are largely removed in respect of stability to hard water and acids, even though Medialan is not quite as good as Igepon. The detergent action is better than that of Igepon and even better than that of

TEXTILE BLEACHING

soap itself, but without the disadvantage of liberating alkali and forming free fatty acid in solution. Neutral scouring is enabled to take place and it is unnecessary to use soda or ammonia for removing soap after scouring woollens. The felting or milling properties with wool are good, as distinct from the Igepons, and the softening action is pronounced.

As the Igepons are alkali salts of strong acids, they are no longer subject to hydrolysis in water. They form molecular solutions and are generally well dissociated; the strong electrolyte does not cause wool to swell so that there is little milling or felting action with the Igepons. As sulphonic acid salts, the Igepons are affected by the basic groups of wool keratin and the free acid has an affinity for wool in acid liquors.

Sulphated Alcohols

SULPHATED FATTY ALCOHOLS represent an attempt to combine detergent powers with resistance to lime. The fatty acids, with their terminal COOH groups are weak in respect of ability to disperse lime soaps, but the corresponding alcohols contain terminal OH groups which may be modified by the introduction of the solubilising SO_3H group to form a sulphuric acid ester which, in its turn, can form a sodium salt. For example, cetyl alcohol can be converted to the acid sulphuric ester and neutralised to form sodium cetyl sulphate $C_{16}H_{33}.O.SO_3.Na$. It is unfortunate that the term sulphonated is often applied to these products, giving rise to some confusion; a sulphate has the formula $R.CH_2.O.SO_3Na$ whereas the sulphonate would have a formula $R.CH_2.SO_3Na$. The conditions of preparation are different, for treatment of the fatty alcohol with chlorsulphonic acid at 30°C. gives a sulphuric ester, whereas the sulphonate is formed by reaction with strong sulphuric acid at 150°C.

The manufacture of the fatty alcohols is a somewhat difficult process and depends on the hydrogenation of the acids and esters at high temperatures and pressures in presence of catalysts. For instance, ethyl laurate is treated with hydrogen at 300 to 400°C. under pressure in presence of finely divided copper to give lauryl or dodecyl alcohol $C_{12}H_{25}OH$ which is readily sulphated.

Cetyl alcohol, $C_{16}H_{33}OH$, may be obtained from spermaceti, and oleyl alcohol, $C_{18}H_{35}OH$, from sperm oil.

The range of suitable alcohols comprises those containing from 10 to 30 carbon atoms in the chain, the most common of these being lauryl alcohol $C_{12}H_{25}OH$, myristyl alcohol $C_{14}H_{29}OH$, cetyl alcohol $C_{16}H_{33}OH$, and the unsaturated oleyl alcohol $C_{18}H_{35}OH$. Whereas those products which are based on the C_{12} alcohol have excellent wetting powers, the maximum scouring and cleansing action is only seen in the products from the C_{18} alcohols. The

WETTING AGENTS 139

detergent action compares favourably with soap, and, in addition, there is a considerable softening power.

Although the sulphated alcohols are relatively insensitive to acid and to lime, this only applies to warm solutions. The lime salts of the sulphated alcohols are soluble in warm water, but are deposited on cooling; nevertheless they do not possess the characteristic sticky properties of ordinary lime soaps.

The sulphated alcohols may be used alone, with soap or even with acid if required; in the recent products, such as the Lissapols, Gardinols, and the sulphated Lorols and Ocenols, there is also an effective action as protective colloids. Gardinol WA is lauryl sodium sulphate.

Wilkes and Wickert (Ind. Eng. Chem., 1937, *29*, 1234) have prepared secondary alcohols of high molecular weight, containing 10 to 21 carbon atoms and also having the hydroxyl groups on carbon atoms near the centre of the hydrocarbon chains. Sulphation and neutralisation of the resulting acid alcohol sulphates produced a group of sodium secondary-alcohol sulphates with remarkable wetting properties, the extent of which appeared to be determined primarily by the size of the non-polar hydrocarbon groups. This class of product is available under the trade-name Tergitol, and is also useful in dispersing or preventing the precipitation of lime soaps. ($R.R'.CH.O.SO_3Na.$)

Although the central position of the hydrophilic groups (compare p. 137) imparts wetting and penetrating powers, the secondary product is less efficient in scouring and dispersing than primary products such as soaps, fatty alcohol sulphates, sulphated fatty acid esters and amides.

Non-anion-active Products

The CATION-ACTIVE assistants arose from the work of the Society of Chemical Industry in Basle on the preparation of diethylamino-ethylamides of oleic acid. The salts of these new substances were found to be of value for increasing the fastness to water and washing of many substantive dyes, and were marketed as Sapamines.

Sapamine CH is $C_{17}H_{33}.CO.NH.C_2H_4.N(C_2H_5)_2.HCl$, Sapamine L is the corresponding lactate and Sapamine A the acetate; their preparation has been covered in B.P. 294,890; 390,533 and 366,918.

Sapamine BCH corresponds to $R\diagdown^{CH_2.C_6H_5}_{Cl}$ and Sapamine MS to $R\diagdown^{CH_3}_{SO_4.CH_3}$ where R represents the base $C_{17}H_{33}.CO.NH.C_2H_4.N(C_2H_5)_2$.

TEXTILE BLEACHING

Another interesting and well-known product, with excellent softening properties is Sapamine KW, which is probably the trimethylammonium-methylsulphate of monostearylmetaphenylenediamine (obtained by heating stearic acid and meta-aminodimethylaniline followed by treatment with dimethyl sulphate).

These Sapamines possess great wetting powers even in acid solutions, but, on the other hand, in alkaline solutions the base is liberated. The Sapamines are not soaps for the solutions of their salts have little detergent power; nevertheless they are excellent emulsifying agents and often superior to soaps, sulphated fatty alcohols and acids.

Whereas most textile assistants which depend for their utility on surface-active properties, carry the fatty chain in the anion, the Sapamines dissociate into positively charged *cations* containing the aliphatic chain, and negatively-charged anions. On this basis, the fixation of dyestuffs may be due to salt formation between the dyestuff anion and the cation of the assistant.

Sapamine FL is an exception to this general statement; it is anion-active, being the sodium phthalate ester of oleyl alcohol.

These cation-active assistants are similar in some respects to the quaternary ammonium salts investigated by Reychler and termed "inverted soaps." A well-known product is Fixanol of the I.C.I., formed by heating pyridine with suitable chlorides, bromides or sulphates of high molecular weight, such as cetyl bromide or dodecyl sulphate.

Fixanol is cetyl pyridinium bromide

$$\bigcirc\!\!-\!\!N\!\!\begin{array}{c}\diagup Br.\\ \diagdown C_{16}H_{33}.\end{array}$$

Another compound of similar properties is Sandofix of Sandoz; alkyl pyridinium sulphates such as lauryl pyridinium sulphate

$$\bigcirc\!\!-\!\!N\!\!\begin{array}{c}\diagup C_{12}H_{25}\\ \diagdown OSO_3H\end{array}$$

have been made by the Bohme Fettchemie.

Whereas the cationic surface-active compounds produce very low interfacial tensions against oils, which is important in detergents, the surface-active cation carries a positive charge which neutralises the negative charge on the dirt particles and may even cause a precipitation of dirt on to the textile material; hence these compounds have little value as detergents. It is not possible to use these cation-active auxiliaries in conjunction with soap, as precipitation takes place.

WETTING AGENTS

Arising from the cation-active assistant, there has been a series of quaternary ammonium compounds which, although not detergents, have found application as stripping agents for dyes, auxiliaries for the fixation of dyestuffs and finishing agents such as the well-known Velan PF.

ANION-CATION SURFACE-ACTIVE compounds have been prepared lauryl pyridinium lauryl sulphate $C_{12}H_{25}N(C_5H_5)SO_3OC_{12}H_{25}$ and lauryl pyridinium laurate $C_{12}H_{25}N(C_5H_5)OOC.C_{11}H_{23}$ being typical examples.

Although these substances are very useful in the preparation of textile emulsions and in dyeing, yet they are practically useless as detergents. The probable reason is that the surface-active cation carries a positive charge which neutralises the negative charge on the dirt particles, and may in fact cause a deposition of dirt on the textile material. Compounds of the type under consideration, therefore, are unable to emulsify the negatively charged dirt particles and separate them from their attachment to the fibre; soap, on the other hand, which is anion-active, is able to peptise the dirt particles.

NON-IONISED SURFACE-ACTIVE compounds have been shown to possess very interesting properties. The condensation of ethylene oxide with octadecylalcohol is capable of producing a compound with the following formula, $C_{18}H_{37}.(OC_2H_4)_n.OC_2H_4.OH$, where n is 15 to 20, that is, 15 to 20 moles of ethylene oxide have combined with one mole of octadecylalcohol; the product is known as Peregal O which is used as a dyeing assistant.

Similar compounds are seen in the Emulphors, which are well-known emulsifying agents of general formula $C_{12}H_{25}(OC_2H_4)_nOH$, where n is greater than 10.

A similar type of condensation is used in the production of the Igepals, which, however, are completely synthetic products and do not rely on natural fats for their origin; they contain no ionic group such as COOH or SO_3H. These unique detergents possess almost unlimited resistance to hard water, salts, acids and alkalis. According to Chwala and Martina (Textilber., 1937, *18*, 998), the general formula is $R.O.(C_2H_4O)_{5-10}.C_2H_4OH$, where R is a hydrophobic aliphatic hydrocarbon chain which fills the same function as the long hydrocarbon chains in soaps and other fatty chain salts. Representing the general formula as $R_1-O-R_2-O-R_2-OH$, where R_1 is the long hydrocarbon chain, from 12 to 16 carbon atoms, and R_2 is the short hydrocarbon chain, usually C_2H_4, it will be seen that the bond between the two groups occurs between an "ether-acid" bridge. The hydrophilic part of the molecule is the terminal hydroxyl group whose solubilising powers alone would be inadequate to deal with such a large molecule; however, the solubility is increased in virtue of the peculiar valency of the oxygen atom which forms the ether bridge. These oxygen atoms can form co-ordinate linkages with one of the hydrogen atoms of water and so bring other solubilising hydroxyl groups into the molecule.

These non-ionic detergents exhibit similar behaviour to the ionic soaps on account of the counter-play of hydrophobic and hydrophilic groups. In aqueous dispersion they form molecular aggregates or micelles carrying an electro-negative charge.

Igepal C has been synthesised for use with cellulosic materials and maintains its detergent action even in presence of metallic salts; the resistance to acid and alkali is good. It has excellent wetting action in neutral and alkaline baths and possesses strong dispersing powers towards lime soaps, when used in conjunction with ordinary soaps, although there is little purpose in employing soap with Igepal C. It cannot be used with aluminium salts for waterproofing and so replace soap, for it will remove the impregnating agent from the proofed material. Igepal C has little softening action.

Igepal W is intended for the treatment of wool and is equal to the Igepons in its scouring powers and superior in stability to hard water and metallic salts. It has no felting action on wool. Excellent scouring may be obtained with the minimum of alkali at low temperatures, and the Igepal is readily removed by rinsing. Igepal W is also a good wetting agent in acid liquors, but, having no sulphonic groups, it is not absorbed by wool as distinct from the behaviour of the Igepons. Strongly acid solutions of Igepal W can remove mineral oils and fatty oils very rapidly at low temperatures.

Igepal L combines the good properties of the Igepals with those of selected organic solvents.

Another non-ionic detergent has been manufactured by Lever Brothers, Ltd. (B.P. 439,435) and belongs to the class of polyglycerol fatty acid esters. Glycerol is polymerised to the penta-glycerol stage and then partly esterified with the fatty acids of coconut oil. The soapy product is mainly a monoester of general formula $CH_2(OH).CH(OH).CH_2.O.CH_2.CH(OH).CH_2O.COR$, where R is the long-chain fatty radical. The product possesses great detergent activity and stability too.

The manufacture and properties of many of these new wetting agents and detergents forms an interesting chapter in modern textile chemistry. The subject has been discussed by Briscoe (J.S.D.C., 1933, *49*, 71), Kertess (ibid., 1933, *49*, 69; 1936, *52*, 42), Dunbar (ibid., 1934, *50*, 309), Hannay (ibid., 1934, *50*, 273), Venkataraman and his co-workers (ibid., 1937, *53*, 91; 1938, *54*, 465, 520; 1939, *55*, 125; 1940, *56*, 503; 1941, *57*, 41) and Nusslein (ibid., 1935, *16*, 325).

A scheme for the chemical identification of many of the newer wetting agents and detergents has been devised by Linsenmeyer

WETTING AGENTS

(Textilber., 1940, *21*, 468), who divides the reagents into the following groups:

(a) Soaps and soaps containing solvents.
(b) Sulphonated oils (Turkey red oil, Monopol oil).
(c) Highly sulphonated oils (Avirol AH, Humectol CX).
(d) Naphthalene sulphonic acid and salts (Nekal BX).
(e) Fatty alcohol sulphates (Gardinol, etc.).
(f) Fatty acid condensation products (Igepon).
(g) Protein fatty acid condensation products (Lamepon A).
(h) Ethylene oxide condensation products (Peregal O? Igepal).

CHAPTER XII
DETERGENCY

THE list of modern auxiliaries includes thousands of compounds, many of which are wetting agents, but by no means all are detergents. For example, ethyl alcohol is an excellent wetting agent (in large amounts), but has no detergent action. Wetting agents as a class will reduce the surface tension of water when used in low concentrations, and also lower the interfacial tension between water and another surface. Wetting power may be estimated by the Herbig method, which consists in immersing a known weight of cotton yarn in a solution of the wetting agent for 5 seconds, removing the superfluous liquor in a hand centrifuge and estimating the liquor retained. Another widely used test is the sinking time method of Draves, which consists in measuring the time necessary for a 5-gram skein of twofold grey cotton yarn to sink in a solution of the wetting agent (Am. Dyes. Rep., 1939, *28*, 425).

Dispersing agents are often confused with wetting agents, but in actual fact a dispersing agent often has little effect on the surface tension of water. For instance, carbon black is wetted in a solution of a wetting agent and settles in large particles, but in presence of a dispersing agent the particles are small and remain suspended. A dispersing agent promotes the separation or deflocculation of particles by overcoming the forces between them.

Many wetting agents have no foaming powers as possessed by detergents; the chief necessity for foam formation is the difference of concentration of the dissolved substance in the surface layer as compared with the concentration in the bulk of the solution. Frothing is only possible when solubility permits, as easily soluble or difficultly soluble compounds will not produce a stable copious foam.

Emulsifying power is a very important factor in a detergent and may be estimated by Baker's method of mixing equal volumes of the solution and light motor oil in a motor-driven drink-mixer and noting the total volume of emulsion. Baker (Ind. Eng. Chem., 1931, *23*, 1026) has also given a useful bibliography of work on the connection between detergent power and emulsifying value of solutions.

King and Mukherjee (J.S.C.I., 1939, *58*, 243) studied soap-stabilised emulsions and found that although soaps form very fine emulsions they are not very stable. Oleates are more efficient than stearates and much better than palmitates.

The de-flocculating value or dirt suspending power has also been studied by Baker (loc. cit.) by shaking 0·15 g. of bone-black in a bottle with 100 c.c. of the detergent and comparing the various results after 60 hours. Finely divided manganese dioxide was preferred by Fall (J. Phys. Chem., 1927, *31*, 801), who has also given a critical summary of the methods utilised for measuring detergent action by McBain, Bancroft and other investigators.

A certain protective colloid action is important in detergents in order to prevent the re-deposition of dirt during scouring. A good protective colloid, gelatin, for example, need not necessarily possess detergent power; the protective action is usually measured by the "gold number" of Zsigmondy (Z. anal. Chem., 1901, *40*, 697).

The "Gold Number" is the number of milligrams of the protective colloid which prevents the colour change when 1 c.c. of 10% sodium chloride solution is added to 10 c.c. of red gold solution (gold chloride) containing 0·0053 to 0·0058% of gold.

Regarded purely as detergents, soaps compare very favourably with the new synthetic detergents. The ability to resist hard water however, is an important requirement in a detergent, and if the water has not been softened, then the detergent must not be sensitive to the constituents of hard waters. It may be better to soften the water than to obviate the trouble by the use of the more costly compounds which disperse lime soaps. The wetting power of sodium oleate, as shown by Herbig numbers, is also of the same order as those for corresponding solutions of the newer wetting agents.

To a limited extent, it is possible to predict detergent behaviour from the molecular structure of the compound under consideration. For instance, soaps and sodium alcohol sulphates have similar hydrophobic groups but different hydrophilic groups; the soaps hydrolyse in water and show an alkaline reaction, but the sulphated alcohol with its stronger acid residue gives neutral solutions. The wetting powers of a compound have been improved by putting the hydrophilic group near the centre, but in general, it appears the terminal position for this group is better in respect of detergency and lathering.

The survival of soaps for cleansing purposes offers some evidence of the intrinsic value of the carboxyl group, in spite of the many ingenious attempts to block it and change it.

Before discussing detergent efficiency, it must be remembered that detergent action is influenced by the nature of the material to be removed by the detergent. Most soiling materials carry a negative charge, so that the surface to be cleansed should also carry a negative charge if possible, in order that they should repel one another. With cellulose in water, there is a negative charge which becomes more pronounced in presence of alkali; Neville (Am. Dyes. Rep., 1936, *25*, 267) has shown that the higher the alkalinity, the more readily is the soiling matter removed. The addition of soap increases detergent efficiency because, among other factors, the fabric assumes a more highly negatively charged condition than in presence of alkali alone. The sulphated alcohols are also more effective detergents in presence of alkali. Wool and silk, on the other hand, are amphoteric, and washing with an anion-active detergent in an acid medium is less effective than in an alkaline medium.

Adam (J.S.D.C., 1937, *53*, 121) examined the relative washing efficiencies on cotton and on wool of six or eight detergents, and found a 10- to 20-fold range on wool, soap being far inferior, compared with a 2-fold range on cotton. Again, with wool there was often a preferential adsorption of the detergent with clean wool to such an extent that weak solutions were exhausted and no longer able to hold the dirt from re-deposition on the fibre; there was a large variation in the amount adsorbed from the different detergent solutions, soap losing considerable amounts. With cotton, the adsorption is more nearly the same for all detergents.

The property possessed by wool of "killing" a soap solution has already been mentioned.

The effect of pH on the detergent powers of soaps and other detergents for cotton has been examined by Rhodes and Bascom (Ind. Eng. Chem., 1931, *23*, 778) and Morgan (Can. J. Res., 1933, *8*, 429), from which it appears that the detergent power of soap is increased by an increase in pH; the work is somewhat complicated by the salt effect (see p. 147). Less work seems to have been devoted to wool, but Palmer (J.S.C.I., 1941, *60*, 60) has found that with detergents of sulphated alcohol type the increase in pH of the detergent solution only had a slight effect on detergency; with sodium oleate the effect is great. With Igepon T, however, an increase in pH causes a decrease in detergent power. In general, a decrease in pH results in a decrease in detergent power of anionic detergents.

The properties of detergent solutions have been studied by Powney and his co-workers (Trans. Farad. Soc., 1935, *31*, 1150; 1937, *33*, 1243, 1253; 1938, *34*, 356, 363, 372, 625, 628), who determined the surface tensions of the sodium and potassium salts of oleic, lauric and myristic acid; small changes in pH produced large changes in surface tension particularly in the case of the laurates. The pH value also has a pronounced effect on the displacement of oil. The presence of unsaturated groups in the alkyl chain decreases the surface activity of the soap.

For degumming silk, Stockhausen (Seide, 1932, *37*, 387) found that soap solutions were better than those of synthetic detergents, as the action of the former was due, in this case, to the alkali formed by hydrolysis.

The effect of salts on detergency has been examined by Palmer

(J.S.C.I., 1941, *60*, 56) in respect of the ability to remove olive oil from wool. The effect of neutral salts on detergent power is not simple, for the detergent power of the solution at first increases as the salt concentration is raised and then decreases at higher salt concentrations. Magnesium, calcium and barium salts increase the detergent activity as readily as sodium salts, but are more effective in lower concentrations; too great a concentration of any salt decreases the detergent power. Experiments were performed with sodium oleate, Igepon T and some fatty alcohol sulphates.

The decrease in surface tension or interfacial tension of a detergent solution on the addition of salts has been discussed in *"Wetting and Detergency"* (Harvey; London, 1937, p. 137) and some isolated washing experiments have been recorded (ibid., pp. 137, 163). Rhodes and Wynn (Ind. Eng. Chem., 1937, *29*, 55) studied the effect of sodium chloride, sulphate, phosphate, borate and acetate on soap solutions and found an optimum salt concentration for a given concentration of soap; this work referred to cotton. The effect of salts on sulphated and sulphonated alcohols with regard to detergent power has been protected in B.P. 352,989.

This type of effect has been employed for years in wool scouring when the suint (largely potassium salts) is extracted with water and added to one of the soap bowls.

The increase in detergent power, which depends on the valency of the added cations in the case of anionic detergents, has been explained by Cassie and Palmer (Trans. Farad. Soc., 1941, *37*, 156) as follows. In the presence of salts, the potential energy of the detergent ions in solution is lowered by the concentration around them of ions of opposite sign, but the potential energy of the detergent ions in the surface is lowered to a much greater extent. The amount of detergent in the surface in equilibrium with a given bulk concentration depends on the difference in potential energy of a detergent ion in the two places; hence the greater lowering of the potential energy of the molecules in the surface will result in a displacement of the equilibrium in favour of the surface, giving greater surface pressures and increased detergence. This is in agreement with the fact that only ions of opposite sign to the detergent ion have an appreciable effect, since for a given cation, change of anion makes little difference; it is also in agreement with the fact that bivalent ions exert the same effect as univalent ions, but at a lower concentration.

Although the relations between the detergent and the soiling material, and also between the detergent and the fibre are of great importance, the adhesion of the soiling matter to the fibre is of some consequence.

Considerable use is made of detergents in the textile industry, particularly in the scouring operation, where the chief difficulty lies in the adhesion of oil or wax to the textile material. Within the same series, adhesion increases with molecular weight, and is further increased by the presence of polar groups.

Speakman and Chamberlain (Trans. Farad. Soc., 1933, *29*, 358) demonstrated that owing to the non-polar character of mineral oil, the interfacial tension is high, rendering emulsification difficult apart from the adhesion phenomena. Mixtures of mineral oil and oleyl alcohol illustrate this point; oleyl alcohol reduces the oil-water interfacial tension, but increases the adhesion to the fibre when present in excess of 7% in the mixture, and makes scouring more difficult in spite of the fact that the interfacial tensions continue to fall. Mixtures containing more than 55% of oleyl alcohol are more difficult to remove than mineral oil alone, on account of the increased adhesion due to the polar groups; oleyl alcohol itself, however, is easily removed as it appears to form an inverted (water-in-oil) emulsion on the surface of the wool.

The scouring of cotton goods with soap and alkali lowers the interfacial tension between fibre and solution sufficiently for the alkali to start its action, but a specific wetting agent is often advisable to hasten the process or to render it more even in its action. The auxiliary must possess a variety of properties which are not always available in soaps. For example, the pH value must be adjusted to stabilise the emulsion formed with the impurities, and to maintain the particles in the fine state of division to avoid deposition. The protective colloid action of soaps is generally not sufficiently marked for emulsification of all the grease, fat, and dirt in grey cotton. On the other hand, mere wetting power in a kier-boiling auxiliary would be a doubtful advantage, for without some emulsifying and cleansing effect the tendency would be to spread the dirt more uniformly on the fibre rather than to remove it completely.

The various criteria suggested for the examination of detergents and auxiliaries have been given by Venkataraman and his associates (J.S.D.C., 1937, *53*, 91). It is concluded, however, that practical trials are often necessary; for kier-boiling, the efficiency of the assistant may be judged by the amount of wax removed, the wetting of the cloth after kiering and the whiteness after bleaching. Substances related to Igepon T were found to be of particular value.

It must be stated that the choice of a detergent is not always decided solely by its cleansing power, for with wool scouring a certain amount of "milling" may be required which may be produced by soap, but not by the non-ionised detergents or the strong soap-like electrolytes.

It is an unfortunate method of classification to group the modern detergents and auxiliaries together under the term wetting agent,

DETERGENCY

or the uglier wetting-out agent, for although wetting is a factor in detergency many wetting agents are not detergents. In the evaluation of detergent efficiency, however, wetting must be considered at least as a preliminary to cleansing.

The reduction of surface tension is not the main factor in detergent action, for most wetting agents will lower the surface tension as much as soaps and other detergents, but will not remove grease from solid surfaces; the surface tension of soap solutions may be increased by the addition of alkali which improves detergent action. Nor does interfacial tension appear to be the fundamental feature, for the effect of cetane sodium sulphonate and cetyl sodium sulphate on the interfacial tension between the solution and oil is almost identical, but the sulphate is a much more efficient detergent.

According to Adam (J.S.D.C., 1937, *53*, 121) the emulsification of a hydrocarbon liquid is not fully understood, but certainly one of the predominating factors is that the interfacial tension must be diminished as far as possible. The dispersion of a solid in water starts as a loosely adherent powder and becomes completely separated into constituent granules; detergent action seems to be a more complicated case of the same type of phenomenon, the essential process being a displacement of the oil from the fibre in fairly large globules, as confirmed also by Palmer (J.S.C.I., 1941, *60*, 56). Detergent action is not a dispersion of the dirt in the ordinary sense, but rather a displacement of the dirt from its adhesion to the surface by the aqueous detergent solution. With cetyl sodium sulphate, for instance, the removal of the soiling material (lanoline) was in large globules; the function of the detergent is to collect the dirt locally on the fibre from which it is then removed by mechanical agitation. The detergent process appears to consist in an alteration of the angle of contact, between the grease-water surface and the solid surface being cleansed, from $180°$ to $0°$, measured in the water. The detachment of the globules has been investigated by Adam, who showed that for this process to occur, not only must the oil-solution interfacial tension but also the fibre-solution interfacial tension be reduced; it is therefore probable that the detergent molecules are adsorbed at the fibre-solution interface as well as at the oil-solution interface. The five actions in detergency (see p. 144) are (*a*) reduction of surface tension of water, (*b*) spreading of water, (*c*) penetration of water, (*d*) emulsification of grease, and (*e*) detachment of grease. Adam suggests that the essential part of detergency is the displacement of oil or grease from the solid surface by the aqueous detergent solution and has deduced a formula showing the relative importance of the different surface tensions and adhesions in producing this displacement.

The assessment of detergent efficiency cannot be made by any simple test, and it has been shown that such criteria as surface tension, drop number, frothing and emulsifying power are not infallible although the drop number against benzene often gives results in accordance with practical washing tests. It is now customary to conduct an actual washing test on fabric soiled with a suspension of lampblack in oil; a typical mixture contains mineral oil, tallow and lampblack in carbon tetrachloride. Hannay (J.S.D.C., 1934, *50*, 273) has criticised many of the extravagant claims made for the new detergents and showed that a more efficient cleansing action might be outweighed by the higher cost. Evans (ibid., 1935, *51*, 233) has graded the detergents according to their efficiency, the esters with C_{12} being inferior to esters with C_{16}. The sulphates surpass the sulphonates in having better solubility, better wetting over a wide range of concentration and temperature under neutral conditions, and better scouring under neutral and alkaline conditions.

PART THREE
SCOURING AND BLEACHING OF THE CELLULOSIC FIBRES

CHAPTER XIII
INTRODUCTION TO COTTON BLEACHING

THE main objects in the bleaching of cotton are the removal of natural and adventitious impurities with the production of pure white material. The extent of this purification is largely determined by the purpose for which the material is intended.

For general sale throughout the world, the plain white material was given the "market bleach," but modifications of this were necessary for coloured-woven goods, particularly shirtings, in view of the susceptibility of the colours to "bleeding" or "marking off" during the cleansing processes.

For subsequent treatment by the dyer and printer, a pure and absorbent finish was necessary in some cases, for example, printing, whereas absorbency was more important than whiteness for the dyer; indeed, in the case of dark shades, the bleaching process proper—the removal of residual colouring matter—may be omitted.

The demands of the dyer, the printer and the market were broadly met by the same sequence of processes, repeated according to requirements. Much of the repetition was due to the absence of favourable conditions for the complete removal of the impurities, rather than to the obstinacy with which they are retained. This incomplete removal of impurities at any one stage was complicated by the fact that thorough washing is essential between most of the stages to avoid reprecipitation of the dissolved impurities and also the partial neutralisation of the various chemical reagents employed. A full bleach was a very prolonged series of operations.

The foreign substances associated with cotton cellulose are the waxes, proteins, mineral matter and the natural colouring material. The waxes are of importance in the spinning operations during which the fibres are made into yarns as they actually assist that process, but when dyeing is considered, the waxes are a hindrance, for owing to their water-repellent nature, they prevent the proper absorption of the dyestuff. The natural colour of cotton and the other impurities also interfere, but to a less degree, with the ordinary processes of dyeing, printing and finishing, so that their removal is practically essential.

In this connection, it must be remembered that not only is Egyptian cotton a darker colour than the American varieties, but it is also more difficult to bleach.

With bleached material, an imperfect removal of the non-cellulosic material beforehand is apt to lead to a poor "white," a dull appearance and to yellowing on subsequent storage. Most colours appear brighter on well-purified material and are easier to apply.

The nitrogenous impurities of cotton are apt to render it susceptible to attack by mildew and bacteria, from which the cellulose itself is comparatively immune.

Many terms used to-day in describing the bleaching processes are only intelligible in the light of the past history of the craft.

The early methods of bleaching cotton and linen goods consisted of "bowking" or "bucking," and "whiteing" or "crofting." The first term is probably derived from the German *bauchen* and refers to the process of boiling in potash-lye obtained from the ashes of plants. Crofting relates to the exposure of the fabric to light and air by "grassing" in crofts, a process which is still carried on in some country districts; it is interesting to note that the word "croft" still persists, and although commonly used to denote a field in many parts of Great Britain, it is also employed to describe that part of the works in which bleaching is done. Between the boiling and the whiteing, the cloth was rinsed and immersed in buttermilk, from which the term "sour" is a survival. The complete process took several months for linen, but it was often considered that the boiled cottons were sufficiently white.

Sulphuric acid replaced buttermilk in 1756 or thereabouts, but although the benefits of lime as opposed to potash had been known for many years, the prejudice against its use did not disappear until about 1764; it seems probable that the prejudice was due to the risk of damage to the fabric. The employment of lime in conjunction with potash led to the occasional use of potassium or sodium hydroxides.

The discovery of chlorine by Scheele in 1774, and the observations of its bleaching action, revolutionised the methods of treating linen and cotton. Berthollet found that chlorine gas could be dissolved in solutions of alkaline hydroxides, and one of these solutions—in potassium hydroxide—was marketed by Javel as Eau-de-Javel; the solution of chlorine in aqueous caustic soda solution was commercially manufactured by Labarraque in 1820.

In the meantime, Tennant had attempted to patent the use of chlorine in milk of lime, but had been forestalled by the Lancashire

INTRODUCTION TO COTTON BLEACHING 153

bleachers, who had already used it. In 1799, however, he patented the manufacture of bleaching powder by the use of dry hydrated lime and chlorine gas, and its commercial production in Glasgow was very successful. Since about 1830, bleaching powder has remained the chief agent for the bleaching of vegetable fibres, but it has many other uses. In textile circles, it was referred to quite simply as "chemical," which has persisted in the form of "chemic"; it may be remarked that its original use was attended with disappointment, for it was hoped that the wonderful "chemical" would obviate the necessity for bowking also.

An interesting account of the bleaching of cotton and linen, from the historical standpoint, has been given by Higgins (J.T.I., 1923, *14*, 209, 277, 319, and 441); parts of the well-known work by Matthews (*Bleaching and Related Processes*, Chemical Catalog Co., 1921) are also devoted to the history of bleaching.

For many years the chief advance in the bleaching of cotton lay in increasing the severity of the scour, in order to obtain a more thorough "bottoming"; the limits of increasing temperature and pressure were soon reached without much real understanding of what was taking place. The boiling was carried out in special iron vessels called kiers, and two rival systems were employed: (*a*) the lime boil, and (*b*) the soda boil; the former and older process consisted of many stages, and is rapidly disappearing. (The word "soda" in connection with the scouring of cotton means sodium hydroxide; sodium carbonate is termed "ash." For water-softening, however, p. 106, "soda" means sodium carbonate.)

The demands of fashion in the way of coloured-woven goods brought with it, on the one hand, the necessity of milder treatment, and on the other hand, a greater scientific interest in the process. Similarly, the advent of rayon rendered modifications of drastic scouring absolutely imperative in the case of the ever-popular cotton warp—rayon weft construction. The closer association of chemistry with the textile industry brought with it, not only a more complete understanding of the bleaching process, but a very satisfactory measure of scientific control. At the same time, new products made their appearance on the market as "assistants" for kier-boiling and as "wetting agents" generally.

The early work of the British Cotton Industry Research Association included a survey of the natural impurities of raw cotton, with methods of estimation, as well as investigations of many aspects of the bleaching process which unless brought under scientific control are apt to lead to damaged material.

Appreciation of modern developments during the rise of Textile Science (1920–40) is best shown against a short background of the "prior art."

Bleaching actually consists of two separate processes:
(a) scouring, or the removal of wax;
(b) bleaching, or the destruction of the colouring matter.

MADDER BLEACH

For many years the standard method of obtaining cotton piece goods in their purest form was known as the "madder bleach." The term originates from the time when madder was used in printing by the dyed style in which the mordant was first printed on the cloth, and the whole piece dyed later with madder. The difficulty was to obtain a clear white ground, for unless the cloth was in a very pure state, the madder stained the unmordanted parts which were very difficult to clear.

As usual the piece goods were stitched together and singed, in order to remove loose hairs and give the cloth a clean, smooth face. On leaving the singeing machine the goods were passed through water in an ordinary washing-machine and allowed to lie overnight, with the result that some of the water-soluble impurities were removed, the materials became thoroughly wet and also a considerable amount of the starch or size was rendered water-soluble in consequence of a fermentation process.

The madder-bleach proper consists of the following steps:

(a) The cloth is impregnated with milk of lime under such circumstances that it takes up between 2 and 4% of CaO calculated on the weight of the goods, and is run into a kier which is whitewashed on the inside in order to prevent stains. The necessary amount of boiling water is then added and scouring conducted for about 8 hours at a pressure of 40 lbs. per square inch. The kier is then allowed to blow-off, and the liquor is replaced as rapidly as possible with cold water in order to avoid tendering owing to local concentration of the lime.

(b) The goods are then "soured" by running through sulphuric or hydrochloric acid at 1 to 2°Tw. and are allowed to lie in the acid solution for about 2 hours at room temperature; this decomposes the lime soaps and removes any iron stains.

(c) A second boil takes place in a solution of 3 to 4% of sodium carbonate and a resin soap equivalent to 1·5 to 2% of resin, all calculated on the weight of the cloth. This boiling operation may take 7 hours at a pressure of 30 lbs. or 3 to 4 hours at a pressure of 45 to 50 lbs. The goods are then washed and soured again, after which process a further washing is given.

(d) A further boil takes place in sodium carbonate solution only, and the amount increases to 5 to 6% of the weight of the cloth; the goods are again washed.

INTRODUCTION TO COTTON BLEACHING

(e) Further souring and washing processes complete the scouring section of the madder bleach.

(f) The cloth is then subjected to the bleaching action proper which consists in immersing the goods in a clear solution of bleaching powder of 0·5 to 1°Tw. at room temperature; the goods are allowed to lie for 2 hours and are then soured in hydrochloric acid of 1 to 2°Tw. and washed for about 1 hour.

MARKET BLEACH

The market bleach for white goods was based on the old madder bleach, but was conducted in low-pressure instead of high-pressure kiers. The sequence of steps was rather different. The lime boil and sour were followed by a short chemicking with 0·5°Tw. calcium hypochlorite solution. The goods were then scoured in soda ash and soap on two occasions with souring and washing between. The final step was that of "chemicking," souring and washing.

As the market bleach was only intended for goods to be sold in the plain white state, it was not necessary to have such a thorough bottoming of the cloth as in the case of material intended for printing.

Another variation of the market bleach is as follows:

(1) Lime boil, 12 hours—wash.
(2) Sour in 2°Tw. HCl, 2 to 4 hours, wash.
(3) Boil in 1% NaOH (calculated on the cloth), 12 hours—wash.
(4) Chemick in 1°Tw. chemic. 2 to 4 hours—wash.
(5) Boil in 1% sodium carbonate solution for 12 hours—wash.
(6) Sour in 1°Tw. H_2SO_4 for 2 to 4 hours—wash.
(7) Tint with blue if necessary.

The Turkey red bleach, for dyeing, does not include a treatment with hypochlorite; the goods are boiled in water, twice boiled in NaOH, soured and washed.

The "half-bleach" is employed for goods to be dyed in medium and light shades. The scoured material is chemicked in 0·5°Tw. bleaching powder solution for 1 to 2 hours, soured and washed.

SODA BOIL

The use of lime was displaced to some extent by caustic soda, and a shorter scouring process may be given as a result. The goods are boiled in either open or closed kiers in a concentration of NaOH varying from 2 to 5% on the weight of the goods: with open kiers the time of boil varies from 8 to 12 hours, but with closed pressure kiers this may be reduced to between 2 and 10 hours, depending on the character of the material and the efficiency of the process.

It is important that all air should be excluded from the kiers during the pressure-boiling processes in order to avoid any possible formation of oxycellulose, and similarly it is important that after the boil the kier liquor should be run off and replaced immediately with water in order to avoid any liquor concentration.

A *general* bleach with caustic soda is sometimes criticised on the grounds that it gives too soft a result as compared with the lime method, but the sequence of steps is as follows:

The goods are scoured for from 8 to 10 hours in NaOH solution of 2°Tw., followed by washing, souring and washing again before a second scour of 6 hours in 1°Tw. NaOH. The goods are then washed and immersed for 2 hours in 2°Tw. bleach liquor at room temperature, after which they are washed, soured and again washed.

COLOURED GOODS

Coloured-woven goods were usually scoured in open kiers which, together with the milder alkaline nature of the scour, necessitated longer periods of boiling. For example, the goods may be scoured for two periods of 10 hours in soda ash and soap, and washed between the two steps. In some cases, a third scour of 5 to 6 hours was given. The concentration of the bleach liquor was usually as low as 0·1 to 0·25°Tw.

A special type of coloured-woven goods was that of mulls with coloured headings which were hung outside the open kier and the bleaching cisterns during the treatment which was approximately the same as that for coloured-woven goods in general.

MODERN METHODS

Many of the older methods were unnecessarily laborious for modern times, and the lime and soda ash boiling process has largely been replaced in favour of a single boil with caustic soda. There are three steps to be followed:

(*a*) A steeping process.
(*b*) A scouring process.
(*c*) A bleaching process.

The term "bleaching" in its broadest sense includes the three steps mentioned above, but in its narrowest sense it relates to the final treatment with the bleaching agent proper. The whitening of the cloth is not necessarily restricted to this final step for a very considerable part of the natural discoloration is removed during the scouring process.

There are other processes which are capable of removing some of the impurities and one of these is the well-known mercerising process;

INTRODUCTION TO COTTON BLEACHING

thousands of yards of shirtings which have been mercerised are merely bleached afterwards; the mercerising process removes sufficient of the cotton wax to render scouring unnecessary in many cases.

Before discussing the modern methods of scouring and bleaching it may be well, briefly to refer to the two chief types of chemical damage which may arise during the bleaching process; the reduction in strength which accompanies chemical attack is often termed "tendering."

The commonest type of tendering is due to oxidation, which may result from the action of air in presence of hot alkali, or from the uncontrolled action of hypochlorite solutions during the bleaching process proper.

Another kind of tendering is due to acid attack with the formation of hydrocellulose; this may result from the uncontrolled use of dilute acids when souring the cotton after the scouring process or after the bleaching process.

CHAPTER XIV
SINGEING AND DESIZING

BEFORE dealing with the actual operation of scouring cotton by the kier-boiling process which generally takes place in cylindrical iron vessels, a short account will be given of the preliminary operations of singeing and steeping processes intended to remove projecting hairs and certain acquired impurities respectively. The steeping process helps the scour to be more efficient by a preliminary removal of water-soluble matter, and also by solubilising the size so that it may be readily removed by washing.

It will be realised, of course, that the fabric is very frequently washed throughout the sequence of stages which make the full scouring and bleaching process; the washing step is interspersed between the various stages of the process, and may be effected by the methods outlined on p. 182.

SINGEING

Singeing is not an essential preliminary to the scouring of cotton piece goods, but it so happens that many cotton cloths are valued for their sheer or smooth appearance; this applies not only to the special satin and sateen weaves but also to poplins, shirtings, handkerchief-cloths, and many others. The smooth and lustrous character would be destroyed by a "fuzz" of cotton hairs projecting from the cloth itself, and it is therefore customary to remove these by singeing. The pieces or "lumps" of cloth, averaging about 120 yards in length, are marked for purposes of identification and record, and then stitched together before passing to the singeing machine in long runs.

With some cotton fabrics, it is sufficient to eliminate the dust and other loose impurities by brushing, and then crop or shear the loose ends of yarn from the surface, followed by a second brushing to make a smooth nap. Singeing, however, is a much more effective alternative method, and is almost essential with goods intended for printing.

Before reaching the singeing apparatus proper, the cloth is carefully dried by passing over steam-heated cylinders, as wet cloth is apt to scorch much more readily than dry cloth. The whole skill of singeing is to burn off the projecting hairs without scorching or otherwise damaging the body of the cloth; care should be taken to ensure that cloths which require singeing do not contain anti-mildew mixtures which may liberate acid on heating and so tender the fabric.

SINGEING AND DESIZING

The types of singeing machine commonly employed are:
   (a) the plate singeing machine
   (b) the rotary-cylinder machine
   (c) the gas-singeing machine.

The plate machine generally consists essentially of one or two curved copper plates, from 1 to 2 inches thick, which may be heated to bright redness by a furnace below them, or by a suitable heating arrangement of gas and air. The cloth passes over and in contact with the plates at a speed varying between 150 to 250 yards per minute, and the travel of the fabric may be arranged so as to singe

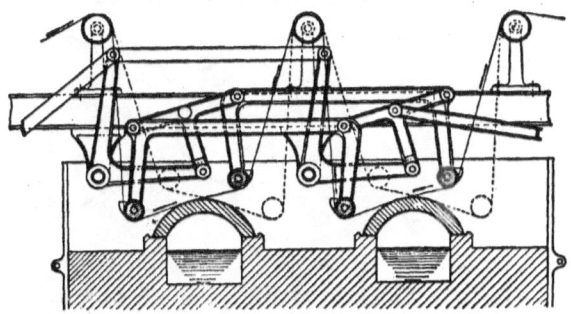

FIG. 44.—Diagram of the plate singeing machine of Mather and Platt.

one or both sides of the fabric during a single passage. An automatic traverse motion is fitted to the machine so that the cloth is brought into contact with a constantly changing part of the plate and this obviates local cooling. The traverse motion also equalises the wear of the plates, for unless the heating is uniform, the singeing is apt to be uneven and give rise to faint stripes when the cloth is dyed later.

The rotary-cylinder machine overcomes these defects, and is a modification of the plate machine, in that the goods are not drawn over fixed plates, but over a hollow cylinder of copper or cast iron, with internal firing, which revolves slowly and so presents a fresh surface to the oncoming fabric. The rotation of the cylinder is in the opposite direction to that of the fabric so that the nap of the cloth is raised, a factor which makes this type of machine particularly suited to the singeing of velvets and other pile fabrics. With two cylinders, it is possible to singe both sides of a fabric in one operation.

The gas-singeing machine is fitted with gas burners so arranged that one or both sides of the fabric may be singed during a single passage of the cloth over the flames. The burners are usually supplied with a suitable mixture of gas and air under pressure; coal gas, anthracite gas or petrol gas may be utilised. It is possible to adjust the width of the sets of burners to suit that of the fabric; it is also possible to adjust the length of the flame and hence the intensity of the heat. At one time it was customary to place small exhaust hoods over the flames, immediately above the cloth, in an attempt to draw the flame through the fabric and thereby increase the singeing power; it now seems to be realised that better results appear when the flame is allowed to spread along the fabric so that special exhaust chambers are no longer considered necessary. Gas singeing machines are available with two, three, four or six burners,

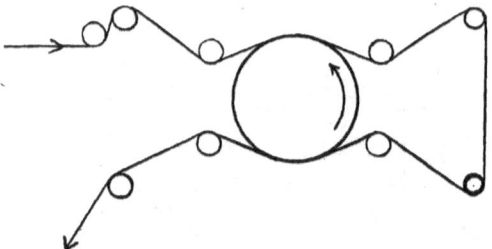

FIG. 45.—Diagram of rotary cylinder singeing machine.

and in many machines the burners are arranged to swivel for singeing at a tangent if required; machines are available in which the general run of the fabric is horizontal or vertical.

For many purposes, a combined gas and plate singeing machine has been found most useful.

No matter which type of machine is used, the singed cloth passes between a pair of draw-rollers or the nip of a mangle revolving in a water-box, so that all sparks are extinguished and risk of fire is reduced to the minimum. Many singeing machines are equipped with a lifting motion whereby, in case of difficulty, the cloth may be raised from the hot plates or flames.

Most singeing machines are also fitted with a device for sucking away the burnt or singed fibres; the simplest arrangement being a hood and exhaust fan over the whole machine.

Within recent years there has been a tendency to prefer the gas-singeing machine instead of the plate singeing machine.

STEEPING

The impurities in raw cotton consist of oil and wax, protein, pectic matter, natural colouring matter, mineral matter and a certain amount of adventitious woody matter consisting of parts of husks,

SINGEING AND DESIZING

seeds and pods, which form the motes, shives, etc., of the yarn or cloth. It is generally stated that cotton yarn loses from 5 to 9% of its weight during bleaching; cotton cloth loses a larger amount, the difference being accounted for by the material added during sizing, which may amount to a further 4 to 8%.

Size is generally composed of farinaceous matter, usually a preparation of starch; in the rare instances where paraffin wax is incorporated in the size, it is now usual to add an emulsifying agent also, as otherwise its removal is not easy.

The three chief methods of removing the size from cotton fabrics are:

(*1*) rot-steeping
(*2*) acid-steeping
(*3*) steeping in enzyme preparations.

In the lime boil, a considerable amount of the starch is left undisturbed and pre-steeping methods in dilute acid, such as 1% H_2SO_4, were suggested. The caustic boil is more efficacious in removing starch, particularly in conjunction with improved construction and operation of kiers, and these advances lessened the value of the preliminary steep. The introduction of the anthraquinone vat colours, however, necessitated a milder boil and restored interest in steeping methods; pressure-kiering in presence of starch, produces degradation products which have reducing properties sufficient to form the leuco state of the dyes, which is dissolved by the alkaline kier-liquor. An added reason for the preliminary steep is that the normal scouring action of the alkali is somewhat hindered by the presence of starch.

The old method of steeping was termed "rot-steeping" and took place in warm water for 24 hours at 35 to 40°C. or overnight at 60°C. The prolonged soaking makes the fibre more absorbent and helps to remove the starch, and also some 2 to 3% of the water-soluble natural impurities present, such as pectins; the organisms present naturally in the water, multiply, and starch-liquefying enzymes are produced so that when the cloth is washed later, most of the starch is removed. The fermentation during rot-steeping must not become too active, or the cellulose itself will be attacked.

The cloth is usually singed before steeping; the singed material is run through a mangle provided with running water to extinguish the sparks and lower the temperature before impregnating with the modern steeping agent which may be either 0·5 to 1% H_2SO_4 or a malt enzyme solution. The acid-steep at 40°C. is preferred by some authorities as it gives a better white after the alkali-boil and also removes mineral matter. The loss in weight with acid-steeping is slightly higher than with water-steeping. The process may be conducted in a continuous manner by suitable choice of temperature and concentration, but care must be taken to avoid acid attack on the cellulose. The acid-steep has been examined by Fargher and his co-workers (J.T.I., 1927, *18*, 29 and 559).

Enzymes

Steeping in enzymic agents may be a very rapid and thorough method of desizing under the most favourable conditions provided that correct conditions of pH and temperature are maintained. Malt extract has been employed for many years and also such preparations as Diastafor, Novofermasol and Polyzime C, which are fairly sensitive to temperature changes from the optimum. Bacterial desizing agents, such as Rapidase, are active over a wider temperature range and appear to have certain advantages, according to Evans (J.S.D.C., 1935, *51*, 318) such as tolerance of variation in pH. Gelatase is a bacterial enzyme for the removal of gelatin sizes as used on rayon. Efficient washing after desizing is essential; if coloured goods are piled during or after the steeping process, precautions should be taken to avoid marking-off through pressure caused by the weight of the cloth, while the colouring matter is in contact with the starch degradation products which have reducing properties.

An outstanding feature of enzyme action is its specific nature, hence with the proper preparation, the steeping process for desizing will have no effect on the cellulose itself nor on the waxes associated with it.

The enzymes appear to be biological catalysts and some of the commoner types are given in the following table.

ENZYMES

Lipases	Split fats into glycerol and fatty acids.
Amylases	Split starches into dextrins and sugars.
Proteases	Split proteins into soluble polypeptides and amino-acids.
Zymases	Convert glucose into alcohol and carbon dioxide.
Catalases	Convert hydrogen peroxide into water and oxygen.
Cellulases	Convert cellulose to soluble degradation products.
Saccharidases	Convert polysaccharides to glucose.

In general, the activity of all the enzymes is affected by concentration, time, temperature and hydrogen-ion concentration or pH. These factors must be specified in any comparisons of enzyme activity. One of the most important factors is that of temperature, and there are very few enzymes which are not inactivated or destroyed by temperatures over 75°C.; further, when destroyed they cannot be revived or reactivated.

SINGEING AND DESIZING

It should also be realised that under the proper conditions, the removal of starch from sized cotton goods by the appropriate enzyme is much more efficient and economic than the use of inorganic acids and alkalis; it is also much safer.

The changes which occur in the decomposition of starch take place in three phases: (a) liquefaction, (b) conversion to dextrin, and (c) conversion to maltose. The liquefaction is generally immediately followed by dextrinisation, and this is readily observed by the gradual change in the coloration with iodine, from blue through violet, reddish-violet to colourless.

Some maltose is formed in the first stage and increased in the second; the ultimate action of amylase on both amylose and amylopectin is the formation of maltose.

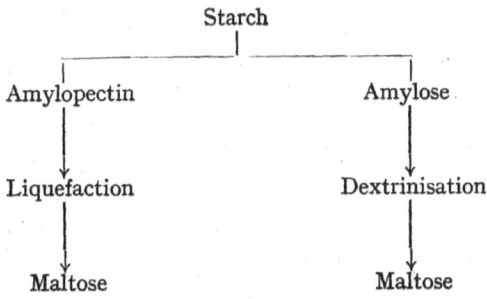

There are two main types of diastase, the animal preparations from pancreas, saliva, blood and liver, and the vegetable preparations which may be sub-divided into the malt diastases like Diastafor, and the bacterial amylases such as Rapidase, Biolase, Arcy and Taka.

The first textile use of malt appears to have been in 1857 when it was employed for the removal of the farinaceous thickeners used in the printing of cotton goods; the printed cloth was steeped for several hours in warm water containing malted barley. About 1900, however, the German Diamalt Co. of Munich introduced Diastafor, which was an extract specially prepared for textile use in a simpler and more effective manner. The malt extract contains two enzymes or diastases, one of which saccharifies starch to maltose, whereas the other converts the starch to soluble starch and dextrin; of the two reactions that of saccharification is the greater.

Until about 1912, the malt extracts were the only diastases commercially available; pancreatic diastases such as Pancreatin, Trypsin, Fermasol, Pancreol and Degomma were then marketed and mainly contained four enzymes, one of which is very sensitive to water. Of the remaining three, trypsin and casease act chiefly on albuminoids and casein, and are of no interest for starch sizes, although they are important for the gelatin size used for rayon; the remaining enzyme is amylopepsin which has great powers of saccharification when allowed to react on starch.

The bacterial diastase was first marketed about 1919 in the form of the well-known product Rapidase; this possesses very rapid action which is mainly liquefying with the production of soluble starch and dextrins. There is practically no saccharification.

The malt enzymes or malt diastases were thus the first to be used on a commercial scale for the removal of starch sizing from cotton textiles; the term "malting" which still persists, is a relic of the terminology of years ago. Malt enzymes are obtained by the extraction of freshly germinated malt, and commercial preparations are available under such trade names as Diastafor, Diastase, Gabalit, Deglatol, Terhyd, Dogmalin, Maltoferment and Maltostase. These preparations act upon starch with the greatest efficiency at a temperature of 60°C. and a pH value of 5·7, but on account of the feeble acidity of cotton itself, the best results for the removal of starch from cotton is at pH 6·7. This is rather fortunate as most croft-waters are slightly alkaline rather than slightly acid, but during the piling of the goods the accumulation of acidic by-products removes this alkalinity; similar effects are observed when the malting liquor is circulated through the grey cloth. As previously mentioned, the optimum temperature for the malt enzymes is 60°C., and this should not be exceeded, for the inactivation is permanent; hence it is better to work at 55 to 60°C. About 5 lbs. of malt extract is required for every ton of cloth.

The pancreatic enzymes are prepared from slaughter-house waste; typical commercial preparations are Novofermasol, Viveral, Degomma, Ultraferman and Dedresan. In general, the maximum activity is reached at pH 7·0 to 7·5 and within a temperature range of 50 to 55°C. It is often claimed that the slight alkalinity required for the best effect offers an advantage over the use of the malt preparations, but this is not necessarily so in actual practice. In presence of fabric, Novofermasol should be maintained between pH 6·5 and 8·0, but requires about 0·5% of common salt to develop the maximum efficiency. Novofermasol develops its maximum activity at 60°C. and again there is a great diminution in activity if the temperature is raised above this point; as the temperature falls however, the loss in activity is less than with the malt preparations.

Bacterial desizing preparations are produced by growing pure cultures of certain micro-organisms in sterilised wort. Typical commercial products are Rapidase and Biolase. The reaction of desizing should be maintained between pH 5·5 and 7·7, so that there is slightly more latitude than with other products. As originally marketed, Rapidase was stated to function well at 90°C. but experience has shown that this temperature is too high and 70°C. should not be exceeded.

Summarising the conditions for the three types of desizing agents in tabular form, gives the following data:

DESIZING PREPARATIONS

	Malt diastase	Pancreatic diastase	Bacterial diastase
Concn. (g./l.)	3 to 20	1 to 3	0·5 to 1
Temperature (°C.)	50 to 60	50 to 60	60 to 70
pH	6·0 to 7·5	6·5 to 7·5	5·5 to 7·5

The Waksman test is often applied for the determination of desizing efficiency, and depends on observations of the thinning of a potato starch paste (J.A.C.S., 1920, *42*, 293). More recently, however, Scott (Ind. Eng. Chem., 1940, *32*, 784) has devised a padding method which approximates to actual practice. At the common temperature of 55°C., the desizing ratings were malt 81, animal 76, and bacterial 70; at optimum temperatures, however, the ratings were malt (60°C.) 83, animal (54°C.) 81, and bacterial (66°C.) 83.

Estimations of the strength of the various diastase preparations have also been outlined by Lenk (J.S.D.C., 1942, *58*, 138), who made use of iodine as an indicator of the degree of conversion; it is possible to determine the end of the liquefaction phase (the decomposition of amylopectin) by finding the starting-point of the dextrinisation phase. In this phase, there are two conspicuous points which may be observed during the series of colour-changes with iodine; these limits are termed $L1$ and $L2$, where the blue colour changes to blue-violet, and the red to yellow respectively.

The following table shows the $L1$ figures obtained with various commercial preparations under the optimum conditions for each class using tapioca starch.

COMPARISON OF DIASTASES

Diastase	$L1$
Degomma	9,100
Diastafor	7,300
Rapidase fluid	1,200
Arcy	250
Vancyme	9,100
Pancreas diastase	10,000

The manipulation of the fabric during the desizing process may follow various alternatives; it is possible to apply to diastase in the trough immediately after the singe but this has the disadvantage that the temperature is apt to be difficult of control. It is also possible to desize on a jig or on some continuous impregnating machine. At least one hour is usually allowed for desizing but more frequently from four to ten hours are taken; with suitable concentrations and control, the operation may be reduced to about 15 minutes and a continuous desizing process employed, with several impregnations to ensure good penetration. The fabric should first be run through hot water to swell the starch and make it more susceptible to hydrolysis.

A common method of using Diastafor is to steep the goods in 0·5% Diastafor solution for 45 minutes at 60 to 65°C. and then rinse with water; alternatively, the goods may be impregnated with the solution, piled, and allowed to lie overnight before rinsing.

The continuous method may be adopted with Rapidase on account of its greater activity under optimum conditions; a simple machine of five compartments with squeezing rollers between each is quite adequate for the purpose. The first and second compartments are for boiling water and the middle compartment for the Rapidase solution (about 10 lbs. per 100 gallons); the fourth and fifth compartments are for hot and cold rinsing waters respectively. The time of treatment in the Rapidase at 70 to 80°C. varies up to 15 seconds according to the type of fabric and other operating conditions. It is, of course, necessary to replenish the middle compartment with Rapidase as it is used.

Before concluding the discussion of desizing processes, it may be remarked that it has been suggested to employ compounds such as the sodium salt of paratoluenesulphochloramide (Aktivin) in the kier; this is quite feasible and efficient. The expense of the reagent has hindered its wide application; further, it is not always convenient to use kiers for this purpose. The various Aktivins are described on page 232.

It is regrettable, but true, that some weavers have a reprehensible habit of lubricating "difficult" warps with paraffin wax when opportunity occurs. The presence of paraffin wax in the fabric creates numerous difficulties; it does not emulsify in the kier but merely melts and deposits itself again on cooling to produce patches which resist water and therefore produce uneven dyeing. If an adequate amount of a suitable dispersing agent were incorporated in the wax beforehand, it might be possible to deal with the difficulty, but wax candles rarely contain sulphated alcohols.

CHAPTER XV
KIERS AND KIERING

THE impurities or non-cellulosic constituents of cotton cannot be fixed with precision, as the amounts vary with the fineness and the variety of the cotton hair, and are influenced by geographical conditions and seasonal factors. Cotton is a seed hair as previously mentioned and contains the highest proportion of cellulose among the natural fibres, in many of which the cellulose is a structural element of the plant, and associated with much greater amounts of "impurities."

Cotton contains a certain amount of hygroscopic moisture which varies according to the relative humidity of the atmosphere but which may be estimated at between 6 and 8% at ordinary humidities; other non-cellulosic impurities amount to between 7 and 11% of the total weight of raw cotton, so that the amount of cellulose is about 80 to 87% of the cotton.

Data collected from various sources indicate the following approximate percentage composition of raw cotton:

COMPOSITION OF RAW COTTON

Cellulose	80 to 85%
Water	6 to 8%
Nitrogeneous matter	1 to 2·8%
Mineral matter	1 to 1·8%
Waxes, etc.	0·5 to 1%
Pectate	0·4 to 1%
Residue of pigment, resin, etc.	3 to 5%

The nitrogenous matter in cotton probably consists of the protoplasmic residues, and the nitrogen content has been examined by many investigators; the work of Ridge (J.T.I., 1924, *15*, 94) showed that the nitrogen content lies between 0·2 and 0·4%, the actual amount being largely determined by the type of cotton. The nitrogenous constituents are readily removed by ordinary scouring but nitrogen estimations are of little value in evaluating the efficiency or the course of the scouring operation. At one time it was assumed that the yellowing of cotton was due to residual nitrogenous impurities, but this defect is now attributed to residual wax, and sometimes to oxycellulose, if unfortunately, such modified cellulose is present.

The mineral matter in cotton exists chiefly in the form of sodium, potassium, calcium and magnesium salts, much of which is soluble in water, but raw cotton direct from the bale is often contaminated with earth and sand. According to Fargher, Hart, and Probert (J.T.I., 1927, *18*, 29)—see page 161—the ash of raw cotton (1·1 to 1·6%), falls to 0·15 to 0·26% after steeping in water, and to between 0·02 and 0·06% after an acid-steep.

The pectic material in raw cotton is present mainly as calcium and magnesium pectates; if the pure pectic acid is isolated from cotton it appears similar to that obtained from other plant sources. The pectates seem to exist in the primary wall of the cotton hair so that the quantity present will vary according to the fineness of the hair; however, when the cotton is boiled with dilute alkali the complex combination is broken down and readily removed.

Before considering the cotton waxes, attention may be drawn to the 3 to 5% of uncharacterised matter which, in addition to resin and pigment, may also include some hemicellulose and modified cellulose. The pigment of cotton is most pronounced in the wild varieties and deepens as the cotton ripens; Egyptian cotton is much more pigmented than the American varieties, with the exception of the well-known "blue bender." The colouring matter of raw cotton is not completely removed by boiling with alkali but needs an oxidation treatment for its destruction.

Cotton wax seems to be the only constituent of raw cotton which is really difficult to remove, and some of it (0·1 to 0·3%) persists through all the scouring and bleaching operations; carbon tetrachloride or benzene will remove the wax and fatty acid, but not the resinous substances, and the best solvent for extraction is chloroform.

Apart from the preliminary work of Schunk in 1868, no systematic examination of the waxy substances in cotton took place until Fargher and Probert (J.T.I., 1923, *14*, 49; 1924, *15*, 337) showed cotton wax to be a mixture of wax alcohols and esters. The material usually estimated as wax, and amounting to between 0·4 and 1% of the raw cotton, is actually a complex mixture of fatty, waxy and resinous substances including wax alcohols, wax esters and wax acids, together with hydrocarbons, fatty acids, resin acids, alcohols and fats. Two alcohols were isolated from the wax of American Upland cotton, gossypyl alcohol, $C_{30}H_{62}O$, and montanyl alcohol, $C_{28}H_{58}O$. Gossypyl alcohol was isomeric with melissyl alcohol, but montanyl alcohol was new. Although these two compounds form the greater part of the cotton wax alcohols, there are also traces of alcohols of the following formulæ, $C_{32}H_{66}O$ and $C_{34}H_{70}O$, together with glycols $C_{30}H_{62}O_2$ and $C_{28}H_{58}O_2$. Most of the alcohols, acids, esters and hydrocarbons of Egyptian cotton are similar to, and often identical with, those of American cottons.

PLATE XI

Fig. 46. Gas singeing machine.

Courtesy of Messrs. Hunt and Moscrop.

To face page 168.

PLATE XII

FIG. 47. Vertical high-pressure kier with multitubular heater and automatic piler.

Courtesy of Messrs. Mather and Platt.

PLATE XIII

FIG. 48. The 'Mather' horizontal kier.
One kier is closed and under pressure, another has the wagons inside but the door is raised; two wagons in the background are being emptied.

Courtesy of Messrs. Mather and Platt.

PLATE XIV

FIG. 49. The Edmeston Bentz continuous kier.

Courtesy of A. Edmeston and Sons, Ltd.

To face page 169.

Cotton wax invariably contains 15 to 20% of free acids—palmitic, stearic and oleic—which can form soaps of high detergent and emulsifying powers; higher fatty acids of low solubility, and resin acids of low detergent powers are also found in the raw cotton wax. The acids are important in connection with the formation of wax emulsions in the scouring liquor and also in connection with the stability of the emulsions. The esters in cotton wax consist partly of fats which are readily saponified, and partly of true waxes which are difficult to saponify; the proportion of unsaponifiable substances in cotton wax is high and consists mainly of the wax alcohols.

Emulsification plays a very considerable part in the removal of wax during the scouring of cotton, and this has been stressed in the work of Knecht and Allen (J.S.D.C., 1911, *27*, 142) and of Scheurer (Bull. Soc. Ind., Mulhouse, 1898, *68*, 24), which mainly dealt with the removal of wax in presence or absence of emulsifying agents.

More direct evidence was provided by Fargher (J.T.I., 1924, *15*, 75 and 120), who compared the properties of the extractive matter in cotton before and after scouring, and also examined the waxy material in the kier liquor for unsaponified waxes. The wax remaining on the cotton still consisted of a mixture of acids, esters and unsaponifiable substances in the same proportions as in the raw material; the spent kier liquor gave a small proportion of esters and contained wax alcohols and acids. Hence scouring appears to be primarily a process of emulsification, but the emulsified material undergoes partial saponification depending on the conditions of treatment.

KIERING

The kiering operation depends on the circulation of hot alkaline liquors, usually under 10 to 30 lbs. pressure, through the regularly packed column of grey fabric for a period of time which may range from 8 to 24 hours.

Before passing to the kiers for the scouring process, the cotton fabric is thoroughly washed to remove those impurities which are, or have been rendered soluble, and so allow the kier-liquor to operate more efficiently. The cloth is manipulated in long lengths which have been turned from the open width into the rope form by passing through the well-known pot-eyes; two strands of cloth generally run together. The cloth is usually impregnated with the scouring liquor before reaching the kiers, and this may be accomplished by passing it through a machine which is very similar to the rope-washing machine described on page 183, but containing lime, soda ash or caustic soda solution according to which of these is intended as the scouring medium. In this manner the cotton is thoroughly impregnated with the liquor before it reaches the kier, and so enables a more even and regular treatment to be given. The old method of filling the kier with cloth was by the plaiting-down operation with the assistance of boys who entered the kier by the manhole at the top and distributed the fabric with a stick, trampling it down with their wooden clogs to form even layers.

The modern method is to use a mechanical piler which gives even and regular piling from top to bottom of the kier, with subsequent improved circulation and freedom from channelling. With the mechanical piler, the cloth may be saturated with alkali and plaited-down in one operation; further, hot liquor may be introduced by the piler with a saving in time. Essentially, the piler consists of a feed pipe and a long oscillating trunk mounted above the kier; the trunk is usually of galvanised sheet steel, but may be made in vulcanite for use with acid solutions. Hot liquor is also fed through the feed pipe and trunking and helps to carry the cloth forward, washing it into the kier. The excess of liquor also runs into the kier by an overflow pipe.

The piling is effected by the simultaneous combination of a rotary motion and an oscillating motion by cams which give even piling and distribute the cloth evenly throughout the kier. The cloth usually passes over a wince of the slipping type which ensures that when the cloth stops, the piler also stops. Even and regular piling is an important contribution to proper scouring.

When the kier has been filled, it is important to see that the goods are kept in position by heavy stones, beams or chains which prevent the cloth from being tossed during scouring; sheeting or sacking is generally placed between the cloth and the weighting devices. Actually, efficient piling methods obviate much of the tossing and also channelling, both of which are largely due to uneven circulation.

Sufficient scouring liquor is admitted to the kier, preferably from below as this sweeps out the air; circulation of the liquor is then started but the valve is left open until all air has been displaced by steam, for otherwise there is a great danger of damage to the cellulose by the formation of oxycellulose which is somewhat rapidly produced by the action of air on warm cotton impregnated with alkali. Hence if the goods are likely to be exposed to the air, as in the open kier, or after draining in a closed kier, it is important that they should be completely immersed in water or the aqueous solution of alkali.

As the kiers are commonly made of iron, it is usual to lime-wash these frequently in order to avoid iron stains on the cotton.

Vertical Kiers

The chief types of kier are cylindrical iron vessels, generally placed vertically, and holding from 500 lbs. of cloth to 5 tons,

KIERS AND KIERING

according to the size of the kier; the two-ton kier is a popular size and is about 9 feet high and 6 feet 6 inches in diameter. The actual capacity of the kier differs slightly from the rated capacity according to the type of fabric. For example, a lightweight fabric will give a denser packing than a heavy cloth; again, a more compact mass of cloth results from hand piling, compared with mechanical piling, although the latter is to be preferred on account of the regularity of the piling.

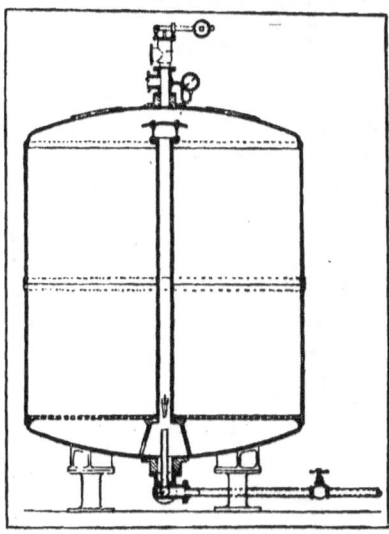

FIG. 50.—Kier with puffer-pipe.

About 80 to 85% of the rated capacity seems adequate, and if this is surpassed there is a danger of over-filling the kier and establishing some hydraulic pressure which will give false readings on the pressure gauge from which the temperature is estimated. Overfilling also compresses the fabric at the top of the kier and hinders circulation, thus increasing the temperature gradient between the top and bottom of the kier. In general one ton of wet cloth containing about 125% of water will require 140 cubic feet of space, but about 12 to 15 inches below the base of the domed top of the vertical kier should be left clear.

Where it may be necessary to kier small amounts of cloth in a large kier, care should be taken to have the kier at least half-full, or the cloth may be turned and almost hopelessly entangled.

The simplest scouring vessel is probably the iron kier, which has a perforated false bottom on which the cloth rests. In the old type

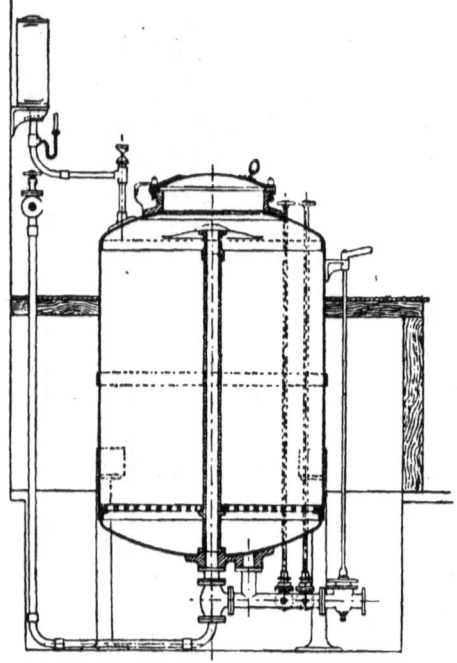

FIG. 51.—Vertical kier with injector heating (Mather & Platt).

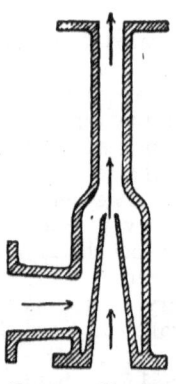

FIG. 52.—Steam injector.

of vomiting kier, a central pipe passed through the grid of the base and an intermittent stream of liquor was forced up the pipe by steam pressure and deflected at the top by a spreader so that it flowed over the goods and down to the bottom of the kier. A diagram of the arrangement is seen in Fig. 50; a vertical puffer-pipe of 4 to 5 inches diameter is attached to a conical base in the centre of which rises a steam-pipe. The liquor rises in the pipe to a height depending on the volume of the liquor in the kier, and the steam discharging into this pipe heats the liquor and expands it until a point is reached where the steam is no longer condensed but blows or vomits the column of liquor up against the spreader, where it is deflected and percolates through the goods to the well of the kier for the process to be repeated.

An alternative method of using steam for circulating and heating the liquor is seen in Fig. 52, which represents the injector system. It will be seen that steam is injected into the pipe supplying the liquor, and as it passes through the nozzle the pressure is turned into velocity and moves the liquor; simultaneously, the steam is condensed so that a column of hot liquor is forced upwards.

In some of the later types of open kier, the pipe supplying the liquor is placed outside the kier, and passes it through a perforated spreader on to the goods.

Although the method of heating by direct steam thus includes circulation, the consequent condensation continually dilutes the liquor and must be allowed for in this particular process. Another disadvantage is that part of the goods comes into contact with very hot liquor and the temperature falls as it percolates through the cloth, tending to give uneven results. Closed steam coils have been used, situated beneath the false bottom of the kier and therefore only in contact with the liquor and not with the cotton. The time of heating is longer, and the liquor has to be circulated by a pump. The modern type of heating is by an external multitubular boiler, the liquor being circulated by a pump.

The open kier is limited in respect of temperature, but closed kiers are available for higher temperatures and hence better scouring; these are made with one or two manholes at the top for entering the fabric; in the earlier types, the injector system of heating and circulation was employed. The early closed kiers are sometimes referred to as "low-pressure" kiers; in these, as in all closed kiers, it is essential to have an efficient safety-valve.

The prototype of the modern high-pressure kier is often referred to as the Walsh kier. The general design permits the constant circulation of the liquor by a centrifugal pump, which withdraws the liquor from the bottom of the kier, passes it through a multitubular heater surrounded by high-pressure steam, and delivers it to the top of the kier where it is showered over the goods. Kiers of this type may work with pressures up to 50 lbs. per sq. in.; they are very suitable for scouring with caustic soda solution as the live steam does not come into contact with the boiling liquor.

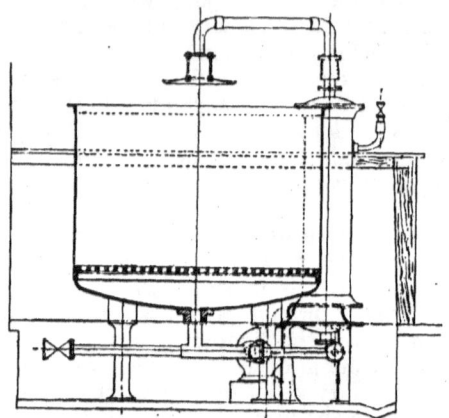

Fig. 53.—Open kier with multitubular heater.

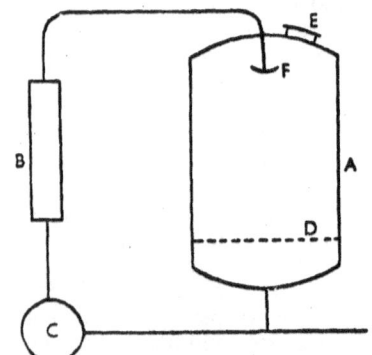

Fig. 54.—Diagram of Kiering operation.
(*A* is the kier, *B* the multitubular heater, *C* the pump, *D* the false bottom of the kier, *E* the manhole, and *F* the spray.)

An interesting variation on the simple closed kier is that in which the boiling liquor is forced in and out of the kier by steam, as for example, in the Pendlebury kier, where a small subsidiary kier is attached to the kier containing the fabric. After the cloth has been piled into the main kier, it is steamed to remove the air; boiling liquor is then admitted and the boiling continued with live steam, after which the liquor is withdrawn to the small kier, and the fabric steamed again. Alternate boiling and steaming were generally continued for from 4 to 6 hours. A modification of this system was seen in the Barlow kier, which comprised two kiers of equal capacity, both of which were filled with cotton; the circulation was effected by steam pressure forcing the liquor out of one kier and into the

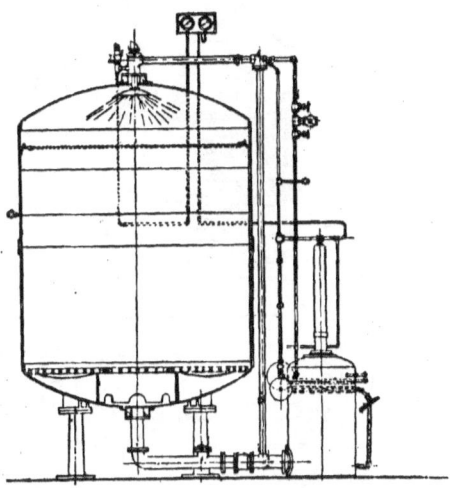

FIG. 55.—The Jefferson-Walker kier.

other every 15 minutes over a period of 4 to 6 hours. A later type is the Jefferson-Walker kier which utilises a vacuum system of circulation. Here, again, there is a small receiver between the kier and the supply pipe; about one-fifth of the liquor in the kier is withdrawn into the small receiver but the supply pipe to the kier itself is closed, and this creates a partial vacuum drawing the fabric together. The liquor is then admitted into the main kier by a rose at the top, when the compressed cloth expands again. In this manner, it is stated that the natural channelling which otherwise occurs, is avoided and a more even treatment is the result. In this system of circulation, the entire liquor in the main kier is circulated every 12 minutes, and the intermittent discharge makes the goods rise and fall with beneficial effects. The filling and emptying of the receiver is regulated automatically.

Reference has already been made to the necessity of regular and even packing in the kier to ensure good circulation of the kier liquor, but a further point of importance is to ensure an even pressure through the kier as far as possible; if the pump is able to pump more liquor than is passing through the cloth, a partial vacuum will occur by the liquor boiling at the bottom of the kier, and some vapour will be pumped. This type of inefficient circulation may be obviated

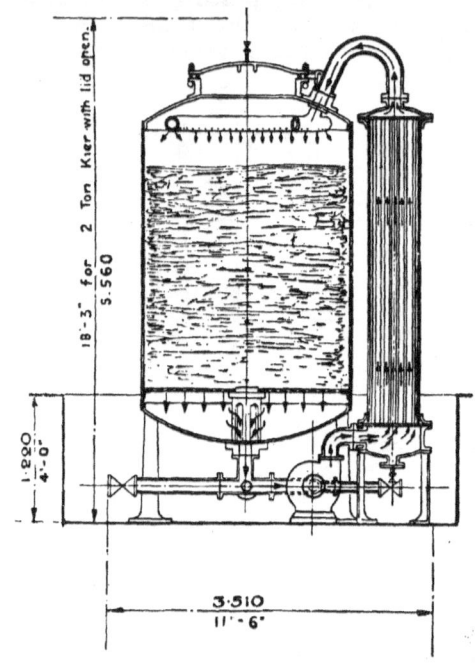

Fig. 56.—Modern vertical kier with multitubular heater (Mather & Platt).

to some extent by arranging pipes between the top and bottom of the kier to equalise the pressure.

Some form of pressure boil is almost essential for cotton goods which are apt to contain particles of cotton seed, husks and leaf, known as motes and shives. These are apt to form dark-coloured particles in the fabric and they are not destroyed by treatment with the bleach-liquors but only by scouring.

The advantages of the pressure boil may be seen by considering the example of a fabric which would require 8 to 12 hours' boiling in an open kier with 2 to 5% NaOH calculated on the weight of the fabric and present to the extent of 74 gallons of liquor per 100 lbs.

of cotton; with a pressure-boil, only 2 to 8 hours' treatment would be necessary.

It may be remarked that the object of heating the kier liquor is not to convert it into steam, but to raise the temperature of the liquid so that the impurities in the cotton are more readily affected.

STEAM PRESSURES AND TEMPERATURES

Pressure Lbs./sq. in.	Temperature °C.	Pressure Lbs./sq. in.	Temperature °C.
0	100	30	134
5	109	32	136
10	115	34	138
15	121	36	139
20	125	38	140
22	127	40	141
24	129	44	144
26	131	48	147
28	133	50	148

Most low-pressure kiers are equipped with a hinged lid, whereas the high-pressure kiers have manholes. Where the mechanical piler is utilised, the manhole must be centrally placed; some high-pressure kiers have two small oval manholes and these kiers must be filled by hand with consequent irregularity in piling.

If, at the end of the boil in the kier, the liquor is run off then the scum accumulates on the top layers of fabric and is apt to adhere quite firmly. Hence it is preferable to "blow-off" the kier before draining and filling with water.

High pressure kiers cannot be treated with impunity and not the least of the difficulties which might have to be encountered is the entanglement of the fabric by "tossing." Most bleachers release the pressure at the bottom of the kier first, and then at the top, when the scum blows off into the drain through a pipe from the top of the kier. When the pressure has been reduced, water may be admitted; many bleachers use hot water and allow it to remain on the cloth for some time before washing by circulation.

An interesting device suggested by Ullman makes use of a jet condenser; at the end of the kier-boil, cold water is forced into the kier at the top and at a pressure not less than the internal kier pressure; at the opposite side of the kier it escapes carrying the scum and sludge with it. As the liquor is still under pressure, it would turn to steam, so it is discharged through a jet condenser as a dirty liquid. The blowing-off is intended to remove the scum with a small volume of water.

Low-pressure kiers generally operate with 5 lbs. pressure and the high-pressure kiers at pressures up to 40 lbs. per sq. in., although there is a tendency within recent years not to exceed 30 lbs.

It will be noted that the kiers previously described make use of a circulation system in which the liquor passes from top to bottom through the fabric; this method is general in England and in the U.S.A. A sectional kier, however, is popular in Germany and utilises a horizontal circulation; it is based on the assumption that irregular results, even with good packing, are due to the pressure exerted by the liquor which is circulated vertically and compresses the fabric, thereby creating difficulties with closely woven goods in particular. With the Gebauer kier (Fig. 57), the cloth more or less

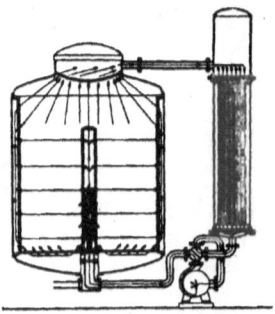 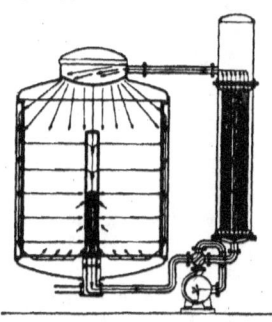

FIG. 57.—The Gebauer kier.

floats in the kier, so the pressure difference between top and bottom is eliminated. It will be noted that the kier is a double-jacketed vessel, the inner jacket being perforated; the liquor passes radially in a horizontal plane and its direction may be reversed if required.

Horizontal Kiers

The horizontal type of kier is well illustrated in the famous Mather kier which differs from all other designs; the kier is made in three sizes to hold 1·25, 2·5 or 3·3 tons of cloth at one time and is constructed to work at 40 lbs. pressure. The horizontal kier is constructed to accommodate two wagons of cloth which are filled and emptied outside the kier; the change of wagons only takes a few minutes and as one set of wagons is under treatment in the kier, the other may be emptied and filled again thus assuring a high output. As the wagons are filled with cloth saturated with caustic soda in the open room, it is possible to obtain better supervision and more even piling than in the confined space of the kier. Further, the horizontal arrangement results in a small depth of fabric and therefore better penetration and circulation. The kier is arranged in a shallow pit so that the wagon lines are level with the floor.

The wagons containing the fabric are generally made of galvanised sheet-iron and provided with a perforated bottom and a telescopic connection to the circulating pipes at the bottom of the kier itself. The circulation of the liquor is generally effected by centrifugal pump, and the heating by an external multitubular heater; in the original models, the older alternatives of injectors or

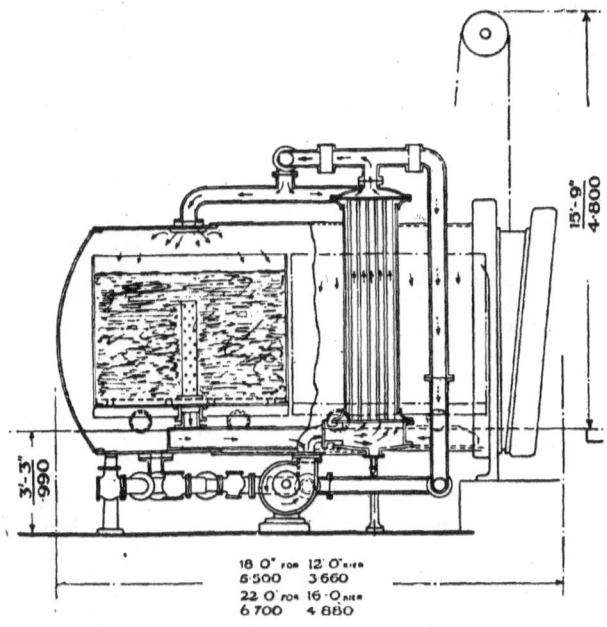

Fig. 58.—Diagram of the Mather kier.

steam-coils were utilised but they are not modern practice. The circulation may be effected equally well in either direction and is usually reversed several times during the boil; the washing after boiling is very effective, for with the circulation in an upward direction it is possible to carry away any accumulated scum or dirt instead of driving them into the fabric.

This type of kier may be used for loose cotton, yarn or fabric; cloths may be treated in rope-form, or in open-width. On account of the fact that it is possible to charge the kier in ten minutes, a high output is assured; even with mechanical pilers, the vertical kiers cannot be filled or emptied at much more than 200–300 yards per minute. The Mather kier is illustrated in Fig. 58.

TEXTILE BLEACHING

Although most cloth is scoured in rope form, two ends running together, yet some fabrics are better treated in open width; heavy, drills, corded fabrics, satins and other weaves are apt to show ropemarks and creases unless scoured in the open form. One of the best known machines for this purpose is the Jackson kier, in which the fabric is maintained free from creases and at full-width on rollers during the scouring process. The fabric is impregnated with alkali and wound on to one of two rollers, mounted on a special frame which goes into a horizontal kier, which is then closed; the ends of the rollers engage with a driving mechanism at the back of the kier so that the fabric may be wound from one roller to another during

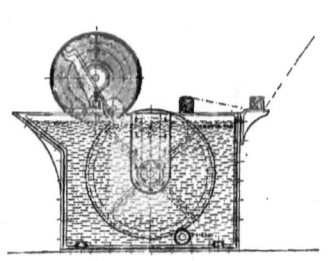

FIG. 59.—Preparing the cloth for the Jackson kier.

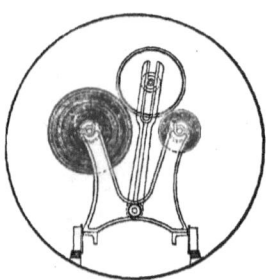

FIG. 60.—Diagram of the Jackson kier.

the boiling process. A boiling solution of 3° Tw. NaOH is forced into the kier by a centrifugal pump and steam-pressure is then applied up to 35 lbs. per sq. in.; the liquor is circulated by the pump, and spread over the fabric as it passes from one roller to the other. (It may be remarked that in the later stages of scouring, washing, chemicking and so forth, the fabric remains on the batch rollers and is wound from one to the other in much the same manner as when in the kier.)

The batches are made to about 40 inches in diameter and the kiers are made in various widths of which 85 inches is probably the most useful; it will deal with 15 cwts. to one ton of cloth in one operation.

It will be remembered that the original "grassing" methods were a form of open-width bleaching; these were followed at a much later date by scouring on jigs, but this was soon found to be dangerous owing to the formation of oxycellulose with 18 to 20% reduction in strength where caustic alkali was used.

General

Various processes have been suggested for scouring in open-width whereby the cloth is plaited or cuttled. One of the best-known systems is that of Tagliani-Rigamonti, sometimes called the Bemberg

KIERS AND KIERING

method in Germany. The fabric passes down a high column of liquor and into the pressure chamber or kier in which is situated a U-shaped transporter made of two conveyor chains with poles; the cloth is plaited into the U and is sprayed on both sides at the entrance and exit. A considerable amount of cloth accumulates in the kier through which it takes about 2 hours to pass; when scoured, the fabric is drawn away through the same column of liquor by which it entered. This method of scouring formed part of a continuous system of cotton bleaching.

Another interesting method of treating cloth in open-width on rollers was suggested by Thies about 1900; the fabric was wound on a drum with supporting bars, but after the thickness of the winding became about 4 inches, fresh bars were inserted to provide openings through which the scouring liquor could be circulated in the ordinary manner.

A well-known form of open-width and continuous bleaching is the Edmeston-Bentz system which originated in B.P. 5,590 of 1889. The process is applicable to cloth in open-width or in rope-form, and also forms part of a continuous system of bleaching.

The kier itself is a strong cast-iron box, at each end of which is a division-plate extending downwards almost to the bottom; the scouring liquor is fed into the bottom of the box and maintained at such a level to form a seal between the inner chamber and the space outside the partitions. Inside the box, arrangements are made whereby the cloth is saturated with caustic soda solution and steamed as it passes continuously through the machine, being subjected alternately to the action of the boiling liquor and steam; only moderate pressures can be reached within the kier on account of the liquid seals. The cloth cannot become dry, and as air is excluded the two chief sources of damage are avoided.

When this apparatus is used for the treatment of cloth in rope-form it is possible to run four ropes side by side at a speed of 30 to 50 yards per minute according to the weight of the fabric. As previously mentioned, this kier forms part of a continuous system of scouring and bleaching; it is illustrated in Fig. 49.

Numerous suggestions have been made for the treatment of special fabrics, such as velvets and cords, in open-width and generally depend on some application of piling-cisterns which may lend themselves to continuous treatment.

Some interesting account of developments in kiers and other machines for the bleach-croft have been given by Grunert (J.S.D.C., 1933, *49*, 285) and by Kershaw and Barrett (J.S.D.C., 1934, Jubilee issue, 90). No account of bleaching would be complete without reference to the work of Matthews (*Bleaching and Related Processes*, The Chemical Catalog Co., New York, 1921).

182   TEXTILE BLEACHING

In addition to the improvements in the design and operation of kiers, some of the steeping, souring and washing processes have been "mechanised" by time-wheels and semi-automatic devices such as the Gantt piler for continuous treatment.

It must be remembered that kiering is not essential for all goods; rayons and coloured-woven cottons are often scoured in rope form with soap and sodium carbonate solutions on a wince machine or rope-washing machine.

The TIME WHEEL may be used in many continuous processes; as a Wheel Kier (supplied by Hunt and Moscrop) it is capable of dealing with the cloth in open-width at a speed of 35 yards per minute to give a scouring period of two hours' duration. The fabric is laid across the space between the inner and outer shell of the wheel, the lower semicircle of which is built up of detachable plates as shown in Fig. 62. Every 10 or 15 minutes one of these plates is removed from the back of the wheel as it emerges over the top of the tank and refixed at the front to keep the semicircle intact. The wheel is not driven by power, but, being balanced, it rotates slowly as the weight of cloth increases at the entrance and decreases at the exit—once filled, the fabric runs continuously.

The DASH-WHEEL must not be confused with the time wheel. The former is a machine for the treatment of towelling and similar material and is driven. The goods are usually charged into the machine in bundles and as the wheel revolves, the bundles move about freely inside the compartment while being sprayed with the scouring liquor (or water when being washed). The wheels are generally situated over a pit to facilitate drainage.

The Gantt piler is not generally used for scouring, but finds employment for continuous souring or chemicking (see p. 207).

During the past few years, attempts have been made to dispense with the kier in favour of a steaming process in presence of caustic soda; these developments come from the U.S.A. and form part of continuous bleaching processes described on page 284.

WASHING

The treatment of piece goods is no exception to the general rule that thorough washing is essential at several stages in the bleaching process; in particular, the well-known dangers of residual acid or hypochlorite necessitate their removal from the goods.

Two main types of washing-machine are generally seen in bleach crofts for the treatment of cloth in rope form, (*a*) that in which the rope is "tight" and the machine has a shallow trough, and (*b*) the slack washing-machine with a deep box or trough. Both of these machines are used for the removal of any alkali, lime, acid, chemic

PLATE XV

Fig. 61. The dash wheel.

Courtesy of Messrs. Mather and Platt.

To face page 182.

PLATE XVI

FIG. 62. The time wheel.

Courtesy of Messrs. Hunt and Moscrop.

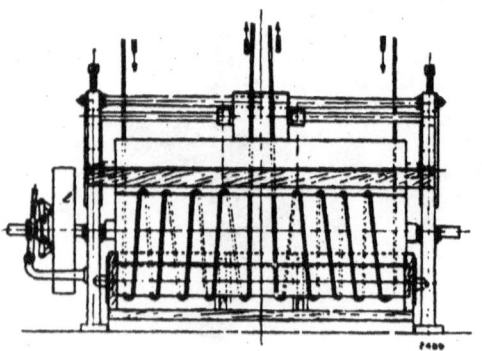

FIG. 63. Diagram of roller washing machine.

KIERS AND KIERING

or acid again, and consist of the usual side frames and a pair of strong wide bowls usually made of wood, although it is possible to have rubber, brass or other suitable material; rope-covered bowls are also common. The machines are very similar to those which may be used for liming, chemicking and souring.

Two strands of cloth are generally washed at the same time, one entering at each end of the machine through a pot-eye into the nip of the bowls and passing under the trough rollers into the nip again being guided by hardwood or porcelain pegs on a rail. The passage through the machine continues in a spiral fashion to the centre, where both strands finally leave the washing process. It is also possible to wash one strand at a time, in which case the goods pass spirally through the machine from one side to the other, along the entire length of the bowls. The peg rail is often fitted with a traverse motion which ensures the cloth passing evenly over the whole surface and eliminates the formation of grooves in the bowls. The loading may be by either lever or spring pressure, the latter being preferred for goods which do not permit heavy pressures. The output of these machines is about 400 yards per minute. A counter-current system of washing may be utilised.

In some cases, roller washing machines are built with a small jockey squeezing bowl in the centre to give an additional squeeze to the cloth before it leaves the machine.

A tensionless roller washing machine is provided by Messrs. Mather and Platt, where the cloth from the nip passes over a wince, under the trough rollers and into the nip again. When the cloth is threaded spirally through the machine, the tension is eliminated by adjusting the sliding wince. Excessive tension or excessive slack can be relieved by the sliding wince.

Whereas the ordinary roller washing-machine has a shallow trough, the slack washing-machine has a deep trough with a sloping base down which the cloth falls in a slack slightly plaited state. The cloth is rippled in a slack and tensionless manner into the box by a wince.

For very delicate fabrics, the elliptical wince machine may be used or, alternatively, a special slack washing machine with hexagonal rollers can be employed in which the wash water is sprayed on to the strands of cloth as they pass from the winces into the trough.

After the final washing process, most cloths in rope form are passed through a squeezing machine which is made of two bowls about 15 inches in width; the two ropes or strands of cloth pass between the bowls which are subjected to heavy pressure by compound levers and weights or by springs for more delicate goods. The bowls are generally made of compressed coconut fibre, but sometimes the bottom bowl is brass and the top bowl sycamore or cotton (see Fig. 111).

TEXTILE BLEACHING

The WINCE MACHINE is often used for the scouring (and washing) of the more delicate fabrics, particularly those containing rayon. A large elliptical wince is usually employed to give a plaiting motion to each strand of cloth as it enters the liquor. According to the width of the machine, several pieces of cloth may be threaded through the machine and treated at one time; each piece passes between pegs and over the wince, after which the ends are stitched together forming a continuous loop with a considerable amount of "slack" which gives an ample period of immersion in the liquor.

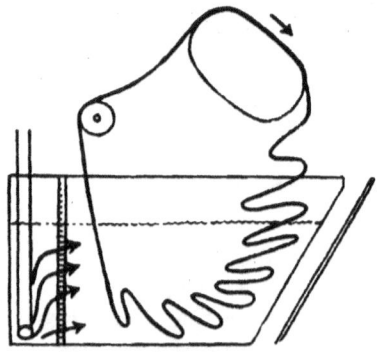

FIG. 64.—Diagram of wince machine.

The pieces are kept apart by hardwood pegs and any tendency to float is overcome by a wooden partition. Entanglement is also avoided by a perforated wooden partition at the front forming a compartment in which additional liquor may be introduced as required and heated by a special nozzle which reduces turbulence to a minimum.

In the Mather and Platt machine the rail on which the hardwood pegs are fitted is arranged to swivel, so that any entanglement will bring into operation an automatic stop-motion between the rail and the driving pulley.

When scouring or washing is in progress, the elliptical wince plaits the material in folds in the liquor, and at the same time the material is withdrawn from the liquor at the opposite end of the beck; the sloping end is open to direct the strands of cloth delivered by the elliptical wince. At the end of the process, the machine is stopped and the pieces separated so that one end of each piece is suspended on the delivery side of the elliptical wince and a draining board brought into position; the machine then withdraws the pieces from the beck by means of the elliptical wince plaiting them on to the draining-board, from which they are later passed over the delivery wince and piled into wagons.

CHAPTER XVI
THE SCOURING PROCESS

A GOOD detergent must wet the greasy surface of the material which has to be cleansed, it must emulsify the fatty or greasy matter, and stabilise the emulsified grease and suspended dirt; the last point is important, for unless the emulsion or suspension is stable there will be a tendency for the impurities to be deposited again.

As already indicated, one of the great textile advantages of cotton lies in its resistance to alkaline liquors; cotton forms the washing fabrics *par excellence*. This property enables the cleansing of the raw material to be undertaken in a manner which is not possible with wool or silk; indeed, the scouring process for cotton would dissolve the woollen material.

The use of alkali in the scouring process for cotton may be considered under three heads:

(*a*) the lime-acid-soda ash sequence;
(*b*) the caustic boil;
(*c*) treatment with soap and soda ash for coloured-woven goods.

The modern tendency is undoubtedly towards the use of caustic soda solutions of 1 to 2% concentration.

The main object of the scouring process is to remove the cotton wax, and in this connection the relative efficiencies of the lime and soda boils have often been discussed. There appears to be little doubt that for a single process, the soda boil has the greater effect; Trotman and Thorp (*Bleaching and Finishing Cotton Goods*; Griffin, London, 1918, p. 95) provided the following comparison:

COMPARISON OF LIME AND SODA BOIL

	I		II		III	
	Soda	Lime	Soda	Lime	Soda	Lime
Ash (%)	0·26	0·52	0·26	0·50	0·42	0·56
Free fat (%)	0·10	0·26	0·20	0·15	0·16	0·11
Fatty acids (as soap)	0·16	0·22	0·13	0·31	0·26	0·56
Nitrogen (%)	0·05	0·07	—	—	0·07	0·07

· A later comparison has been made by Fargher and Higginbotham (J.T.I., 1926, *17*, 233), from which the following data are taken:

COMPARISON OF WATER, LIME, SODA ASH AND CAUSTIC SODA FOR SIX HOURS

	Loss in wt.	Wax content	Methylene Blue	Copper Number
Grey	—	0·49	5·49	1·15
Water (25°C.)	2·4	0·49	5·40	0·34
Water (125°C.)	4·2	0·48	1·85	0·17
Lime (0·7% at 100°C.)	5·0	0·40	1·22	0·08
Na_2CO_3 (1·3% at 100°C.)	5·1	0·49	1·17	0·08
NaOH (1% at 100°C.)	5·2	0·36	1·10	0·03
Lime (0·7% at 125°C.)	6·1	0·28	1·12	0·09
Na_2CO_3 (1·3% at 125°C.)	6·6	0·31	0·87	0·07
NaOH (1% at 125°C.)	7·0	0·20	0·86	0·01

The temperature 125°C. corresponds to 20 lbs. pressure.

The superiority of the caustic soda scour is clearly established, and it is estimated that about 75% of modern kiering processes employ it. The pioneer work of Scheurer (Bull. Soc. Ind. Mulhouse, 1888, p. 399) emphasised the efficiency of NaOH for a single process, but drew attention to the rapidity of the action of lime, and stated that the complete saponification of the wax could only be effected by a subsequent acidification and boiling with sodium carbonate solution.

Scheurer stated that saponification of the fatty constituents of cotton could be effected by (*a*) a single treatment with NaOH and rosin, or (*b*) by the lime-acid-soda ash sequence, the latter being preferable on account of its greater elasticity and certainty.

Higgins (J.T.I., 1916, *7*, 30) has pointed out that the alleged efficiency of the lime boil is not a mere survival of the days of empiricism, but may be the best treatment for certain classes of goods. During this boil, a lime soap is formed with the saponified portion of the wax and this soap is decomposed in the subsequent acidification, with the result that the goods proceed to the boiling with soda ash in a state where the remaining cotton wax is closely associated with free fatty acid. The alkali of the boiling liquor produces a soluble soap in the immediate neighbourhood of the wax which is therefore readily emulsified. This view is associated with the experiments of Shorter (J.S.D.C., 1918, *34*, 137), who showed practically instantaneous emulsification by the addition of an alkali to a mixture of a mineral oil and a fatty acid.

Higgins (*Bleaching*, 1919, p. 39) gives the following figures in support of the preference of the lime method in certain cases; the data refer to treatments of hanks of linen yarn:

Wax Removal

Treatment	Residual wax
Lime, lye, lye	0·06%
Lye, lime, lye	0·046%
Lye, lye, lye	0·16%

Although the lime boil has given place to the caustic boil, yet it has not entirely disappeared. It is customary to use about 2 to 4% of lime (on the weight of the cloth) as $Ca(OH)_2$ where kiering pressure of 30 lbs. per square inch are concerned. The actual consumption of lime appears to be about 2% of the weight of the cloth, but a further 1% is necessary to maintain saturation. The spent liquor dissolves more lime than does water at the same temperature.

It is usual to estimate for the use of 45 to 50 lbs. of calcium oxide per ton of cloth, but more than this is necessary with low-pressure kiering.

The "lime" is generally applied by a liming machine, and the multi-nip type is to be preferred, on account of the better and regular impregnation; it must be remembered that the cloth may act as a filter for undissolved lime so that an even distribution before the cloth goes into the kier is highly desirable. The liquor is prepared so as to contain 5 to 10 grams of CaO per litre.

Questions of efficiency apart, it should not be forgotten that the lime boil is only the first stage in the lime-acid-soda ash sequence which occupies a great deal of time; kiers are slow to fill with fabric and take a long time to heat to the temperature of the boil. Hence the caustic boil has a great advantage.

Within recent times, the caustic boil has been examined in some detail, with particular reference to the fact that the aim of the process is to remove the natural fats and waxes and leave a relatively pure and absorbent product.

Some interesting figures have been provided by Fargher and Higginbotham (J.T.I., 1927, *18*, 283), who employed chloroform for the estimation of fats, waxes and resins in cotton—the standard method being to extract for three hours in the Soxhlet apparatus. The data refer to cotton *yarn*.

Wax Removal

Treatment	Chloroform extract (%)
None (grey)	0·46
3% NaOH, 40 lbs.	0·25
1% NaOH, 20 lbs.	0·30
1% NaOH, open kier	0·34
Acid steep, 1% NaOH, open kier	0·22
Acid steep, 1% NaOH, 20 lbs.	0·21
1% NaOH, 0·6% soap, open kier	0·14
1% NaOH, 0·3% monopole soap, open kier	0·17
1% NaOH, 0·3% coconut oil soap, open kier	0·05

TEXTILE BLEACHING

The chief variables in the process are the temperature of boiling, the time of boiling, and the concentration of NaOH. These factors have also been investigated by Fargher and his co-workers, and some of the results are shown in the following table:

SCOURING WITH CAUSTIC SODA

Treatment	American Scouring Loss %	Wax %	Egyptian Scouring Loss %	Wax %	Indian Scouring Loss %	Wax %
1% NaOH, 6 hours at 50°C.	4·1	0·49	4·7	0·53	6·0	0·42
100°C.	5·2	0·36	6·8	0·40	7·7	0·37
116°C.	6·9	0·21	7·7	0·25	8·1	0·29
125°C.	7·0	0·20	8·0	0·26	8·5	0·28
134°C.	7·2	0·18	8·6	0·20	8·6	0·24
141°C.	7·1	0·17	8·7	0·21	9·0	0·22
1% NaOH, 125°C., 2 hours	6·6	0·30	7·8	0·29	8·1	0·34
4 hours	6·7	0·22	7·9	0·26	8·2	0·28
6 hours	7·0	0·20	8·0	0·26	8·5	0·28
12 hours	7·1	0·23	8·3	0·25	9·0	0·27
6 hours at 100°C., 1% NaOH	5·2	0·36	6·8	0·40	7·7	0·37
2% NaOH	6·3	0·26	7·3	0·26	8·0	0·26
3% NaOH	6·4	0·19	7·5	0·21	8·1	0·26
125°C., 0·5% NaOH	6·7	0·20	7·8	0·29	8·1	0·31
1·0% NaOH	7·0	0·20	8·0	0·26	8·5	0·28
2·0% NaOH	7·0	0·28	8·2	0·28	9·0	0·29
3·0% NaOH	7·3	0·24	8·5	0·26	9·1	0·30
141°C., 1% NaOH	7·1	0·17	8·6	0·21	9·0	0·22
3% NaOH	8·1	0·32	9·6	0·33	10·4	0·29

From the above data, it appears that the greater part of the non-cellulose matter is removed by open kiering at 100°C.; the effect of increasing temperature over the range of 100 to 140°C. is slightly to increase the scouring losses. The small changes, however, are apt to be very significant in the final product.

It will also be seen that the loss in weight increases slowly as the concentration of the NaOH is raised; this is of interest in view of the popular conception that an increase in the concentration of NaOH is objectionable as it leads to a considerable loss in weight. The effect of increased concentration of NaOH is comparable with that of increased temperature.

Some later work by Kollman (Textilber., 1937, *18*, 994) bears on the question of the scouring efficiency of various alkaline solutions. Sodium carbonate is definitely inferior to caustic soda in equivalent concentration; the results with milk of lime are also interesting.

THE SCOURING PROCESS

WAX REMOVAL

Treatment	Wax content
1% CaO for 2 hours at 20 lbs.	0.39%
As above, soured, and boiled in 1% Na_2CO_3, 2 hours at 20 lbs.	0.15%
Boiled 2 hours without pressure in 0.5% NaOH and 0.5% CaO	0.28%

One of the chief reasons for removing cotton wax during the scouring treatment is to render the material absorbent; in many cases, absorbency is just as important as whiteness in the bleached product. Beadle and Stevens (J.S.C.I., 1913, *32*, 174) measured the time of wetting of cotton which had been subjected to various scouring processes.

ABSORBENCY

Treatment	Wetting time
None	Over 24 hours
Bleach, without scour	31.3 secs.
Boil in 1% NaOH	12.3 ,,
Boil in 2% NaOH	5.7 ,,
Boil, bleach and boil	4.0 ,,
Extract with ether and alcohol	0.5 ,,

Kollman (loc. cit.) has compared the effects of various scours in respect of wax removal and absorbency; the most absorbent fabric is not always that with the lowest wax content—apparently, the distribution of the cotton wax is as important as the amount when considering the quantities remaining after scouring.

WAX REMOVAL AND ABSORBENCY

Treatment	Time of wetting	Wax content
Kiering in 1% NaOH		
2 hours without pressure	8.8 secs.	0.36%
2 hours at 30 lbs. pressure	3.1 secs.	0.28%
4 hours at 30 lbs. pressure	2.4 secs.	0.26%
6 hours at 30 lbs. pressure	2.0 secs.	0.22%
Kiering in 0.66% NaOH and 0.43% Na_2CO_3 for 2 hours without pressure	>1 min.	0.24%
Kiering in 1.33% Na_2CO_3 for 2 hours without pressure	>1 min.	0.27%
Kiering in 1% NaOH, once	<1 sec.	0.19%
Kiering in 1% NaOH, twice	<1 sec.	0.12%

WAX REMOVAL AND ABSORBENCY

Treatment	Time of wetting	Wax content
Kiering for 2 hours at 125°C. in—		
1% NaOH	3·1 secs.	0·28%
1% NaOH+0·1% Monopol soap	>1 min.	0·38%
Sodium aluminate	>1 min.	0·34%
0·9% NaOH, 0·1% Monopol soap and 0·1% sodium aluminate	1·8 secs.	0·12%

After comparing the various treatments in respect of their effect on the cotton wax and also on the absorbency of the cotton, Kollman arranged them in order of decreasing efficiency, as shown in the table.

WAX REMOVAL AND ABSORBENCY

Wax removal		Absorbency
1. Alcoholic NaOH	0·05%	Two kierings in 1% NaOH
2. Alcohol	0·10%	As (4) opposite
3. 0·9% NaOH, 0·1% soap, 0·1% aluminate	0·12%	As (5) opposite
		One kiering in 1% NaOH
4. 1% NaOH following Br vapour	0·14%	As (6) opposite
5. 1% NaOH following chlorine water	0·15%	Kiering in 1% NaOH after first chemicking
6. Lime and soda ash	0·15%	

From these results it seems that the scouring process for minimum wax content does not necessarily produce a fabric of maximum absorbency.

It must be remembered that the presence of fats and waxes affects the handle of the goods, and it may be that the alleged superiority of the lime scour was due to the efficient removal of waxy matter; where the wax content is reduced to 0·1% or less the goods have a crisp handle. More often, however, scouring is only taken to the stage where about two-thirds of the wax is removed, i.e. the wax content of the scoured fabric is about 0·25%, and this gives a more acceptable handle to many goods.

Although the saponifying powers of lime appear to be very remarkable and there is some saving in the cost of chemicals, yet the so-called lime boil actually consists of several processes so that the costs for time, labour and manipulation are greater.

CAUSTIC BOIL

Within the last twenty years, the caustic boil has superseded the lime-ash sequence; for the lime boil about 56 lbs. of lime was required

THE SCOURING PROCESS

per 100 gallons, but with the caustic boil it is common to use about 1% NaOH. In both cases about 60 gallons of liquor are required per 100 lbs. of cotton.

It is important to have adequate caustic soda in the kier, for during the process some of the alkali is absorbed by the cotton and some is neutralised by the material extracted from the raw cotton; if the concentration becomes too low, then the impurities are no longer maintained in colloidal solution and precipitation occurs. It is customary to use 2 to 4°Tw. NaOH and ensure that the concentration has not fallen below 0·7°Tw. at the end of the boil. Whilst it is not possible to examine the fabric during the kiering process, yet the aqueous extract may be tested; titration with potassium dichromate solution enables the carbohydrates removed to be expressed as glucose, and a constant value denotes the end of the operation.

Similarly, it is possible to estimate the alkalinity of the kier liquor during the scouring process.

To some extent, there is a tendency to vary the concentration of the kier liquor according to the type of cloth. American practice, for instance, favours 2 to 3°Tw. NaOH solutions (0·9 to 1·35%) for lightweight fabrics, and solutions of 4 to 5°Tw. (1·75 to 2·18%) for heavier material. These values usually refer to the strength of the kier liquor as prepared for application to the cloth; as the fabric is wet, however, there will be dilution by the water in the cloth, so that the true initial strength of alkali is not easy to determine until the liquor has been circulated for some time. For instance, with fabric containing 150% of its weight of water the concentration of the kier liquor may be reduced from 4°Tw. to 2·7° Tw.

Estimating the concentration of NaOH on the weight of the fabric may be more convenient for many purposes; there is a general tendency to keep the amount of alkali to 3% or less, as it is feared that the loss in weight will increase considerably if the concentration of NaOH is high. Clark (*Finishing Materials*, W. R. C. Smith Publishing Co., Atlanta, Georgia, 1939) states that at least one bleach croft uses 4% NaOH on the weight of the goods. There seems to be an impression amongst many bleachers that the lowest possible concentration of NaOH should be used, as the minimum of alkali will mean the minimum loss in weight. It will be seen, however, from the table on page 188, that there is little difference between 0·5% and 3% NaOH in this respect. Many English bleach works use between 0·5 and 1·5% NaOH; the upper figure gives an improved product, but if the concentration falls below the lower limit, the results are definitely inferior.

German practice seems broadly similar; kier liquors of 8 to 12 grams of NaOH per litre are common, corresponding to between

3 and 5% NaOH estimated on the weight of the goods. The liquor to cloth ratio is usually 4 : 1 or 5 : 1.

The time of boiling appears to follow no set rule, and is determined to some extent by the temperature. Rapid and efficient wetting of the fabric is of great importance in connection with the time of scouring and a thorough desizing treatment beforehand will result in more rapid kiering. It should be realised that a kier may take from 1·5 to 2 hours to come to the boil at atmospheric pressure; the air is allowed to escape through a pipe or safety-valve during this heating and is then closed. The establishment of the higher temperature (e.g. 130°C.), as indicated by pressure, takes further time, so that true boiling under pressure may not begin until 2 to 3 hours after the start of the process. Here, again, a thoroughly desized fabric is an advantage as it permits more rapid circulation and therefore more rapid heating.

Temperatures of 100 to 125°C. are regularly used and the higher temperature is preferred because the action is more rapid and the wax and non-saponifiable matter is dispersed more readily. The velocity of reaction is doubled for a rise of 10°C. in temperature.

The time of boil after the required pressure has been reached may vary from 4 hours upwards; liquors stronger than 1% NaOH do not appear to shorten the time of actual boil, but the addition of kiering assistants or wetting agents is of value in this connection. The time of boil will naturally vary according to the type of goods, but a period of 8 hours seems a common one.

In the Mather kier, there are two wagons in which the cloth is piled, so that in effect there are two loads of cloth; hence a shorter period of time is required than with the same weight of fabric in a vertical kier.

The average rate of circulating the liquor appears to be of the order of 12,000 gallons per hour for a 4-ton kier; there is little advantage in higher rates of circulation, but the efficiency of the scour is reduced at lower rates of circulation.

The improved rate, however, brings difficulties with it, and channelling may occur, for circulation is still a problem with some kiers; automatic pilers are an advantage. On account of the large amount of cloth in the kier, there is a temperature gradient down the side and time is necessary to complete the kiering.

Some fabrics may be scoured by a single kier boil with caustic soda, but it is often preferable to give two boils and "turn the cloth over" between the boils by running from one kier to another; in this manner the fabric which was at the top in the first kier becomes the lower part in the second kier. This method also provides an opportunity for a more thorough wash than may be given in the kier itself. Again, although the souring process is an essential part

THE SCOURING PROCESS

of the lime-ash sequence, and is not essential with the caustic boil, it is common knowledge that a better colour is obtained if the goods are soured after a caustic boil.

A single boil is rarely satisfactory for goods which are to be dyed with the vat colours.

It must be confessed that even modern kier-boiling cannot be regarded as a highly efficient process, and is still the dearest part of the bleaching routine; kiers form the bottle-neck in the bleach works and prevent continuous methods being developed to a greater extent. Many attempts have been made to improve the efficiency, and hence the rapidity of the kiering process.

Auxiliaries

Within recent times, there has been great activity in the provision of assistants or auxiliaries for almost every textile process, and naturally the bleaching process was no exception.

Actually, the use of kier-assistants is old, for rosin soap was introduced by Sykes of Edgley into the caustic boil about 1845. Soap has very powerful emulsifying properties; for instance, Knecht (J.S.D.C., 1911, *27*, 142) showed that the amount of wax removed in four hours' boiling with 2°Tw. NaOH solution was doubled when 5% Marseilles soap (on the weight of cotton) was added to the kier liquor. Unfortunately most soaps are not stable to hard water and difficulties are apt to arise; a scoured cloth may appear white, but the presence of lime soaps renders it in a poor condition for subsequent dyeing. Turkey red oil, which is the sodium salt of the sulphuric acid ester of ricinoleic acid, is more stable to hard water, but its scouring action and emulsifying power are not very good; even the highly sulphonated oils lack detergent power although they are not precipitated by dilute acids.

Rosin soap may be prepared by boiling caustic soda and rosin or colophony until the latter is completely saponified. A typical example is to use 10 parts of rosin, 2·5 parts of caustic soda and 100 parts of water and boil gently for about 2 hours.

Other recipes are 4 parts of sodium carbonate and 1 part of colophony, or 3 parts of caustic soda and 1 part of colophony.

Rosin soap is not utilised to as large an extent as formerly, because of the use of caustic soda at high temperatures in suitable kiers, and also on account of the growth of other textile auxiliaries.

Rosin soap is still used to some extent, as it possesses good lathering and detergent powers, but it necessitates a very thorough washing with soft water; Trotman (loc. cit.) prefers not to use rosin soap for the best quality of white production, and states that it is very difficult to remove with hard water. Nowadays many other assistants are available.

A satisfactory kier-assistant must not only be stable to hard water, but it must have good wetting power, good detergency and be capable of emulsifying the impurities in cotton and maintaining these in the dispersed state during scouring and washing. For many years no such substances were available, but about 1924 a number of preparations were marketed to assist penetration and act as fat solvents, during scouring. Many of the products consisted of hydrogenated aromatic hydrocarbons or mixtures of these with the sulphonated oils. Hexalin (cyclohexanol), Methylhexalin (a mixture of three isomeric methyl cyclohexanols), Tetralin (tetrahydronaphthalene), Decalin (decahydronaphthalene) formed the basis of many trade products. These solvents, however, seem to have little value in scouring under pressure.

The alkyl derivatives of naphthalene sulphonic acids were utilised in products such as Nekal, Leonil and Neopermin; whilst emulsified pine oil formed part of Perminal K.B. Many of these preparations were very useful in their wetting power, enabling more thorough penetration of the kier liquors to produce better scouring action. Apart from the pine oil preparations, the dispersion of lime soaps remained a difficulty until about 1930 in so far as textile assistants were concerned; the situation changed, however, with the introduction of auxiliary products with high detergent power and good dispersing action, as described elsewhere (see p. 133).

The two chief types for use in scouring cotton materials are the sulphated fatty alcohols, such as the Gardinols, Lissapols, Lorols and Ocenols, and the fatty acid amides of which Igepon T ($C_{17}H_{33}.CO.NH.C_2H_4.SO_3Na$) is the best-known example. The uses of these substances, generally 0·5 lb. per 100 gallons, have been described on numerous occasions; interesting comments have been made by Briscoe (J.S.D.C., 1933, *49*, 71), Kertess (ibid., 69; 1936, *52*, 42), Dunbar (ibid., 1934, *50*, 309), and Hannay (ibid., 273); other papers have been contributed by Kling (Dyer, 1931, *65*, 315), Nusslein (ibid., 320), Evans (J.S.D.C., 1935, *51*, 233) and others.

Hannay (loc. cit.) states that the new agents are of superior efficiency and offer a great advance in respect of wideness of application, stability and adaptability, but not to the great extent which is sometimes claimed. Scholefield and Ward (J.S.D.C., 1935, *51*, 172) found that Lissapol (0·2%) when added to 1% NaOH solution for scouring was definitely superior to rosin soap in respect of the white cloth produced and as good for wax removal.

The difficulties of calcium and magnesium soaps may be dealt with by a water-softening plant, or by the use of Calgon (sodium hexametaphosphate); Trilons A and B (J.S.D.C., 1940, *56*, 473) appear to be the organic counterparts.

Whilst many auxiliaries are classed together as wetting agents,

THE SCOURING PROCESS

their mere wetting power is not particularly valuable, as shown by Kollman (Textilber, 1938, *18*, 994), when solutions of Nekal BX and Hydralin were little superior to water alone. It must also be remembered that kier-assistants are used at high temperatures, with a low liquor to cloth ratio, whereas most tests of wetting power are made at room temperature with a very high liquor to cloth ratio. Detergent value is of great consequence, and in this respect the sulphated products $R.O.SO_3Na$, are definitely superior to the sulphonated compounds, $R.SO_3.Na$, but probably the most important requirement is the ability to maintain the emulsified impurities in the dispersed state and not allow them to be reabsorbed on the cloth before their removal by water.

Kier-assistants have been examined by Munch (Textilber., 1934, *15*, 558), Meyer (Am. Dyes. Rep., 1932, *21*, 90), Hart (ibid., 1934, *23*, 646), Brandt (Melliand, 1931, *3*, 409) and Clark (Cotton, 1931, *95*, 1318), among many others; Igepon T seems a popular favourite.

Victoroff (Textilber., 1925, *6*, 333) has given an account of the use of naphthalene sulphonic acids, including Kontakt and Episol. Kontakt is a Russian product and may be a sulphonated naphthenic acid.

Where soft water is available, soap is a very valuable and cheap kiering assistant; Kollmann (loc. cit.) obtained his best results in respect of wax removal and absorbency with 0·9% NaOH, 0·1% Monopol soap and 0·1% sodium aluminate. Jambuserwala (Am. Dyes. Rep., 1937, *26*, 799) found Turkey red oil to be a valuable addition to NaOH, not only for wax removal, but also for the final white produced.

Both Tschilikin (Textilber., 1928, *9*, 397) and Kornreich (Textilber., 1938, *19*, 61) attach considerable importance to the dispersing powers of the pectic and nitrogenous impurities of cotton when treated with alkali and dissolved in the kier liquor.

Pine oil appears to be a favourite kier-assistant in many quarters; it is often sold in admixture with soaps or sulphonated oils as a proprietary preparation. It is also possible to mix the pine oil with soap and add the product to the caustic soda kier liquor; when this is practised, about 0·025 to 0·25% of pine oil in the solution may be considered adequate.

There seems to be little doubt at the moment that soaps are pre-eminent among the cheaper scouring assistants; they enable the rate of penetration of the liquor to be increased and therefore the time of heating throughout the kier. Similarly, the length of time necessary for actual boiling may be reduced somewhat. A more uniform product is also obtained on account of the improved rate of scouring. As previously mentioned, many bleachers prefer the use of Igepon T in concentrations of 0·02 to 0·05%, but the British

Cotton Industry Research Association has shown that good results may be obtained by using castor-oil soap.

Many impressions of the value of these kiering assistants are largely subjective, and some of the experimental work has been criticised by Gotte (Kolloid Z., 1933, *64*, 222) on the grounds that maximum scouring efficiency is not reached with the sulphated alcohols until an alkalinity of pH10 is established. However, an undoubted advantage of the assistant lies in the retention of scouring power in hard water, but even this may be dealt with by the use of Calgon.

Some of the scouring assistants are dear, for 0·5 grams per litre becomes a considerable quantity in a two-ton kier which may hold 9,000 gallons of water.

The cheapness of ordinary chemicals militates against the use of the new assistants, but they have been found advantageous in the saving of time and enable two boils to be performed in 24 hours by reducing the time from 12 hours to 8 hours. The actual manipulation of tons of cloth requires considerable time with vertical kiers, and this, of course, is largely saved with the Mather kiers.

Some chemical manufacturers supply sodium carbonate containing sodium hydroxide and this forms a useful scouring agent.

Trisodium phosphate has been shown to be a useful auxiliary in scouring, according to Venkataraman (J. Ind. Chem. Soc., Ind., 1939, *2*, 81), particularly for the open kiering of coloured goods.

Sodium silicate is often employed in kiering as the soap which it forms is sometimes superior to pure soap; this may be attributed to the mild alkalinity of this reagent, its emulsifying power, and also to the adsorptive capacity of the hydrated gels formed by reaction with metallic salts, such as calcium and magnesium. Silicate also possesses definite wetting powers and increases the stability of the foam produced by soap. Both the phosphates and the silicates cause the soaps formed to function better in hard water. Of the various silicates, that known as water-glass, $Na_2Si_4O_9$, has been used not only to assist the detergent action of the kier liquor, but also on account of its deflocculating action in diminishing the danger of kier stains. Mixtures of sodium silicate, soap and solvents are often the basis of proprietary preparations sold as kier-assistants, and are employed to the extent of 0·5% on the weight of the fabric. Care must be taken thoroughly to wash the goods before souring or insoluble deposits will be formed on the fabric and affect its handle.

Sodium meta-silicate, $Na_2SiO_3.5H_2O$, has also been suggested as a kiering assistant, not only on account of its deflocculating powers, but also because of its emulsifying and dispersing properties. The buffer action is good, and this is an additional advantage in its use.

When sodium silicate or meta-silicate are used with sodium

hydroxide for kiering, in concentrations of the order of 0·5%, there is a perceptible improvement in the colour of the fabric.

Where the special assistants are utilised, it is advisable to carry out a preliminary boil in an empty kier, as otherwise they may cause deposits on the cloth by affecting the surface films established on the walls of the kiers by the well-established and previous methods of kier-boiling.

A kiering process by Ullmann (B.P. 339,850) rendered possible the simultaneous use of lime and caustic soda in presence of one of the newer kier-assistants which is capable of dissolving or dispersing metallic soaps; in this manner the high rate of saponification of lime is utilised without an additional boiling process after souring. The Sirial scouring compounds (Textilber., 1931, *12*, 577) were developed on this basis, Sirial B being a combination of aluminium with a derivative of an aliphatic alcohol sulphonate for addition to caustic soda solution. Without the dispersing agent, the lime soaps formed in the kier would envelop the cotton and prevent access by the sodium hydroxide.

Coloured-woven Goods

Reference has already been made to the scouring of coloured woven goods such as shirtings, when the reducing solution formed in kiering is sufficiently powerful, in presence of the alkali, to cause a return to the soluble leuco state with vat dyes, and so bring about bleeding and marking-off. (The strongly reducing oxycelluloses exert a similar effect.) Auxiliaries to prevent this trouble appeared on the market about 1928, and consisted mainly of substances such as p-nitrosophenol, nitroso-m-cresol or p-nitrosodimethylaniline; Ludigol and Kieropon are well-known trade products. Ludigol is the sodium salt of m-nitrobenzene sulphonic acid.

The use of mild oxidising agents in the kier led to more venturesome suggestions, sometimes with disastrous results, for obtaining the maximum whitening effect during the alkali boil. Activin (p-toluenesulphochloramide) was found by Haller and Seidel (Z. angew. Chem., 1928, *41*, 698) to produce purer whites without deterioration of the cotton and so necessitate a milder final bleach. Peractivin (p-toluenesulphodichloramide) was also quite effective. Biancal, which is an organic derivative of Caro's acid, $C_{10}H_7.SO_2.O.ONa$, has also been suggested as an addition to kier liquors and again appears to have no deleterious action on the cotton (J.S.D.C., 1933, *49*, 373). The suggestions for complete bleaching in kiers will be discussed at a later stage (see p. 288).

General

The modern method of scouring coloured-woven goods is to use sodium carbonate solution (about 2·5% on the weight of cotton) or soap or both on wince machines, or in open kiers but low-pressure kiers are occasionally employed. Goods to be sold in the white, dyed, or printed state are generally boiled in NaOH solution in low- or high-pressure kiers. In the light of modern knowledge and scientific control, caustic soda appears to be the best scouring agent as it saponifies and emulsifies the waxes, hydrolyses the proteins and eliminates the pectins; it may also be used with reasonable safety. Even with modern kiers of improved circulation and the improved packing of the kier by means of mechanical pilers which distribute cloth and alkaline solution together, it is still advisable to kier the goods twice. Temperature is of greater consequence than concentration of alkali, and the temperature falls as the liquor passes through the kier; on the second scouring, however, the position of the cloth is reversed as it has been passed from one kier to another with washing between. Some absorption of NaOH takes place during kiering and Muir (Melliand, 1937, *1*, 14) suggests that for maximum efficiency the kier effluent should be at pH 7·5 to 8; extravagant washing or neutralising should not be necessary.

The soda-boil, in replacing the lime-acid-ash sequence, not only saves a process and therefore time and cloth-handling, but may often provide a means of utilising the dilute caustic soda washing liquors from the mercerising ranges.

Efficient kiering is the secret of good bleaching in accordance with the saying, "well-boiled is half-bleached," but in spite of the great improvements in kiering methods and improved circulation which enable the entire liquor to be circulated every twenty minutes, it is not possible to point to any one method as being the best; cloth constructions and the purpose for which the material is intended have still to be considered. Although the methods vary, there is a general tendency not to exceed 30 lbs. pressure. A single good caustic boil for about 9 hours is adequate for most cottons intended for printing, and sodium carbonate may be used for goods which are to be dyed; sometimes two boils are given. Most goods which are to be sold in the white state are given two boils, both of which may be with sodium hydroxide, although one with the hydroxide and one with the carbonate is quite common. Lime is sometimes added to the sodium carbonate scouring liquor, but the lime-sour-ash sequence is rapidly disappearing.

Before leaving the discussion of kiering, it may be well to repeat that even and regular packing of the goods in the kier is of the greatest importance. Where the older types of kier are still utilised, care should be taken to ensure that the steam pipes do not come into contact with the cotton, and also that no pockets are formed in which live steam may collect and overheat the material.

Complete immersion of the cotton is essential, for in presence of air and alkali and high temperatures, there will be a great danger of oxycellulose formation; hence when steam is admitted into the kier at the start of the scouring operation it is essential the kier liquor should be admitted from below to sweep out the air, and when the steam is admitted to the kier at the start of the scouring operation, it is essential to keep the valve open for some time to allow the air to pass out. With the old-fashioned type of kier, using direct steam, it was sometimes difficult to ensure a proper supply of liquor owing to the condensation of the steam in the kier.

When the scouring process has terminated, the alkaline liquor should be removed in absence of air, that is, it should be replaced by hot water and then by cold water. Again, this precaution is to avoid the formation of oxycellulose. However, the cloth should not be allowed to cool in the kier or the impurities suspended in the hot liquor will be re-deposited on the fabric during cooling.

It may be considered good practice to wash the goods with water in a washing machine, even after washing in the kier; an exception to this general rule may be found in the treatment of soft-spun yarns and of knit-goods whose open nature permits better cleansing.

Knit-goods may be treated in one vessel throughout the whole of the processes of scouring and bleaching; wooden kiers have been used with liquors consisting of 1·5 to 2·5% NaOH. About 560 gallons are required for a ton of fabric. Trotman (loc. cit.) has recorded a method of steeping overnight followed by boiling for 7 hours at low pressure; the goods may then be washed in the kier, and treated with "chemick" for 4 hours, followed by washing, souring, and again washing. Finally the goods may be taken from the kier and soaped in the dolly machine.

Coloured goods are generally scoured with sodium carbonate solution instead of caustic soda, although there is also a tendency at present to utilise soap and water together with a modern wetting agent. Both temperature and time are less than with plain goods, and on this account it is general to give a very thorough desizing treatment before the scour.

Where kiers are used, it is customary to hang coloured headings and borders outside the top of the kier and moisten them from time to time; alternatively, the headings may be collected and wrapped in a bundle of old cloth and disposed at the top of the kier. Many coloured-woven goods with the coloured yarns distributed throughout the fabric, may be scoured on a wince or roller machine with a weak lye in presence of a protective agent; the continuous process may be employed or the goods may be thoroughly impregnated with the hot liquor and allowed to lie in a pile which must not be allowed to dry.

CHAPTER XVII
THE HYPOCHLORITE BLEACH

THE removal of cotton wax and other impurities leaves the material in a more absorbent condition than that of the raw cotton; there is also a considerably cleaner appearance which is adequate for some purposes, and, indeed, at one time it was not usual to treat cotton with hypochlorite solutions for the final bleached effect, but only linen.

The final bleaching process is essential for a good white effect, and may be carried out in two ways:

(a) bleaching with dilute hypochlorite solutions at room temperatures;

(b) bleaching with peroxide solutions at temperatures of 80 to 85°C.

It is important to appreciate that *cold* hypochlorite or *hot* peroxide solutions are employed; if hot solutions of hypochlorite are used, then the cotton is rapidly disintegrated, and even warm solutions are apt to cause considerable damage by the formation of oxycellulose. It should also be appreciated that whereas the hypochlorite bleach follows the scouring operation, modern developments with peroxide bleaching are directed to the elimination of the scouring process as such, and to the achievement of a simple bleaching method in which the scour and bleach are combined in one process.

HYPOCHLORITES

Since the introduction of bleaching powder on a commercial scale about 1830, dilute hypochlorite solutions still form the most popular means of bleaching cotton goods. When properly applied, with some understanding of the principles involved, it is possible to obtain good results with very simple plant, practically no expense for heating, and the chemical reagents employed in very dilute solution are both cheap and easily procurable.

An efficient bleaching process must impart to cotton goods a pure and permanent white, and level-dyeing properties without "tendering" or diminishing the tensile strength. It is only within recent times that precise and elegant methods of testing, developed by the B.C.I.R.A., have enabled a satisfactory method of control to be established—copper number, viscosity or fluidity, and solubility number—and raised the general technical level of bleaching in all its branches.

PLATE XVII

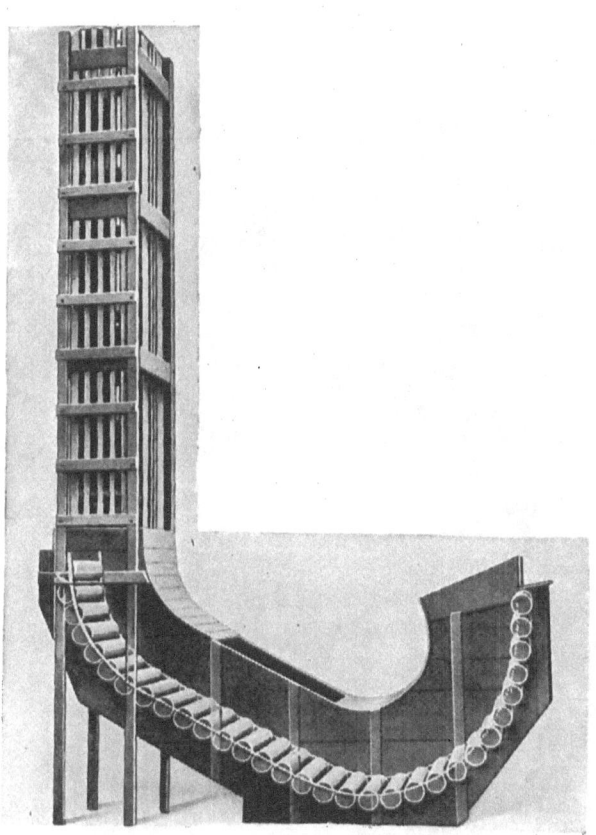

FIG. 65. The Gantt piler.

Courtesy of Messrs. Hunt and Moscrop.

PLATE XVIII

FIG. 66. Bleaching cisterns with automatic pilers.

Courtesy of Messrs. Mather and Platt.

To face page 201.

THE HYPOCHLORITE BLEACH

The principle chlorine compounds used in commercial bleaching are calcium and sodium hypochlorites in the form of dilute solutions termed "chemic." Both bleaching powder and sodium hypochlorite solution are available commercially, although within recent times the use of liquid chlorine has extended, and many bleach works now make their own bleach liquors by passing chlorine from cylinders into milk of lime. Sodium hypochlorite solution may be prepared by passing chlorine gas into dilute caustic soda solution, but cooling devices are necessary on account of the heat developed; it may also be prepared electrolytically where cheap power is available or by mixing solutions of "chloride of lime" and sodium carbonate, allowing the calcium carbonate to settle and withdrawing the clear liquor.

Bleaching powder was first made by Tennant; in 1798 he tried to patent the manufacture of a solution made by passing chlorine into lime water, but this had already been used by Lancashire bleachers. However, the following year Tennant used dry hydrated lime for absorbing the chlorine as described in B.P. 2,312 of 1799; bleaching powder was first made in Glasgow and soon became available in very large commercial quantities. At the beginning of the twentieth century, however, the production suffered a decline on account of the availability of liquid chlorine; the lack of stability of bleaching powder in hot climates was a disadvantage.

For very many years, the composition of bleaching powder was assumed to be $Ca(OCl)Cl$ as postulated by Odling in 1861, but this formula was regarded with doubt after the work of O'Connor in 1927 revealed the presence of a basic hypochlorite. By adopting phase rule and X-ray methods of investigation, Bunn, Clark and Clifford (Proc. Roy. Soc., 1935, *151*, 141) showed that the first stage in the production of bleaching powder was the formation of the basic hypochlorite $Ca(OCl)_2.2Ca(OH)_2$ and the basic chloride $CaCl_2.Ca(OH)_2.H_2O$. On further chlorination, the former compound was converted into a mixed crystalline substance, the chief constituent of which was calcium hypochlorite.

Ordinary bleaching powder, which contains about 35% of available chlorine, is a mixture of the hypochlorite mixed crystal and the basic chloride; further chlorination converts the basic chloride into the tetrahydrate $CaCl_2.4H_2O$. The presence of the basic chloride $CaCl_2.Ca(OH)_2.H_2O$ accounts for the difficulty of producing bleaching powder with more than 38% of available chlorine; the basic chloride, however, prevents deliquescence.

The term "available chlorine" means the chlorine available for bleaching; the chlorine in calcium chloride is not available chlorine.

The classical chamber process for the manufacture of bleaching powder is now almost obsolete; slaked lime of high quality was spread to a depth of 3 to 4 inches on the floors of the chambers by shovels and rakes, after which the chlorine gas was passed through, and the lime turned by hand occasionally. The chambers were generally operated in sets of four, the concentration of the gas varying according to the amount of chlorine in the bleaching powder.

Pure chlorine cannot be used without dilution on account of the heat of reaction rising to the region of 55°C., where bleaching powder decomposes.

A later process depended on staggered shelves for the lime arranged in a tower through which the chlorine gas pursued a circuitous route.

The well-known Beckman chamber method of manufacture utilised a series of superimposed circular chambers, on the floors of which the descending lime and chlorine gas were moved by ploughs on rotating arms so that the mixture passed from the centre to the rim and back again; the fall of the lime was in the opposite direction to that of the ascending chlorine gas.

The Ridge tube process makes use of a rotatable cylinder about 60 feet in length fitted with lifting devices causing the lime to travel in the opposite direction to the oncoming chlorine gas; as the cylinder is made of iron, it can be cooled to prevent any undue increase in temperature.

The stability of bleaching powder is generally improved by the addition of some quicklime which reduces the amount of water present in the free condition and also as water of crystallisation. Perfectly dry bleaching powder was produced in 1925 (B.P. 246,000); some improvements in bleaching powder made by the I.C.I. (B.P. 317,572 and 344,012) are also of interest. An Italian method is to use hydrated lime in CCl_4, chlorinate, and then remove water and the solvent by distillation; the product was termed Sichlor.

As previously mentioned, ordinary bleaching powder generally contains about 35% of available chlorine, and occasionally 38 to 39%; the stabilised or tropical bleaching powder contains less available chlorine.

Calcium hypochlorite $Ca(OCl)_2$, in crystalline form, has been made by the I.G.; it was sold as Perchloron and contained 80 to 90% available chlorine. The I.C.I. marketed a product termed Maxochlor and containing 75% or more of available chlorine. Another compound is H.T.H. (high test hypochlorite) of the Mathieson Alkali Works; it contained over 60% of available chlorine.

The pure hypochlorite seems to be fairly stable: apparently the presence of calcium chloride creates difficulties on account of its hygroscopic nature.

As will appear later, it is possible to make solutions of calcium

THE HYPOCHLORITE BLEACH

hypochlorite for use at the bleach-works itself by passing chlorine into milk of lime; the temperature must not be allowed to rise to 40°C., but the evaporation of the liquid chlorine may be used for cooling by means of a cast-iron vessel.

Calcium hypochlorite solution is generally prepared from bleaching powder, which is first mixed into a smooth paste with water and then diluted until sufficiently mobile to flow easily. The liquor is next passed through a sieve into a large concrete cistern and diluted with water until there is about 1 gallon of water per pound of bleaching powder; this suspension is thoroughly mixed and allowed to settle. The clear supernatant liquor is generally about 12°Tw. and contains about 3·9% available chlorine. The stock solution may be syphoned to another tank and diluted to the required strength. Some works make their own calcium hypochlorite solution by passing chlorine into a suspension of lime.

Sodium hypochlorite may be purchased from chemical manufacturers, but this involves paying for the transport of a considerable amount of water. It is possible to prepare sodium hypochlorite solution by passing chlorine gas into an alkaline solution; this method appears to be popular in the U.S.A., according to the American Cotton Handbook (1941, p. 650). The solution of alkali should not be too concentrated, nor should the chlorine be admitted at a high speed, or the temperature will rise on account of the heat of reaction; the solution should never exceed 27°C. or sodium chlorate will be formed. Solutions of sodium carbonate absorb chlorine more rapidly than sodium hydroxide solution; the concentrations of alkali must be adjusted to absorb the chlorine easily and must not be too dilute. With sodium hydroxide, 4 parts of chlorine may be used for every 5 parts of sodium hydroxide and the concentration of the latter should be between 3 and 4%; with sodium carbonate, the ratio should be 1 lb. of chlorine to 3·5 lbs. of sodium carbonate. Mixtures of hydroxide and carbonate may also be used, the stability of the hypochlorite solution varying with the ratio of Na_2CO_3 and NaOH.

The following table gives some proportions which have been employed in the U.S.A.:

PREPARATION OF SODIUM HYPOCHLORITE SOLUTION

Water	500 galls.	500 galls.	500 galls.
Chlorine	100 lbs.	100 lbs.	100 lbs.
NaOH	125 lbs.	—	80 lbs.
Na_2CO_3	—	350 lbs.	200 lbs.

It is possible, of course, to prepare sodium hypochlorite solution from calcium hypochlorite solution by the addition of sodium carbonate, 70 lbs. of which should be added for every 100 lbs. of bleaching powder in solution. The calcium is precipitated as the carbonate and the clear liquor used for bleaching; the bleaching powder originally contained some calcium hydroxide which is replaced by sodium hydroxide, and this has a stabilising effect.

The so-called electrolytic bleach actually refers to the use of hypochlorite solutions produced by electrolysis, although in its original form nascent chlorine was formed in presence of the cotton. Very exaggerated claims were made when the method was first devised about 1880; many salts were used, such as $MgCl_2$, $CaCl_2$ and $AlCl_3$, but finally common salt was realised to give equally efficient results. Early electrolysers only produced unstable solutions containing 3 to 5 grams of chlorine per litre, but the method became more economic about 1900 when solutions containing 15 to 30 grams of available chlorine per litre were obtained. The yield of chlorine, however, was far from theoretical (see Reuss, J.S.D.C., 1911, *27*, 110).

The use of the electrolyser depends on the decomposition of sodium chloride, with the production of sodium and hydrogen at the negative pole or cathode whilst chlorine and the hydroxyl ions are formed at the positive pole or anode; the products of electrolysis are allowed to combine in the electrolyser so that the chlorine and caustic soda form sodium hypochlorite.

$$2\ NaOH + Cl_2 \longrightarrow NaOCl + NaCl + H_2O$$

The hydrogen evolved during the process reduces some of the hypochlorite which impairs the efficiency of the process; it is also important to keep the temperature below 24°C. or chlorate will be formed, and there will be a waste of electric power in addition. Most electrolysers are equipped with cooling devices.

The three factors in the cost of producing electrolytic bleach are the power, the salt and the electrodes; the last-mentioned were a considerable source of trouble with the early types, but modern electrodes, if kept in good and clean condition, hardly enter into the operating costs of the electrolyser.

The relative costs of salt and electric power vary in different places and the operating conditions should be varied accordingly. The greater the strength of the brine, the greater will be the cost of salt and the less the cost of power; hence with cheap power and expensive sodium chloride it may be better to work with 10% solutions of salt instead of 15%. It should also be realised that the current consumption rises as the salt concentration falls when the hypochlorite is forming; this is shown in the following table (Industrie Textile, 1924, *40*, 339):

THE HYPOCHLORITE BLEACH

Electrolysis of 10% NaCl Solutions

Hypochlorite	Kilowatts/hour	Salt
15 g. Cl_2/l	5·6	6·6%
20 g. ,,	6·1	5·0%
25 g. ,,	6·7	4·0%
30 g. ,,	7·8	3·3%
35 g. ,,	10·0	2·8%

Electrolysis of 15% NaCl Solutions

Hypochlorite	Kilowatts/hour	Salt
15 g. Cl_2/l	5·3	10%
20 g. ,,	5·5	7·5%
25 g. ,,	5·8	6·0%
30 g. ,,	6·4	5·0%
35 g. ,,	7·2	4·3%

It is generally found satisfactory to circulate the brine through the electrolyser until a solution containing 12 to 15 grams of available chlorine per litre is obtained, although many electrolysers are capable of providing solutions containing 15 to 30 grams of available chlorine per litre. These solutions have a characteristic odour.

Whatever type of electrolyser is used, it is essential rigorously to follow the instructions.

Owing to the presence of salt and the neutrality of the solution, electrolytic bleach is much more active than the ordinary solutions, but we now realise that such activity is apt to be dangerous and little is heard to-day of the extraordinary properties of these solutions except as a warning (see p. 224). Electrolysers are not very common in bleach-works, but if the solutions are adjusted in respect of pH there is no reason why the bleaching action should not be quite readily controlled.

In spite of the greater cost, solutions of sodium hypochlorite are being used to a greater extent in preference to calcium hypochlorite. One of the difficulties of the latter is the "sludge" which is formed; this is less when the bleach-liquor is formed by passing chlorine into lime water than with the dissolution of bleaching powder, but with sodium hypochlorite there is no sludge. Another point in favour of NaOCl is that no insoluble products are formed during bleaching, and this obviates the necessity of souring with HCl to remove the lime salts from the calcium hypochlorite bleach, salts which are apt to give a harsh handle to the goods; a favourable aspect is that sodium hypochlorite solutions give more uniform results as they diffuse better and have superior powers of penetration. At one time extraordinary claims were made for the superiority of the white produced, but for the same consumption of active chlorine under optimum conditions there is little difference between the effects of sodium and calcium hypochlorite solutions in this respect.

One important factor which will appear later, is that sodium hypochlorite solutions lend themselves to controlled bleaching by buffering. In the initial stages of bleaching, the alkalinity of calcium hypochlorite solutions falls rapidly owing to precipitation of calcium carbonate and the calcium salts of the acids formed during bleaching; this results in increased activity of the calcium hypochlorite as compared with sodium hypochlorite as speed of bleaching depends to some extent on alkalinity. Increased activity, however, is apt to lead to difficulties of control.

Many practical bleachers, as distinct from chemists, prefer sodium hypochlorite solutions as they obviate many minor difficulties on the works' scale.

Calcium hypochlorite solution may be made at the bleach-works as previously mentioned; the process is quite simple and merely consists in passing chlorine, from cylinders of liquid chlorine, into milk of lime. The chlorination vessel is filled to about two-thirds of its capacity and the chlorine is admitted at the bottom until the liquor has a density of $13°$ Tw. corresponding to 40 to 44 grams per litre of available chlorine. The temperature should not be allowed to rise about $35°$ C. or decomposition to chlorate occurs.

Convenient quantities appear to be about 6,000 to 6,500 gallons of liquor containing 3,500 lbs. of hydrated lime; this can be converted into calcium hypochlorite by the use of 2,750 lbs. of chlorine in between 90 minutes and 2 hours.

The following table shows the relation between the density of calcium hypochlorite solutions and the available chlorine:

CALCIUM HYPOCHLORITE SOLUTIONS

Density	Available chlorine
$0.125°$ Tw.	0.35 g. per litre
$0.25°$ Tw.	0.70 g. ,,
$0.36°$ Tw.	1.00 g. ,,
$0.50°$ Tw.	1.40 g. ,,
$0.72°$ Tw.	2.00 g. ,,
$0.75°$ Tw.	2.05 g. ,,
$1.00°$ Tw.	2.71 g. ,,
$1.08°$ Tw.	3.00 g. ,,
$1.40°$ Tw.	4.00 g. ,,
$1.80°$ Tw.	5.00 g. ,,
$2.00°$ Tw.	5.88 g. ,,

As is well known, the density of the solution may be a misleading guide to its chemical composition owing to the presence of impurities;

THE HYPOCHLORITE BLEACH

for this reason it is preferable to express the concentration of bleach liquors in grams per litre of available chlorine. This is particularly so with sodium hypochlorite solutions, whose density, for the same chlorine content, is apt to vary with the method of preparation.

Available chlorine is usually determined by titration with a standard $0 \cdot 1\ N$ solution of sodium arsenite, using an external indicator of filter paper moistened with potassium iodide and starch solutions. This has the advantage of obviating estimations of chlorine which may be present as chlorate and without bleaching value; the usual iodine-thiosulphate method would include such chlorine and therefore not estimate the "available" chlorine.

In most commercial bleaching processes it is customary to employ solutions containing from 1 to 3 grams of active chlorine per litre; the solutions are used at room temperatures.

BLEACHING WITH HYPOCHLORITES

The scoured fabric intended for bleaching in hypochlorite solutions should be thoroughly rinsed before the bleaching treatment and so have approximately a neutral condition. The fabric, of course, is still wet, but it should be squeezed as thoroughly as possible so that excessive amounts of water are removed, as these would otherwise dilute the bleach liquor or "chemic" into which the cotton is led.

There are two main methods for bleaching with hypochlorites: (*a*) machine chemicking, and (*b*) the circulation method. In the former, the goods are impregnated with the solution on a machine which resembles the rope-washing machine, and are then piled in boxes until the bleaching action has taken place; during impregnation the liquor is constantly replaced. For the circulation method, the goods are piled in cisterns with false bottoms; the hypochlorite is sprayed on to the goods, percolates through the cloth to the well below, from which it is pumped up to be sprayed over the goods again. For circulating through the cisterns, calcium hypochlorite solutions of $0 \cdot 5$ to $1 \cdot 0°$ Tw. may be employed for periods of two hours or so; with machine chemicking a stronger solution is utilised, often $2°$ Tw., as the ratio of liquor to cloth is less, and the period of bleaching may vary from 4 to 12 hours according to circumstances.

In some of the larger modern plants it is customary to employ devices, such as the Gantt piler, which permit a definite time of action for the hypochlorite solution and allow at least part of the bleaching process to be conducted in a continuous manner.

The GANTT PILER is made of wood and consists of a vertical trunk with a curved foot or chute which is provided with a series of rollers over which the cloth passes when a sufficient weight of it has filled the bottom of the trunk. These rollers help the cloth to pass through the foot. The cloth is fed into the trunk by an overhead wince with a traversing pot-eye, and when the piler is filled, the cloth is taken from the front and passed over a wince to the next operation.

The Gantt piler may be used for souring as well as for bleaching, in which case one machine is sufficient to give the required steeping time, but for bleaching, as the cloth only takes 20 minutes to pass through the machine, it is usual to have two in tandem, and run the cloth direct from one to the other. In this manner, each yard of cloth receives the same period of treatment, whereas with the cistern method of bleaching, the first end down into the cistern is the last to be removed; this may mean a difference of several hours with large-scale bleaching.

Whatever method is employed, it is essential to shield the goods from contact with bright sunlight or "tendering" will result; similarly, it is important that the hypochlorite liquor should not come in contact with metal during the bleaching process or it will be activated locally, and the surrounding cotton damaged.

Bleach liquors vary in their activity with the temperature and are adjusted in concentration to meet the variations of summer and winter temperatures; dilute and warm (up to 40°C.) bleaching is occasionally suggested to hasten the effect, but is apt to give irregular results as well as tending to form oxycellulose. Warm hypochlorite solutions form chlorates which have no bleaching power.

Between winter and summer conditions, the temperature may vary from 5°C. to 15°C.; an increase of 8°C. is sufficient to double the velocity of the reaction.

Machine chemicking, followed by piling on stillages, is sometimes employed for the production of "printers' whites," but is not considered to be good practice; there is always a tendency towards over-bleaching and slight damage to the cellulose.

The circulation or cistern method is common practice for the normal bleaching routine.

The cisterns may be constructed of stone, slate or cemented brickwork, and are large enough to accommodate two tons of fabric; many of the cisterns are lined with slats of wood, but all of them have a false bottom which assists drainage and supports the fabric. The cisterns have two outlets, one to the drain and the other to the chemic well which lies beneath the cistern, and from the chemic well, the bleach liquor may be raised by a pump to perforated trays which lie loosely on top of the cistern; the trays are generally shaped to distribute the liquor evenly over the fabric.

Generally the cisterns are arranged in pairs, and the fabric is "turned over" from one to another, with an intermediate washing, touring and washing, so that the upper parts of the cloth goes to

THE HYPOCHLORITE BLEACH

the bottom of the cistern for the second stage of the bleaching treatment. The souring cisterns are arranged in a similar manner.

The rope of fabric is generally fed to the cistern by an overhead wince and is plaited down into the cistern by a mechanical piler which supplies the "chemic" at the same time; when the cistern is

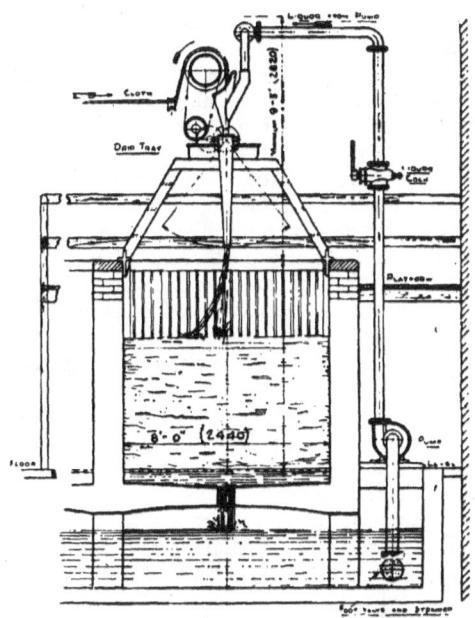

FIG. 67.—Diagram of bleaching cistern equipped with mechanical piler (Mather and Platt).

filled, the hypochlorite liquor is circulated for 1 or 2 hours according to circumstances, after which the goods are washed in a rope-washing machine, soured for an hour, and again washed.

Circulation of the bleach liquor by compression and exhaustion is sometimes adopted in special machines for the treatment of yarn in the form of cops and cheeses, as well as for loose cotton.

When bleaching by the circulation method, there is a rapid initial fall in the concentration of the hypochlorite liquor, but this is mainly due to dilution of the chemic by the water retained in the cloth after leaving the kier and washing machines; it must not be assumed to be caused, to any large extent, by the consumption of active chlorine in bleaching. The apparent fall in the amount of available chlorine is much slower when the bleach liquor is thoroughly mixed with the water entrained by the cotton.

When the bleaching operation has been completed, any chemic retained by the cotton will be removed by washing and go to the drain; hence, on grounds of economy, it is wise to arrange that the concentration of the bleach liquor at the end of the bleaching operation is as low as possible. As it is not possible to bleach by the circulation method in such a manner that the liquor is completely exhausted of its available chlorine, some loss of hypochlorite by washing must be expected. This may be minimised by intelligent use of measurements of the strength of the liquor at the end of the bleaching process; the standard arsenite solution, together with Soluble Blue 2RS as an indicator, may be of value in this connection. Hence it is comparatively simple to find the lowest final concentration available of chlorine which is consistent with efficient bleaching.

In general, bleach liquors are alkaline when formed, and it was customary to employ them in the alkaline state, for "standing baths" were used and it was found that an acid bleach liquor was unstable. During the bleaching process the acidic products formed tend to neutralise the liquor and led to observations on the "activity" of old liquors; in some cases it was also customary to "sharpen" the bleach by the addition of acid. As the solution passes from the alkaline side, first hypochlorous acid is formed and then chlorine itself, and later, when moderately acid, chlorates are formed and these have no bleaching action.

Little is known about the colouring matter which exists in cotton, and a great deal of the information on the action of hypochlorites comes from investigations with organic dyestuffs, and a very large number of those which are affected by hypochlorite solutions are much more rapidly bleached by acid hypochlorite solutions and are only slowly affected by alkaline hypochlorites. The reaction may be one of chlorination rather than oxidation.

Some work on this subject has been published by Oparin (Textilber., 1930, *11*, 644) and Schilow (ibid., 1936, *17*, 425).

It is usually considered that bleaching with hypochlorites is a process of oxidation, but Taylor (J.C.S., 1910, *97*, 2541) put forward the view that the actual bleaching agent was free chlorine; Higgins (J.C.S., 1912, *101*, 222), however, showed that free chlorine had little effect on the colouring matter in linen, and also stated (J.S.D.C., 1912, *28*, 30; 1914, *30*, 85) that in textile bleaching operations hypochlorites produce oxygen directly and a little chlorine at the beginning of the reaction. The discussion is of interest, but Taylor worked on unprepared cloth with laboratory reagents and Higgins with scoured material and commercial bleaching liquors under industrial conditions (see Higgins, *Bleaching*, p. 54).

Taylor (J.S.D.C., 1922, *38*, 93) also showed that free hypochlorous acid was a less vigorous bleaching agent than chlorine water, but neither has much effect on the colouring matter of linen, which is readily bleached with hypochlorites. Coward (ibid., 1922, *38*, 97) has suggested that the difference between hypochlorous acid and its salts may be due to low ionisation of solutions of the former compared with the latter. As will be discussed later (see p. 289), chlorine can play a part in the bleaching of linen as distinct from cotton.

After the bleaching action of the hypochlorite solution is complete it is customary to wash the goods or, with calcium hypochlorite, to pass the goods through weak acid—the so-called white sour—in order to decompose the last traces of hypochlorite, and to decompose any insoluble lime salts which may be deposited on the fibres; hydrochloric acid is therefore preferable to sulphuric acid. In some cases, a whitening is observed as the alkaline bleach liquor passes through the neutral point.

Where souring is employed, it is essential that the goods are thoroughly washed in order to remove the last vestiges of acid and avoid hydrocellulose formation on subsequent drying; sometimes the goods are "sweetened-off" with dilute sodium carbonate solution in order to produce a safer but slightly alkaline final state.

The souring process is discussed further on p. 230.

Much of the early scientific work on bleaching processes suffered from lack of sufficiently accurate methods of testing both the solution and the bleached product. Prior to the development of the elegant methods of estimating damage on a chemical basis, the somewhat crude measure of tensile strength afforded the only estimate of chemical attack. Hübner and Pope (J.S.C.I., 1903, *32*, 70) made one of the earliest examinations of the scouring of cotton and found that the treatment led to an increase in tensile strength. O'Neill (see Higgins, *Bleaching*, p. 32) studied the effect of bleaching on the strength of yarn and found an increase, as also did Jecusco (J.S.D.C., 1917, *33*, 34). Where the scouring process was so restricted as to require severe hypochlorite treatment, then the total effect is apt to show as a diminution in tensile strength. Greenwood (J.T.I., 1919, *10*, 274) made the interesting observation that the hair strength was actually reduced by bleaching, and the increase in yarn strength was due to the increased cohesion of the fibres in the yarn on account of the removal of the wax which normally acts as a lubricant.

Improved testing methods and better control have made some of these observations of historic interest only, as indicating the efficiency of bleaching at a particular period.

One of the earliest scientific tests for the efficiency of the bleaching process, i.e. a good white without damage, was the "copper number." This depends on the reduction of copper from the cupric to the cuprous state by modified cellulose, and was put on a satisfactory basis by Clibbens and Geake (J.T.I., 1924, *15*, 27). In their investigations they found that the original impurities of raw cotton possess reducing properties which are increased to a slight extent by the addition of size. Scoured cottons gave low copper numbers, the lowest being from the lime-sour-ash sequence, but very low values were also obtained from the caustic soda scour. Practically all technical bleaching processes gave a rise in copper number after the hypochlorite treatment. Birtwell, Clibbens and Ridge (J.T.I., 1925, *16*, 15), however, produced samples of technical processing in which the scour had been so thorough and the bleach so mild that an excellent white was obtained without any significant rise in copper number. Such cases were rare in 1925, and the general position with regard to copper numbers showed values between 0·01 and 0·3. It was suggested tentatively that a maximum value might be set at 0·25 to 0·3. The use of better tests and the well-known anomalies of copper numbers as indices of damage have minimised the value of this suggestion. Copper numbers were also used to follow the changes in the oxidation of cellulose with hypochlorite solutions, which may be considered to contain three oxidising agents—hypochlorous acid, hypochlorite ion and chlorine; the ratio of the concentrations varies with hydrogen ion concentration so that hypochlorite ion is the main agent in alkaline solutions, chlorine in acid solutions and hypochlorous acid near the neutral point.

Solutions of pH 10 to 11 contain, for practical purposes, hypochlorite ions as the only oxidising agent and these produce a maximum absorption of methylene blue on their modified celluloses; the concentration of free chlorine appears negligible at pH 4·6 and becomes of importance at pH 2·7 when there is a fall in copper number which continues in still more acid solutions with rapidly increasing concentrations of free chlorine. Maximum copper numbers were found in solutions of pH 4·6 to 2·7 where hypochlorous acid is the main oxidising agent.

Oxycelluloses—the product of the oxidative attack on cellulose—were divided into two main types: (*a*) high reducing power and loss of weight in alkali, and (*b*) great affinity for methylene blue and alkali, on a broadly similar basis to that suggested by Witz (Bull. Soc. Ind. Mulhouse, 1883, *43*, 334). It was also found, however, that the determining factor in the production of these oxycelluloses in "over-bleached" material was the acidity or alkalinity of the hypochlorite solution, and that the properties of the oxycelluloses were greatly affected by very slight changes in acidity or alkalinity of the solution which produces them, if the latter is near the neutral point. The high-reducing oxycelluloses are formed on the acid side of neutrality and regain their original chemical properties if treated with alkali but suffer a loss in weight. A chemical test in such cases, therefore, gives no indication of oxidative attack, but this may always be revealed by the fluidity of the material in cuprammonia. The fluidity determination has been described frequently (see J.T.I., 1936, *27*, 285), and properly bleached cotton should have a fluidity (0·5% solution) not exceeding 5, higher fluidities indicating degradation; viscose yarn (2% solution) should not exceed 12·5 units. For linen material the Solubility Number is preferred (J.T.I., 1928, *19*,

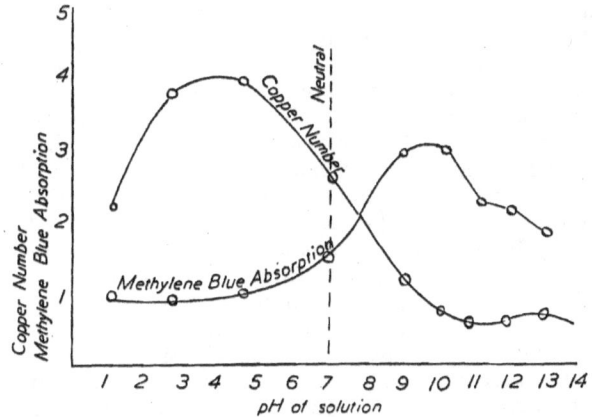

Fig. 68.—Effect of the pH of hypochlorite solutions on the copper number and methylene blue absorption of oxycellulose.

349; 1931, *22*, 416) and this should not exceed 10; with cotton, values as low as 3 are common, but in this case the preliminary boil in 2% NaOH solution may be omitted when carrying out the test (see page 468).

Effect of pH

The composition of some commercial bleach liquors was examined by Birtwell, Clibbens and Ridge in connection with their work on oxycellulose (J.T.I., 1925, *16*, 15), the properties of which depend on the exact acidity or alkalinity of the hypochlorite solution, as previously stated.

With machine chemicking, fresh hypochlorite solution is fed continuously to the machine at the same rate as it is withdrawn by the cloth, and therefore remains alkaline (pH values of 9·5 to 10·5 were given as typical); in the circulating process, the "old liquor" at the end of one operation is strengthened with fresh hypochlorite for the next operation, and as the acidic oxidation products accumulate, the solution tends to become neutral or acid (pH values of 9·5, 6·0 and 5·5 are given as representative).

Fresh 1°Tw. bleaching powder solution cannot be more alkaline than pH 11·5.

This preliminary investigation established the range of commercial bleach liquors as lying between pH 5 and 10, and a more detailed investigation of the effect of pH was made by Clibbens and Ridge (J.T.I., 1927, *18*, 135), when it was clearly established that the maximum rate of oxycellulose formation occurs at the neutral point. The rate of oxidation of the neutral solution is roughly ten times that of a similar solution at pH 9, or at pH 4·6. At these two values, the effective oxidising agents are presumably the hypochlorite ions and hypochlorous acid respectively, and the rapid rate of action of neutral solutions may be explained by the relatively high concentrations of both hypochlorite ions and hypochlorous acid existing simultaneously; the formation of chlorate in hypochorite solutions is thus most rapid near the neutral point. In these particular experiments, fully bleached cotton was used in order that over-bleaching could take place without great changes in the concentration of the solution. With hypochlorous acid solution at pH 4·6 the rate of oxycellulose formation was beyond control when circulating methods were employed and the rate of self-decomposition of the liquor was also greatly increased, so that a steeping method had to be utilised. Although buffered solutions were used, and it was shown that slightly acid solutions are less rapid in their action than neutral solutions, yet these acid solutions are very susceptible to influences which are not easily controlled and which greatly increase the activity of the solution.

The presence of small amounts of chromium in cotton was found considerably to increase the rate of oxidation by hypochlorite solution.

In these experiments, the time to consume half the available chlorine was taken as a measure of the rate of oxidation of the cellulose, as shown in Fig. 69. Now the test of efficient bleaching is to obtain a good white without sacrificing other textile qualities, so that the rate of change of copper number, methylene blue absorption and viscosity with pH are also of interest; these are shown in Figs. 70, 71 and 72, which serve to emphasise the great activity of neutral bleach liquors.

Blakeley (J.S.D.C., 1934, *50*, 306) has measured the oxidation potential of hypochlorite solutions at various pH values and established a strong maximum near pH 7, thus confirming previous experience of great oxidising properties at this particular value. The question was examined in somewhat greater detail by Turner, Nabar

THE HYPOCHLORITE BLEACH

and Scholefield (ibid., 1935, *51*, 5) in connection with other work (see p. 228); curves relating fluidity of bleached cotton with the pH of the solution were shown to be similar to those relating oxidation potential with pH. The dangerous neutral region was again clearly established, as shown in Fig. 73.

From a different standpoint, Kauffmann (Ber., 1932, *65*, 179)

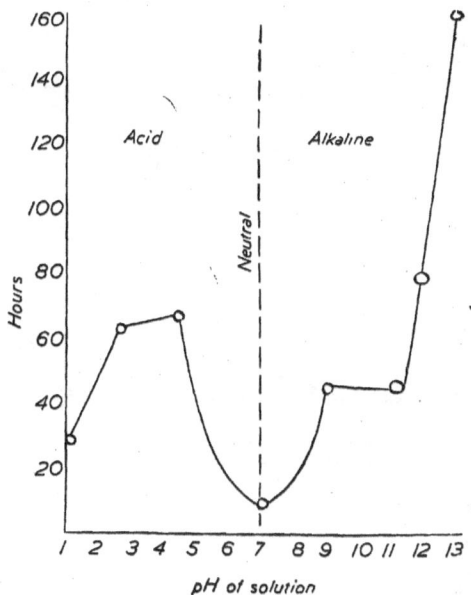

FIG. 69.—Rate of oxidation of cellulose estimated in terms of the consumption of half the total available chlorine.

calculated theoretically that the maximum oxidation by hypochlorite bleach liquors takes place in the pH range of 7·0 to 8·0. Serious injury to the cotton material will be caused as the hypochlorite (the oxidising agent) and hypochlorous acid (the accelerator) are present in the same equivalent amount (see p. 229).

The exposition of the dangers of neutral hypochlorite bleach liquors is undoubtedly a great contribution to the progress of efficient bleaching technique. It is almost impossible to overemphasise the dangers of neutral hypochlorite solutions on cotton; reference to Fig. 73 reveals that below pH 9 the activity of the solution increases to a manifold degree as it approaches neutrality,

216   TEXTILE BLEACHING

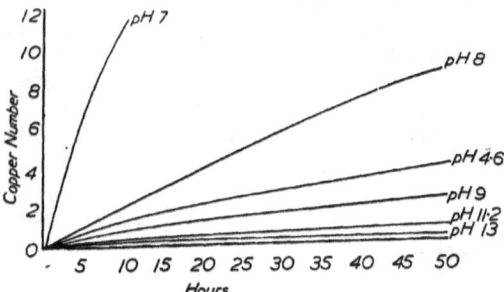

Fig. 70.—Rate of increase of copper number.

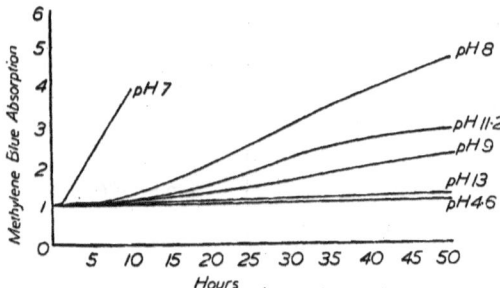

Fig. 71.—Rate of rise in methylene blue absorption.

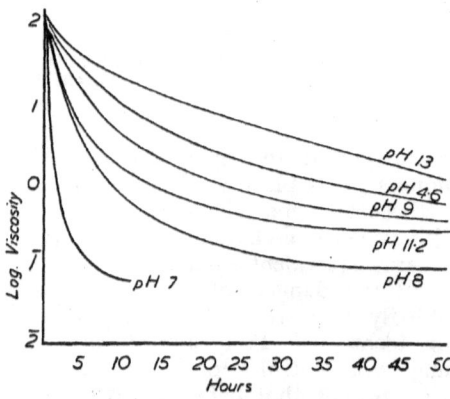

Fig. 72.—Rate of fall in viscosity.

THE HYPOCHLORITE BLEACH

so that small changes in pH have a much greater effect on the oxidation of cellulose than was at first supposed.

Attempts have been made to utilise this increased activity to produce a rapid bleaching action which, however, is apt to be superficial and impermanent in the case of cotton (see also p. 291); less active liquors can be safely used for longer periods to give a thorough and complete bleaching effect without degradation of the cellulose.

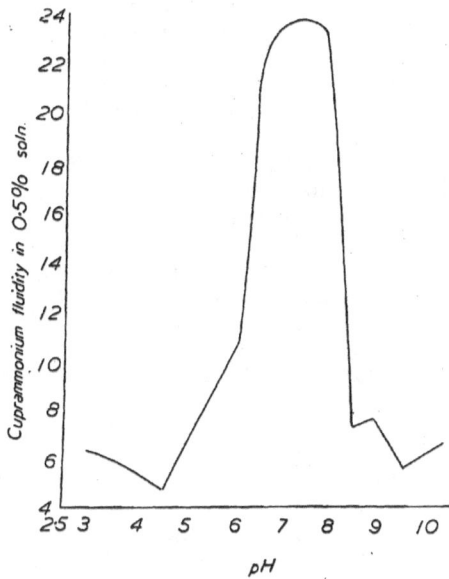

FIG. 73.—Degradation of cotton with self-buffered hypochlorite solutions (3 g. available chlorine per litre) after 5 hours' treatment.

The activation of bleach liquors may be realised in an indirect manner through insufficient diffusion which produces local reduction of alkalinity by the formation of acidic products of oxidation. Where diffusion is retarded, therefore, the liquor is more apt to enter the dangerous neutral region and cause local tendering if all the active chlorine has not been utilised. This state of affairs is obviated by efficient circulation of the liquor.

From the consideration of the action of neutral hypochlorite bleach liquors and the damage they are capable of producing on cellulose, it is obviously of interest to discuss what takes place with uncontrolled hypochlorite solutions in presence of the cotton, or other forms of cellulose, and further to find what steps may be taken to avoid the possibility of the hypochlorite solutions becoming neutral and causing tendering.

Buffered Solutions

The course followed by the pH of various bleach liquors in contact with the cellulose has been examined in some detail.

Utilising the glass electrode, Davidson (J.T.I., 1933, *24*, 185) carried out determinations of the hydrogen ion concentration of hypochlorite solutions. The dissociation constant of hypochlorous acid was also determined as 3.7×10^{-8} at 18 to 20°C,; this is in agreement with the value found by Sand (Z. physikal. Chem., 1904, *48*, 610), Britton and Dodds (Trans. Farad. Soc., 1933, *29*, 537) and Yorston (Pulp and Paper Mag., Canada, 1931, *31*, 374).

Cotton cloth was scoured with sodium carbonate solution and experimental bleaches were conducted utilising as bleaching agents (*a*) calcium hypochlorite, (*b*) sodium hypochlorite, and (*c*) sodium hypochlorite buffered with 5 g. per litre of sodium carbonate. The consumption of hypochlorite was very rapid in the early stages of the bleach, the available chlorine concentration falling from $0.05\ N$ at the beginning to an average value of $0.018\ N$ at the end of half an hour; the concentration was then increased by $0.025\ N$ through the addition of fresh hypochlorite when it fell more slowly, again reaching a value of $0.018\ N$ at the end of the bleaching period of 2.5 hours. The partially spent liquor was stored in the cistern of the apparatus overnight, when the chlorine content fell to about $0.005\ N$.

The lower free alkalinity of the calcium hypochlorite solution gave a value (pH 11·05) which was less than that of the sodium hypochlorite solution (pH 11·55), but in these two cases the circulation method of bleaching caused a rapid fall in the early stages of the process; the pH rose again on the addition of fresh hypochlorite and then fell more slowly as bleaching proceeded. When the partially exhausted liquor was allowed to stand overnight, there was a considerable fall in pH, and on making up the liquor for the second bleach, the pH was lower than at the start of the first bleach. In general, the second bleach was similar in respect of the course of the pH values to the first, but lower throughout and without any rapid fall at the beginning. The main difference between the calcium and sodium hypochlorite solutions was that the pH values for the former were consistently lower than those for the latter, the difference amounting to 1 unit after three bleaches. With the buffered sodium hypochlorite solution the fall in pH is much less rapid, and there was no great rise on the addition of fresh liquor.

pH OF BLEACH LIQUORS

	Calcium hypochlorite		Sodium hypochlorite		Soda ash and hypochlorite	
Before circulation	11·05	8·55	11·55	9·35	11·2	10·1
5 mins.	9·25	8·45	10·5	9·25	—	—
15 mins.	8·95	8·3	9·7	9·05	10·95	10·0
*30 mins.	8·7	8·15	9·35	8·9	10·8	9·95
34 mins.	8·85	8·4	9·8	9·2	10·75	—
45 mins.	8·7	8·2	9·4	9·1	10·65	9·9
60 mins.	8·5	8·0	9·2	8·9	10·5	9·85
90 mins.	8·25	7·65	8·8	8·6	10·3	9·8
120 mins.	8·1	7·45	8·5	8·45	10·2	9·75
150 mins.	7·95	7·35	8·35	8·3	10·15	9·7
23 hours	6·6	6·0	6·75	7·0	9·8	9·6

* Addition of hypochlorite immediately before this result. The bleach liquor was circulated through the cloth.

The effect of buffering is very pronounced. In view of the popular assumption that the carbon dioxide of the air plays an important part by neutralising the free alkalinity of bleach liquors, this point was also investigated by Davidson when it was found that its effect was small, and quite inadequate to explain the rapid fall which takes place during the first 15 minutes of bleaching.

In the early stages of bleaching, the acidic products of the bleaching process reduce the pH of the liquor, but below pH 9·7 or thereabouts this action is supplemented by the acids produced from the hypochlorite itself; this produces a buffer system of partly neutralised hypochlorous and carbonic acids which is effective between pH 8·5 and 6·5 and as it is gradually acted on by the acids produced during bleaching, and by hypochlorite consumption, the pH falls gradually. When the partly exhausted liquor is allowed to stand, organic matter from the cloth is further oxidised, thus accounting for the fall in available chlorine which is somewhat rapid.

The variation in pH of liquors containing sodium hypochlorite and sodium carbonate is confined to above pH 9·7, and is therefore due to the effect of the acidic products of bleaching and to any atmospheric carbon dioxide. At pH 10 and above, all the carbon dioxide remains in solution, and at lower pH values it diffuses into the atmosphere.

The slight solubility of calcium carbonate suggests that a precipitate would be formed during bleaching with calcium hypochlorite, and some turbidity is frequently observed; as acidic products are formed some of the precipitate dissolves but it seems probable that there is appreciable filtering action by the cloth itself—hence the necessity for souring after bleaching with calcium hypochlorite.

The results of Davidson's investigation show that with calcium or sodium hypochlorite, the pH falls into the danger zone where attack on the cellulose itself is rapid, but this can be prevented by using sodium hypochlorite solution containing 0·5% of sodium carbonate. Yorston (Pulp and Paper Mag., Canada, 1932, *33*, 74) has shown that calcium hypochlorite bleach liquor may be buffered at about pH 9 by the addition of magnesia.

Williamson and Oakes (J.T.I., 1936, *27*, 197) have recorded pH values for some of the commoner alkaline sodium salts used in textile processes; most of the data refer to 0·2% solutions.

pH Values of Solutions Measured at 20°C.

Substance	Concn. %	pH
NaOH	0·2	12·69
$Na_2SiO_3.5H_2O$	0·2	11·94
$Na_3PO_4.12H_2O$	0·2	11·67
Na_2CO_3	0·2	11·17
Na_2CO_3 (commercial)	0·2	11·12
$Na_2CO_3.NaHCO_3.2H_2O$	0·2	10·11
$Na_2B_4O_7.10H_2O$	0·2	9·15
$NaHCO_3$	0·05	8·45
$NaHCO_3$	0·2	8·43
$NaHCO_3$	0·5	8·41
$NaHCO_3$	2·0	8·39

Estimations of the pH of bleach liquors may be effected with B.D.H. Universal Indicator by an indirect means owing to the destruction of the colour by chlorine; a little of the bleach liquor should be warmed with a few drops of neutral hydrogen peroxide solution and then cooled, when the indicator may be added to show the pH according to the colour developed. The active chlorine is destroyed by the peroxide with the liberation of oxygen.

A somewhat less reliable method of testing, but nevertheless one which is satisfactory in experienced hands, is to add the Universal Indicator to the bleach liquor and carefully observe the colour change; if the chemic is over pH 10 then the violet colour persists for some seconds, and if near 10, then there is a momentary flash of violet colour which rapidly disappears. Where the liquor has a pH value of less than 10, there is no violet coloration, but there may be a momentary green or greenish blue colour.

Another method has been described by Lynch and Nodder (J.T.I., 1932, *23*, 309); a filter paper is folded four times so that eight thicknesses are obtained along the final fold, and the folds lie tightly together. A drop of the bleach liquor is placed on the eight-fold edge and allowed to soak in, when four or five further drops are similarly added. After the last drop has been absorbed, a small drop of B.D.H. Universal Indicator is placed on the damp spot, and the colour produced noted, and compared with standard colorations.

For safe bleaching, the pH of the liquor should not be less than 9; not only are fluidities of the bleached cellulose satisfactory but the permanence of the white is better than with samples bleached at lower pH values. The safety of the bleaching operation is controlled by the pH value of the liquor, for within reasonable limits, variation in the concentration of available chlorine is of less importance.

An important contribution to the control of pH in hypochlorite

THE HYPOCHLORITE BLEACH

liquors has been made by Ridge and Little (J.T.I., 1942, 33, 33; 59), recognising that under uncontrolled conditions of processing with alkaline hypochlorite solutions, changes occur which cause a rapid fall in pH.

By consideration of the equilibria

$$HOCl + H^+ + Cl^- \rightleftharpoons Cl_2 + H_2O$$

which is predominant above pH 5, and

$$HOCl \rightleftharpoons H^+ + OCl^-$$

which predominates below pH 5, the ratio of hypochlorous acid to total available chlorine in the hypochlorite solution was calculated for all pH values within the range of 1 to 10. The fraction of hypochlorite as hypochlorous acid exceeds 96% over the pH range of 6 to 3, but outside this range it falls progressively as the pH increases or decreases, becoming negligible at pH 1 and 10.

Hypochlorous acid is a weak acid, and therefore in mixtures with its salts it exerts a buffering action of its own, which is greatest over the range of pH 6·5 to 8·2; but during bleaching, the components of the system decompose and liberate hydrochloric acid, causing a marked fall in pH. The ability of various substances to buffer the pH of various hypochlorite systems was determined in the absence of textiles, by potentiometric titrations at 18 to 20°C. of sodium hypochlorite and bleaching powder solutions containing added buffers. The usual method was to determine the pH of the solution with a glass electrode, after increasing additions of HCl, although in some cases the effect of increasing amounts of buffer material on the pH of the hypochlorite solution was examined.

BUFFER REAGENTS

pH	Buffer	Concentration
11	Lime (milk of lime)	1–1·5 g. per litre
10–11	Sodium carbonate	5–10 ,,
10	Sodium sesquicarbonate	5+ ,,
7·5–8·5*	do.	8–10 ,,
8·6	Sodium bicarbonate	10+ ,,
6·8–7·3*	do.	3–7·5 ,,
6·9	Carbon dioxide	Saturated
6 –6·3	Calcium carbonate	8–10 ,,
8 –8·5	Boric acid	5+ ,,
9 –10	Sodium borate	3 ,,
5	KH_2PO_4	3–12 ,,
6 –8	Na_2HPO_4	3–12 ,,
7 –8	Calgon	10–15 ,,
6 –7	Na pyrophosphate	5 ,,
3·5–5	Acetic acid and acetates	Varies with ratio
3·8–4·5	Aluminium sulphate	2 g. per litre

* denotes pH with bleaching powder solution.

The concentrations given in the table refer to hypochlorite solutions containing 3 to 5 g. per litre of available chlorine, but they will tend to vary with the available chlorine content.

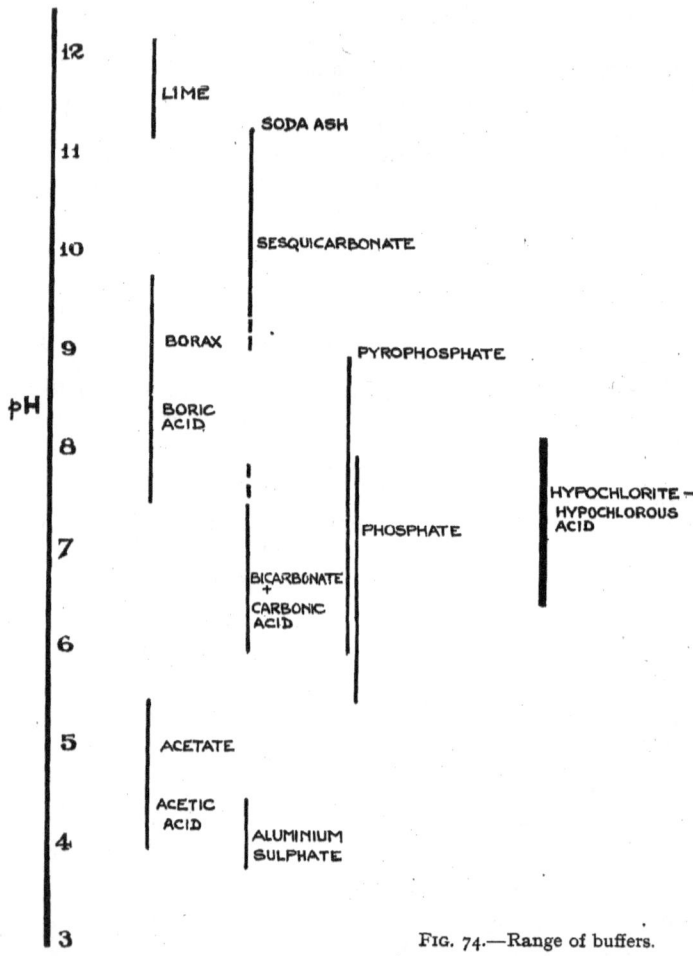

Fig. 74.—Range of buffers.

In addition to the changes in pH which may occur in unbuffered solutions due to atmospheric carbon dioxide, even buffered solutions will change in pH during the actual bleaching process. The changes are least for cotton and rayon, which only contain small amounts of

THE HYPOCHLORITE BLEACH

non-cellulosic matter to react with the bleach liquor, but there is considerable alteration with jute.

Tests were made with sodium hypochlorite and bleaching powder solutions, buffered with various reagents, of the changes in pH

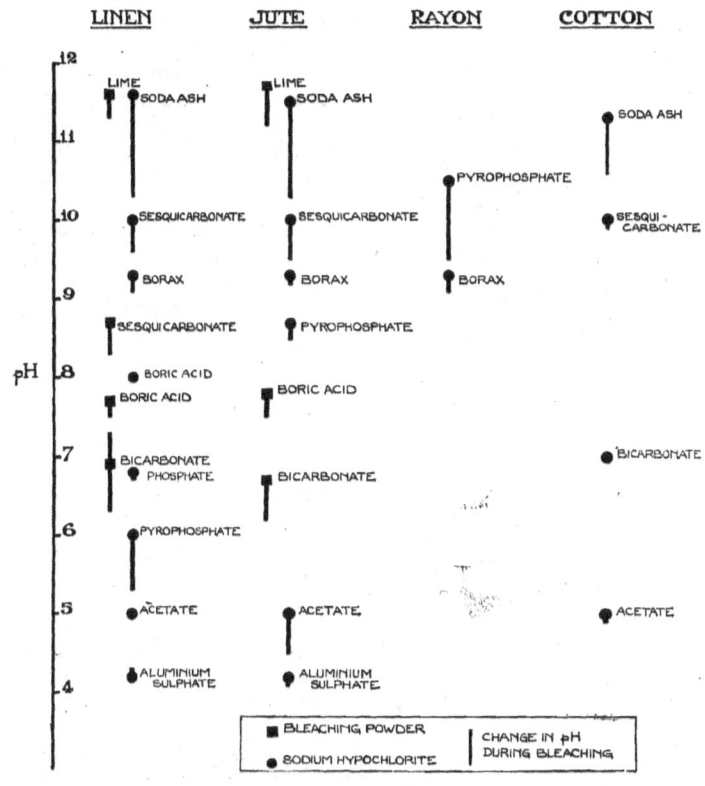

FIG. 75.—Change in pH of buffered hypochlorite solutions during bleaching.

during the bleaching of linen, cotton, rayon and jute; the ratio of liquor to textile was 15 : 1, the temperature was 19 to 19·5°C., and the yarns were scoured before bleaching.

The general effect is shown in Fig. 75 over a period of one hour's bleaching.

The chlorine consumptions were recorded at intervals during the bleaching process and indicate that with jute, the percentage consumption of available chlorine increases steadily as the pH falls, and in acid liquors the consumption is sometimes so rapid that the solution is exhausted before the end of the assigned period.

With linen, the consumption also increases as the pH approaches 7, and then falls on the acid side, but the changes are much less than with jute. The buffering appears to be less satisfactory with jute than with the other fibres, but is nevertheless good in most cases, excepting sodium carbonate and pyrophosphate.

The conditions of bleaching shown in Fig. 75, are not necessarily those which would be recommended for the various fibres, but show to what extent the non-cellulosic constituents affect the buffer properties of the various hypochlorite solutions. For instance, borax is not likely to be used to give a pH of 9 for jute or linen, but might be used for bleaching viscose rayon.

It will be noted that the buffering effect of soda ash is not very good above pH 11, but the capacity is greater between pH 11 and 10. Sodium sesquicarbonate in sodium hypochlorite solutions buffers in the region of maximum stability of the carbonate-bicarbonate system, but in a region where the hypochlorite-hypochlorous acid system is unbuffered; hypochlorite solutions of different concentrations all have the same initial pH in this locality, but as the hypochlorite is consumed during bleaching, a fall in pH takes place depending on the chlorine consumption. Borax stabilises the solutions just above pH 9 and little change takes place in bleaching.

The activity of old bleach liquors has already been mentioned and is now comprehensible in the light of modern knowledge of the effect of the pH of the hypochlorite solution on its oxidising powers.

Effect of Salt

Higgins (J.S.D.C., 1912, *28*, 30) examined some old bleach liquors and found they contained about 5% of calcium chloride; this had little effect as a stimulant, but when replenished with fresh bleach liquors the free lime of the latter would be removed and hence result in an increase in oxidation.

The acidification of bleach liquors to "sharpen" them is also comprehensible in terms of the pH of the liquor. Higgins (J.C.S., 1913, *29*, 302) realised what was happening, for he pointed out that an excess of boric acid produced an energetic bleach liquor because it liberated hypochlorous acid from the hypochlorite, whereas an excess of hydrochloric acid produced free chlorine and a solution of weak bleaching power. If, however, calcium carbonate is added to the latter, then hypochlorous acid is regenerated and active bleaching properties are restored.

Higgins also prepared an active solution of sodium hypochlorite by precipitating bleaching powder with sodium carbonate and bicarbonate; the addition of bicarbonate to bleaching liquor was patented by Macilwaine (J.S.C.I., 1915, *34*, 75). As pointed out by Higgins, similar results are obtained by impregnating the fabric with either dilute hydrochloric acid or sodium bicarbonate and then running through bleach liquor.

Although the use of acidified bleach liquors is now known to be dangerous because of the approach to the neutral point, and the activity of strengthened "spent" liquor is due to the same reason, numerous attempts were made at one time to utilise activated bleach liquors. One of the first of these was due to Lunge (D.R.P., 31,741 of 1884), in which acetic or formic acid was used to liberate the hypochlorous acid from solutions of bleaching powder. D.R.P. 107,093 describes the Hadfield-Sumner process in which the fabric was impregnated with 1° Tw. calcium hypochlorite solution and then passed through acetic acid vapour. The Mather-Thompson process described a similar impregnation of the cotton followed by a passage through a chamber containing carbon dioxide. Lagache (Deutsche Färb. Z., 1902, *13*, 313) has suggested passing carbon dioxide into fresh solutions of bleaching powder, which seems the scientific method of producing the same neutral bleach as obtained empirically through "working" a fresh bath with old material to get it into the proper active condition.

The great defect of these old methods was that they brought the bleach liquors into what is now known to be the dangerous neutral region; a further difficulty was associated with the uncontrolled nature of the liquor and the bleaching action.

The bleaching action of hypochlorite solutions may be stimulated by the addition of various salts; Higgins (J.S.C.I., 1913, *32*, 350; J.S.D.C., 1913, *29*, 85) has examined the action of sodium chloride, nitrate and sulphate and found the effect to be small. The activation of bleach-liquor by common salt has been mentioned in the *Manual of Dyeing* by Knecht, Rawson and Loewenthal (Griffin and Co., London, 1925).

The effect of adding sodium chloride to potassium hypochlorite solution used for bleaching linen fabric is shown by the exhaustion of the chlorine in the bath, according to the following data from the work of Higgins (supra).

EFFECT OF SALT ON HYPOCHLORITE SOLUTION

	Potassium hypochlorite	
	Alone	With salt
Original	5·5	5·5
After 15 mins.	4·8	4·6
30 mins.	4·6	4·35
45 mins.	4·4	4·2
60 mins.	4·35	4·1
3·5 hours	3·65	3·45
46 hours	1·25	1·05

It will be noted that the activation produced by the presence of salt occurs only in the first fifteen minutes, after which there is a similar rate of reaction from the two solutions, as shown by estimations of the available chlorine. The effect may therefore appear pronounced on a laboratory scale but is of little moment in industrial bleaching, for the sudden initial action is of little real consequence. Similar effects were observed with bleaching powder solution and calcium chloride, and also with hypochlorous acid solutions and sodium chloride.

Higgins has put forward the following explanation of the stimulus given by common salt. According to the equation

$$HOCl + NaCl = 2\,Cl + NaOH$$

the production of the nascent chlorine give a sudden stimulus, but when the new equilibrium is reached the action ceases, and bleaching resumes its normal course. Hence the action of salt is temporary and only confined to the first stage of bleaching.

Activated Hypochlorite

A somewhat recent use of activated hypochlorite bleach liquor has been described by Minaev (Textilber., 1936, *17*, 219) and relies on working at the usual temperatures with solutions containing one gram of active chlorine per litre and activated with sodium bicarbonate. In this manner, the duration of the bleach is only 2 to 3 minutes so that it is possible to adopt a continuous process. The chlorine consumption is stated to be one gram of chlorine per kilogram of cotton, and the amount of oxycellulose formed was only 0·18 to 0·2%.

The effect of various amounts of $NaHCO_3$ in sodium hypochlorite solution (1 g. of active Cl per litre) used for 10 minutes at room temperature may be seen from the following table.

ADDITION OF $NaHCO_3$ TO $NaOCl$ SOLUTIONS

Concn. of $NaHCO_3$	Cl consumption per 100 g. cotton	Purity of white	Oxy-cellulose
0	0·11	78%	0·21%
2·37 g. per litre	0·11	81·4%	0·21%
4·73 g. ,,	0·12	82%	0·20%
10 g. ,,	0·12	82·1%	0·21%

On a large scale, however, it was found that compared with ordinary alkaline hypochlorite solutions, the time of bleaching could be reduced to one-fifth of that usually given, and also that the chlorine consumption was only 60% of that with alkaline hypochlorite.

The acceleration of bleaching may be due to the liberation of HOCl from the NaOCl, or to hydrolysis $ClO^- + H_2O \rightleftharpoons ClOH + OH^-$.

There is also the possibility of double decomposition:
$$NaOCl + NaHCO_3 \longrightarrow Na_2CO_3 + HOCl$$
or
$$NaOCl + 2NaHCO_3 \longrightarrow Na_2CO_3 + HOCl + NaHCO_3$$

If the mechanism of bleaching is $NaOCl+H_2O \to HOCl+NaOH$, then Minaev considered that the $NaHCO_3$ would be beneficial in partially neutralising the accumulating NaOH and preventing the reversion of hydrolysis. It was found that solutions of NaOCl containing $NaHCO_3$ (in absence of cotton) decomposed slowly and lost about 50% of their chlorine in 200 to 250 hours.

The dangers of neutral bleach are, of course, well known, and it was immediately pointed out by Elöd (Textilber., 1937, *18*, 64) that the bleaching may only be superficial on account of the short period of treatment so that the fabric would acquire a false whiteness. Wasser (ibid., 225) drew attention to the four types of bleach liquor which were being discussed: (*a*) alkaline NaOCl at pH 9·6, (*b*) NaOCl neutralised with acetic acid to pH 6·9, (*c*) NaOCl with 4·8 g. per l. of $NaHCO_3$ giving pH 8·3, and (*d*) neutralised NaOCl with $NaHCO_3$ giving pH 7·5. His findings were that the quality of the white followed the pH, 9·6 giving the best results and 6·9 the worst; there was no diminution in strength at pH 9·5, but lower values resulted in damage to the cellulose.

A practical bleacher, Schmidt (ibid., 1936, *17*, 878) also disputed the results of Minaev, and stated that the use of bicarbonate was well known, but the treated goods tend to turn yellow, particularly if allowed to stand in the bleach for several hours, and that tests showed that alkaline hypochlorite gave a good stable white whereas acid hypochlorite gave a poor white and caused yellowing. Kornreich (ibid., 1937, *18*, 304) pointed out, however, that although the use of bicarbonate was known, yet Minaev was the first to use it in the correct way, i.e. with dilute hypochlorite for a short time. It may be remarked that the recent work of Butterworth (see page 290) with neutral hypochlorite solutions is of some interest in this connection.

Processes suggested at various times for the use of activated bleach liquors, are apt to exert too vigorous an oxidising action and form oxycellulose to the detriment of the physical properties of the cellulose including the permanence of the white effect; poor colour may also be due to the presence of soap, lime, iron and impurities in the wash-waters.

Coloured-woven goods should contain dyestuffs which are fast to bleaching and in this connection the pH of the bleaching solution is also important, for dyes which are resistant to hypochlorite at pH 11 may not be so resistant in solutions of lower pH values.

Accelerated Oxidation

Some peculiar effects in the bleaching of coloured goods have been investigated between 1930 and 1940 with very interesting results. It is common experience with many textile technologists that some of the yellow vat dyeings on cotton and rayon produced tendering when exposed to light under conditions which were without effect on the undyed material. Scholefield and Patel (J.S.D.C., 1929, *45*, 175) found that with dyed material—certain vat yellows and oranges —treatment with hypochlorite in presence of light brought about tendering which did not occur in the dark. Derrett-Smith and Nodder (J.T.I., 1932, *23*, 293) investigated the action of hypochlorite solutions over a wide range of pH values, on cotton dyed with certain vat and azoic dyes; yellow, orange and red dyes were active and tendered the cellulose, the greatest effect taking place in the neutral region. It was suggested by Scholefield and Turner (ibid., 1933, *24*, 330P) that in presence of light, these active dyes can be reduced to their leuco forms, which, during oxidation (by air or other oxidising agent) bring about a strong simultaneous oxidation of the cellulose. Oxidations of this type are known to occur with a variety of organic compounds (Traube, Ber., 1881, *15*, 659, 2421, 2434; 1883, *16*, 123; Englers *et al.*, see Chem. Soc. Ann. Rep., 1924, p. 109).

Turner, Nabar and Scholefield (J.S.D.C., 1935, *51*, 5) confirmed this activating effect of reduced vat dyes on cotton in hypochlorite solutions, and found that the effect depends to a large extent on the concentration of dye on the fibre. The increased rate of oxidation of cellulose in sodium hypochlorite solution at pH 7 in presence of leuco-vat dyes is so great that for the same fluidity a number of seconds is adequate in place of the hours required in absence of dye. In presence of a third substance which is oxidised, cellulose is not protected from oxidation but is attacked with increased vigour. The suggested addition of oxalic acid to hypochlorite solutions in 1877 (G.P. 2148) certainly produced "a stronger bleaching effect" and mixtures of dichromate-oxalic acid have been used (Clibbens and Ridge, J.T.I., 1927, *18*, 135) in recent times for the purpose of forming oxycellulose, which is produced much more rapidly than with dichromate-sulphuric acid.

Accelerated oxidation with hypochlorites has been examined in some detail by Kauffmann (Ber., 1932, *65*, 179) who showed that the destruction of certain readily oxidisable dyes by hypochlorite is increased by the addition of small proportions of hydrogen peroxide. It was also established that cellulose itself was attacked by the accelerated oxidation of the activated hypochlorite; other agents capable of acting as activators are ammonia, chloramine, acid amides and amino-acids. This is a departure from the usual opinion that hypochlorous acid is the agent which is responsible for attack on the fibre during uncontrolled bleaching, and is formed according to the scheme

$$NaOCl + H_2O \rightleftharpoons NaOH + HOCl$$

or

$$OCl^- + H_2O \rightleftharpoons OH^- + HOCl.$$

Actually, Kauffmann found that there was only slight attack on cellulose by hypochlorous acid, alone.

A number of dyestuffs are readily bleached by dilute sodium hypochlorite solution but the bleaching action is slower in presence of NaOH; the addition of hydrogen peroxide to hypochlorite solution produces rapid bleaching, but with ammonia there is a certain period of induction during which chloramine is formed, similarly with urea, when dichlorurea is formed. Chemical compounds of very different types can produce this effect, as for instance, dimethylglyoxime. The presence of these activators in hypochlorite bleach liquors may produce tendering of the cellulose.

Kauffmann is of the opinion that in bleaching, two substances are necessary, the oxidising agent and the activator; with hypochlorites the oxidising agent is the hypochlorite ion and the activator is hypochlorous acid which should not exceed about 1 mg. per litre. Bleach liquors containing higher proportions of activators are injurious to the cellulose.

This effect is interpreted on the basis of the Haber-Willstätter theory of chain-reactions, the activators being substances which form hydroxyl ions in solution. With hydrogen peroxide, for instance, the first stage is

$$H_2O_2 + ClO^- \longrightarrow HO_2 + Cl^- + OH$$

followed by

$$OH + ClO^- \longrightarrow ClO + OH^-$$

This generates hydroxyl ions:

$$ClO + ClO^- + OH^- \longrightarrow 2 Cl^- + O_2 + OH.$$

The process continues until the hydroxyl is used completely in the reaction

$$OH + HO_2 \longrightarrow H_2O + O_2.$$

With neither catalyst nor activator, the following reaction occurs:

$$HClO + ClO^- \longrightarrow ClO + Cl^- + OH.$$

The decomposition of hypochlorite is a second order reaction and calculations show that maximum activity takes place at pH 7 to 8.

It will be seen, therefore, that a very considerable amount of theoretical and practical evidence is available on accelerated oxidation and activated bleach liquors, as a result of which the bleaching of cellulosic material with hypochlorite solutions has been brought under control. Although the dangers of the neutral region with hypochlorite solutions are now very well known, the last word has not been written on this topic, for with knowledge of the data available it has been suggested to use the active liquors in a special manner for commercial bleaching of linen as described on page 290.

SOURING

As will have been gathered from previous remarks, the term "souring" means a treatment with dilute mineral acid to neutralise any alkali which may be in the fabric, and which on drying may concentrate and give rise to oxycellulose. Another reason for "souring" due to the bleaching process, as distinct from the scouring treatment, is associated with the use of calcium hypochlorite.

It has already been mentioned that calcium carbonate is formed during bleaching with calcium hypochlorite solutions, and the greater part of this compound is deposited on the cotton fabric from which it is very difficult to remove by washing with water. If these calcium deposits are allowed to remain on the cotton, it acquires a harsh feel and objectionable handle on drying; hence dilute acid is employed to change the carbonate into a soluble salt which, therefore, can be rinsed from the fabric with water. Hydrochloric acid is commonly employed for this purpose, and is employed at room temperatures in concentrations of 1 to 2° Tw.

The dangers of treating cellulose with mineral acids are well known on account of the formation of hydrocellulose (see page 43) with the accompanying tendering of the cotton. The acid must not be allowed to dry on the cotton and so become sufficiently concentrated to damage the material; as long as the cotton is kept completely covered with the cold dilute acid solution there is no danger of damage.

The manipulation of the fabric during the souring operation is similar to that which takes place during the chemicking; the rope of cloth may be circulated through the dilute acid and led into a cistern with the acid liquor where it is allowed to lie for about 30 minutes, after which a thorough rinsing with water is essential.

One of the commonest faults in the bleaching of cotton goods is due to incomplete removal of the sour, which concentrates and damages the cotton by acid hydrolysis. As previously mentioned, some bleachers adopt the precaution of rinsing the fabric with dilute sodium carbonate or ammonium carbonate (the so-called "sweetening-off") to ensure that the fabric is not in the acid state when dried.

It might be expected that the use of sodium hypochlorite instead of calcium hypochlorite would obviate the necessity of souring, for with sodium hypochlorite only soluble salts are formed and these

THE HYPOCHLORITE BLEACH

can be washed from the cotton directly. Under somewhat ideal conditions this procedure may be followed, but unfortunately, many bleach-crofts leave something to be desired and the cloth is apt to acquire various stains in its passage through the plant; some of these stains come from drops of condensed moisture from the roof and pipes, and contain traces of metal which do not become apparent until the fabric has been dried. Hence it is sometimes a wise routine procedure to sour fabrics which have been bleached with sodium hypochlorite solutions; in such a case, however, it is possible to use dilute sulphuric acid which is cheaper than hydrochloric acid.

It should be realised that the function of the sour is *not* to decompose the hypochlorite in the fabric, which should be thoroughly rinsed before souring. This washing step avoids the heat of reaction on neutralisation. and also helps to decompose the residual hypochlorite by the dissolved and entrained carbon dioxide in the water.

With efficient rinsing there should only be a trace of hypochlorite in the fabric when it is ready for souring; otherwise there may be a tendency to over-bleach and damage the fabric because of the accelerated oxidation produced by neutral oxidation with hypochlorite. In presence of a large excess of acid and only a slight trace of hypochlorite, the latter is rapidly decomposed to free chlorine, a slight odour of which is sometimes apparent.

It may, perhaps, be emphasised that the decomposition of residual traces of hypochlorite by acid is a secondary and very minor consideration; no attempt should be made to "clear" the white or complete the bleaching action in the souring treatment with dilute acid.

Souring after the bleaching operation is sometimes referred to as the white sour to distinguish it from the souring after the scouring process—sometimes termed the grey sour. Attention has been drawn to the fact (see page 154) that the grey sour originated in the lime-acid-ash sequence of scouring, where the function of the acid is to decompose the lime soaps; here again the use of the sodium compound in place of the calcium compound (caustic soda in place of lime water) does in theory obviate the necessity for a sour, but nevertheless the fabric is cleaner and of better colour after a sour subsequent to a caustic boil.

It has been suggested by Dubeau and Vincent (*Textile World*, 1943, *93*, No. 5, 79) that sodium bicarbonate solutions may be preferable for souring, particularly in the grey sour. The goods are stated to be cleaner, whiter, softer and more absorbent; the souring and rinsing take less time than with acid and there is no need for very careful chemical control.

CHAPTER XVIII
AROMATIC CHLORAMIDES AND CHLORITE

For many years, solutions of calcium or sodium hypochlorite have been used for the bleaching of cotton and linen materials; they are cheap and simple to use, particularly in view of the large amount of information now available as to the correct methods of utilising them to their best advantage and avoiding the dangers which they may possess. Nevertheless, suggestions have been made for the application of other compounds which contain chlorine and which exhibit mild oxidising properties; although they are not cheap, yet their properties are of interest in several directions.

CHLORAMIDES

CHLORAMINE T and DICHLORAMINE T have been mentioned on page 197 as substances which may be added to the kier liquor during the scouring process, and assist the purification of the cotton on account of the bleaching action they possess at high temperatures. Chloramine T or Activin is the sodium salt of paratoluenesulphochloramide

$$CH_3-C_6H_4-SO_2.N(Cl)(Na)$$

Chloramine TO is the corresponding ortho-derivative, and Chloramine BX is a somewhat similar meta compound in which the methyl group is replaced by a carboxyl group; it gives an acid reaction in solution. These products are described in B.P. 241,579, 251,580.

Peractivin or Dichloramine T is a similar product in which both the hydrogen atoms of the amino-group are replaced by chlorine; it is therefore capable of providing twice as much chlorine:

$$-SO_2.N(Cl)(Cl)$$

Activin is a white powder which is very soluble in water; the aqueous solution is quite stable and only decomposes at the boil:

$$-SO_2.N(Cl)(Na) + H_2O \longrightarrow -SO_2.NH_2 + NaCl + (\dot{O}).$$

CHLORAMIDES AND CHLORITE

Decomposition is slow in the absence of compounds which may be oxidised, but in presence of acid it is rapid; hence Activin is a moderate oxidising agent in neutral or alkaline solution.

The expense of these reagents is a disadvantage when large scale work is considered, for hypochlorite is very cheap.

The use of the aryl sulphodichloramides in textile bleaching has been outlined by Feibelmann in U.S.P. 1,892,548. For instance, p-toluenesulphodichloramide, which is insoluble in water, may be used in suspension or in alkaline solution. Compared with Activin or the monochloramide, the bleaching effect is greater on account of the two molecules of available chlorine.

$$CH_3.C_6H_4.SO_2.N\genfrac{}{}{0pt}{}{Cl}{Na} + H_2O = CH_3.C_6H_4.SO_2.NH_2 + NaCl + O$$

$$CH_3.C_6H_4.SO_2.NCl_2 + 2NaOH = CH_3.C_6H_4.SO_2.NH_2 + 2NaCl + O_2$$

The dichloramide dissolved readily in hot water in presence of alkali metal hydroxides. For example, 1 to 2 parts of p-toluenesulphodichloramide may be dissolved in 1,000 parts of hot water containing 3 to 6 parts of soda ash, and utilised for the bleaching of rayon over a period of one to two hours. Alternatively, the dry dichloramide may be mixed with an equal part of soda ash and kept as a stock reagent. One part of this mixture may be added to 1,000 parts of water and the temperature raised to 70 or 80°C.; the dichloramide does not entirely dissolve but forms a suspension which dissolves as the chlorine is consumed in the bleaching action. More rapid action may be obtained by dissolving 1 to 2 parts of the dichloramide-soda ash mixture in 10 to 20 parts of 5% caustic soda solution and diluting with 1,000 parts of water. The solution may be heated to about 70°C. and neutralised by the addition of acetic or hydrochloric acid.

The use of the dichloramide in the kier is seen in the addition of 1 to 2 Kg. of the dichloramide, separately or with 1 to 2 Kg. sodium carbonate or trisodium phosphate, for every 1,000 Kg. of cotton; scouring may take place for 4 to 6 hours at a pressure of 1 to 2 atmospheres.

CHLORITE

The use of chlorites has been suggested by the Mathieson Alkali Works (B.P. 380,488) as safe bleaching agents for cotton as they can be used over a wide range of conditions of time, temperature, acidity and alkalinity, the deleterious action being negligible. Sodium chlorite, $NaClO_2$, dissolves in water at 20°C. to an extent of about 40% by weight, and forms a stable solution. The anhydrous material which is marketed does not cake on storage.

TEXTILE BLEACHING

Now in the case of bleaching with solutions of hypochlorites, a number of precautions have to be taken and careful control exercised to render the process safe, but with chlorite bleaching the need for control is almost eliminated as the concentration, temperature and acidity are not particularly important in respect of damage; bleaching can be effected under acid conditions without fear of degradation for chlorine dioxide (ClO_2) is produced, whereas chlorine is formed from hypochlorites.

According to Dubeau, MacMahon and Vincent (Am. Dyes. Rep., 1939, *28*, 590), the activity of chlorite bleach liquors is determined by the rate at which chlorine dioxide is liberated and this increases with (*a*) a decrease in the pH value of the solution, (*b*) increase in temperature, and (*c*) increase in concentration. It has been established that even in boiling solutions of pH 3, with a concentration of chlorite in excess of any commercial requirement, there is no significant degradation of cellulose on prolonged treatment. Any attack from solutions of lower pH values is due to the effect of the acid and not the oxidising action of the chlorite.

The sodium chlorite is marketed under the trade name of Textone, and as the metals used in textile plants have no catalysing action on acidified Textone, there is no danger of degraded cellulose on this account, rendering it unique among common textile bleaching agents. Acidification of sodium chlorite solutions, in absence of any oxidisable matter, gives chlorine dioxide and chlorate, but no chlorine except with hot concentrated hydrochloric acid. Treatment with sodium hypochlorite above pH 11 to 12, gives chlorate and chloride but at lower alkalinities some chlorine dioxide is formed. When sodium chlorite solutions are treated with chlorine, there is formation of chlorine dioxide especially if the production of chlorate is hindered by the removal of the chlorine dioxide as it forms.

Mixtures of sodium chlorite with chlorine or sodium hypochlorite form good bleaching agents, as discussed later, as there is little deleterious action on cellulose particularly at pH 8 to 9 where little chlorine dioxide or chlorate is formed.

Sodium chlorite (Textone) with its remarkable properties, offers certain advantages over the usual and less expensive bleaching agents.

The reaction with various carbohydrates has been examined and found to be slow in neutral solution but rapid in acid solution; quantitative measurements show that chlorous acid is the oxidant and the reaction, in the main, may be represented by

$$R.CHO + 3 HClO_2 \longrightarrow RCOOH + 2 ClO_2 + HCl + H_2O$$

A short and simple bleaching operation may be utilised in which scouring and bleaching are combined in a single treatment involving the addition of a synthetic detergent to the hot acid Textone bath. Full scale trials on spun rayon cloths gave good results in 45 minutes at 80 to 90°C., 1,000 yards of cloth requiring 1·5 lbs. of Textone, 1 lb. of synthetic detergent and 4 quarts of 28% acetic acid. This type of treatment is covered by B.P. 535,107.

Oxidation potentials of sodium chlorite solutions containing 1 g. per litre available chlorine (1/142 M) showed 0·79 v. at pH 4 and 0·66 v. at pH 9, compared with 1·2 to 0·95 v. for similar sodium hypochlorite solutions in the range of pH 7 to 10; although hypochlorite under alkaline conditions is a stronger oxidising agent than Textone under acid conditions, the absence of factors producing degradation of the cellulose renders the latter superior. Similar considerations apply in comparing peroxides with chlorites.

Advantage may be taken of the slow oxidising action of Textone on the alkaline side of neutrality, to utilise it as an addition to kier liquors. Trials are quoted in which the use of 1·5 lbs. of Textone per 1,000 lbs. of cloth eliminated the acid steep and the double scour; no changes in the operation of the kier were necessary and there was no corrosion even with pressure scouring.

The use of chlorine in conjunction with a chlorite in aqueous solution for bleaching purposes is suggested in B.P. 519,561. Vincent, Dubeau and Synan (Am. Dyes. Rep., 1941, *30*, 358) recommend a ratio of hypochlorite to chlorite of 1·5 : 1 in terms of available chlorine and a pH value of between 8·7 and 10·0. The usual hypochlorite equipment and procedure may be employed but a considerably smaller quantity of total available chlorine is required; cotton goods were stated to have a more even and permanent white, a higher tensile strength and lower oxycellulose content than with hypochlorite alone.

For bleaching cellulose fibres, Textone may be used under acid conditions at elevated temperatures (60°C.), or with hypochlorite at room temperatures; both reactions are characteristic of chlorite in that no chlorine is liberated.

$$5\ NaClO_2 + 4\ HCl \longrightarrow 4\ ClO_2 + 5\ NaCl + 2\ H_2O$$
$$2\ NaClO_2 + NaClO \longrightarrow 2\ ClO_2 + NaCl + Na_2O$$

Hypochlorite may therefore be used to activate the chlorite even when slightly alkaline. The speed of bleaching depends on pH and is more rapid as pH falls; this is shown by the following data on properly "bottomed" cloth.

Speed of Bleaching

Time	pH
30 mins.	8·8– 9·0
2 hours	9·2– 9·4
6–12 hours	9·7–10.0

The best method of maintaining pH is by buffering; for example, sodium carbonate and bicarbonate (30 lbs.) may be added to the bleach liquor (1,000 galls.) and the range of pH adjusted by varying the ratio of carbonate to bicarbonate.

BUFFERING OF CHLORITE-HYPOCHLORITE

Bicarbonate	Carbonate	pH
3	1	8·7–9·0
2	1	9·2–9·4
1	2	9·7–9·9

As previously mentioned, the ratio of hypochlorite to chlorite was 1·5 : 1 in terms of available chlorine; in these circumstances the saving in chlorine consumption over hypochlorite alone may be 33 to 50%.

CHAPTER XIX
GENERAL CONSIDERATIONS

THE somewhat complicated series of processes, with their various repetitions, which go to make up the full madder bleach and which depend on the lime boil, have been discussed on page 154. As previously mentioned, the lime boil is rapidly declining in popularity in favour of the caustic boil, in spite of the fact that some authorities maintain that the lime boil is still the best for certain classes of goods; the reasons for this opinion are not always clear, but if they are based on the "crisp" handle of goods which have been subjected to the lime boil, and which is due to the low wax content of the fabric, then it may be stated that an equally thorough bottoming with caustic soda leaves little to be desired.

In view of the present position, it may be sufficient to restrict the consideration of the sequence of operations in various common bleaching processes to treatments in which the lime boil does not feature.

In comparing the two stages of processing, scouring and bleaching, it will be realised that the latter is much more efficient than the former in actual practice; scouring still remains somewhat unsatisfactory as a modern process, whereas the bleaching effect is relatively simple. With regard to attack on the cellulose, however, careful scouring may be carried out without increasing the fluidity, but even the most careful treatment with hypochlorite is accompanied by an increase in fluidity of the cellulose if a good white is to be obtained.

In considering the scouring operation, it is important to remember the importance of suitably soft water and the necessity for even packing of the fabric and regular circulation of the liquor. As previously mentioned, a popular size of kier will accommodate 2 tons of cloth, and as most kiers require a liquor to cloth ratio of 6 : 1, some 2,500 gallons of liquor will be required. The alkali present must be adequate, not only to remove the impurities from the fabric, but also to maintain them in a dispersed state. Further, with high-pressure kiers, it is essential that all air should be excluded before the boiling starts and also that the scouring liquor should be removed from the fabric immediately the scour has ceased, and that this removal should take place in absence of air, that is to say, the kier must not be drained, but the cotton should be washed with hot water and then with cold water.

A good scour is essential for the success of the subsequent bleaching operation, which must rely on a uniform impregnation of the fabric and an even distribution of the liquor. Any wax or fatty matter which remains in the cloth after the scour will tend to inhibit permeability, and the presence of calcium or magnesium soaps also renders bleaching difficult and irregular for the same reason. Wetting agents are of assistance in both scouring and bleaching.

Between the various stages of the scouring and bleaching operation, the fabric is washed with water; it is almost impossible to over-emphasise the importance of the washing process. Products which are rendered soluble by the actual scour or bleach are removed by washing with water, for if allowed to remain the fabric would be unsatisfactory; further, a copious supply of water is required to remove the various reagents which are employed in purifying the fabric, for if these were allowed to be carried over from one stage to the next, not only would they often interact with each other, but sometimes they would bring about very substantial damage to the cotton. It has been estimated that in the scouring and bleaching of cotton, it is necessary to use about 4,000 gallons of water per 100 lbs. of cotton, compared with 10,000 gallons per 100 lbs. of linen.

Most cotton fabrics are treated in rope-form and are drawn from one part of the works to another through the well-known pot-eyes. In passing from the washing machines, however, the final squeeze should be adequate to ensure that no great excess of water is entrained in the fabric to dilute the next liquor to an unnecessary extent. With ropes of fabric in different stages of treatment being drawn about the works, it is essential that no undue excess of liquor should accompany them, or else damage may occur, as for example, by acid from the soured cloth dripping on to a lower rope of washed fabric on its way to the white box for subsequent drying.

Modern technique relies on automatic pilers for entering the fabric into the kiers, chemicking cisterns and souring cisterns so that even piling is assured. In the chemicking and souring operations, it is customary to circulate the liquor through the fabric, and lead pipes are often preferred for the feed; it is important that there should be no iron in the souring system. The cisterns for chemicking and souring may be made of slate, stone or cemented brickwork, although some bleachers prefer to line the base and sides with strips of wood. Separate cisterns are provided for grey and white chemicks and also for grey and white sours.

Cloth for printing, as previously mentioned, is sometimes impregnated with chemic on a mangle or rope machine and then piled on wooden stillages until bleaching is complete; with this method there is generally a tendency to over-bleaching.

GENERAL CONSIDERATIONS 239

BLEACHING ROUTINES

Before considering some of the various bleaching routines, it may be of interest to point out that many of the variations on the old lime boil have been recorded by Trotman and Thorp (*The Principles of Bleaching and Finishing of Cotton*, Griffin, London, 1927).

In the following accounts of the treatment of cotton, it is assumed that the fabric has been desized before passing to the kiers; the summarised descriptions, of course, should be taken in conjunction with the detailed discussion of scouring and bleaching previously given. For the moment, the accounts are limited to the normal bleaching routines, and discussion of some specialised methods of scouring and bleaching are to be found on page 279.

The purification of cotton goods has been classified by Trotman and Thorp (*supra*) under four headings:

(a) Full bleach which denotes the highest standard of purity.
(b) Half-bleach for goods which are to be weighted or assisted in finishing.
(c) Bleach for dyeing and printing where a high degree of absorbency is of more consequence than a perfect white.
(d) Bleach for coloured-woven goods where the chief object is to bleach the undyed part of the cloth without altering the colour; in this case, it is obvious that the coloured portion has already received some purification before the dyeing of the yarn.

From the standpoint of manipulation, one of the simplest methods of scouring and bleaching is the open-width system of Edmeston-Benz (see page 181), which merely involves a low-pressure scouring treatment, after which the goods are piled for a time and then washed, chemicked, soured and washed.

The Jackson method is also very simple, the cloth being passed from one roller to another in the special kier, after which it is removed and may be treated with water, chemic, water and dilute acid by winding from batch to batch as in the kiering operation.

Many curtain materials, nets and other goods of low quality may be given a simple treatment consisting of one high pressure boil with caustic soda, followed by washing, souring, washing, chemicking, washing, souring, and washing. The scouring may take place in a 2% solution of NaOH (estimated on the cotton) and the chemic may be used in a concentration of approximately 4°Tw. where calcium hypochlorite is employed.

Lightweight cotton goods for printing are required to have a high degree of absorbency rather than a perfect white; hence it is often customary to give the fabric two scours. The cloth should first be desized and then boiled in 2% NaOH solution for about 8 hours under 30 lbs. pressure; the goods may then be washed in the kier for about 2 hours, or alternatively, they may be washed in a washing machine and so turned over before passing into the second kier where they may be boiled in 1% sodium carbonate solution. The washing in the kier, of course, will save a considerable amount of time. Following upon the second scour, the goods may be washed, and then chemicked in 2°Tw. calcium hypochlorite solution, with the liquor circulating for 2 hours, followed by washing, souring and washing.

A popular process for many plain white cotton goods is to boil the desized fabric in a solution containing about 0·75% of NaOH and 1% of soap estimated on the fabric, gradually raising the pressure to about 20 lbs. per square inch over a period of 2 hours and then maintaining the conditions for a further 4 to 5 hours. The cloth is then withdrawn from the kier and washed in the rope-washing machines and passed to the cisterns to be soured for an hour in 1°Tw. hydrochloric acid solution, after which it is turned over into a second cistern for a further sour. The material is then washed again, and run into a kier for a second scour with 0·5% soda ash and 0·5% of soap for about 5 hours, after which the goods are washed, chemicked, washed, soured, and washed.

A more thorough process for light and medium-weight goods depends on a high-pressure boil with caustic soda (1%) for 8 hours, followed by washing, chemicking for 2 hours in 2°Tw. calcium hypochlorite solution, washing, souring and washing; the goods are then scoured again for 6 or 7 hours in 2% sodium carbonate solution, under low pressure, followed by washing, chemicking as before, washing, souring and washing. It will be noted that this method includes two scouring stages and two chemicking stages; the scour with soda ash takes places between the two chemicks.

Heavier cotton goods may be processed by scouring in a high-pressure kier for 8 hours in 2·5% of NaOH, calculated on the weight of the cotton, followed by washing in the kier and then through the washing machines, after which the goods are again scoured in a mixture of caustic soda and sodium carbonate (75 : 25) corresponding to 1·5% Na_2O on the weight of cloth. The material is then washed in the machines, soured, washed and chemicked for 2 hours in 2°Tw. calcium hypochlorite solution. After washing, souring and washing again, the fabric may be scoured in 1·5% of sodium carbonate solution for 4 hours at low pressure, followed by washing, chemicking, washing, souring and washing.

Where coloured-woven goods are to be scoured and bleached, a great deal must obviously depend on the fastness of the colours, but with the modern tendency towards colours of the highest degree of

GENERAL CONSIDERATIONS

fastness, it is possible, in many cases, to conduct the scouring in the kier. With high-quality shirtings, the singed and desized goods may be entered into a kier and treated at 50 to 60°C. for 8 hours in a solution containing 0·5% soap and 0·5% soda ash; the goods are then washed and chemicked for an hour, followed by washing, souring, and scouring in a kier again under similar conditions to those of the first or grey scour. A second washing and chemicking complete the process, followed, of course, by the usual washing, souring and washing.

An alternative method for dealing with coloured-woven goods is to treat them on a wince machine or rope-soaping machine; solutions of 1% soda ash and 0·65% soap may be employed at 80°C., for owing to the freedom of the fabric from heavy pressure against itself there is less danger of the colours "marking-off." The more concentrated scouring liquor and the higher temperature are accompanied by a shorter time of treatment which may be only 20 to 30 minutes with the rope of cloth running through the machine at about 30 yards per minute.

OPENING

In dealing with cloth in rope-form, it is necessary after the scouring, bleaching and washing operations, to restore the cloth to the full width or open state. This is effected by an opener or scutcher. The apparatus consists of a revolving brass beater and two spiral brass scrolls which help in the opening.

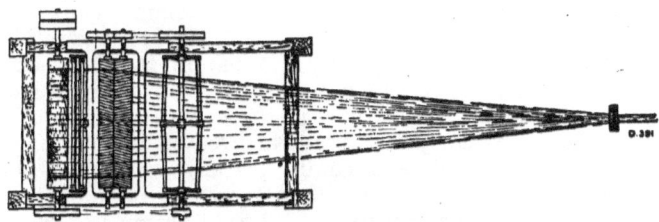

FIG. 76.—Diagram of scutcher for opening cloth from rope-form.

The cloth, in rope-form, passes through suitable guides, such as pot-eyes, which must be situated at about 30 feet or more from the opener; when the cloth comes to the scutcher it first meets a rapidly revolving beater which opens it into loose folds. It then passes between the scroll rollers which are covered with copper twigging, and these act on the cloth still further and complete the opening action; the beater and scrolls are driven in the opposite direction to the cloth.

TEXTILE BLEACHING

In order to keep the cloth centrally positioned as it passes through the scutcher and to correct any tendency to wander, a governor, which comprises an oscillating or swivelling set of bars pivoted on the centre, is placed in front of the draw rollers.

Although the scutcher is generally arranged in a horizontal position, a vertical arrangement is also possible.

The scutcher may be arranged to work with a plaiting mechanism or arranged to run continuously with a drying machine.

RAW COTTON, SLIVER AND YARN

The fundamentals of cotton bleaching are not seriously affected by carrying out the process on raw cotton, sliver, yarn, hosiery or piece goods; the main differences in treatment are connected with the handling of the material, that is to say, with mechanical manipulation rather than chemical treatment.

Raw Cotton

Some of the poorer qualities of cotton and also cotton waste are occasionally bleached before spinning into yarn of medium quality for use as wefts in coloured-woven goods. The general method of treatment is in a machine of the type employed for the dyeing of loose cotton, that is, a machine which consists essentially of a closed vessel and a pump for circulating the liquor. In many cases, the raw cotton is packed into the machine and then "wet out" with hot water and a wetting agent; when the cotton has cooled, a solution of hypochlorite is circulated, followed by water, dilute hydrochloric acid, water and an antichlor such as 0·5% sodium thiosulphate solution. The cotton is finally washed and soaped. It will be noticed that there is no scouring operation in this process, as it is considered necessary to retain the greater part of the wax so that the cotton may easily be spun into yarn. Motes and shives are not affected by this process as they seem to require treatment with alkali under some pressure. Where scouring before bleaching is required, the same type of machine may be used and the raw cotton scoured with caustic soda solution. After the bleaching process, the raw cotton may be lubricated with soap solution or one of the new synthetic products.

Absorbent cotton for surgical purposes usually comes from the waste of combed Egyptian cotton, but ordinary wadding is made from gin waste or the ordinary waste from spinning mills.

The bales of waste are opened and as much as possible of the dust and dirt is removed by mechanical means, such as given in an opening willowing machine, which also opens the material, as its name implies, and affords an opportunity of blending different types of cotton waste. The cotton is then placed in a kier by hand labour and sprinkled or wetted from time to time with warm water and a wetting agent, every effort being taken to ensure that the packing is firm and uniform. When the kier is full, hot water is introduced and the cotton steeped for some time. After the water has been drained off, the scouring liquor is introduced; the common lye consists of 1% NaOH and 1% Na_2CO_3, together with some soap, sulphated fatty alcohol or other kiering auxiliary. Scouring may continue for from 6 to 10 hours at 30 lbs. pressure, according to the nature of the cotton undergoing treatment; the type of kier usually employed for this work is equipped with pump and multitubular heater. After scouring, the cotton is washed in the kier and allowed to drain.

The loose cotton is then removed by hand to a large cistern with a perforated or slotted wooden floor, the spaces being covered with cloth to prevent loss of the fibres; here it is washed again and then bleached in the ordinary manner with hypochlorite which circulates through the cistern. Washing, souring and washing complete the bleaching process, but the cotton is finally passed to a machine of the "harrow" type (see p. 301), where it is soaped and blued if required. Hydro-extraction is followed by drying on a lattice machine, and the dried material then passes to a scutcher and to a carding engine, the laps from which, when superimposed, comprise the well-known "cotton-wool."

Sliver

The bleaching of SLIVER may be carried out in a simple manner if required, the chief difficulty being so to arrange the material that any channelling is avoided and even circulation of the liquors assured. Package dyeing machines are easily adapted for this type of work, and the various liquors circulated in the usual manner.

As with the case of raw cotton, the material must be carefully packed into the machine, and when it has been thoroughly wetted with warm water and a wetting agent, the liquor is replaced by 1 to 1·5% NaOH solution which is circulated at the boil for 2 or 3 hours, after which the cotton is washed in the machine with hot and then cold water. Bleaching in the same apparatus follows, utilising sodium hypochlorite solution of 2° Tw., which is circulated for an hour, after which the cotton is washed, acidified with dilute hypochloric acid and washed again.

Yarn

YARN, in the form of cops and cheeses, may also be scoured and bleached, but the process is more difficult with cops, which may be placed in bags and kier-boiled. Alternatively, it is possible to use horizontal kiers with their wagons where large quantities are required. A vacuum kier is sometimes used for cops of hard-spun yarn; inside the kier shell is a secondary perforated wooden container in which the cops are placed. The air is removed from the kier and the scouring liquor admitted from the top, when it rapidly penetrates the cops and is drawn off at the bottom. The same principle is used for the various solutions at the different stages of the bleaching process.

Although the scouring of yarn in hank form is generally easier than the treatment of woven goods on account of the open nature of the material and the absence of size, yet it may be necessary to treat the yarn in the stocks to remove dirt.

Where it is required to bleach yarn in hank form, several possibilities present themselves. Small quantities may be treated on sticks or poles in wooden becks as in the dyeing of hanks; they must be turned from time to time, as also in dyeing. The time of boiling is determined by the nature of the yarn, but somewhere between 1 and 3 hours may be taken as illustrations of treatment in 2°Tw. NaOH solution or 5°Tw. Na_2CO_3 solution. Larger quantities are sometimes treated in bundles; the ties are removed and the bundles loosely bound with one hank which is placed round the bundle at right angles to the length of the bundle; one end is then slipped through the loop formed by the other end and then over the entire bundle. Such bundles may be treated in bags in kiers or in package machines. An alternative method is to thread the hanks into a long chain and treat them in rope form. In both cases, the knotted portions tend to tighten and so resist penetration by the circulating liquors; it is advisable therefore to keep the knots as loose as possible and to examine them from time to time.

WARPS may be scoured by running through tanks fitted with guide rollers and containing solutions of soda ash from 5 to 10°Tw. concentration; they are washed in similar tanks. The bleaching proper is usually carried out in a bleaching cistern through which the hypochlorite solution is circulated. Warps on the beam may be scoured and bleached on a special perforated back beam, which is placed horizontally in a special tank, and one end connected to a powerful pump which circulates the scouring and bleaching liquors through the beam and the warp.

HOSIERY

The scouring and bleaching of knitted cotton goods is much more readily effected than the treatment of piece goods; the soft-twisted yarns and the open nature of the knitted structure afford easy penetration by the scouring liquors and the bleaching solutions.

GENERAL CONSIDERATIONS

With tubular knit material, the fabric may be plaited into kiers in the ordinary manner, but it is sometimes preferable to treat them in the roll or in bundles on account of their spongy and extensible character. Manufactured hosiery and garments may also be treated in bundles, and so may lace and net goods whose structure is apt to be distorted if drawn about the works in rope form under some tension. These bundles may be wrapped in net or in bags of knitted webbing.

Because of their open nature which facilitates all wet processing, it is possible to treat the goods in one kier, and without removal, during the whole sequence of operations from kiering to bleaching and washing. In these circumstances, the kiers are often made of wood and the bundles are carefully packed to ensure an even treatment. Low-pressure boiling for 6 or 7 hours is generally adequate for removal of the natural wax and fats, but it is often necessary to treat some of the knitted material specially to remove adventitious dirt; this is done in the tom-tom machine (see p. 349), which is referred to in the Midlands of England as the dolly, and must not be confused with the dolly of Yorkshire, used for the scouring of woollen and worsted piece goods, as described on p. 331.

An alternative to the dolly is seen in the stocks, but often the bundles may be treated in the dash-wheel, or indeed by merely boiling them in an open kier with a high ratio of liquor to cloth.

The stocks and the tom-tom machines are very efficient in promoting detergency, and also in removing the dirty liquors from the scoured materials; soapy liquors are used for scouring knit-goods in the stocks or the dolly.

After washing, excess of liquor may be removed on the centrifuge, after which the goods may be returned to the kier for bleaching in the usual manner. The final removal of the rinsing water also takes place on the centrifuge.

STAINS

Stains of various types are apt to occur from time to time in the treatment of goods in the bleach-croft. The commonest types are due to mineral oils from contamination with machinery, metal stains chiefly from iron, mildew stains which appear on the grey goods occasionally, stains which appear on kiering through uneven treatment of one sort or another, and a general yellowing of the fabric which may occur on ageing if the cotton has not been properly scoured.

The presence of mineral oils may be shown by their fluorescence in ultra-violet light, but this property may be inhibited to some extent by the presence of metallic powders. The vegetable oils.

exhibit no fluorescence, but tallow and paraffin wax can often be recognised in this way.

The oil stains themselves may be difficult to remove in the kier, but most of them yield to treatment by hand with a mixture of potassium coconut oil soap and trichlorethylene; the chief danger of the mineral oils in bleaching is not the difficulty of removal, but the fact that they are often associated with metals which catalyse the hypochlorite or peroxide and cause local damage.

The presence of iron stains or rust is easily detected and may be removed by hand-treatment with oxalic acid, potassium oxalate, or hydrochloric acid solution; the acid should not be allowed to dry or tendering will occur. Copper stains, which sometimes occur on knit goods, may be removed with 4% potassium cyanide solution. After the removal of any of these stains, the cloth should be thoroughly rinsed around the treated areas.

Where cotton goods have been stored for some time under moist and warm conditions with poor ventilation, mildew may form on the cotton, unless it has been protected; the presence of mildew is often recognised by the odour, long before the spores are visible on the cloth. Where there is doubt, the goods should be kiered as soon as possible; alternatively they may be dried on steam-heated cylinders to destroy the spores and then treated as soon as possible.

Scoured fabrics from the kier sometimes show stains or marks which may be of two main types, the first of which is termed "channelling," and is probably due to uneven attack of the alkali on the natural colouring matter in the cloth. This is in agreement with the fact that the "stains" are often warm when the cloth is removed from the kier showing that the water, and hence the scouring liquor, has not been in regular contact with these areas. The marks due to channelling are removed in the subsequent chemicking or bleaching process.

The second type of stain is called "stone-marks" because they occur at the bottom of the kier where the fabric is in contact with the drainage stones; the shape of the stain is often similar to that of the stones. It has been suggested that these stains are due to the filtering of some of the impurities by the fabric itself; the liquor is coolest at the bottom of the kier and towards the end of the scour when the alkalinity is reduced there is a tendency for some of the impurities to be precipitated. The remedy is to have adequate alkali and good circulation. The tendency for stone-marks to form is aggravated by the intermittent working of the circulating pump, coupled with uneven circulation, when the suction at the bottom of the kier produces vapour and precipitates the solid matter. If these precipitated materials are allowed to dry on the fabric, or if the cloth becomes overheated, the stains are difficult to remove; normally however, they are removed during the second scour, or in the later washing and bleaching operations. It has been suggested to introduce filters into the circulatory system of the kier, but they do not appear to absorb the colouring matter, and further, they are apt to become clogged and hinder circulation of the liquor and so reduce the efficiency of the kier.

Yellowing

At one time considerable difficulty was caused by the yellowing of bleached cotton goods with time. Various suggestions were put forward to account for this phenomenon, and it appears to be established that one contributory factor is the gradual decomposition of the calcium or magnesium soaps which may be present when hard water has been utilised in the bleaching process. The yellowing, however, also occurs in the absence of these soaps.

It was also suggested that the nitrogenous impurities were responsible for the yellowing, for the chloramines formed would gradually decompose, but here again it was shown that yellowing occurs in the absence of nitrogenous matter, which, after all, is readily removed in a good scour.

The chief cause appears to be due to an excess of cotton wax, for Trotman (*Bleaching and Finishing of Cotton*, Griffin, London, 1927, p. 299) states that in nearly every case of yellowing which he examined, wax has been present which could be removed by extraction with benzol with an improvement in colour. Cotton wax varies in colour from yellow to dark green, so this substance is quite capable of causing discoloration if it works its way to the surface during storage. Hence the better the scour, the less the danger of yellowing.

The tendency to yellowing on storage may be estimated by ascertaining the ease with which the fabric "scorches" under a hot iron, or, alternatively, by exposing a sample to steam under pressure in an autoclave or "ager" for an hour or so.

Yellowing is also caused by the formation of oxycellulose, but this is accompanied by tendering.

CHAPTER XX
THE PEROXIDE BLEACH

THE use of hydrogen peroxide as a bleaching agent has aroused great interest within recent times, particularly in the U.S.A.

There are very definite advantages in its use, in certain circumstances; it is a universal bleaching agent, whereas hypochlorites cannot be employed with wool and silk. Another advantage is that the manipulation of the fabric is less, so that although chemical costs may be higher, if we are approaching an era of high labour costs, then it may be preferable to use more expensive reagents which entail less labour. It has also been stated, and with reason, that the losses in weight in bleaching with peroxide are less than with hypochlorite, and this may be a great asset in the treatment of towels and other goods where condenser yarns are used.

Less water is also required with the peroxide bleach and there is no necessity for souring.

The fact that hydrogen peroxide is used at a raised temperature has brought forth many attempts to utilise this product in one stage which combines both scouring and bleaching; further, as will appear later, the use of hypochlorite and peroxide has some interesting possibilities.

The earliest methods of bleaching with peroxides relied on the application of sodium peroxide solutions, but owing to the later manufacture of stable and concentrated solutions of hydrogen peroxide, the sodium compound is rarely utilised.

Bleaching with peroxides seems to have passed through three phases in its history, the first being the early use of barium peroxide and the solutions of hydrogen peroxide which could be made from it; this came to an end about 1912, when sodium peroxide became available in commercial quantities. Since about 1927, however, the concentrated solutions of hydrogen peroxide made from barium peroxide and phosphoric acid, and the later developments utilising potassium persulphate and persulphuric acid, have given a great impetus to the use of hydrogen peroxide.

It must be remembered that the aim of most bleachers is to produce the ideal process—a one-bath treatment. The first use of hypochlorite solutions, many years ago, was attended with great disappointment because it was believed that the "chemic" would not only obviate the "grassing" but also the "bowking." In considering the possible uses of hydrogen peroxide, the prospect of a single process is much brighter, for there appears to be no theoretical

THE PEROXIDE BLEACH

reason why a hot alkaline solution should not scour and bleach in one operation.

It is not always fully realised that the whitening stage of the bleaching process is much simpler than the scouring treatment for the vegetable fibres; any method of shortening or eliminating the scour will receive some consideration provided it is safe and economic. Hence the fact that peroxide solutions could be used in the warm alkaline state aroused great hopes.

SODIUM PEROXIDE

Sodium peroxide is generally sold in the form of a light-yellow, anhydrous powder whose oxygen content is almost 20%; for bleaching purposes, the powder, as marketed, usually contains a little trisodium phosphate to act as a stabiliser when the peroxide has been dissolved. Sodium peroxide absorbs moisture on exposure to air and becomes white, but this does not seem to interfere with its bleaching action; a more serious property, however, is the rapid oxidising action on paper, straw, cotton-waste, wood shavings and other similar packing materials, and this is often sufficient to cause combustion, sometimes with great violence.

In the manipulation of sodium peroxide, it is well to avoid contact with copper, brass and other metals with the exception of tin; hence tinned scoops or enamelled iron utensils should be used.

When sodium peroxide is added to water, it dissolves rapidly with generation of heat to give a strongly alkaline solution, and even for the bleaching of cotton goods, the alkalinity is excessive and must be neutralised. The general method of working is to prepare a stock solution containing about 40 lbs. of sodium peroxide per 100 gallons of water; this is from 2 to 4 times as strong as required for the actual bleaching operation. The dissolution may be brought about in a wooden vat, taking the precautions mentioned on p. 252. Care must also be taken to ensure that neither the water nor the acid used for making the solution should contain dissolved metal, particularly copper or iron. Either sulphuric or hydrochloric acid may be used for partial neutralisation, 12·75 lbs. of the former or 9·5 lbs. of the latter per 10 lbs. of sodium peroxide which is slowly sprinkled on to the diluted solution of acid with stirring; it may be necessary to allow the solution to cool during the addition of the peroxide.

The actual bleach liquor is prepared by diluting the stock solution until it contains about 15 lbs. of sodium peroxide per 100 gallons.

The cotton goods should be thoroughly scoured before treatment with the peroxide liquor which may require 12 to 16 hours; an alternative to the steeping method is simply to saturate the cotton with the peroxide solution, and pile the goods in a warm place to complete the bleaching action. Washing with warm water and then cold water completes the process.

Brandt (Text. World, 1931, *79*, 1100) has given the following prescription for sodium peroxide: 104 lbs. of sodium peroxide, 130 lbs. of sulphuric acid, 100 lbs. of sodium silicate, 20 lbs. of sulphonated oil made up to 1,400 gallons with water. This gives a solution containing 1·63 grams per litre of available oxygen, an alkalinity corresponding to 0·13% Na_2O and a pH value of 10·3. Curves are given showing the changes which occur during the boil, up to 15 hours.

Methods of bleaching hosiery with sodium peroxide have been described by Hand (Text. World, 1933, *83*, 740). Suitable proportions appear to be 7·5 lbs. of sodium silicate (82°Tw.), 8 lbs. of sulphuric acid (168°Tw.) and 6·75 lbs. of sodium peroxide; the bleaching is conducted at 80 to 85°C. for an hour.

It seems to be generally agreed that better results are obtained if the goods are kiered beforehand.

A method of bleaching scoured cotton goods with sodium peroxide solution, used in some French bleach-works, is to make a solution in the proportion of 1 Kg. of sodium peroxide to 100 litres of water, together with 1·5 Kg of sulphuric acid of 66°Be (170°Tw.); the solution is then made alkaline with sodium silicate and used at 40 to 50°C. for several hours.

The concentration of sodium peroxide is generally of the order of 1·5% Na_2O_2 corresponding to 3 grams of available oxygen per litre. As is well known, the action of neutral peroxide is too slow for practical bleaching, but, on the other hand, if the solution is too strongly alkaline, there is some danger of the cellulose being attacked.

As hydrogen peroxide is much safer and more convenient to use than sodium peroxide, the latter is very rarely employed for bleaching nowadays.

HYDROGEN PEROXIDE

Hydrogen peroxide was discovered by Thenard in 1818 and was used for the bleaching of silk as early as 1878; such use was very limited and even the improved methods of manufacture in 1920 did not greatly extend its application except for the more delicate and expensive textile materials. One of the chief difficulties was the instability of the peroxide solutions, as then manufactured. About 1925, however, very great improvements were made in the production of hydrogen peroxide (B.P. 252,768) based on the reaction between barium peroxide and phosphoric acid, giving products of 100 and 130 vol. strengths of greater purity and stability than the old 10 vol. strengths. Hydrogen peroxide is usually sold in 10, 12, 20, 100 and

130 volume strengths; the 10 vol. strength gives ten times its own volume of oxygen, the 100 vol. strength will give 100 times its volume of oxygen, and so forth. The 10 vol. strength contains 3% H_2O_2 by volume, the 100 vol. strength contains 30% and the 130 vol. strength contains 40% of H_2O_2 by volume. The system of volume-strengths is convenient for many purposes; for instance, 1 gallon of 100 vol. strength will give 100 gallons of 1 vol. strength H_2O_2.

Impurities of all types should be avoided in the solution which should be stored in a cool place and away from sunlight; an acid solution of hydrogen peroxide is much more stable than an alkaline solution, so that all commercial solutions are stabilised with acid up to the time of actual use in bleaching.

The production of these concentrated stable solutions more or less coincided with the difficulties in connection with the bleaching of coloured-woven goods discussed on page 153, and aroused renewed interest in the possibilities of peroxide as a bleaching agent.

As previously mentioned, hydrogen peroxide is sold on a 'volume' basis, so that 10 vol. H_2O_2 will give ten times its volume of oxygen. Other convenient quantities are seen in the table.

Hydrogen Peroxide

	Oxygen liberated	H_2O_2 (%)
10 vol. H_2O_2	10 volumes	3·04
20 vol. H_2O_2	20 volumes	6·08
100 vol. H_2O_2	100 volumes	30·40

Hydrogen peroxide is occasionally defined by its active oxygen available, and in this connection it is sufficient to remember that 100 g. of 3% H_2O_2 solution will afford 1·4 g. of active oxygen, or that 10 vol. H_2O_2 contains about 14 g. of available oxygen per litre.

For estimation of the concentration of peroxide, the iodine method is preferred to the permanganate method. The solution of hydrogen peroxide is diluted to about 1 vol.; in the meantime, a solution is prepared of 2 g. of potassium iodide in 200 c.c of water, mixed with 30 c.c. of sulphuric acid solution (1 in 2) and allowed to cool. About 10 c.c. of the dilute hydrogen peroxide are added to this solution and allowed to stand for a little while in order that the reaction may take place, liberating the iodine which is then determined with 0·1 N thiosulphate solution. (1 c c. is equivalent to 0·0017 g. of H_2O_2.)

An important point in connection with the use of hydrogen peroxide solutions for bleaching is that the method is only practicable within a certain temperature range, roughly 80 to 85°C.; below 80°C. the bleaching action is inadequate for commercial purposes and above 85°C. the loss of oxygen is too rapid.

The stability of the peroxide solution varies with the pH, as shown by the time required to reduce H_2O_2 of 1 vol. concentration to 0·5 vol. concentration.

TIME OF FALL FROM 1 VOL. TO 0·5 VOL.

pH	Time
6·8	3 hours 10 mins.
7·1	2 hours 50 mins.
7·9	2 hours 10 mins.
8·9	1 hour 10 mins.
9·9	25 mins.

In addition to the control of stability by pH of the solution, it has been found that the addition of certain compounds appears to exert a specific action; one of the best known is sodium silicate. For instance, at 85°C. 1 vol. H_2O_2 solution containing silicate takes 90 minutes to fall to 0·5 vol. strength, but in absence of silicate, although at the same pH, there is complete decomposition in less than 5 minutes. Mixtures of silicate and carbonate accelerate decomposition, but mixtures of silicate and trisodium phosphate have a great stabilising action.

Peroxide solutions are also rapidly decomposed by many metals, of which copper appears to be the worst, and even copper alloys should be avoided. The use of iron, lead and aluminium is practicable in presence of silicate on account of the layer of the metallic silicate which is formed; stainless steel appears to be suitable, particularly that type with 18% chromium and 8% nickel.

Decomposition of peroxide is accelerated by the presence of moulds or mildew in the cotton fabric and may be shown by the great frothing which takes place.

Peroxide Bleaching

Although excellent results may be obtained with up-to-date apparatus, it is not so simple to produce a satisfactory bleach when utilising old kiers with a vomiting puffer-pipe operating on low-pressure steam; the cleaner and more modern the plant, the better the results. Attention has already been drawn to the catalytic action of most metals, and where a local action is allowed to develop in this way, the goods will be tendered. For this reason, some bleachers prefer to use a bleaching vat made of white hardwood, but even here care must be taken to ensure that the joints are made with hardwood pegs and not with nails, screws, or nuts and bolts. The vat is provided with a perforated false bottom below which is placed a steam-coil which may be made of tin or lead; some bleachers have used rubber hosepipe, but the low conductivity of this material is apt to present difficulties. A wooden cover for the vat prevents undue loss of heat.

Where efficient kiers are available, it is possible for peroxide to be used in the least expensive manner. The goods are scoured in the usual way, which would be utilised as a preparation for the hypochlorite bleach, but instead of withdrawing the cloth from the kier, it is thoroughly washed, and then the peroxide is introduced and circulated. A modification of the ordinary circulation has been stated by Smolens (Am. Dyes. Rep., 1939, *28*, 495) to give better results; the peroxide is withdrawn from the top of the kier and pumped into the well from which it is caused to rise through the cloth; a closed heating coil in the well is an advantage. The upward circulation tends to conserve heat and is beneficial in obviating channelling.

For bleaching cotton goods, a concentration of 0·5 to 1 vol. H_2O_2 is adequate, the process being carried out at or near boiling-point. Hydrogen peroxide is particularly sensitive to traces of many metals, as previously stated, and its activation will produce oxycellulose and considerable degradation; for this reason, therefore, where ordinary kiers are employed it is customary to line them with cement and silicate, or a similar composition, taking care that the surface is well covered. Weber (J.T.I., 1933, *24*, 178P) suggests the following procedure: four parts of Portland cement, one part of burnt lime, one part of magnesium oxide, are made into a thin paste with a dilute solution of sodium silicate of approximately 20°Tw. and sprayed or painted on to the inside of the kier. This lining is allowed to dry for 24 hours, sprayed with 20°Tw. sodium silicate solution and again allowed to dry, after which the coating is painted with 5% HCl and again allowed to dry when the kier is ready for use.

The pH of the solution has a great effect on its stability and efficiency, and peroxides are used under alkaline conditions; caustic soda and sodium silicate are added to the bath in a ratio to give sodium metasilicate and produce an alkalinity of pH 11·5. The addition of pyrophosphate is also commonly practised, alone or with ammonia. An American prescription given by Weber (loc. cit.) comprises 500 gallons of water, 40 lbs. of sodium silicate, 5 lbs. of sodium hydrate and 18 lbs. of 100 vol. hydrogen peroxide.

For yarn it appears preferable to bleach at about 85°C. for 90 minutes or so and then raise the temperature to 90 to 95°C. for a further period of 30 minutes. Knitted fabrics and hosiery may be bleached in winch machines, but piece goods are usually treated in the kier after a preliminary open boil; about 6 to 7 hours are required for the bleaching process, but with thoroughly scoured goods this may be reduced to about 2 hours. Wetting agents are beneficial in the liquor.

Weber suggests that where special apparatus is made for peroxide bleaching, it should be constructed of wood and lined with aluminium, the heating coils, valves, pipes and false bottoms for the bleaching cistern should also be made of aluminium, which is the most suitable of all metals for the purpose. Reference has been made to the dangerous effects of contact with heavy metals, and in this connection it is interesting to recall the experience of Kershaw (J.T.I., 1933, *24*, 193P), who found that traces of copper in the water supply had a detrimental action on the hydrogen peroxide.

Hydrogen peroxide bleaching has not been examined with the same detailed thoroughness as hypochlorite; for instance, there is no series of researches comparable with those carried out by the B.C.I.R.A. Weber (loc. cit.), however, states that the fluidity of goods bleached with peroxide is less than 5, and that the white is good. If the peroxide treatment is preceded by an alkali boil, the bleaching cost is greater than with hypochlorites, but as all the operations can take place in the kier there is a full utilisation of the plant and economy of labour. When the hydrogen peroxide and silicate treatment is the sole process, then there is considerable conservation of weight, but although the handle is softer than with normal bleaching, the white is not so good and the absorbency is poorer; a second bleach is advisable. In England, where hypochlorite bleaching has reached its highest pitch of efficiency, peroxide bleaching is less common than in Germany and the U.S.A.

The weakening of cotton with hydrogen peroxide has been examined by Kollmann (Melliand Text. Monthly, 1933, *5*, 162), who impregnated scoured cotton which was then dried and stored; loss in strength decreased with concentration of H_2O_2 and was negligible below 0·015% retained by the cotton. Hence appreciable tendering only occurs if the goods are insufficiently rinsed; after thorough washing no tendering need be expected.

The Kauffmann process for bleaching with peroxides is outlined in B.P. 352,690; the goods are first subjected to a pressure boil with sodium carbonate and caustic soda for a few hours, after which the liquor is run off, and replaced by hydrogen peroxide solution containing sodium silicate and a soluble oil. The second treatment is carried out at 80°C. If an iron kier is used it must be coated with calcium silicate beforehand to prevent decomposition of the peroxide; the process saves time in that all operations may be effected in the same vessel.

Weber and Laporte Ltd. claim the addition of sodium metaphosphate, another metaphosphate or metaphosphoric acid as stabilising agents in B.P. 434,599; the metaphosphate may be added alone to the peroxide bath or in addition to sodium silicate or carbonate or caustic soda.

The addition of salts of metaphosphoric acid to the peroxide bleaching solution is stated to have special advantages according to B.P. 435,465; 435,475; 435,562 and 435,710 by Henkel of Dusseldorf:

THE PEROXIDE BLEACH

the goods should be desized and scoured before bleaching. Better and clearer whites are obtained with economy in the use of the peroxide; about 1 to 3 lbs. of sodium metaphosphate is sufficient for 100 gallons of the peroxide bleach liquor. In an actual example, a bleach liquor was prepared by diluting hydrogen peroxide to 0·5 vol. concentration, and sodium silicate added until the liquor was neutral to phenol-phthalein. The bath was then divided into two parts, to one of which was added 0·125% of sodium metaphosphate; cotton was bleached in the two baths for 90 minutes, at 80 to 85°C. until the peroxide was half-exhausted, but that in the liquor containing the metaphosphate was a much better white than the other sample. Comparisons of the degree of white showed that for the same time and temperature, it was necessary to use 2 vol. peroxide for the same result produced by 0·5 vol. peroxide with metaphosphate.

It has been suggested that peroxide alone may be used to effect scouring and bleaching in one operation, but the volume of practical experience is against this course, preferring clean goods before the bleaching step; it is not sufficiently realised in many cases that cotton goods are often dirty and need a scouring process for cleansing before destroying the natural colouring matter, as well as removing the natural impurities. Motes, neps and shives also require attention.

Most processes for bleaching with peroxide follow the usual course of employing a dilute solution in large volume, and circulating the liquid through the fabric. An alternative method has been described in B.P. 351,217 by Adolf, according to which it is possible to apply to the scoured goods a sufficiently strong solution of peroxide that the actual amount of liquor absorbed by the cotton is adequate for bleaching; the circulation of further liquor is not necessary. For example, cotton fabric is impregnated with 0·2% of hydrogen peroxide solution containing sufficient sodium carbonate and sodium silicate to act as alkali and stabiliser. The liquor is then maintained at 80°C. for 4 hours; this is sufficient to produce a half-white, but a full white may be produced by repeating the treatment for 3 hours with 0·3% of hydrogen peroxide and sodium silicate without the sodium carbonate.

B.P. 436,268 by Danzinger describes the use of still higher temperatures. The cotton fabric is first kiered at 3 atmospheres pressure, 2,500 Kg. of cloth necessitating the use of 8,000 litres of a liquor containing 25 Kg. of caustic soda and 25 Kg. of soda ash. The goods are then washed and bleached in the same kier by circulating again under 3 atmospheres pressure, a similar volume of liquor at 110°C., but containing 15 Kg. of 30% hydrogen peroxide to which has been added 30 Kg. of 38°Bé sodium silicate to act as a stabiliser and also as an alkali.

It will be noticed that hydrogen peroxide may be employed on cotton goods at temperatures of the order of 80 to 90°C., whereas with wool the usual range is about 40 to 50°C. (see p. 362).

It now seems to be generally agreed that the best way of using hydrogen peroxide for the bleaching of cotton goods is first to scour the cotton; the two chief disadvantages in a single stage process or combined scour and bleach are there is some danger of oxidising the cellulose if sufficient alkali is added to the peroxide for efficient scouring, and also the soluble impurities are apt to bring about an increased consumption of peroxide. The peroxide bleach on scoured cotton will enable the minimum amount of somewhat costly peroxide to be employed.

In spite of the general tendency to use peroxide after the kiering operation, suggestions have been made to employ it *before* kiering. For example, in B.P. 444,059 of Weiss and von Reich, it is stated that a mild treatment with peroxide modifies the impurities in such a manner that the scouring is rendered more effective and produces such a clean cotton that the final bleach need only be of a mild type. The goods are singed in the usual manner, after which they are worked in a peroxide liquor for 3 to 4 hours; for every 1,800 Kg. of cotton there is required about 5,000 litres of water at 85°C. and 7·5 Kg. of sodium peroxide or its equivalent of sodium hydroxide and hydrogen peroxide.

It is possible to saturate cotton with hydrogen peroxide solution and then allow a steeping period to take place during which bleaching occurs; this forms the basis of some of the continuous bleaching processes. "Steep-bleaching" at room temperatures takes a considerable time, however, and two treatments may be necessary. The time of steeping may be reduced by raising the temperature, either by heating the impregnating solution or steaming the cloth after impregnation; both methods may be combined if necessary.

A special continuous peroxide bleach has been devised by Du Pont, and has been applied with success to a variety of medium and light-weight cotton goods. The equipment is simple and consists essentially of a specially designed heating tube and a series of *J*-boxes or Gantt pilers. The fabric is first padded in open width with a solution of hydrogen peroxide at a strength of 1·4 to 1·6 vol. and at pH 10·2 to 10·8; the temperature is maintained at 38°C. In addition to the peroxide and silicate, the bath also contains a small proportion of a suitable wetting agent. After leaving the padding mangle the goods are passed through a pot-eye to convert them into rope-form, and then enter the heating tube which is maintained at 80 to 95°C. by steam at about 30 lbs. pressure. The cloth is drawn through the tube and piled into the *J*-box, where it remains for an hour; the box is made of stainless steel and well insulated in order to conserve the heat. Speeds of 100 yards per minute are possible, and the goods are stated to be somewhat more absorbent than those bleached in a kier with peroxide; the uniformity and permanence of white, the tensile strength and viscosity in cuprammonia are equal to those of goods treated with peroxide in kiers. The process is completed by the usual washing treatment.

H_2O_2 or NaOCl

The relative merits of hypochlorite and hydrogen peroxide bleaching have often been discussed without any very definite decisions; hypochlorite bleaching is undoubtedly very cheap and capable of excellent results without damage when properly controlled. American experience (Am. Dyes. Rep., 1942, *31*, 61) sometimes in the same mill, points to a superior softness in the finished product which is an advantage for knit goods and towels; minor advantages are that less water is required for washing and the goods are stated to be more absorbent. On the other hand, some bleachers obtained a better "bottoming" from an alkali boil followed by hypochlorite bleaching. It has also been stated that peroxide bleaching is safer, as it is less liable to cause damage, but this is largely a matter of careful control, although it may be argued that in such a case, labour costs would be higher for hypochlorite bleaching.

There is undoubtedly a place for each of these processes.

Modern American practice may be represented by the following series of operations. First, sour in 1·5°Tw. sulphuric acid for 30 minutes, followed by a thorough rinsing; secondly, scour in the kier with 4% NaOH, 1% Na_2CO_3, 0·3% of a wetting agent and 0·3% of sodium silicate, all calculated on the weight of the goods; finally, the cloth is washed in the kier and bleached in a sufficient volume of hydrogen peroxide to operate the kier—using 1% of H_2O_2 and 2% of sodium silicate (82°Tw.) for 4 hours at 82°C.

An alternative to the use of peroxide depends on 1 to 1·24°Tw. sodium hypochlorite at room temperatures, followed by souring in 0·75 to 1°Tw. sulphuric acid and rinsing in the usual manner.

Mecheels (Textilber., 1935, *16*, 725) also refers to the softness of cotton and rayon bleached with peroxide, and states that a better white may be obtained than from hypochlorite; if the goods are thoroughly scoured, however, the white from subsequent hypochlorite treatment is improved, but at the expense of tensile strength. References have also been made to comparative losses in weight; according to *Textile World* (1935, *85*, 1866) the records for 15 years show 6% for kier-boiling followed by hypochlorite, and 5% for scour and peroxide, while Schramek (Leipzig Monats. Text. Ind., 1931, *46*, 313) gives similar figures for high-quality yarn, but a difference of 2% for lower quality material. Schramek's work, however, utilised peroxide without a scour, but as a two-bath process in alkaline solution at 80 to 90°C. Minimum loss in weight resulted from a single peroxide bath followed by a mild hypochlorite bleach.

Mecheels (*supra*), however, finds that the "combined" bleach has the greatest effect on strength and viscosity of the cellulose, although the white is very good. Similar views are expressed by Baier and Hundt (Textilber., 1937, *18*, 301) in their examination of the bleaching of a cotton-staple fibre cloth containing 84% cotton; samples treated twice by the simple peroxide method gave products with the highest viscosity. The loss in weight was greatest, as would be expected, in those processes which included an alkali boil under pressure. Joubert (Bull. Soc. Ind. Rouen, 1938, *66*, 349) also favours peroxide for goods with a high proportion of rayon, and reserves the open boil, chlorine and peroxide method for fabrics with low proportions of rayon or containing cotton that is difficult to bleach.

It seems probable that the use of hydrogen peroxide solutions for bleaching will increase in the future.

Some specialised applications of hydrogen peroxide in conjunction with hypochlorite bleach liquors are given on p. 288.

PERBORATES

Other per-salts have been suggested for bleaching from time to time, and some attention has been devoted to sodium perborate, which is a constituent of many household bleaching powders and also finds application in laundries. The per-salts have been produced because of a demand for solid compounds of the peroxide type, but capable of being handled with safety. Sodium perborate $NaBO_3 4H_2O$ is a white powder of high stability which dissolves in water with absorption of heat; it may be handled with safety, but is decomposed by metallic salts in a somewhat similar manner to peroxide.

When dissolved in water, the perborate is much less alkaline than the peroxide and may be used immediately, but it is important that complete dissolution should take place or intense local action by undissolved particles will cause damage.

The use of perborates with cotton is rare, but they are sometimes employed for the bleaching of rayon materials, as discussed on p. 277.

PERACIDS

Organic peracids and their salts have recently been suggested as bleaching agents for cotton; the monoperacids are characterised by the presence of the perhydroxyl group, —OOH, as part of the molecule, and may be regarded as being derived from other acids by replacing the hydroxyl group containing the ionisable hydrogen atom, with the perhydroxyl group. Du Pont, in B.P. 550,490, outline a simple process for making aqueous solutions of peracids by reacting an organic acid anhydride with an alkaline solution (at least pH 10) of hydrogen peroxide whose concentration corresponds to 3-vol. strength. Typical peracids or salts are

$$\begin{matrix} CH_2\text{—}CO\text{—}O\text{—}OH \\ | \\ CH_2\text{—}COOH \end{matrix}$$
Persuccinic acid

$CH_3COOONa$
Sodium peracetate

[benzene ring with COOOH and COOH substituents]
Perphthalic acid

The use of these reagents has been outlined in B.P. 550,491, also by Du Pont; it is not necessary to isolate the pure compounds for use in bleaching. Their use in concentrations of not more than 2 vol. is stated to give results superior to those obtained with hydrogen peroxide. Cotton goods are not damaged nor degraded, and the residual bleaching is readily removed by simple rinsing. Although these reagents may be used in acid, neutral or alkaline solution, it has been found preferable to adjust the aqueous solution to pH 8 for the most rapid and satisfactory effect; this value is not very critical, but it is better to keep within the range of pH 6 to 9. The lower portion of this range is valuable for coloured goods; even smaller values down to pH 3 may be employed if volume concentrations less than 2 are used, such as 0·2 vol., for example.

A wide temperature range is also available, but 50 to 95°C. is advisable in order that the action may be complete within a reasonable time; the duration of the process is generally from 10 to 30 minutes, according to the nature of the goods requiring treatment. Occasionally, goods containing large amounts of motes or non-cellulosic material may need bleaching in two stages.

In actual examples, scoured cotton goods were immersed for 15 minutes in a solution of monoperphthalic acid at pH 8 and equivalent in active oxygen content to 2-vol. hydrogen peroxide; the temperature of the liquor was maintained at 82°C. At the end of the process the goods were rinsed and dried. Dyed cotton goods may be similarly treated in a solution of pH 7 to 8, and corresponding to a 1-vol. concentration.

An interesting comparison was made of the bleaching action of hypochlorite, peroxide and peracid; a piece of scoured cotton muslin was divided into three portions which were immersed in separate solutions corresponding to (a) 5 g./l. of available chlorine at pH 11·6 at room temperature, (b) 0·25 vol. H_2O_2 at pH 10·3 and 82°C., and (c) peracetic acid of 0·0825 vol. concentration at pH 3 to 4 and at 82°C. In all three experiments the time of bleaching was 15 minutes. After bleaching, all samples were rinsed and then divided into two parts, one of which was then treated with peracetic acid solution of

0·16 vol. concentration at pH 3 to 4 for 1 hour at 82°C. All six samples were finally rinsed and ironed; the two-stage treatment produced similar results in all cases, but the single peroxide treatment did not give as good a bleach as the single hypochlorite or single peracid treatment. The effect of the peracid and hypochlorite was the same in respect of bleaching efficiency.

PERMANGANATE

The well-known properties of potassium permanganate as an oxidising agent have, from time to time, directed attention to the possibility of its use as a bleaching agent. The general method appears to rely on acidified permanganate solutions, and at least 0·5 lb. of potassium permanganate is required for 100 lbs. of cotton. One method suggested is to treat 1,000 lbs. of scoured cotton fabric on a jig for 2 to 3 hours in a solution of 14 lbs. of potassium permanganate acidified with 3 lbs. of sulphuric acid of 170°Tw., followed by a second bath consisting of 300 lbs. of sodium bisulphite solution (65°Tw.) and 100 lbs. of sulphuric acid (170°Tw.). The function of the bisulphite is to reduce the dioxide formed in the first treatment; a thorough washing completes the process.

Not only is potassium permanganate a relatively expensive reagent to employ, in spite of the oxygen content, but the manipulation of the fabric is more complicated than with simple reagents such as hypochlorite and peroxide on account of the oxides of manganese which are deposited on the fabric and have to be removed by chemical means. Hence, although two molecules of permanganate contain five atoms of available oxygen, this advantage is more apparent than real.

For instance, when cotton is soaked in acidified permanganate, it is not only bleached but coloured brown by the oxides of manganese which cannot be removed with water or dilute acid, but require the assistance of sulphurous acid which reduces them to the soluble manganese oxides. When the "brown" cotton is placed in sulphurous acid, the oxides are removed, but some potassium hydroxide is produced and there is a danger of oxycellulose formation by the alkaline oxidising agent.

Various methods have been suggested to overcome these difficulties, and they have been fully described by Trotman and Thorp (*The Principles of Bleaching and Finishing of Cotton*, Griffin, London, 1927, p. 517). The first, or neutral, method depends on the addition of magnesium sulphate which replaces the potassium hydroxide by magnesium hydroxide. The permanganate is dissolved in water and magnesium sulphate added (316 parts of permanganate require 246 parts of magnesium sulphate); the fabric is allowed to soak in the solution for 3 to 4 hours, and is then immersed in dilute sulphurous acid or hydrogen peroxide.

The second method depends on keeping the bath acid throughout, and it may be remarked that hydrochloric acid should not be employed or chlorine will be formed. In the "acid" method, a solution of 0·25 to 0·5% permanganate is prepared, to which is added the necessary amount of sulphuric acid ($2KMnO_4$ requires $3H_2SO_4$). The wet fabric is soaked in the solution for 3 or 4 hours at room temperatures, and the oxides are still deposited because there is not enough acid present to keep them in solution. However, when the fabric is placed in acidified hydrogen peroxide solution, or in dilute sulphurous acid, decomposition takes place, and in presence of acid the manganous oxide is dissolved.

$$MnO_2 + H_2O_2 \longrightarrow MnO + H_2O + O_2$$
$$Mn_2O_3 + H_2O_2 \longrightarrow 2\,MnO + H_2O + O_2$$

OZONE

The use of ozone for the bleaching of textile materials was suggested by Siemens and Halske about 1890, but although the term "ozone-bleaching" is often employed it actually refers to the use of ozonised air. It was suggested that 12 to 24 hours in ozonised air was adequate to bleach cellulosic materials, and that one day in ozonised air was equivalent to 3 days' grassing in summer and 14 days' grassing in winter.

There are numerous other suggestions for the use of ozone, and one of the more recent has been put forward by Crespi in B.P. 247,738. The cotton fabric is first treated with a dilute acid such as hydrochloric acid, and then passed through a current of ozonised air charged with water vapour at low temperatures. The gaseous mixture may be circulated through the material and its activity may be maintained by adding fresh ozonised air or by returning to the ozoniser.

The action of ozone on cellulose has been described by Dorée (J.S.D.C., 1913, *29*, 205; J.T.I., 1938, *29*, 27).

CHAPTER XXI
THE SCOURING AND BLEACHING OF LINEN

BEFORE discussing the bleaching of linen, it is perhaps of interest to recall that flax is a bast fibre which has to be separated from the woody matter, associated with the flax stems, by the process of retting. This is essentially a fermentation process which has been practised for thousands of years, and is still done in a somewhat primitive manner. Retting may be carried out in pools or in rivers; Courtrai flax is retted in the sluggish waters of the river Lys, but within recent times there has been a movement towards tank-retting where the process can be brought under better control. Many chemical methods of retting have been suggested, but none is of great industrial value; the basis of most suggestions is a hot or boiling treatment with solutions of oxalic acid, or with solutions of caustic soda or sodium carbonate. Some chemical retting methods have been described by Ruschmann (J.T.I., 1924, *15*, 61 and 104).

Various methods have been suggested to replace the old retting processes, but commercial production has not been accomplished as yet. The chemical methods include boiling at 100 or 110°C. with water or with soap or dilute sodium carbonate.

Now in the bleaching of cotton the chief object is the removal of all impurities, but this is not so with linen, which contains a much higher proportion of non-cellulosic matter than cotton. With a full bleach there may be a loss in weight of 25 to 30%, so that most of the bleaching processes are directed to the conservation of some part of the non-cellulosic material. In addition to the larger amount of impurities, linen is much more easily attacked by chemical reagents than is cotton; this is probably due to differences in structure of which the absence of any cuticle may possibly be the most important. Complete removal of the intercellular matter in flax causes the disintegration of the fibre-bundles—the so-called "cottonisation" of flax.

The susceptibility to acid, alkali and oxidising agents necessitates a longer and more tedious bleaching treatment than is required for cotton; a single severe treatment is apt either to tender the fibres or to "set" the colour, the latter being very difficult to rectify.

The bleaching of linen, therefore, is a milder process than that employed for cotton, but as it is more difficult to bleach, then the process becomes longer, generally involving the repetition of the sequence scour-sour-chemic, the number of repetitions being determined by the effect required. The bleaching process for linen is used as a means of producing the final type of cloth in a manner which is not possible with cotton.

The impurities in flax are mainly classed together as pectic matter and hemicellulose, but there is also present about 0·5 to 2% of wax. It has been suggested that this could be removed by solvent processes without any deleterious effect, and such processes have been protected by Higgins (B.P. 102,892) and Lumsden, Mackenzie, Robinson and Fort (BP.. 165,189 and 221,296); the operation of one of these processes on a large scale has been described by Fort (J.S.D.C., 1923, *39*, 42).

The colour of the wax varies according to the type of flax, but it is characterised by an unpleasant odour similar to that of flax itself; the melting-point of the wax is about 61·5°C.

The wax contains 81% of unsaponifiable matter and 19% of saponifiable compounds, including caproic, stearic, palmitic, oleic and linolenic acids; the unsaponifiable material resembles ceresin.

The bleaching of linen yarn is different from that of cotton, mainly in that the process is decided by the type of cloth for which the yarn is intended; the removal of the whole of the 25% or so of impurities would involve a serious loss in weight and render the ultimate product much dearer; yarn bleaching is graded by the loss in weight into creamed yarn (8 to 12% loss), half-white (10 to 16% loss), three-quarter white (12 to 20% loss) and full white (15 to 25% loss). The cloth from the creamed yarn is usually lightly bleached after weaving. The finer linens and damasks are woven from boiled yarn which has been scoured to lose up to 20% of weight, and the bleaching is completed after weaving. The commoner types of linen cloth are woven from the "green" yarn and the whole bleaching process takes place after weaving.

Boiled yarn, for the finer quality goods, is prepared by kiering in 3 to 4°Tw. (1·5 to 2%) sodium carbonate solution for about 4 to 6 hours under 5 to 7 lbs. pressure. Creamed yarn should suffer the minimum loss in weight and is sometimes merely steeped in hot alkali overnight; if kiered, however, then 45 to 60 minutes is adequate in 5 to 7% sodium carbonate solution. As in all kiering operations, whether for cotton or linen, the goods must be well washed in the kier, first with hot and then with cold water to prevent the resinous impurities settling on the material. The yarn is bleached by reeling for 30 to 40 minutes in calcium hypochlorite solutions of from 2 to 4°Tw., the higher concentration being used for the coarser yarns. During the reeling of the hanks their lower ends are only submerged to a depth of about one foot so that the linen is exposed alternately to air and solution; if the liquor becomes too active, then slaked lime is added as required, but the use of buffered sodium hypochlorite solutions obviates this necessity. The usual washing, souring and washing operations complete the process, a little sodium bisulphite being added in the final wash.

Half-white yarn is "creamed" as outlined above, and after a thorough rinsing, is "scalded" or boiled with a weaker alkaline solution than that employed at first. (The term "scald" is peculiar to linen-bleaching and refers to any boiling process other than the first.) The chemicking is carried out by steeping instead of reeling.

Three-quarter and full-white yarns are generally given one lye boil and two or more scalds together with the intervening souring, washing, chemicking, souring and washing operations.

Clayton (J.S.D.C., 1923, *39*, 31) has given an interesting account of the bleaching of yarns made from the bast fibres.

A method of bleaching linen yarn, sponsored by Laporte, is first to boil the material in soda ash solution (8 to 10% Na_2CO_3 on the weight of the yarn); the strength of solution usually starts at 2%, but is diluted by the introduction of the live steam used for heating. The time of boiling is about 3 hours, at the end of which the yarn is dark brown in colour.

The alkali is then removed by rinsing and the yarn is reeled in 2°Tw. calcium hypochlorite solution containing 5·6 grams of available chlorine per litre; the reeling may take 1 to 1·5 hours. The yarn is then rinsed with water and soured in 0·1 to 0·2% hydrochloric acid solution for 30 to 60 minutes, followed by further rinsing until the linen is free from acid.

The yarn is then bleached in 0·5 vol. hydrogen peroxide solution to which has been added 2% soda ash and 3% of sodium silicate (100°Tw.), both estimated on the weight of the yarn. The ratio of liquor to linen is 5 : 1, so that for 1,200 lbs. of yarn, 600 gallons of solution would be required containing 3 gallons of 100 vol. H_2O_2, 24 lbs. of soda ash and 36 lbs. of sodium silicate. The temperature of the liquor is brought to 70 to 77°C. over a period of one hour, and maintained at that temperature until the peroxide is exhausted; this is generally 4 to 5 hours. The yarn is then rinsed in hot water and in cold water.

Finally, the linen is soured in hydrochloric acid (1 gallon of hydrochloric acid per 200 gallons of water) for 30 to 60 minutes, followed by a thorough rinse to remove the acid.

The above process gives a good "half-white."

For a "three-quarter white," the peroxide treatment is followed by rinsing and then immersing for 10 hours in sodium hypochlorite solution containing 0·3 to 0·4 grams of chlorine per litre. It is then rinsed, soured, and rinsed again.

For a full white, a further process of treatment with hypochlorite (0·1 to 0·2 grams of chlorine per litre) for 6 to 10 hours, follows the process for the "three-quarter white." Finally, the linen is rinsed, soured, and thoroughly rinsed again.

It is perhaps wise to point out that all treatments with hypochlorite solutions and with acid are at room temperatures.

Where kiers are used for the peroxide bleach with linen, they should be lined with cement; the kier is first scoured with boiling caustic soda solution to remove any grease or paint. It is then treated with a boiling solution containing 2 gallons of 80° Tw. sodium silicate over 100 gallons of water; after 30 minutes treatment, the liquor is drained but the kier is not rinsed. Portland cement is then made into a slurry with water and brushed on to the sides of the kier, and circulated through the pumps, the pipes, and the spreader. As the cement is setting, it should be brushed with 10° Tw. sodium silicate solution and then allowed to remain for 2 days. The kier is finally treated with dilute sodium silicate (2 gallons of 80° Tw. solution per 100 gallons of water) at the boil.

The bleaching process for linen piece goods is more severe than for linen yarn, but is much less severe than for cotton materials, particularly in respect of pressure boiling which is usually limited to 5 to 10 lbs. Lime or caustic soda may be used for the first boil, and in the former method, is followed by the usual sour and a sodium carbonate boil. The actual bleaching or whitening is rarely effected by "grassing" alone, but it is still fairly common to expose the material in fields during the process, generally between the boil and the chemick. Grassing is rarely practised in Ireland and the U.S.A. and Higgins (Textile Mercury, 1939, Jubilee No.) suggests it may have publicity value.

The old method of bleaching linen comprised about twenty or more different operations, and even more recently (Irish Textile J., 1937, *3*, 6) the usual process involved a lime boil, six alkali boils, four hypochlorite treatments and the series of interspersed acid treatments. As previously indicated, the susceptibility of linen to chemical reagents and the greater amount of impurity necessitate a larger number of milder processes than for cotton. Kiering assistants are of value in boiling of linen goods.

Naturally there are many variations of the bleaching process, but it seems fairly general to start with a lime boil for 8 to 12 hours in a low-pressure kier, followed by washing, souring, washing and boiling in sodium carbonate solution, again in low-pressure kiers; sodium carbonate and 10% sodium hydroxide is often used. These second boils or "scalds" are repeated from two to six times, according to the weight of the material and the closeness of the weave. At certain stages it may be necessary to rub the cloth to remove "sprit" or the small particles of straw which are contained in the flax. The goods are steeped in soap and passed between heavy corrugated rubbing boards, arranged in two or three pairs, with the lower board remaining stationary while the upper boards move backwards and forwards.

If required, the goods may be grassed for from 2 to 7 days after the boiling processes. The next stage is a series of alternate treatments with hypochlorite and mild boiling with sodium carbonate solution, interspersed with the usual washing and souring steps; each sequence is termed a "turn" and from 2 to 5 "turns" may be given until the requisite colour is obtained.

Where the cloth has been woven from partly bleached yarn, the processes are reduced in number and severity accordingly.

Although the bleaching of linen is fundamentally similar to that of cotton, it is nevertheless more complicated and tedious. Attempts to accelerate the removal of impurities with due regard to the sensitivity of the material are apt to discolour the fibre—the so-called "setting" of the colour—presumably by resinification of the decomposition products through prolonged single treatments, high temperatures or strongly alkaline liquors. A further difficulty is the maintenance of the high lustre of linen which necessitates the use of weak solutions repeatedly rather than fewer treatments with strong solutions; the concentration of hypochlorite solutions, for instance, is generally one half of that employed for cotton.

Some data on the bleaching of linen have been given by Higgins (J.S.C.I., 1911, *30*, 1295).

BLEACHING OF LINEN YARN

	Weight	Loss (%)	Ash (%)
Brown linen	92·1 g.	—	1·28
Steep	88·7 g.	3·8	—
Lime	77·15 g.	16·2	0·18
Lye	70·93 g.	22·9	0·08
Chemic	69·53 g.	24·5	0·08
Full bleach	67·52 g.	26·7	0·07

BLEACHING OF LINEN FABRIC

	Loss in weight	Ash	Breaking Load
Brown Linen	—	1·28	1,270 g.
Steep	3·8%	—	—
Lime	16·2%	0·18	960 g.
Soda	22·9%	0·084	1,070 g.
Chemic	24·5%	0·08	930 g.
Full bleach	26·7%	0·074	910 g.

It will be noted that the bleached product shows 28% decrease in breaking load.

Butterworth and Elkin (J.T.I., 1933, *24*, 10) have provided some modern data on the scouring of linen previously desized and washed. Samples were boiled in solutions of sodium hydroxide of various concentrations from 0 to 2·5%; preliminary experiments showed that approximately 18% loss in weight was obtained by boiling for 4 hours in 0·5% NaOH solution, and a further 4% by increasing the concentration to 2%. These residual impurities are much more resistant than those removed in the early stages.

Samples containing various amounts of the non-cellulosic components were examined and it was established that the copper number decreases progressively with increase in loss in weight; the fluidity in cuprammonia is independent of the non-cellulosic content. The loss in weight is proportional to the alkali used in the process and depends on the amount of alkali remaining at the end of the boil; there is strong evidence for sorption of the alkali by the cellulose complex.

Some details of a special method of bleaching linen are given later on page 290.

Chloramines

It has been known for many years that it is exceedingly difficult to remove the last traces of chlorine from linen by washing alone; a positive potassium iodide-starch reaction may be obtained months after the bleaching process and storage will produce signs of chemical degradation. Raschig (Chem. Zeit., 1907, *31*, 920) showed that in dilute solutions, chlorine was able to combine with amino-compounds to form chloramines of the NH_2Cl type and this was recognised by Cross, Bevan and Briggs (J.S.C.I., 1908, *27*, 260) as having an important bearing on the bleaching process where protein-chloramines may be formed. Particularly in linen bleaching, chloramines are likely to accumulate in the old standing bleach liquors as the process proceeds and there is a tendency for the fibre to absorb them in much the same manner as mordants. Cold hydrogen peroxide does not decompose chloramines and this fact may be used to differentiate between the chlorine of hypochlorites and the chlorine of chloramines.

The possible role of the chloramines has been a matter of discussion ever since the work of Cross, Bevan and Briggs (loc. cit.); Tschilikin (Textilber., 1928, *9*, 592), Auerbach (ibid., 1928, *9*, 769), and Kornreich (ibid., 1936, *17*, 227; 1938, *19*, 61) have published data and views on this topic which is becoming of renewed interest on account of modern work on "combined bleaching processes" which are mainly centred on linen and bast fibres generally. Two phases may be distinguished in the treatment of linen with hypochlorite solutions; first, a rapid whitening or brightening of the fibres within a few minutes, and secondly, a slower bleaching until the full white is obtained. The first is assumed to be caused by chlorination which builds up the chloramines which are light yellow in colour as against the brown colour of the proteins. Alkaline treatments restore the dark colour to a large extent. Unscoured cotton will give similar results to a minor degree.

In the normal processes of purification, the series of boiling processes removed the proteins from the fibre almost completely so that the question of protein-chloramines did not arise to any appreciable extent, but with acceleration of the whole procedure and the tendency towards mild scouring, and even no scouring, the role of the chloramines has to be reconsidered.

Kornreich (loc. cit.) has pointed out that if the protein-chloramines are treated with hot alkaline solutions of hydrogen peroxide, then the active chlorine liberated is destroyed by the peroxide and the brown substances which are rebuilt are bleached by the oxygen; in this manner the chloramines are destroyed. From this standpoint the "combined" bleach is based on protein-chloramines, which are built up by a short process of chlorination and then destroyed by treatment with hot alkaline peroxide solution. Often there is no previous scouring, but Kornreich regards the scour-chlorine-peroxide sequence as the best method of purifying cotton or linen. Questions of efficiency apart, other workers do not consider that chloramines play any important part in the combined bleach. Butterworth (J.S.D.C., 1940, *56*, 355) suggests that the role of chloramines in the hypochlorite bleach liquors has been exaggerated; nevertheless chlorination does play some part in the bleaching of linen, but there is insufficient available data to explain the chemical reactions.

Any possible damage on storing bleached linen which may be caused by hydrochloric acid from the chloramines may be obviated by treatment with sodium bisulphite or thiosulphate solutions.

[The present work contains no account of the bleaching of jute. A useful modern paper on this topic has been given by Ridge, Little & Wharton (J.T.I., 1944, *35*, 93), to which the reader is referred.]

CHAPTER XXII
THE SCOURING AND BLEACHING OF RAYON

FUNDAMENTALLY, the processes for purifying rayon textiles are similar to those used with other types of cellulosic materials. It must be remembered, however, that natural impurities are almost non-existent with rayons and further, that the regenerated celluloses and cellulose derivatives are more sensitive to many chemical reagents than is the case with the parent cellulose. Hence the bleaching methods used with cotton, for example, are modified because of the greater sensitivity to chemical influences and the smaller amount of impurities present; these modifications generally comprise the utilisation of similar reagents for scouring and bleaching but for shorter times and at lower concentrations and temperatures.

All types of regenerated cellulose are weaker than the native cellulose fibres, and, moreover, the wet-strength is greatly diminished so that great attention must be paid to any wet processing. The fact that some rayons, such as the cuprammonium varieties, have a high wet-strength relative to other rayons does not mean that they can be treated with impunity, for their filaments are generally so fine that friction and tension should be avoided to the utmost possible extent.

In addition to the dangers of manipulation in the wet state, which apply particularly to rayons of regenerated cellulose, care must also be taken on account of the sensitivity of the material to many chemical reagents. Rayon is a dispersed form of cellulose and is therefore much more readily affected by chemical treatment; both hydrocellulose and oxycellulose are formed more easily from rayon than from native cellulose, and hence special care must be taken with the time of treatment and the concentration of the reagents. Acetate rayon, on the other hand, although it does not swell greatly in water and so become ultra-sensitive to physical and chemical influences, is peculiar on account of its susceptibility to alkali and also to hot aqueous solutions.

There is practically no bleaching of rayon in the hank form, for the two chief varieties of rayon, viscose and acetate, are both bleached by the rayon manufacturers and delivered to the dyer, weaver or knitter in the bleached state.

Hence, in so far as bleaching is concerned, it is only woven and knitted fabrics which require attention. Of these, the materials consisting of 100% rayon rarely require bleaching if the fabric is

270   TEXTILE BLEACHING

intended for dyeing; with whites, only the mildest bleach is necessary. On the other hand, many fabrics contain natural fibres in addition to the rayon, and here, of course, some form of bleaching is generally necessary. Similar considerations apply to the question of scouring, for there are no natural impurities to be removed from rayon, but only the acquired impurities; these fall into two main classes, the size which has been applied to the warps of the woven goods and the oily lubricant which has been employed in knitting. It should also be remembered, however, that rayon is apt to become rather dirty during weaving or knitting, so that some scouring treatment is generally necessary to cleanse the goods before dyeing and finishing.

SCOURING

The general process for cleansing woven or knitted rayon materials is sometimes termed scouring and sometimes boiling-off; the latter has come from the treatment of real silk, but it must be realised that there is nothing similar to the silk-gum to be removed from rayon.

Most rayons may be treated in a solution of soap containing about 3 to 5 lbs. per 100 gallons of liquor (i.e. 0·3 to 0·5%), to which has been added a little sodium carbonate or trisodium phosphate; ammonia is often added to scouring solutions employed for cuprammonium rayons. There are numerous wetting agents and detergents available as "assistants" for the treatment of rayons and an account of these substances has been given on page 133. Pine oils, sulphonated pine oils, and many terpene preparations with wetting and penetrating properties seem to be popular for scouring rayon goods.

For the regenerated cellulose rayons, it is satisfactory to treat the goods at a temperature of 80 to 90°C., but lower temperatures should be employed for acetate rayons.

The application of scouring assistants has undoubtedly been of great value in the treatment of rayon goods, not only in obviating the difficulties which occur when lime soaps are formed, but also in connection with the drawbacks of strongly alkaline liquors which are well known with acetate rayons, and present to some extent with the regenerated cellulose rayons. Hence, if the fibre is alkali-sensitive it is now possible to use neutral scouring agents, or the bath may be made only slightly alkaline by the addition of ammonia.

An important factor in the purification of rayon goods is the type of size that has been used; at one time, it was considered that simple mixtures of starch, dextrin and so forth would prove adequate, but defects were soon apparent and these are rarely used to-day. One of the best-known preparations is the so-called Boyeux mixture, which is mainly an emulsion of boiled linseed oil, which suffered from oxidation to an insoluble film; this has been overcome by the addition of retarding agents. Another suggestion was the use of triethanolamine soaps, and further products include gelatin with sulphonated oils, gelatin with dextrin, and emulsions of olive oil with various soaps.

Sulphonated oils, gelatin and dextrin may be removed quite readily by hot water, and many of the emulsion sizes can be dispersed or emulsified again by mild alkali or even soaps and modern synthetic detergents. Proteins and starchy preparations need treating with the appropriate enzyme such as malt extract, peptase, etc. (see page 162). Activin (see page 166) may be employed for the removal of starch sizes; it also exerts a bleaching action on the rayon.

Linseed oil sizes are not always easy to remove evenly, particularly when the size has "aged" or oxidised on the yarn. The addition of solvents such as carbon tetrachloride or trichlorethylene to the soap bath is very helpful, and there are special preparations (see page 135) which include solvents and detergents for the purpose of scouring rayons which have been sized with linseed oil or which have been oiled before knitting with mineral oil emulsions. A subsequent treatment with hot soap and sodium carbonate solution may be necessary to complete the cleansing of the rayon.

The methods adopted for the desizing of rayon, therefore, are determined by two factors: (*a*) the type of size present, and (*b*) the type of rayon to which the size was applied.

It appears convenient to group rayon sizes into four main classes: (*1*) linseed oil, (*2*) glue or gelatin, (*3*) starch, including dextrins, and (*4*) glue and starch mixtures. It is fairly simple to recognise the type of size which has been employed; for example, linseed oil sizes generally contain some lead salts which act as driers, and their presence may be revealed by sodium sulphide solution, gelatin is readily detected by the biuret reaction, and starch and dextrin by the starch-iodide test.

In so far as the type of rayon is concerned, it is generally only necessary to know whether the goods are made of regenerated cellulose or of cellulose acetate rayon; the latter is soluble in acetone. With all goods containing acetate rayon, the desizing bath should not be very alkaline or the yarns will be adversely affected.

In general, linseed oil sizes may be removed by scouring with soap and soda ash solutions containing a little solvent for fats, such as the cyclohexanol type. A suitable scour for acetate rayon containing linseed oil size is Marseilles soap and a solvent such as Laventin KB at 75°C. The gelatin and glue sizes are often readily removed by hot water in a jig or wince machine; obstinate material may be treated with a proprietary preparation such as 0·3% of Degomma DL. Starch and dextrin sizes yield to treatment with enzymes in the ordinary manner, Diastafor being a popular preparation. Many enzymes, such as Degomma, remove starch as well as glue.

Both viscose and acetate rayon are often sized with gelatin and it is important to realise that the methods for removing starch sizes are not very satisfactory with gelatin. Atkinson (J.S.D.C., 1931, *47*, 5) has described the ideal method for the removal of gelatin size. The goods are wetted in water and allowed to stand for a short time and then worked on the jig at a temperature rising to 80°C. The goods are next padded in Courtauld's Desizing Compound which consists of 4·5 lbs. of Monopole Soap No. 1, 4 pints of 40°Tw. NaOH and 2·75 pints of xylol, after which they are allowed to lie overnight and then given the usual scour.

Although this is a lengthy process, it is advisable to approach these conditions as nearly as possible for the best results on goods of high quality.

A method suggested by the I.C.I. for the scouring of regenerated cellulose rayons is to treat them at 50 to 60°C. for 30 to 60 minutes in a liquor consisting of 1 lb. of Lissapol C, 5 to 10 pints of Astol A, and 1 lb. of soda ash per 100 gallons. (Astol is trichlorethylene.)

In so far as the manipulation of the goods is concerned, treatment on a jig may be employed, or alternatively the wince machine. The jig is sometimes preferable with acetate rayon goods of a woven structure to avoid the formation of crease-marks; the treatment of knitted acetate rayon goods is discussed on page 273.

Delicate structures may be treated as for crepes, in book-fold or loop form; alternatively they may be treated on the star or spider frame.

Cuprammonium rayon, with its fine filaments, may be scoured with boiling soap, ammonia, and a little dispersing agents. The scouring is usually at its best when the fabric is fed into the warm and prepared bath. With plain fabrics, it appears usual to scour for about one hour with frequent motion of the material in book-form; for knit goods, a winch treatment in the boiling liquor for about 30 minutes is often adequate.

Acetate Rayon

In dealing with acetate rayon, it should be remembered that this material is alkali-sensitive and also susceptible to aqueous solutions at temperatures above 85°C.; it may also be useful to remark that acetate should not be dried at high temperatures.

A common method of cleansing acetate goods is to use a neutral soap of high quality, such as an olive oil soap, together with ammonia; about 0·5 to 1 lb. of soap and 1 quart of ammonia per 100 gallons will be satisfactory at 65 to 75°C. for 30 minutes. Where a starch size has been employed, it may be removed in the same manner as for regenerated cellulose rayons, but it is sometimes advantageous with acetate to allow the material to soak overnight because of the low degree of swelling of cellulose acetate in water compared with ordinary rayons.

With all-acetate fabrics, it sometimes happens that the goods contain crease-marks when they are received at the bleach works; in this case, it is better to wet the material thoroughly on the jig, remove the water by hydro-extraction and dry the cloth on a hot-air stenter to remove the creases. Alternatively, a blanket finishing machine of the Palmer type may be found of use in removing these marks.

Fabrics containing a viscose warp and acetate weft are often desized by soaking in water overnight and then working in water at 50 to 75°C. for 1 to 2 hours to remove the gelatin size. Scouring may be effected on the jig with a liquor consisting of 0·1 to 0·25% soap and 0·05 to 0·01% of ammonia (d 0·88); from 1 to 2 hours at 50 to 75°C. is generally adequate.

A very popular fabric consists of an acetate warp and a viscose crepe weft. These are generally manipulated by stringing the pieces from the selvedge in book form, or framing, and then immersing in the trough so that they are fully suspended. The scouring liquor may contain 0·1 to 0·25% olive oil soap in which the goods are wetted, after which the temperature is raised to 75 or 80°C. over a period of 2 hours. This allows the crepe effect to appear uniformly before the scouring action starts; it may be wise to ensure that all the size has been removed, in which case the desizing bath mentioned on page 272 may be utilised.

Knitted Fabric

Knitted material containing cellulose acetate rayon demands greater care in some respects than silk or ordinary rayon, as it is less able to withstand the action of alkaline solutions and high temperatures on account of possible saponification of the acetate. Hence it is usual to scour in soap and ammonia solution (2 lbs. of soap and 2 pints of 0·88 ammonia in 100 gallons of water) at 60 to 70°C. for 30 to 45 minutes. In cases of dirty oil on the goods, a solvent scourer such as Astol A may be added; 1% of xylol has also been suggested.

The major difficulties with knitted goods of acetate rayon are generally of a manipulative nature; the material tends to become plastic at high temperatures and it is essential to prevent or to minimise the formation of creases which may become permanent on cooling. In passing through rollers in the hot wet state, for example, it is advisable to have the minimum of pressure on the fabric; it is also wise to cool the fabric rather gradually as sudden cooling leads to cracking or surface markings. On no account should the fabric be allowed to lie piled in folds in the wet state. It is customary to remove excess of water by hydro-extraction, but this should not be prolonged or vigorous, as creases and cracks will develop in the goods.

Considerable quantities of knitted material are made in the form of Milanese fabric which is flat-knit in a strong, compact and elastic texture which is also ladder-proof. The original Milanese material was made from real silk but it is now made from rayon and also from acetate rayon.

One of the defects of this material is that the selvedges show a pronounced tendency to curl, so that instead of remaining flat as in a woven fabric they form rolls to the extent of about nine inches on each side of the 60 to 70 inches wide fabric. Hence with wet treatments there is great difficulty in obtaining a uniform and regular effect. For this reason, therefore, it is usual to convert the fabric into a tubular form before wet-processing and this may be accomplished on special sewing machines developed for the purpose.

It must be realised that the Milanese fabric readily forms creases during its wet processing and when these have once been formed they are very difficult to remove; curiously enough, although viscose rayon fabrics crease more readily than those of acetate rayon, it is much more difficult to remove creases from the latter. On account of the smooth and sheer appearance for which the Milanese material is prized, it is essential to avoid the formation of creases.

It must not be assumed that the formation of permanent creases is merely due to the bending of the yarns, for it has been found that shrinkage plays a large part. For example, if the fabric is allowed to shrink in warm soapy water when in the flat state then no creases occur, but if the shrinkage takes place when the fabric is in a creased or crumpled condition then the creases which form are practically permanent. This is due to maintaining the fabric in a creased state during shrinkage.

During the shrinkage which takes place, the strains in the original material are released so that it passes from an unstable to a stable condition, and this stable state tends to make any creases permanent. Hence if efforts are made to remove the creases, they must depend on some return to an unstable state which will disappear with time or on wetting restoring the creases.

The effect of temperature is important in connection with the formation of these crease-marks, for although wetting in moderately warm water or soap solution will bring about some shrinkage this will not be complete until the temperature has been raised to near boiling-point. With this in mind, it is easy to realise that permanent creases may still be formed by passing from warm liquors to hotter liquors; hence it is advisable to ensure that the fabric is completely relaxed in a crease-free state as soon as possible in the sequence of wet-processing, and thus avoid the formation of creases in the later stages of bleaching and dyeing.

As the real origin of the crease-marks has been shown to be due to the fabric shrinking whilst in a creased or crumpled state, care must be taken to bring about uniform shrinkage without creasing. With the ordinary wince machine, it is not possible to avoid a certain amount of creasing as the fabric passes over the wince and falls into the liquor, but the use of a wide (from back to front) and shallow machine with an elliptical wince enables the minimum number of folds or plaits to be formed. The wetting operation generally starts in a lukewarm solution of soap containing about 1 lb. per 100 gallons, with careful plaiting guided by hand if necessary. When the two ends have been stitched together, the fabric may be run for a little while to ensure thorough wetting, with acetate rayon it may be advisable to leave the goods for several hours in order to give them the necessary time to swell, but this is rarely necessary with regenerated cellulose which wets and swells very rapidly and easily.

Care should be taken to avoid running too much material in the wince machine for the weight of fabric, even in the large folds, may be sufficient to cause creases in the wet state on account of the pressure involved.

After the preliminary steeping in the soap solution, the temperature should be raised gradually with the wince in motion, it will be found that the highest practicable temperature in a shallow wince machine is about 85 to 90°C., for there is considerable heat loss by evaporation and radiation. During the whole of the period of scouring in the wince machine, care must be taken to avoid tight folds, and it is sometimes advisable to add further quantities of soap after the preliminary steeping as this lubricates the goods and facilitates slippage of the layers of fabric over one another.

The scouring process generally requires about 30 minutes at the higher temperature and this is adequate not only to remove most of the impurities but also thoroughly to shrink the fabric.

It may be useful to remark again that the first stages of wet processing are most important in connection with avoiding the formation of permanent crease-marks.

A typical method of dealing with fabrics composed of 100% acetate is to treat them on the jig as far as possible and so obviate the formation of the residual crease-marks which are so difficult to remove. The fabric may first be desized with an enzyme preparation (0·125%) when 4 passages or "ends" at 60°C. is usually adequate;

the goods are then lightly scoured in a solution consisting of 1% soap and about 0·125% of a sulphated alcohol at 50 to 60°C. Four to six ends is often sufficient, but it is very important that the temperature should not be allowed to approach 80°C. in the case of the bright acetate rayons or 90°C. with the delustred acetate rayons. Instead of the solution of soap and sulphated alcohol, it is possible to use xylene, Turkey red oil and soap in an emulsion; other similar products may also be utilised, the I.C.I. having suggested a solution of 1 lb. of Lissapol C, 5 to 10 pints of Astol A and 1·5 pints of ammonia (0·91) per 100 gallons. This I.C.I. recipe requires a treatment of about 30 to 45 minutes at 50 to 60°C.

BLEACHING

The bleaching proper may be effected by about 6 ends in dilute acid hypochlorite at pH 5·5; a suitable solution may be made by dissolving 3 pints of NaOCl (20% available chlorine) and 4 pints of 40% acetic acid in 90 gallons of water. In dealing with fabrics containing acetate rayon, it must be remembered that this material is sensitive to high temperatures and also sensitive to alkaline attack. Hence it is usual to employ hypochlorite solutions on the acid side of neutrality; bleach liquors containing 0·3 to 0·6 g. of available chlorine per litre may be used at pH 4·5 to 5·5, for a period of about 45 to 90 minutes according to the degree of bleaching required. Where hypochlorites are used with acetate rayon, there is sometimes a little difficulty in getting rid of the last traces of chlorine from the acetate, and it may be necessary to follow the washing with a treatment in sodium bisulphite solution; a solution of 1 to 3·5 g. per litre for 30 minutes at 50°C. is adequate.

Where it may be preferred to use hydrogen peroxide for the bleaching of acetate rayon, care must again be taken on account of its susceptibility to hot alkaline liquors; the usual requirements for the peroxide bleach are pH 8·5 to 9·5 at 70 to 80°C. After the bleaching process, the fabric should be washed with hot water, and then thoroughly rinsed with cold water.

A recipe suggested by Laporte, is to prepare a solution containing 5 c.c. of 30% H_2O_2 solution, 3 g. of soap and 1 g. of a stabiliser per litre. The stabiliser should not be alkaline like sodium silicate; sulphated alcohols are suitable. The acetate rayon may be treated for 45 to 60 minutes at 45°C.; this temperature should not be greatly exceeded. Where the articles are difficult to bleach, it is preferable to use two successive baths, rather than one prolonged or severe treatment.

The dangers of alkaline bleaching of cellulose acetate, with attendant risk of saponification, may be obviated by the use of

Textone or sodium chlorite in acid solution (see page 233). Hence this method may be of value in dealing with cotton and acetate mixtures where the degree of alkalinity necessary to remove motes in the cotton would almost ruin the acetate yarn. A cotton and acetate interlining may be bleached in 2 hours at 85°C. in 1,000 gallons of water containing 8 lbs. of Textone, 8 lbs. of a synthetic detergent, 12 quarts of 56% acetic acid, 0·5 pints of 85% phosphoric acid and 1 lb. of sodium pyrophosphate; this liquor is sufficient for 1,200 yards of fabric. The treatment is concluded by washing in hot water and rinsing in cold water.

Both viscose and cuprammonium rayons may be bleached with very dilute alkaline hypochlorite solutions; viscose may also be bleached with hydrogen peroxide, but the traces of residual copper in cuprammonium rayon may cause difficulties if this material is treated with hydrogen peroxide solutions.

A suitable peroxide bleaching process, suggested by Laporte, is to prepare a liquor containing 2·5 to 4·0 c.c. of 100 vol. hydrogen peroxide, 0·05 to 0·15 g. of sodium silicate, 0·1 to 0·2 g. of a wetting agent and 0·1 to 0·2 g. of a sulphated fatty alcohol per litre. The bath should be adjusted to pH 9·4 to 9·8 by the addition of sodium silicate before the addition of the wetting agent and sulphated fatty alcohol. The viscose rayon is placed in the solution at room temperature, but the liquor is gradually raised to 70°C. and maintained at that temperature for 2 hours. The bath may become slightly acid at the end of the process. If the bleach is inadequate, the pH of the liquor may be brought to 9·8 again, and the temperature raised to 77°C. for a further 30 minutes. The goods are then washed with cold water, rinsed with hot water (50°C.) and again with cold water.

Another bleaching bath for viscose rayon may be made by using 1·5 to 2·0% of 30% hydrogen peroxide, 0·8 to 1·2% of sodium hydroxide, 2 to 3% of trisodium phosphate, and 1 to 1·5% of sodium silicate, estimated on the weight of the rayon.

It is generally considered unwise to treat viscose rayon at 80°C. for longer than 1 hour.

Sodium perborate appears to have aroused some interest in the U.S.A. where it has been utilised for the scouring and bleaching of rayon goods because of the ease of application and the economy of time and labour.

A typical method is to make a solution of about 2 to 3% of sodium perborate, 2 to 5% of a modern sulphonated detergent and a small amount of a wetting agent; the pH of the solution should be about 10·2. The goods may be treated in this liquor for about 30 minutes at 70°C. after which the temperature may be raised to 95°C. for a further period of 30 minutes; the fabric is then thoroughly rinsed in hot water and finally in cold water.

Hydrogen peroxide may be utilised in a similar manner; 1 lb. of hydrogen peroxide of 100 vol. strength corresponds in bleaching power to 1·3 lbs. of sodium perborate, but it is necessary to add alkali to the bath where hydrogen peroxide is used.

Where the goods have been scoured beforehand, sodium perborate may be used in the proportion of 1·5 lbs. per 100 gallons for bleaching only; the goods should be treated for an hour at 70°C. followed by rinsing in hot water and then in cold water.

CHAPTER XXIII
SPECIAL BLEACHING PROCESSES

THE bleaching processes for cotton and linen as commonly practised include the two main steps of scouring and bleaching; these may be described as the standard methods of treatment although there are many variations in the actual procedure and the chemicals employed as, for instance, hypochlorite or peroxide, for the final treatment.

There are, however, certain special processes which do not follow the ordinary course of scouring and bleaching; one of these is to bleach and then scour, another is to use hypochlorite alone, whilst a third is the combined bleach in which a treatment with hypochlorite is followed with peroxide.

In addition, continuous bleaching processes have been devised.

The Greisheim "intermediate boiling" process is one in which the normal sequence of bleaching operations is reversed; the goods are first treated with alkaline hypochlorite under such circumstances that the available chlorine is all utilised and then scoured in the still alkaline "spent" liquor. A final bleach may be given in fresh hypochlorite if necessary.

A reversal of the normal procedure has also been suggested for the treatment of coloured woven goods in particular; desized material is treated in hypochlorite solution (which may be at a temperature of 40°C.) for about two hours, followed by washing and then boiling in soap and caustic soda solution.

PRE-CHEMICKING

The application of hypochlorite before the scouring process has been examined by Clark (Cotton, 1932, *96*, No. 7, 28), who worked with cotton sheetings. The general procedure was to wash the fabric, then treat it with 1°Tw. chemic, followed by washing, souring and washing, after which the sheeting was kier-boiled in NaOH solution, washed, chemicked in 0·5°Tw. hypochlorite, washed, soured and washed. With the hypochlorite solutions employed, 1°Tw. corresponded to 1·25 grams per litre of available chlorine.

Clark was of the opinion that the fabric was more thoroughly bottomed as a result of the pre-chemic treatment; the motes were softened and partly bleached, and the colour of the final fabric was equivalent to that treated with two scouring stages and one chemic.

The permanent nature of the white obtained was tested by steaming in the ager for 1 hour at 5 lbs. pressure and found to be satisfactory.

A comparison of samples of the fabric extracted at each stage of the two processes has been given.

Chemic—Boil—Chemic

	Grey	Chemic	Boil	Chemic
Ounces/sq. yd.	3·50	3·04	3·06	3·11
Tensile strength warp	64·4 lbs.	72·4 lbs.	69·9 lbs.	67 lbs.
Tensile strength weft	62·9 lbs.	80·7 lbs.	86·2 lbs.	80 lbs.
CCl_4 extract	0·61%	0·48%	0·29%	0·28%
C_2H_5OH extract	0·64%	0·32%	0·18%	0·22%
Size	5·34%	1·81%	0·92%	0·33%
Ash	0·93%	0·34%	0·25%	0·08%

Boil—Boil—Chemic

	Grey	I boil	II boil	Chemic
Ounces/sq. yd.	3·13	3·49	3·47	3·49
Tensile strength warp	70·2 lbs.	67·7 lbs.	69·8 lbs.	68·6 lbs.
Tensile strength weft	73·6 lbs.	72·4 lbs.	75 lbs.	74 lbs.
CCl_4 extract	0·66%	0·43%	0·11%	0·11%
C_2H_5OH extract	0·6%	0·30%	0·23%	0·24%
Size	8·2%	1·0%	0·63%	0·34%
Ash	0·9%	0·6%	0·11%	0·98%

A comparison of the matter removed at various stages of these two processes is also of interest.

Effect of Pre-chemic Treatment

Removed	Pre-chemic	I boil	II boil	Chemic	Residue
Fat and wax: pre-chemic	20·2%	31·1%	—	0·6%	46·4%
ordinary	—	33·8%	49·6%	0·4%	16·1%
Pectic matter: pre-chemic	49·7%	21·2%	—	5·2+%	34·3%
ordinary	—	48·1%	12·2%	0·7+%	40·3%
Size: pre-chemic	66·1%	16·6%	—	11·0%	6·1%
ordinary	—	87·8%	4·5%	3·5%	4·1%
Ash: pre-chemic	63·8%	9·8%	—	25·4%	8·4%
ordinary	—	34·6%	53·7%	1·0%	10·5%

The above table shows that the removal of wax is better by the standard process with the two scouring treatments; on the other hand, there is a better extraction of pectic matter by the pre-chemic process than by the double boil. The residual size and ash are approximately the same in the two processes.

SPECIAL BLEACHING PROCESSES

Clark is of the opinion that a pre-chemic treatment can be used with some advantage; there are, however, no details of the fluidity or other chemical characteristics of the materials, and it is generally supposed that a pre-chemic treatment for cotton is apt to affect the fluidity to a marked degree.

Some interesting data of the effect of chemic before scouring have been supplied by Kollmann (Textilber., 1938, *18*, 994). From experiments with cotton having a wax content of 0·46%, it was established that treatment with chemic beforehand was not in itself sufficient to give the cotton either a low wax content or a high rate of wetting; nevertheless, it does appear to enable the subsequent kiering process to effect remarkable changes.

EFFECT OF PRE-CHEMICKING

Treatment	Wetting time	Wax content
Bleach for 3 hours with 5 g./litre of chlorine	22·6 secs.	0·36%
As above, followed by kiering without pressure for 2 hours in 1% NaOH solution	<1 sec.	0·20%
Steeping for 6 hours in 1 g. per litre of chlorine, anti-chlor with thiosulphate, and kier 2 hours without pressure in 1% NaOH	<1 sec.	0·15%

The wetting time was that taken for a small piece of fabric to sink in water after being placed on the surface.

It is also interesting to note that Kollmann found that a pre-treatment with hydrogen peroxide was much less efficacious than hypochlorite.

Freiberger (Textilber., 1931, *12*, 697) has outlined a process in which the desized material is chemicked and then subjected to an after-cleansing process in which the goods are treated with a solution containing soda, soap and water-glass which is continuously decolorised by sodium peroxide. The peroxide does not exert any bleaching action, as the duration of the treatment is less than one-fifth of that for a peroxide bleach, the temperature is much lower (55°C. as against 85°C.), and the amount of peroxide is only a fraction of that employed for bleaching. The method also forms the basis of a continuous process.

CONTINUOUS BLEACHING

Continuous bleaching processes have not received attention for many years, but latterly there has been a tendency to investigate these more closely, mainly by way of the peroxide bleach. Very large quantities of cloth are necessary to ensure the economic value of such processes, and it is preferable to have a series of fabrics about the same width and weight running together; the chief difficulty in continuous processes lies in the scouring part of the treatment.

In the Edmeston-Bentz system, as discussed on page 181, the cloth runs through a tank containing sets of rollers where it is impregnated with the boiling alkaline liquor and then steamed.

In another method, the cloth is run slowly through a series of tanks by means of guide rollers; the various tanks contain the different solutions necessary for the various operations of the bleaching process.

It will be realised that such methods lend themselves to open-width treatment, which partly compensates for the lower temperatures and shorter times in the scouring stages; stronger solutions are generally used in the continuous processes than in the discontinuous treatments.

Open-width washing, souring and chemicking machines are made by many manufacturers of bleaching machinery, and these permit continuous treatments. Some ingenious devices in the way of time-wheels, J-boxes, such as the Gantt piler, offer economies of space and lay-out.

As previously stated, the peroxide bleach lends itself to continuous processing more readily than the scour-hypochlorite method. A recent process sponsored by Du Pont is of particular interest (Text. World, 1939, *89*, June, 63). Mercerised cloth in the open width is impregnated in hydrogen peroxide solution of 1·4 to 1·6 volume and pH 10·2 to 10·8 at a temperature of 35 to 40°C.; the bath contains sodium silicate and a wetting agent. The goods, containing 85 to 125% liquor, are passed through a pot-eye to convert them to rope form, and then enter a specially designed heating tube which is an essential feature of the plant. Here they are heated with steam to 85 to 95°C. and fed into a J-box where they remain for an hour, and are then drawn through an ordinary continuous rope-soaper. Rinsing and drying complete the process. The chemical costs are stated to be about 60% of the older method of peroxide bleaching; tests indicate that the goods are more absorbent than those bleached with peroxide in a kier and are also equivalent in uniformity, permanence and degree of whiteness, tensile strength and fluidity in cuprammonia.

On the basis of two boils and one chemic, it is possible to represent the purification of the cotton diagrammatically. Data collected from various sources by Scholefield and Ward (J.S.D.C., 1935, *51*, 172) form the basis of Fig. 77.

Voronkov (Investia Textilnov Promushlenosti i. Torgovli, 1930) has reviewed the bleaching process as consisting, in the main, of three stages:

SPECIAL BLEACHING PROCESSES

(a) Starch removal by desizing which may take 6 to 12 hours by ordinary means.

(b) Removal of fat and wax in the kier which may take from 7 to 20 hours according to the means adopted and the type of material.

(c) Colour removal by bleaching which may take 30 to 60 minutes.

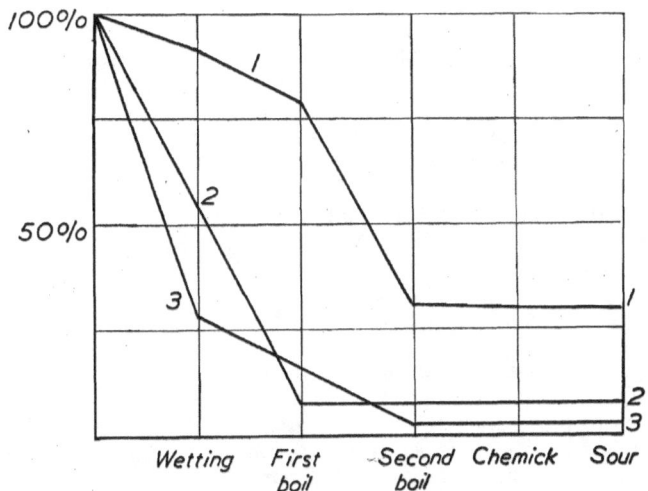

FIG. 77.—Impurities remaining in cotton after the sequence of bleaching operations. (*1* denotes fats and waxes, *2* nitrogenous material, and *3* mineral matter.)

From this Voronkov concluded that the kier boiling processes represent the main obstacle to rapid and continuous bleaching.

The first attempts at a continuous process were to replace the desizing and scouring treatments by a single process in a J-box with 5°Tw. NaOH, together with some sodium silicate and a wetting agent of the Nekal type.

A rapid and continuous bleaching process was finally produced by Fillipov and Voronkov (Chim. et. Ind., 1930, *24*, 943; 1931, *25*, 1498), in which the usual steeping and lengthy scouring treatments are replaced by a short process lasting for 30 to 40 minutes so that the complete bleach is reduced to about 1·5 to 2 hours. Continuity of operation is provided by a suitable arrangement of the apparatus in series. The scouring is carried out in open kiers containing a solution of the following composition—15 to 35 g. of sodium hydroxide, 5 to 15 g. of Contact T, 12·5 to 25 g. of potassium sulphite

(72° Tw.) and 7·5 to 10 g. of sodium silicate (74° Tw.) per litre; the goods are treated for 30 to 40 minutes at 90° C. and then washed with warm and cold water before bleaching for 30 minutes, in sodium hypochlorite solution of 0·5° Tw. followed by souring, washing and drying. It may be remarked that Contact T is a wetting agent derived from a special fraction of Russian petroleum (see page 195).

An important feature of the manipulation of the fabric is the heavy squeezing to which it is subjected between the various stages of the process. It is stated that the bleached cloth contains 0·18% of fat and wax, 0·009% of ash and that the nitrogen content is 0·01%. (The nitrogen content of the proteins present in cotton is 0·232%.)

Continuous scouring and bleaching processes have attracted considerable attention in the U.S.A., whose enterprising chemical manufacturers have sponsored developments which are now on the works' scale. The different methods have one common stage, in that the kier boiling is replaced by steaming in presence of caustic soda.

The Du Pont process has been described by Campbell (Am. Dyes. Rep., 1944, *33*, 293); it is a high-speed operation requiring 2 hours from the grey cloth to the fully bleached product, compared with 15 to 21 hours by usual methods. Units now in operation are producing at the rate of 1,500 lbs. of cotton fabric per hour, and over 500 million yards of cloth have been treated.

The cloth first passes through a solution of caustic soda and after uniform saturation, it is squeezed to contain 100% of liquor; the temperature of the alkali is 30°C., but the impregnated cloth is rapidly heated to the operating temperature of 100°C. by passing through a specially designed heating-tube (U.S.P. 2,304,474) at a rate of 100 yards per minute. The average length of the tube is 40 feet, so that the cloth is only in the heater for about 8 seconds. Where the cloth has to be treated in open width, the cloth heater takes the form of a stainless-steel box like a modified *J*-box or Gantt piler. The ratio of liquor to cloth is 1 : 1, compared with the usual kier ratio or 4 : 1 or 5 : 1, and less than 750 lbs. of steam is required per 1,000 lbs. of goods, compared with 2,500 to 3,000 lbs. in the kiering system.

The usual concentration of NaOH solution is 3%, but it may vary between 2·5 and 4% depending on circumstances; the consumption of alkali is therefore 3 lbs. of NaOH per 100 lbs. of cotton, and this is in interesting agreement with the amounts employed in kier-boiling. The heated and impregnated fabric is stored in an insulated chamber for the necessary time during which the alkali reacts with the impurities, i.e. one hour; the storage chamber is based on *J*-box construction.

Following upon the alkali treatment, the fabric is rinsed and

SPECIAL BLEACHING PROCESSES

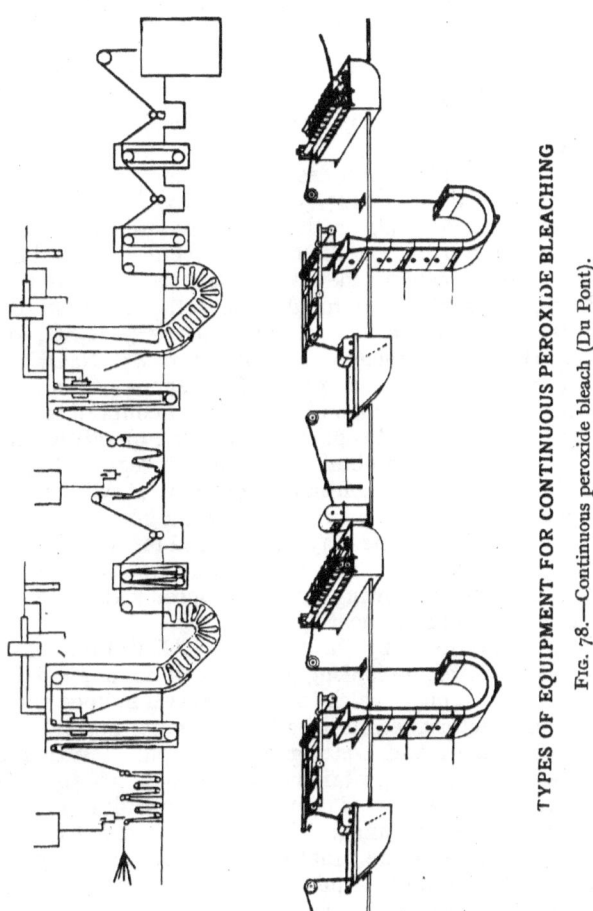

TYPES OF EQUIPMENT FOR CONTINUOUS PEROXIDE BLEACHING

FIG. 78.—Continuous peroxide bleach (Du Pont).

impregnated with 0·5 vol. H_2O_2 solution at pH 10·5 to 10·8, using silicate as a stabilising and buffering agent; after impregnation, the fabric is again heated to about 100° C. and stored for an hour or so. Some fabrics are more difficult to bleach than others, and may require H_2O_2 of 0·7 to 1 vol. concentration. A second rinse completes the basic process.

With coloured-woven goods, there is also a two-stage process, but both are with H_2O_2 solutions; the first treatment is in 0·5 vol. H_2O_2 and the second in 0·5 to 1 vol. H_2O_2, the storage temperatures being 95° C.

The Becco method has been described by the staff of the Buffalo Electro-Chemical Co. (ibid., 345, 365, 380, 385, 401, 402, 405). The flow of goods again depends on the use of the stainless J-box which permits uniform heating by direct steam, but it may be varied according to the character of the material and the degree of bleaching required. For example, a thorough preparation may entail a preliminary treatment for a few minutes in 0·2% H_2SO_4 or HCl, followed by rinsing and then saturating in 3 to 4% NaOH solution before steaming in the J-box at 100° C. for an hour. After a washing process, the final bleach is given by a solution containing hydrogen peroxide, caustic soda, and sodium silicate, the quantities and proportions being adjusted to obtain the required degree of bleaching. A final rinse completes the continuous process. It will be seen that the basic unit consists of a saturating device, a heated J-box, and a washing machine; these units may be arranged for continuous steam scouring and peroxide bleaching of goods in rope form or in open width, the dry steam flowing in the opposite direction to the material to ensure uniform heating.

The method sponsored by the Mathieson Alkali Company (ibid., 536) also utilises steam-scouring with alkali in a specially constructed steamer, through which the goods travel in open width on a conveyor. Satisfactory results in alkali-steaming depend on penetration of the alkali and wetting agent, rapid elevation to the required temperature, maintaining the goods at the elevated temperature, the prevention of cooling before washing, and washing with hot water after steaming. The goods so treated are suitable for bleaching by any of the usual processes, or they may be dyed in shades which normally required an absorbent but unbleached fabric. The Mathieson steamer may be used for bleaching with hypochlorite-activated Textone, or with acidified Textone, thus ensuring a continuous scouring and bleaching process. It is also possible to add Textone (sodium chlorite) to the alkaline liquor with which the cloth is impregnated before the first steaming treatment, a typical recipe being 6·5% NaOH, 0·5% Igepon, 0·1% Nekal and 0·25% Textone. After impregnating at 80° C., the fabric is steamed for 45 mins., washed with hot water, and bleached in NaOCl solution (2·5 g./l.) in a continuous manner.

The replacement of kiering by caustic steaming is stated to give a good white, and it is also stated that the removal of motes is good. It is, of course, possible to get a good white without thorough bottoming, but the removal of motes by alkali-steaming is rather surprising. The fluidity of the bleached fabrics is quite satisfactory, being of the order of 4, but more data should be forthcoming about the extent of wax removal, ash content, and the permanence of the white against yellowing before these methods commend themselves to the more exacting demands of the British bleachers.

COLD BLEACHING

Cold bleaching processes are frequently discussed in the textile literature; the term does not refer to the use of cold hypochlorite solutions, which is the normal procedure, but to methods of bleaching in which there is no previous kier-boil or scour. In most of these processes, the cotton is first steeped and then treated with the hypochlorite solution; the chlorine consumption is greater than for properly scoured goods and may require solutions containing 6 to 8 g. per litre, but the white is not so permanent, nevertheless the process is fairly satisfactory for certain classes of goods, including loose cotton and yarn intended for dyeing. Wetting agents are sometimes added to the bleach-liquors in order to assist penetration. The cold bleach has also been suggested as a means of avoiding the bleeding of coloured woven goods which is apt to occur in the kier. Damon (Textile World, 1940, *90*, 122) has given an account of an American process in which the desized goods are given two treatments in 1°Tw. sodium hypochlorite solution, first for one hour and later for 30 minutes with thorough washing between the two processes; the purification is less than that obtained in the standard treatments.

For raw cotton, it is claimed that the retention of the wax as a result of cold bleaching gives a product of better spinning properties. The presence of the waxes and other impurities, however, hinders wetting, penetration and diffusion; on this account it has been suggested to circulate the hypochlorite under pressure or, alternatively, to add soaps to the bleach-liquors possibly in the hope of obtaining a little scouring action. If soaps are utilised, however, then sodium hypochlorite solution is essential to avoid precipitation of lime soaps.

These treatments of unscoured materials with strong chemic are apt to give uneven results, and are generally regarded as poor methods of processing cotton.

An improved cold bleaching process is described in B.P. 399,320 according to which the cotton is first treated with hypochlorite, then given an intermediate treatment with a solution of an alkali, and, finally, a further oxidising treatment with hypochlorite.

Enzymatic removal of starch and size is common practice in the preliminary operations of bleaching; this has been extended by Rohm in B.P. 100,224, which describes the use of pancreatic enzymes for the elimination of nitrogenous matter and fats; other suggestions have been made to effect a complete bleaching process by similar means but without any commercial success.

THE COMBINED BLEACH

Combined bleaching processes usually refer to the use of the two oxidising agents—hypochlorite and peroxide—in succession; if it be assumed that this process of chlorination followed by oxidation, then a process of Zellstoff-fabrik Waldhof, as outlined by Ristenpart (Textilber., 1922, *3*, 363) is of interest. Cotton is first subjected to the action of acid hypochlorite solutions and later to alkaline solutions containing about 1 g. of available chlorine per litre. Good whites and excellent tensile strength are claimed.

The peroxide bleach has always been fairly popular in Germany, and within recent years the "combined bleach" has come to the fore. In one variation of this method the goods are kier-boiled, chemicked and then treated with peroxide. A permanent white is obtained in a manner which is not possible with a single kier boil followed by a chemic, and the necessity of a second kier boil for many goods is thus avoided. It may be that the peroxide treatment is responsible for the permanent white by removing any of the chlorine compounds such as chloramines which may have formed; although the subject has not reached finality, it has been suggested some of the yellowing which may appear on ageing with fabrics that have not been thoroughly bottomed is due to the presence of a chlorine compound.

The Mohr process was one of the first combined bleaching treatments but also made use of the "intermittent" effect which is obtained in pressure kiering. The suggestion is that wax removal during kiering may be due in part, to the mechanical disturbance brought about by pumping the liquors through the kier; temperature may not be the chief factor. On this basis, apparatus was built in which the cotton goods could be placed, and cold liquors could be pumped through under 30 to 40 lbs. pressure. Trials were first made with cold alkali for one or two hours, followed by washing, souring, washing and the circulation of chemic; later developments, however, utilised the combined bleaching process. Hypochlorite is first circulated through the goods under pressure, followed by water, acid, and

SPECIAL BLEACHING PROCESSES

water; peroxide completes the bleaching process (Textilber., 1925, 6, 909). The method appears to be covered by G.P. 410,106.

Many variations are possible in the plant which is of Zittau construction, some fifteen at least, having been erected in Germany and Holland. The simplest variation is the use of warm peroxide solution in the second main stage of the process. On a large scale, the used alkaline peroxide liquor may be circulated at the boil under pressure to give a preliminary scouring and whitening effect before washing and introducing the cold hypochlorite solution.

In the Mohr process there is no re-packing in kiers as all operations take place in the one lead-lined vessel thus saving a considerable amount of time. The general procedure appears to be first, a treatment in rope-form in the kier with exhausted liquor from the bleaching process containing the protein degradation products from the fabric. The cloth is then washed, and air removed from the kier, after which cold hypochlorite solution is pumped into the kier and circulated under pressure; when the necessary time has elapsed, the liquor is withdrawn and the fabric is washed and soured in the kier. After another wash, the goods are treated with hydrogen peroxide solution at 60 to 70°C. with circulation of the liquor. The whole sequence of operations may take 10 hours.

The Mohr bleach has been shown to give remarkably soft results.

Dahlenvord (Leipzig Monat. Textilind., 1929, 44, 262) described the "Osmotor" apparatus in which cotton is treated for 2 to 3 hours with cold hypochlorite solution, washed for 45 minutes with water, giving a three-quarter white; a full white is obtained by a further treatment in the same apparatus for 3 hours with a hot solution of sodium peroxide and sodium silicate.

The Ce-Es process may also be described as chloramine bleaching, as the primary action is the chlorination of the protein substances. B.P. 401,199, by T. H. Bohme, is considered to apply to this method, in which the goods are impregnated with hypochlorite and then treated directly, without an intermediate bath, with peroxide. Presumably the Kauffmann acceleration effect comes into play, although it is stated that when the goods are treated with hypochlorite and a sulphonated oil-wetting agent, the fibres are rapidly saturated, and in a short time the active chlorine is almost entirely used so that the peroxide contains practically no free chlorine. The chloramines formed are more readily decomposed and dissolved in the alkali of the peroxide bath than the original albuminous substances.

The unboiled goods are first impregnated in hypochlorite containing 2% of active chlorine, 2% NaOH and 1% Na_2CO_3 for one hour at room temperature; Floranit VP or Avirol AH may be used as wetting agents. The goods are then squeezed and passed, without washing, into an alkaline hydrogen peroxide bath where the chloramines dissolve and act as colloidal dispersing agents. The action may be increased by the addition of Homogenit B and the stabilising of the hydrogen peroxide bath may be improved by the addition of Ondal W 20.

The bleaching loss is smaller than that experienced in the scour-hypochlorite processes, and the goods are stated to have a full white appearance, a soft handle and exceptional absorptive capacity.

The Korte process, as developed by the I.G., starts with a preliminary treatment in alkali to remove the pectins and proteins; the lignins and some of the wax are chlorinated by hypochlorite at pH 3 to 4, using a weak solution for a short time. The chlorinated products may then be removed by hot alkaline peroxide which also bleaches the cellulose, or, as an alternative, the chlorinated products may be removed by an alkaline boil and the process completed in dilute alkaline hypochlorite solution of half the usual concentration of active chlorine.

This process is particularly applicable to the bast fibres, and Kayser (Textilber., 1938, *19*, 725) states that linen bleached in this way has a lower solubility number than that from the ordinary oxidation treatments.

Kling (Z. ges. Text. Ind., 1933, *36*, 496; 1934, *37*, 325) describes the effect as mainly chlorination of the lignified substances in an acid medium (pH less than 5) whereby they become soluble in alkali. The treatment of linen starts with an open boil for 2 hours in sodium carbonate solution containing 10 g. per litre, followed by immersion for 2 hours in a solution containing 8 g. of active chlorine and 7 g. of hydrochloric acid per litre; the goods are then scoured for one hour with soda (5 g. per litre) and bleached in a solution containing 4 g. of active chlorine per litre.

Modern Developments

Butterworth's process, which has been developed with Ridge, and in conjunction with Messrs. Frazer and Haughton and the I.C.I., appears to be the only British contribution to combined-bleaching. It has the additional novelty of departing from the modern attachment to alkaline hypochlorite solutions and using the highly active and usually dangerous neutral solutions; conditions are controlled, however, in the light of recent developments and also in subdued light.

In general, there is little modern published work on the bleaching of linen, so that the contribution by Butterworth (J.S.D.C., 1939, *55*, 589) is all the more valuable on that account. Butterworth points out that the extensive work on the oxidation of cellulose, published between 1920 and 1940, has gone far to clarifying the chemistry of the process, but much of the data refers to carefully purified and bleached material subsequently treated under much more severe conditions than would obtain in normal practice. The problem from the standpoint of the bleacher is not the degradation of the cellulose, but the bleaching of the colour without attack on the cellulose and in presence of its impurities. These two viewpoints are, of course, closely related, but the range of conditions is smaller in bleaching than in laboratory production of oxycelluloses.

The position of linen is different from that of cotton in that it is customary to retain some proportion of the non-cellulosic constituent;

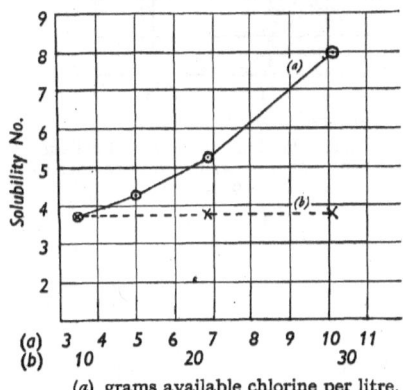

(a) grams available chlorine per litre.
(b) ratio of liquor (2·8 g. Cl/l.)

FIG. 79.—Linen bleaching.

the ease of bleaching, therefore, corresponds inversely with the degree of scouring. The final colour produced may be obtained by adjusting the strength of the hypochlorite, and concentrations up to 10 g. per litre of available chlorine may be used. Butterworth states that this adjustment is largely empirical, and is determined by the final shade of the bleached material; the reaction with hypochlorite can be pressed too far, since after a certain limit, the degradation of the cellulose is not accompanied by a corresponding whitening.

It is suggested, therefore, that for commercial bleaching processes, if a safe method of using neutral hypochlorite liquors could be devised, any drift from the neutral region would retard the action on the cellulose. In these circumstances, the question arises whether the acid, neutral, or alkaline hypochlorite is most specific in the action on the colouring matter. It must be recalled that the colour of flax is more pronounced than that of cotton, and the high degree of whiteness often demanded has led to various degrees of degradation of the cellulose. One bleaching step, even if drastic, only produces a yellow colour, and the subsequent alkali treatment darkens the fibre

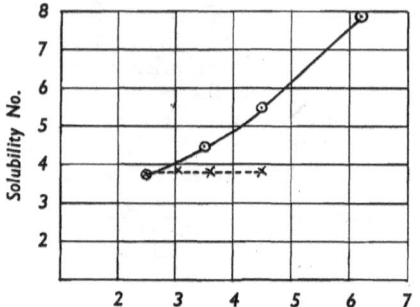

Fig. 80.—Linen bleaching (full curve for liquor of increasing concentration; broken curve for increasing ratio of liquor.)

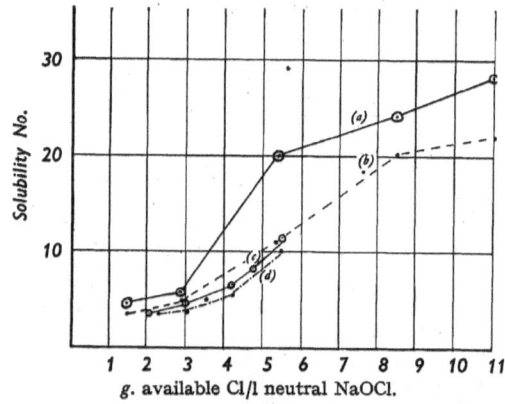

Fig. 81.—Scoured linen yarn reeled 30 mins. in liquor to show effect of increasing concentration of neutral hypochlorite (a) and (c) in diffused light, (b) and (d) in the dark at 18·5° C. (a) and (b) and at 14°C. (c) and (d).

so that further oxidation is necessary, the sequence of these steps usually leads to a certain amount of degradation.

Experiments in presence of the colouring matter revealed that, over the useful range of available chlorine, the attack is a function of the concentration and is almost independent of the total amount of chlorine present. For the same chlorine consumption, the attack is less in liquors of lower concentration. The bleaching effect was rapidly produced on yarns scoured to lose 5 to 6% in weight, when subjected to concentrations of 3 to 3·5 g. of chlorine per litre in solutions near the neutral point, and shows little improvement up to 10 g. per litre although the attack rises accordingly.

The effects of temperature and sunlight during the bleaching process are well known, but Butterworth also showed that strong diffuse light has a pronounced effect in degrading cellulose in presence of hypochlorite solution.

As previously indicated, one bleaching step is not sufficient to destroy the colouring matter of linen, and a further alkaline treatment produces a greyish-brown colour which in its turn requires bleaching; a modern alternative is the treatment with warm peroxide solutions, in which the degradation is less. Nevertheless, temperature and pH have an effect on degradation, but this is determined to some extent by the previous history of the material.

After the chemicking of linen it is impossible to remove all the chlorine by washing with water, and Butterworth has shown that in a standard peroxide bath, the degradation increased with the amount of "fixed" chlorine. There is, however, no simple relation between the amount of chlorine and the extent of degradation in the peroxide bath.

Material which has been treated with hypochlorite and washed before entering the peroxide bath, showed minimum degradation for pH 11 to 12 in a bath containing 1·2 g. of H_2O_2 per litre after 3 hours at 75°C. The colour differences were also prominent as the acid, neutral, and slightly alkaline solutions showed poor bleaching properties.

The treatment with hot alkaline peroxide solutions may possibly be regarded as combined alkaline scouring and bleaching, but if the two aspects are separated by using first, an alkaline scald, and then a slightly alkaline bleaching bath, there is greater degradation and less bleaching; the same phenomenon is observed if the peroxide bleach precedes the alkaline scald.

Now the reaction with alkaline peroxide cannot be taken too far without producing degradation, so the course to follow is that of repetitions of the hypochlorite-peroxide sequence. It was found that there seems to be a threshold value for hypochlorite below which there is only slight degradation even with repeated treatments.

These investigations point to a method of bleaching in which the non-cellulosic component is oxidised and the cellulose is not attacked. If such a treatment follows a lime boil and alkali boil, a very good white is obtained. The amount of degradation is small and Solubility Numbers of less than 5 and even less than 4 may be obtained on linen; the estimation refers to the method of Nodder (J.T.I., 1931, 22, 416). [See page 468.]

The preliminary treatment comprises a thorough wetting of the material by a known wetting agent, or, preferably, an alkaline scour with sodium carbonate solution. The hypochlorite treatment takes place in a neutral solution containing 1·5 to 5 g. per litre of available chlorine, buffered with sodium bicarbonate. (3, 5 and 8 g. per litre of $NaHCO_3$ are required for solutions containing 1·5, 3 and 5 g. per litre of available chlorine prepared from bleaching powder.) The time of treatment should not exceed 30 minutes for solutions containing more than 3 g. of chlorine per litre, and bright light should be excluded by screening the vessels in which the treatment is carried out; as an example of permissive light for linen, a normal electric light bulb of 60 c.p. should not be placed less than 6 feet from the material undergoing treatment. After the hypochlorite treatment, the goods are washed in cold water and then soured (0·5% HCl for 30 minutes) followed by a further wash. The goods are next treated in an alkaline peroxide bath of alkalinity equivalent to 0·15 to 0·4% Na_2O and containing 0·05 to 0·2% H_2O_2; the time of treatment should be about 3 hours at a temperature of 65 to 85°C.

The severity of the treatment with hypochlorite varies inversely with the severity of the initial scour.

With most linen yarns, a half-white may be obtained as outlined above and if higher degrees of bleaching are required, then further treatments with hypochlorite and alkaline peroxide may be given, but of reduced severity. A final washing process completes the treatment in all cases.

These methods have been covered in a series of patent specifications by Butterworth, Ridge, Frazer and Haughton, Ltd. and I.C.I. Ltd. The main document is B.P. 489,496, variations being covered in B.P. 497,346; 499,873; 500,121 and 500,473.

B.P. 497,346 covers a preliminary treatment with reducing agents, whilst B.P. 499,873 discloses that the preliminary scour may take place in the liquor from the peroxide bleach. B.P. 500,121 refers to the addition or carbonates of alkaline earth metals to the hypochlorite solutions and B.P. 500,473 gives details of determining the pH value of the hypochlorite liquor from time to time in cases where no buffer has been added to maintain neutrality.

The neutral hypochlorite solution, buffered with bicarbonate, had previously been utilised by Minajev (Textilber., 1936, *17*, 218) as a method of bleaching in a sufficiently rapid manner to form a continuous process. The use of either warm chemic, or neutral chemic, is normally regarded as dangerous to the cellulose so that it is usual to employ very dilute solutions for a short time. In any case, it is obvious that most careful control of conditions is essential to avoid damage, and a number of these conditions have been outlined by

SPECIAL BLEACHING PROCESSES

Butterworth; in addition, as previously stated, the circumstances for bast fibres are different from those surrounding scoured cotton.

In connection with the above reference to the use of warm hypochlorite solutions, it may be remarked that one such process has been suggested by Freiberger (Farber-Zeit., 1919, *30*, 89) in which cotton is treated with hypochlorite solutions containing 1 gram per litre of available chlorine for 35 minutes at 35°C., in contrast to the use of 3 grams per litre of available chlorine at 15 to 18°C. The use of warm "chemic" cannot be advised.

PART FOUR
THE SCOURING AND BLEACHING OF ANIMAL FIBRES

CHAPTER XXIV
SCOURING RAW WOOL

THE impurities present in raw wool may be divided into two classes:
(a) adventitious impurities
(b) natural impurities.

The adventitious impurities are mainly composed of particles of dried grass and straw, thistles, burrs, seeds and other substances of a cellulosic character, together with a certain amount of dust, dried earth and excremental material. In particular, the straw and other cellulosic material become very firmly entangled with the wool fibres on account of their waviness and scaly surfaces so that they cannot be removed mechanically but need a special process of chemical destruction known as carbonising:

The natural impurities are due to the growth of the wool hair, the follicle of which secretes an oil which is supplied to the hair during its growth and imparts some lubrication; two types of gland surround the follicle, the suint glands and the sebaceous glands. The suint glands secrete a complex mixture of the potassium salts of various fatty acids which give raw wool a disagreeable odour; it has been suggested that they prevent wool being damaged by exposure to sunlight, because the original suint is transformed from sheep-sweat by fermentation in presence of light. The suint itself possesses some detergent power which can be utilised in the scouring process. The wool fat is secreted by the sebaceous glands and forms a protective coating on the wool hairs, preserving them from injury, hindering entanglement and exerting considerable water-repellency. From the fatty matter, it is possible to extract and purify the product known as lanoline.

Suint is generally assumed to consist mainly of the sodium and potassium salts of the lower fatty acids, together with a small amount of other substances. The first examination seems to have been made by Vauquelin (Phil. Mag., 1804, *19*, 33), who considered it to be a soapy substance containing potassium carbonate, potassium acetate, potassium chloride and compounds of animal origin.

Buisine (J.S.C.I., 1886, 5, 539) made a more detailed investigation and found suint to contain potassium, sodium and ammonium, carbonate, and all fatty acids from acetic to octoic, oleic, stearic, and cerotic acids, oxalic, succinic, lactic, benzoic and uric acids, together with glycine, leucine, tyrosine and some phenol, present as the potassium sulphonate.

A more recent examination of suint by Freney (J.S.C.I., 1934, 53, 131) has shown that the product is normally neutral or slightly alkaline; suint samples vary greatly in ash and nitrogen contents. Sulphates, carbonates and chlorides are present in small amounts, and chiefly as the potassium salt. Suint was found to contain up to 20% of saturated organic acids, including the lower fatty acids from acetic to palmitic acid; succinic acid was definitely isolated and small quantities of lactic and hippuric acids were found.

Many unidentified organic materials are also present, but there are only small amounts of urea and ammonia; there was no positive result from a test for proteins and it may be that these compounds have suffered bacterial decomposition.

Before wool is carded it is scoured; this process is sometimes termed washing, but it should also be remembered that wool is washed on the sheeps' backs, and confusion between these two types of washing should be avoided. In the scouring process for raw wool there is a loss in weight amounting to about 40%, and of the material removed in this manner about 40% consists of natural fatty matter. Hence the effluent from the washbowls is extremely rich in organic materials.

Grease Recovery

An interesting account of the recovery of wool fat has been given by Brock (J.T.I., 1937, 28, 30P) and also by Wontner-Smith and Campbell (Paint Mfr., 1942, 12, 40).

About 20% of the world production of wool is washed in the Bradford area and the Corporation has established an immense plant to deal with the greasy suds from the works. Of 64 wool washing and combing factories, only 16 extract their own wool grease; this is known as Yorkshire Brown Grease, whereas that passing to the Esholt works is termed Recovered Grease, but both qualities contain the fats derived from the soap used in scouring wool.

The extraction of the grease generally depends on precipitation by sulphuric acid followed by hot pressing. Where the suds are dealt with on the mill premises, they are usually collected in a series of wooden tanks where they are "cracked" with acid; the fatty layers of magma form partly by precipitation and partly by scum concentration on the surface. The magma is collected, heated and pressed.

As the Brown Grease is produced from various sources, there is a great variety of crude wool fats depending on the nature and amount of the soap used in scouring. Where olive oil soap has been used, a pale, soft, sweet-smelling grease may be extracted, but the wool fats from the dirtier types of wool are dark in colour, and some may be fairly acid. The free acidity of the recovered wool fat may be represented by about 13% estimated as oleic acid, but the figure may vary between 10% and 20%. The remaining portion of the fat consists of wax alcohols and fatty acids; the latter are completely combined as wax esters, but there is also a certain proportion (10 to 15%) of the wax alcohols present in the free state.

The combined fatty acids of wool fat are latent in the wool fat, whereas the free fatty acids are derived from the scouring soaps. The former seem to be of an unusual type for they exhibit a tough waxy character and belong to a special series of acids designated by the prefix "lano-"; they appear to be saturated for their iodine value is very low. These acids range from C_5 to C_{26}; the latter typifies wax acids for cerotic acid exists in beeswax in the free state.

The molecular weights of some of the acids correspond to palmitic and stearic acid, but the lano-stearic and lano-palmitic acids seem to have a special hydroxy construction for they form internal anhydrides on heating; the nature of the soap from these acids suggests a somewhat unusual constitution for they remain translucent, soft and viscous for years.

There is also present about 20 to 25% of a hard waxy acid (m.p. 80°C.) with a molecular weight of 460; this appears to affect the physical properties of the mixture of acids by conferring toughness and wax-like features.

The unsaponifiable matter extracted from saponified wool fat exhibits the characteristics of the wax-like fatty alcohols, and resembles cholesterol.

Cholesterol and isochloesterol are two compounds belonging to the series of polycyclic polyterpenes; the melting-point of cholesterol is 127°C. and that of its isomer is 115°C.; the former is laevorotatory and the latter dextrorotatory.

The empirical formula for cholesterol is $C_{27}H_{46}OH$, but it may be more fully represented by the following formula.

As previously mentioned, wool fat has found an outlet in the manufacture of commercial lanolin; it may be used as a rust preventive and for the production of super-fatted soaps. The neutral fat may be extracted by alcohol and bleached with hydrogen peroxide. Lanolin is actually the neutral ester of the fatty alcohol and the fatty acids, and should be termed wool wax, but as it is soft and greasy, the term wool wax has come to denote the extracted wax alcohols derived from pure wool fat.

The wool-wax alcohols have been described in some detail by Lower (Manufacturing Chemist, 1943, *14*, 231).

The general method of producing wool wax is to saponify the wool fat with alcoholic caustic soda and extract the alcohols with volatile solvents; the tough waxy product, sometimes termed cholesterol, possesses remarkable powers of emulsification.

The Recovered Grease from Esholt is somewhat high in acidity and dark in colour; it does not lend itself to refining. It was found, however, that it could be dry-saponified by heating and stirring with powdered caustic soda. The unsaponifiable matter was shown to possess drying properties and may be used in paint manufacture, an account of which has been given by Wontner-Smith and Campbell (loc. cit.).

General

The amount of the different impurities in raw wool varies from 30 to 80%; merino wool often contains 40 to 70% impurity, crossbreds from 20 to 45%, whilst English wools only contain 10 to 20% of impurities.

Skin wools contain certain impurities additional to those found in the fleece wools, depending on the method of removal from the skin. In the sweating process, the skins are moistened and hung in a closed chamber at 16 to 32°C. until the hairs are loosened by bacterial action and may be "pulled"; this sweating may take 3 to 4 days. A quicker process is to paint the skin with a 30% solution of sodium sulphide on the flesh side and then pile the pelts, wool to wool and skin to skin—like sandwiches—for a day before pulling the wool. Slipe wool is obtained in a broadly similar manner using a cream of slaked lime instead of sodium sulphide; as will appear later, this use of lime causes difficulties in the scouring process unless some special precautions are previously taken.

SCOURING

Now the great cleansing materials generally available for most purposes are soap and hot water; unfortunately wool is very sensitive to alkali and also to temperature. The methods employed for scouring cotton, for instance, would rapidly destroy wool. In addition, any mechanical pressure in the wet state must be reduced to a minimum to avoid felting the material.

Wool may be scoured at various stages during its manufacture, as raw wool, in sliver form for backwashing during the preparation of tops, as yarn, and in the form of woven or knitted cloth.

Unlike cotton, which needs its wax in the mechanical operation of spinning, raw wool is most difficult to spin and must be washed or scoured before it can be manipulated by the spinner. Good and careful scouring is also of importance as affecting other subsequent operations such as fulling or milling and dyeing, as the greasy wools are more difficult to process in these chemical treatments. Hence, although the scouring of raw cotton fibres is comparatively rare, and only for special purposes, the scouring of loose wool is a standard process which is applied to the greater part of wool before spinning.

There are various methods of purifying loose wool and these are determined to some extent by the type of material to be cleansed, and the nature and amount of the impurities; the chief methods are:

(*1*) Emulsion scouring
(*2*) Suint scouring
(*3*) Solvent extraction
(*4*) Freezing.

EMULSION SCOURING

Scouring with soap solutions is by far the commonest method of cleansing loose wool, for although wool fat is difficult to saponify under conditions which would not damage the wool substance, yet it may readily be emulsified, particularly at temperatures slightly in excess of its melting-point (40 to 45°C.).

As a wide generalisation, it is customary to use about 2 to 4% of soap and 2% of sodium carbonate, calculated on the weight of the wool; the scouring of the raw wool is popularly known as "wool-washing."

A great deal of wool is dyed directly after scouring, so that any residual alkali will be washed away or even neutralised as most of the wool colours are acid colours. Even where the wool is not dyed before spinning, it is customary to use a much greater proportion of oil in spinning woollen yarns than worsted yarns. The fibres for the latter are rarely dyed in the loose state and it is customary to have the material well scoured before carding and combing. It is not always appreciated, however, that a small proportion of oil or fat —about 0·75%—should be left in the wool to avoid brittleness with consequent damage in the mechanical processing.

Before discussing the scouring of wool in detail, reference may be

SCOURING RAW WOOL

made to neutral scouring with some of the auxiliary products or soapless soaps. For example, the burrs of certain wools, such as some of the New Zealand qualities, become red with alkali and stain the wool; hence neutral scouring is a great advantage. Again, with limed or "limey" wools, although it is usual to remove the lime with hydrochloric acid and rinse before scouring, this involves extra processes. As previously indicated, some of the synthetic detergents are of value here and the limed wools may be scoured directly.

Machines

Wool-washing machines are made in two main types:
- (a) the Harrow machine
- (b) the Swing Rake machine.

In the Harrow machine, the rakes are mounted on one frame and move together through the scouring liquor, whereas in the Swing Rake machine each rake may move separately.

Although the Harrow machine will wash practically all classes of wool, it is specially suited to the fine and medium wools; the rakes in the Swing Rake machine, however, bring about a greater agitation of the wool, and whilst this may be beneficial in loosening heavy dirt and in opening matted wool, yet it is apt to be detrimental for the finer wools.

With the long lustre wools, it is usual to employ the swing rake system of propelling the loose wool through the liquor. These rakes can be adjusted to act in unison or may be timed to exert an individual sweeping movement by operating at different periods. This irregular motion brings about more agitation of the wool and so produces a better cleansing action but is only suited to the coarser wools, for the finer wools would tend to felt if severely treated, and cause breakages in the subsequent mechanical processes of spinning.

At the end of the bowl, the wool has to be carried forward and upwards to the squeezing rollers; this may be done by the Belgium lift which operates like a small grab, some of its forks dipping into the wool and lifting it on to a small conveyer which carries it to the nip of the rollers. Alternatively, the slide lift may be utilised, and this consists of a series of slides with projecting spikes which carry the wool to the rollers.

The long merino wools may also be scoured by the swing rake method but this is too drastic for the finer wools so that the alternative Harrow fork system is used. These forks are fitted to a framework which passes along the greater part of the length of the bowl, the prongs projecting through the frame so that, in the lowest position they almost reach the perforated bottom of the bowl. The forks are moved by a cam and crank arrangement in such a manner that the prongs first descend into the liquor and press the wool down, they then move along horizontally carrying the wool through the liquor and at the end of the sweep they rise vertically out of the liquor and are returned to their original position. The usual length

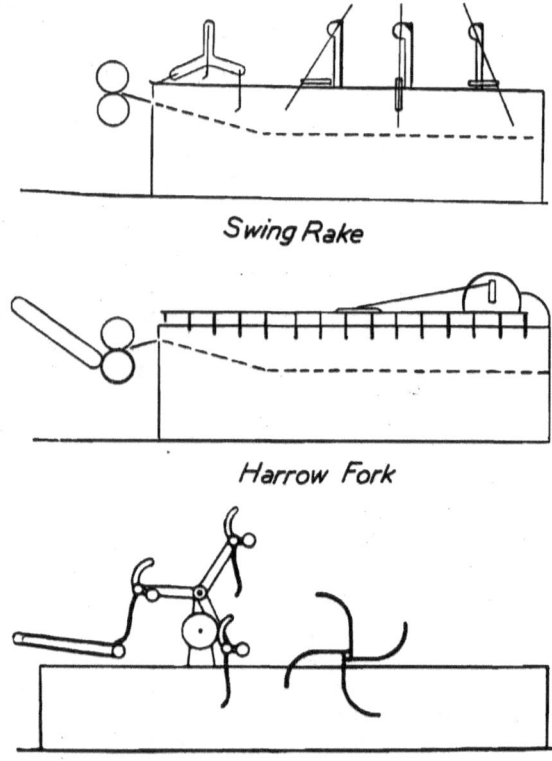

Fig. 82.—Diagrams of wool washing machines.

of sweep is about 9 inches but a 12-inch sweep is sometimes used; the harrow generally operates about ten times per minute. A gentle movement with no felting is the chief characteristic of the harrow forks.

Some systems of scouring raw wool of medium quality include both methods, first the swing rake and then the harrow.

The feeding devices for the bowls generally consist of some form of endless apron made of laths or lattice wire on to which the wool is evenly spread; as the greasy wool tends to float some device in the form of an immersion roller or box, popularly termed a "ducker," is fixed at the entering end of the bowl. The hollow roller or box is perforated to maintain the level of the liquor under which the wool is forced, and moved forward.

The continental systems of scouring wool are based on the fact that they had to cope with the shorter and dirtier wools; in general, more and larger bowls are used than in England. Many scourers follow the suint method of cleansing the wool but the paddle machine is often used. Here the paddle action gently "swishes" the wool along the bowl in which it has a longer period of immersion at a much lower temperature than in the English methods. It has been suggested that the continental system is a closer approximation to true emulsification than the English method in which the wool fat is partly removed by melting and washing away.

The English systems operate at temperatures of 50 to 55°C. for a total period of 10 to 15 minutes, whereas the continental methods use temperatures of 28 to 38°C. but for longer periods of 20 to 25 minutes in all.

In all these systems, however, the wool is scoured in troughs or bowls with perforated false bottoms which permit the heavy impurities to escape by settling into the external compartment of the bowl. At the end of the bowl, side compartments are also provided to receive the liquor from the squeeze-rollers, which are usually covered with a special roller-lapping top, prepared from long English wool. As the dirt settles and the fats rise, the intermediate purer liquor may be fed back into the bowl for re-use in scouring.

Wool-scouring machines are generally made in standard widths of 28, 36 and 48 inches, and in lengths of 30 to 40 feet to suit requirements. The bowls may be self-cleaning or not, but the self-cleaning type is recommended for the first bowl of the set in nearly all cases. Whenever possible, the bowls of the washing machine are arranged in one straight line, but if space does not permit then they can be arranged at right angles, or alternatively, one or more bowls can be parallel to the others. This arrangement does not necessarily interfere with the continuity of the operation, for the wool may be carried from one line to the other by brattices or water-carriage conveyors.

There are generally three or more of these bowls arranged in line to scour the wool. Many types of bowl are available and in some of them it is possible to impart additional motion to the rakes, opening the wool and agitating it during the scour. The process is naturally gentle to avoid felting and owes a great deal of its efficiency to the open nature, or state of division, of the loose wool which assists penetration and permits the escape of the scouring liquors. The various bowls of Messrs. Petrie and McNaught, Dawson, Hughes and Lancaster, and Taylor and Wordsworth are well known. In most of the bowls, there is a constant circulation of the liquor which assists in the movement of the wool through the bowls. In normal use the scouring liquor becomes too dirty for further use and has to be "run off," but self-cleansing bowls are available which reduce the number of stoppages for cleaning. A semicircular trough runs through the lower part of the bowl and contains a spiral conveyor which revolves slowly and carries the deposited dirt to a hopper fitted with a valve which is regulated to open at intervals.

Broadly speaking, whatever the arrangement of the bowls, the first bowl is responsible for the major part of the cleansing, and the later bowls for the washing and rinsing of the cleansed wool. It is thus possible to arrange the bowls to operate to some extent on the counter-flow principle. When the scouring liquor in the first bowl becomes too dirty for further use, it is withdrawn; it is possible to do this continuously. Alternatively, the dirty liquor may be clarified by a centrifuge and used again.

The quantity of water and soap is determined to some extent by the nature of the wool, but assuming that an output of 1,000 lbs. of wool per hour is required and the wool contains about 15% of grease, the first bowl will require an inflow of approximately 200 gallons per hour; the second bowl may require a constant supply of 100 gallons of fresh water, but the third bowl may be supplied with water from the fourth bowl which in its turn will need about 250 gallons of fresh water per hour. When starting the scouring operation, it will not be necessary to make any additions for some hours; the additions of soap and/or alkali, of course, will need to be made from the start. The necessity for adding soap may be seen from the state of the surface froth or lather in the bowl.

When non-self-cleaning bowls are used, it may be necessary to empty the first bowl every 24 hours; it may then be filled with the liquor from the second bowl and brought up to the required strength.

The Washing Process

In the scouring process it is important to use soft water and a soap of high emulsifying power; soft water is important as the formation of metallic soaps is avoided by its use. The scouring liquor has not only to remove the dirt and grease, but it must also prevent them being deposited on the fibre again and therefore has to possess powers of suspension as well as emulsification. There are many different opinions as to the best scouring liquor.

A neutral soap is important on account of hydrolysis and the preferential absorption of alkali by the wool, which would cause discoloration, harsh feel and alkali damage. Soft soap appears to be preferable for wool scouring but is dearer than the sodium soaps.

The *exact* neutrality of a soap for scouring, as distinct from milling, is sometimes over-stressed in considering raw wool; small quantities of alkali have beneficial effects on the scour, as will appear later, but it is important that no free alkali should be left in the wool at the conclusion of the scour. The very dilute nature of the scouring liquor also provides a safeguard in itself.

It is frequently assumed that the best soap for scouring is a potash-olive oil soap but some cheaper alternatives are available. The hardness or softness of a soap is not solely decided by its alkali constituent; the fatty acid also plays its role. Palmitic and stearic acids are solids at ordinary temperatures and their soaps are not very soluble in cold water and tend to gelatinise. Oleic acid, on the other hand, is liquid at room temperatures and yields soaps which are very soluble and do not gelatinise. The harder soaps are more difficult to remove from the wool and their residues tend to give a harsh feel to the scoured material. A satisfactory substitute for the potash-olive oil soap is a soda soap made from oleine which is free from the solid fatty acids and sufficiently free from stearine in particular to have a low melting-point; this may be tested in a simple manner by observing the tendency to gelatinise on cooling the soap solution or, alternatively, by precipitating the fatty acid from the soap and observing the melting-point which should not exceed 27 to 30°C.

Part of the detergent action of soap is due to the reduction of the interfacial tension between the scouring liquor and the wool grease; the surface tension effect increases rapidly with concentration up to 0·4% after which there is little further improvement. A second factor is the stabilising effect due to its action as a protective colloid, and this is at a maximum around 0·2 to 0·3%. It would appear, therefore, that there is no advantage in exceeding a concentration of 0·4% soap; actually 0·7% soap is frequently used in the first bowl.

Now an aqueous solution of soap hydrolyses, and in presence of purified wool there is a preferential absorption of the alkali with the production of a slightly acid soap; with raw wool, there is a similar phenomenon and as wool grease is frequently acidic, the effect is greater. The detergent value of the solution suffers on this account and an adequate alkalinity must be maintained, sufficient to neutralise the free fatty acid produced, and also to improve the detergent value of the soap. With insufficient alkali, there is apt to be frothing or foaming, and the free fatty acid may be deposited on the wool during the final rinse. Soaps containing a slight excess of alkali have

306   TEXTILE BLEACHING

a better detergent action than when neutral, the surface tension effect reaching a maximum in soap of 0·2% concentration; with sufficiently strong alkali, however, the soap may be precipitated in the form of curds. The addition of alkali, therefore, improves the detergent action of the soap and saponifies the fatty acids in the grease, converting them into soaps, but it is usual to add the alkali in the form of sodium carbonate; in order to exercise the necessary careful control over the alkalinity of the scouring liquor it is preferable to start with neutral soaps. The amount of sodium carbonate which may be added is apt to vary from works to works, and with the type of wool undergoing cleansing, but 0·25% is common.

Some wool scourers use only a small quantity of soap in the first bowl, and the amount varies again not only from wool to wool, but from works to works. It seems desirable, however, to have sufficient soap in the first bowl to start the emulsification process, after which the natural impurities in the wool are to some extent self-emulsifying. Although in theory, the best results come from soap and water alone, it must be remembered that all water is not perfectly soft, and the use of soda may be a necessity from this standpoint too.

Three of the chief functions of a detergent are to wet the greasy surface, to emulsify the grease, and to stabilise the emulsified grease and suspended dirt. Although the lowest surface effects are exhibited by soap at pH 8 to 10, and solutions of 0·1 to 0·2% of soap have excellent emulsifying properties at pH 10, yet the emulsification and suspension power falls off rapidly below pH 9 unless the concentration of soap is increased.

Priestman (*Woollen Spinning*, Longmans, Green; London, 1924) gives the following concentrations of soap and alkali for scouring raw wool, utilising bowls of 1,500 gallons capacity.

CONCENTRATIONS OF SOAP AND ALKALI

	Soap	Alkali
First bowl	0·80%	0·27%
Second bowl	0·40%	—
Third bowl	0·37%	—

Schofield (*Finishing of Wool Goods*) gives the following data:

CONCENTRATIONS OF SOAP AND ALKALI

	Soap	Alkali
First bowl	0·75%	0·25%
Second bowl	0·50%	0·10%
Third bowl	0·25%	0·10%
Fourth bowl	Water only	

Schofield has also suggested a system of scouring in which alkali alone is employed in the first bowl, and soap with weak alkali is not used until the second bowl.

American practice may be seen in the following table taken from the American Wool Handbook (page 274):

CONCENTRATIONS OF SOAP AND ALKALI

	Soap	Alkali
First bowl	0.08%	0.24%
Second bowl	0.01%	0.2%
Third bowl	0.01%	0.1%
Fourth bowl	0.017%	—
Fifth bowl	Water only	

It will be noticed that the above figures are much lower than those given by British authorities; it may be that more frequent additions of soap and alkali are made to the bowls according to American practice.

The time of immersion of the wool varies from about 3 minutes in the first bowl to about 90 seconds in the last bowl.

The temperature of the scouring liquor is important not only with regard to the efficiency of the scour, but also in respect of minimum wool damage. The wool substance is affected by boiling water, and there is also some chemical evidence that water at $50°C$. may break the disulphide linkage.

The susceptibility of wool to alkali is well known and severe degradation may easily be brought about, even in dilute solutions at high temperatures. With the very dilute solutions employed in wool scouring there is less risk of damage; sodium carbonate in solution up to 5% concentration gave no appreciable tendering in one hour at $50°C$. Strong solutions of soap have tendering actions equal to that of about one-fifth of the equivalent of caustic soda, and appreciable tendering is produced in an hour at $50°C$. by either 1.5% of sodium oleate or 0.04% sodium hydroxide. An average concentration of 0.5% soap, as commonly employed, is well below these figures, and even a soap containing 4% of alkali will not produce tendering provided it is neutralised before the wool is dried. Sodium carbonate, of course, is safer than the hydroxide which is completely ionised in aqueous solution at room temperatures.

From consideration of damage to the wool substance, the melting-point of the wool fat (40 to $45°C$.), and the commercial efficiency of the scour, a temperature range has been established varying from 40 to $70°C$. according to the type of wool. Schofield (loc. cit.) gives temperatures of the following order:

TEMPERATURE OF SCOURING

First bowl (alkali and soap)	50–55°C.
Second bowl (alkali and soap)	43°C.
Third bowl (soap)	38°C.
Fourth bowl (water)	32°C.

The alkali degradation of the wool protein has been examined by Crowder and Harris (Am. Dyes. Rep., 1936, *25*, 264); the effect with increasing time is shown in the following table:

ACTION OF 0·05 N.NaOH AT 65°C.

Time	Loss in weight	Sulphur content	Cystine
0 minutes	0%	3·72%	13·40%
15 minutes	2·27%	2·91%	6·91%
30 minutes	3·52%	2·56%	4·85%
45 minutes	4·67%	2·35%	5·13%
2 hours	6·40%	2·24%	4·41%
4 hours	9·38%	2·13%	3·70%
8 hours	15·21%	2·03%	2·64%
40 hours	61·50%	2·28%	2·65%

With the 40-hour experiment, the samples became gelatinous and some of the wool was lost in washing.

More precise indications of alkalinity on the pH basis enable a closer control to be established. Speakman and Stott (Trans. Farad. Soc., 1934, *30*, 539) have demonstrated that the absorption of alkali by wool increases rapidly at pH values of 10 and over, and having reached equilibrium with these solutions, the physical properties of the wool are so modified as to show high swelling, minimum resistance to extension and reduced power of recovery.

At pH values from 6·5 to 10·4 at 20°C., there is no damage to the wool fibre, but yellow discoloration and tendering occur at greater pH values; at 50°C. there is rapid tendering and yellowing above pH 10·4, and at 70°C. the fibre is disintegrated above pH 8·4. With normal soap scouring, the greater part of the grease and dirt is removed in the first bowl which mainly determines the efficiency of the whole scouring process. It has been found by Phillips (J.T.I., 1936, *27*, 208P) that whereas the optimum pH of the first bowl should be of the order of 10, this is rarely the case in practice; the suint salts have a high buffering action (i.e. they resist changes in pH on the addition of acid or alkali) so that substantial amounts of alkali should be added to maintain the desired pH, particularly in view of the neutralisation of some of the alkali by the fatty acids in the wool fat. The liquors of the later bowls are much cleaner and therefore more sensitive to alkali; the wool itself is cleaner and hence not protected by the grease of the raw wool, so that care must be taken to ensure that the pH does not rise above 10 on account of the alkali carried over from the more highly buffered first bowl. The subsequent bowls generally function as rinsing bowls, the last of which serves as a final rinse. With the exception of the last bowl, sufficient soap must be added to prevent grease and dirt being deposited on the fibre again, during what is mainly a rinsing process; the pH varies from 8·5 to 9·8 during the course of a run and may even be greater where excess of alkali has been carried forward. The final bowl contains water only and is sometimes supplied with running water in order to maintain pH 8, but if a lower value is attempted then there is danger of the deposition of acid soaps on the fibre.

The great danger of damage by alkaline attack is when the wool is practically clean, so the pH of the third and subsequent bowls must not be allowed to attain too high a value.

Good rinsing is an essential conclusion to the scouring process and care should be taken to ensure an absence of alkali which, if dried into the wool, causes the well-known yellowing effect.

General

Owing to the great variation from wool to wool it is not possible to formulate a standard scouring process. For instance, with slipe wool it may be considered advisable to wash with acid before scouring in order to remove the lime; alternatively, sodium metaphosphate may be added to the scouring liquor to disperse the lime soaps or sodium hydroxide and carbonate may be used in the first bowl. In general, where only three or four bowls are used for scouring, the first often contains 0·5 to 0·7% of soap, and sodium carbonate is added at a later stage; this increases the emulsifying power but the emulsion is less stable. With the finer wools, the conditions should be as mild as possible consistent with efficient scouring; less carbonate should be used and the temperature should not exceed 50°C. Where five or six bowls are employed, the first occasionally contains sodium carbonate only, and those later bowls containing soap and carbonate operate on the counter-current system by "Blowing-back" as it is termed in wool scouring. In all cases, the temperature of the liquors is reduced in the later bowls concluding with water at 30°C.

Details of wool-scouring sets are given in the Textile Recorder Year Book, the Wool Year Book, the Textile Manufacturer Year Book, and also in the standard works of reference such as the *Finishing of Wool Goods* (Schofield).

"Cardonia" of the Textile Mercury (1941, *105*, 31, 81, 145, 210) has supplied the following data:

Lustre Wools

No.	Capacity	Soap	Alkali	Temp.
1	1,000 galls.	0·3%	0·2%	52°C.
2	900 galls	0·22%	0·1%	49°C.
3	750 galls.	0·06%	0·0%	43°C.

Merino Wools

No.	Capacity	Soap	Alkali	Temp.
1	1,000 galls.	0·50%	0·50%	54°C.
2	900 galls.	0·26%	0·13%	52°C.
3	750 galls.	0·10%	0·0%	46°C.
4	600 galls.	0·0%	0·0%	43°C.

As the scouring of raw wool is a continuous process, some soap and alkali will be carried forward from one bowl to another; the counter-flow returns some liquor, so that having "set" the first bowl and established emulsification, it is necessary to maintain emulsifying power by the addition of soap and alkali as required as shown to some extent by the state of the liquor itself.

A variation of the usual emulsification process originated in the treatment of the dirtier types of wool; it must be realised that a great deal of dirt is embedded in the grease and drops out during the emulsification process. It is also possible to remove the dirt by washing with warm water and this method was developed in continental Europe in order to cope with the shorter and dirtier types of wool; the scouring methods generally entail the use of more and larger bowls than in England.

The wool suint is also soluble in water, so that a preliminary steep not only helped to clean the wool from adhering dirt, but also provided a solution of suint which could be used for scouring later. The steeping process may be carried out in various ways, one of which is to place the wool in a series of vessels at various heights and allow the warm water to flow down in a slow cascade; another method was to arrange a series of tanks at different levels and lift cages of wool from one tank to another.

Perhaps the best-known method is that due to Malard; the wool is carried on a travelling lattice over a bowl divided into compartments; water at 27°C. is sprayed on to the wool and passes into the tanks from which it is pumped again to operate on a countercurrent principle whereby the dirtiest wool is treated with the strongest solution and the last compartment receives clean water.

SCOURING RAW WOOL 311

The suint solution is generally concentrated and evaporated, for subsequent use.

The further treatment of these shorter wools, which are not very strong, is to remove the grease by soap emulsification at a lower temperature and over a longer time than that described in the English system where some grease removal is by melting as well as by true emulsification; the poorer wools are too sensitive for such a treatment however.

SUINT SCOURING

It has been known from time immemorial that the potassium salts present in the impurities of raw wool and originating from the perspiration, possess great detergent value. The old-fashioned method of treating wool was to steep it in water and use the liquor as a detergent, often re-activating it with lant or weeting (stale urine) and pig-dung, when it became impoverished. Oriental peoples still use the aqueous extract from raw wool as a detergent, as it contains potassium salts of fatty acids.

Modern suint scouring by machinery originated in France through the work of Duhamel, and the method is often called the Duhamel system. The general course of treatment is to remove the suint and yolk separately, and use the suint for the emulsification of the yolk. One of the chief difficulties originally encountered was the removal of the dirt from the raw wool and from the suint. Precautions should be taken to ensure that the wool is well dusted and opened beforehand.

The process generally operates in five stages: (*1*) de-suinting, (*2*) grease removal, (*3*) rinsing, (*4*) soaping, and (*5*) rinsing.

Cold suint is used in the first bowl for treating the dusted wool, and the liquor is withdrawn to be purified by sedimentation or centrifuging, for if the dirty suint is used in the second bowl in the warm state, the dirt deposits itself on the fibres which still contain their natural fat. Care must be taken, therefore, only to use the suint solution for scouring when it has been clarified.

Grease removal is brought about in the second bowl by the use of the clear suint solution at a temperature of approximately 60°C.

Suint scouring is sometimes referred to as the natural emulsion process, for it makes use of the wool suint to emulsify the fats. At least four bowls are necessary, the clarified suint being used in the first two or three bowls; subsequent bowls contain soap with the exception of the last bowl which holds water only. The suint concentration is maintained by a continuous system of circulation.

Suint liquors vary from pH 5·5 to 8·4 so that higher temperatures may be used in the bowls containing suint only; in these circumstances 70°C. may be employed without damage to the wool. The bulk of the grease and dirt is removed in the first two bowls, and soap is only required at a later stage to remove the small amount of remaining impurity. This means that the soap bowl is not rapidly contaminated and may be used over long periods.

The bowls for the Duhamel or suint washing process are rather smaller than for the average English plant and are specially designed to separate the grease-contaminated liquor, recovering the lanolin, and returning the scouring liquor to a reservoir for use again.

The suint scour is very economical and gives a brighter and whiter wool which is more open and lofty than that from the soap scour.

Although the optimum dirt-suspending power of suint is at pH 7, the maximum grease-suspending power occurs at pH 10, so that the suint scour could be improved by the addition of alkali; this would mean a reduction in working temperature to obviate alkali attack on the wool. Recent developments show a tendency to combine the suint and soap scours by employing suint and alkali in the first bowl with a little soap, and suint and soap in the second bowl with a little sodium carbonate.

It has been established that, for the same pH, the emulsifying power of suint is similar to that of 0·6% soap solution.

Where suint scouring is utilised on a large scale, it is preferable to select wools having the minimum amount of dirt, but with a high suint content. Although it is possible to use temperatures of the order of 70°C. in the initial stages without damage to the wool, this would not be satisfactory on a commercial scale as the wool fat would melt; the commercial success of suint scouring depends to a large extent on the order of removing the impurities. In many cases, it is customary to dust the wool before it is allowed to enter the first bowl, and in this manner the desuinting liquor is kept as clean as possible. The suint is removed at a temperature of 16°C. and then the wool fat at about 55°C.; rinsing, soaping and rinsing are carried out at 40 to 32°C.

One of the great problems of suint scouring is the removal of the dirt which is apt to foul the desuinting liquor and deposit itself on the hairs containing the wool fat and stain them; the pumping of suint liquors tends to disperse the dirt throughout the liquor and ample precautions must be taken to allow the dirt to settle. The separation of the dirt and the grease from the second bowl is a still more difficult matter, and it is customary to withdraw this liquor and subject it to centrifugal action so that the dirt settles to the bottom and the wool fat rises to the top; the suint solution may then be drawn off and allowed to settle in tanks before being used again for scouring. It is worthy of mention that the wool fat obtained in this manner is purer, and therefore more valuable, than that obtained from the emulsion method of scouring.

PLATE XIX

Fig. 83. The swing rake machine.

Courtesy of Messrs. Petrie and McNaught.

To face page 312.

PLATE XX

FIG. 84. The Leviathan harrow machine.

Courtesy of Taylor, Wordsworth and Co., of Leeds.

FIG. 85. The Klauder Weldon machine.

Courtesy of Sellers and Co. Huddersfield).

PLATE XXI

Fig. 86. Single-bowl yarn washing machine with squeezing press.

Courtesy of Messrs. Petrie and McNaught.

PLATE XXII

Fig. 87. Yarn washing with three-bowl set.

Courtesy of Messrs. Petrie and McNaught.

The third bowl of the set is used for rinsing purposes in order to free the wool from any traces of potash from the suint and from residues of wool fat; the fourth bowl gives the material a good soaping and the last bowl, a rinse in water only.

Where the bulk of the grease is emulsified by suint, the pH value is determined by the suint itself, and Phillips (J.T.I., 1936, 27, 208P) has found that the pH value of the water-extract of Australian wools varies from 5·5 to 8·5. If the natural pH values of these liquors is raised by the addition of alkali, then greater quantities of suint and grease would be removed, by converting the free fatty acids in the raw wool into soaps. Suint has been estimated to contain 20% of its dry weight of fatty acid, and grease contributes between 4 and 7·5%.

Phillips (*supra*) has determined the amount of alkali required to bring raw merino wool into equilibrium with solutions of pH 10, and found that for 1,000 lbs. of raw wool 5·5 to 9 lbs. of soda is necessary. In the case of the suint only 2·2 to 3·4 lbs. of soda is necessary according to the pH value of the fleece.

Suint liquors contain many constituents not found in ordinary soap solutions, and act as efficient emulsifiers of wool grease at pH values below 10. As previously stated, at the same pH, the emulsifying power of suint is similar to that of 0·6% soap solutions.

It is possible to use suint alone by raising the temperature and so extracting more suint and hence emulsifying more grease; for maximum grease removal, however, it is better to increase the pH and so convert the fatty acids present in raw wool into soaps.

The harrow system or the paddle method may be used for moving the wool through the bowls to the squeeze rollers, which, as usual, are covered with a long, strong, wool, such as Lincoln hog wool. "Cardonia" (loc. cit.) has given the following data for suint scouring:

SUINT SCOURING

No.	Capacity	Process	Soap	Temp.
1	1,000 galls.	Desuint	—	16°C.
2	1,000 galls.	Fat removal	—	60°C.
3	800 galls.	Rinse	0·1%	38°C.
4	1,000 galls.	Soap	0·3%	41°C.
5	800 galls.	Rinse	0·1%	32°C.

It will be noted that the third bowl contains a little soap although the main object is to rinse the wool; the wool coming from the second bowl carries liquor containing some grease and dirt, and precautions must be taken to prevent this being deposited on the fibres again which would be the case if water alone was used for rinsing.

314   TEXTILE BLEACHING

Scoured wool should not contain more than 0·5 to 0·75% of fat or grease and any excess is liable to cause difficulties in the subsequent mechanical processes of spinning, such as carding. On the other hand, "over-scouring" impairs the handle of the wool. Similarly, an excess of soap in the final product creates difficulties in spinning and also in dyeing. The drawbacks of metallic soaps are well known, particularly calcium and magnesium soaps which come from the use of hard water. Calgon (sodium hexametaphosphate) may be used to remove the undesirable metallic soaps, as already explained. The use of sulphated fatty alcohols in the rinse bowls affords advantages on account of their stability to hard water, their lower pH values and high dispersing powers.

COMBINATION SCOUR

It has been realised that a suitable use of the first bowl in a set of scouring bowls enables some of the better features of the emulsion method and the suint method to be utilised in one process; there is, therefore, an increasing tendency to use suint and soda in the first bowl, together with a little soap, the subsequent bowls containing soap with the minimum of soda.

In the ordinary emulsion scour, the only suint in the first bowl is that taken from the wool which passes through it, but in the combination scour various alternatives present themselves. It is thus possible to give the wool a preliminary steep in water and feed the suint solution into the first bowl, where it will meet the soap and alkali solution. Alternatively, it is possible to add alkali only to the suint solution, or even to rely on the liquor flowing back from the second bowl. Many wool scourers consider it best to add the cleansed suint to the first bowl, where the addition of alkali brings the pH to the optimum figure, and the emulsifying power is improved by the presence of soap.

It will be seen that the behaviour of the first bowl is very important for it is here that the greater part of the dirt and grease is removed. In the later bowls, the wool is washed clean with soap and soda, and finally rinsed with water. High concentrations of alkali in the first bowl would not damage the wool, on account of the buffering action of the suint, but it must be remembered that wool-scouring is a continuous process, and it would not be safe for large amounts of alkali to be carried forward to the second bowl which is less highly buffered.

SOLVENT EXTRACTION

In spite of the fact that the removal of the oils and grease from wool is principally effected by soap and alkali, interest has been aroused in the use of solvents, particularly in Belgium and in the U.S.A. Wool fat may be removed from the wool by benzine, carbon tetrachloride, solvent naphtha or white spirit; both the wool fat and the solvent may be recovered.

The general procedure is to pack the wool into kiers which are then evacuated of air and the solvent introduced; the extraction is generally effected in three stages, first with solvent which has been utilised on two previous occasions, secondly, with solvent which has been used once, and lastly, with clean solvent. The two last quantities of solvent are stored for re-use, but the first solvent is distilled. It is important to leave about 0·5 to 0·75% of fat in the wool, as otherwise it becomes harsh and brittle, giving rise to difficulties in spinning.

The solvent remaining in the wool is removed by blowing warm air through the kier, and then passing it through a condenser where the solvent is recovered.

The extracted wool, of course, still contains suint and this may be removed by warm water in a special bowl or de-suinting machine where a series of troughs is arranged so that water flows through them by gravity, the wool being floated along by the water.

There are certain drawbacks to the use of solvents, such as fire risks and special plant, but it is claimed that this method gives material of superior quality, strength and softness; all danger of alkali damage and felting is eliminated, and the wool is delivered in an open state more suitable for carding.

FREEZING

The partial cleansing of wool in a novel manner has been described in B.P. 395,471 of the Frosted Wool Process Company; the moistened material is subjected to low temperatures and a large part of the impurities may be removed by mechanical means. Some information about the process has been given by Wig (Am. Dyes. Rep., 1935, *24*, 270) and Barker (Text. World, 1935, *85*, 1085); a comprehensive account has also been given by Townend (J.T.I., 1936, *27*, 219P).

When the raw greasy wool is submitted to a sufficiently low temperature, the fats freeze and become hard and brittle but the wool fibre itself is not affected. For regular production the temperature should be maintained at $-30°C$. or less, and actually -35 to $-45°C$. is commonly reached in the freezing chambers. After sufficient time has elapsed for the wool to come to equilibrium with the surrounding atmosphere, it may be subjected to vigorous agitation when the frozen grease breaks up into a fine powder and falls away together with substantial quantities of vegetable and mineral matter.

The necessary plant comprises a refrigerating unit and a refrigerating chamber; the raw wool is fed into the top of the cold chamber through a pair of rollers which act as a seal and exclude the warm air. Inside the chamber, the wool falls on to a conveyor, about 20 feet in length, where it remains for 3 to 7 minutes and is then passed on to the feed sheet of a specially designed intermittent batch duster or cleaning machine, which comprises a cylinder about 36 inches in diameter fitted with steel pegs 5 inches in length on its surface.

Experience has shown that it is possible to remove 10 to 18% of the weight of the greasy wool or 30 to 57% of the actual grease; in these circumstances, the capacity of the scouring bowls may be increased by 30% or so, and the scouring liquors may be used over longer periods before renewal. Alternatively, fewer bowls are found necessary. Vegetable matter is also removed to a large extent and perhaps the most striking example cited is that in which 8% of vegetable matter present before processing was reduced to 0·68%. Although the carbonisation process cannot be eliminated by using the freezing technique for the finest felt wools, yet it is often not essential with other wools.

It would appear that any mechanical process must result in some breakage of the fibres, even to a small extent, but on the other hand, very large quantities of wool have been treated by this novel method. After the milder scouring process, however, the wool has a softer handle and a more lofty appearance together with a better colour than material treated by the usual scouring processes.

CHAPTER XXV
SUBSIDIARY PURIFICATION PROCESSES

BEFORE passing to the consideration of the scouring and bleaching of fabrics, it is proposed to discuss briefly certain subsidiary processes of purification which sometimes take place before the weaving or knitting of fabrics.

Most wools are apt to contain greater or less amounts of cellulosic impurities which may be removed mechanically or chemically; the chemical process is that of carbonising and relies on the relative immunity of wool to acids in contrast to the susceptibility of vegetable matter. The cellulosic impurities have to be removed sooner or later, and it is customary to remove them before spinning worsted yarns; actually more wool is carbonised as piece goods than in the raw state, but it is convenient to discuss the process at this stage.

The preparation for worsted spinning involves combing during which the fibres are opened to a considerable extent and appear dirty; a mild scouring treatment known as "backwashing" may therefore be applied to the material.

Finally, considerable quantities of yarn are required in a moderately high degree of purity for hand-knitting and also for dyeing; the amount of yarn-dyed wool for weaving is proportionally much higher than in the cotton trade. Hence, a special yarn-scouring process is often necessary.

CARBONISING

One of the most unwelcome cellulosic impurities in raw wool exists in the form of burrs or the prickly seed-cases of various plants which have become entangled with the wool; if these are not removed after the scouring process, they are apt to be broken into small particles during mechanical operations such as carding, and often cause damage to the card clothing and combs. Yarns containing burrs are difficult to spin evenly and in the final fabric the presence of particles of cellulosic matter is shown by a difference in affinity for dyes.

It is possible to remove a part of the burrs by mechanical means depending on the type of wool and its "burriness." One type of machine operates by burr rollers in conjunction with the opening rollers or cards; the burr rollers are about 4 inches in diameter, and carry steel blades set close to the card rollers where the wool is held while the burrs are flicked off. Another machine operates differently by crushing the burrs between smooth and fluted rollers, whose pressure is adjusted according to the type of burr. There are also special machines for cleaning wools which contain an excessive amount of burrs.

The mechanical removal of the grosser cellulosic impurities enables the wool to be spun and woven, so that the carbonising process for the complete removal of vegetable matter may take place more conveniently with the wool in fabric form. Although there is a tendency to prefer the carbonising of woollens in fabric form, the worsted industry relies on carbonisation of the wool before spinning.

As pointed out by Schofield (*The Finishing of Wool Goods*), the carbonising of wool in the loose state is essentially a large-scale process, requiring scribbling and carding plant, machines for the mechanical removal of burrs, the carbonising machines proper, washing machines and drying machines.

As previously mentioned, the essential function of the carbonising process is to turn the vegetable matter or cellulose into the friable hydrocellulose; the chemistry of hydrocellulose formation has been discussed by Marsh and Wood in *An Introduction to Cellulose Chemistry* (Chapman and Hall, London, 3rd Edition, 1945). Some data are also given on pages 43 and 473.

The methods of carbonising wool fall into two main classes:

(1) dry carbonising, with hydrochloric acid gas
(2) wet carbonising with either

   (a) sulphuric acid solution, or
   (b) aluminium chloride or magnesium chloride solutions.

WOOL AND ACIDS

The treatment of wool with acids (apart from dyeing) is mainly for the purpose of removing any vegetable matter with which the fleece has become contaminated during the time it was on the sheep's back; particles of grass and straw, thistles and burrs, are very frequently encountered in the wool itself.

The treatment with acid may be given to the loose wool immediately after scouring or it may be applied to woollen fabrics. The principle of carbonising relies on the degradation of the vegetable matter by acid to give a powdery hydrocellulose which is readily removed mechanically.

The affinity of wool for mineral acids is very great and it is not possible to remove sulphuric acid, for instance, even by repeated washing with boiling water. The tenacity with which the acid is held is such that it is very difficult to estimate the quantity retained

SUBSIDIARY CLEANSING PROCESSES

by the usual methods of titration. With many acids it has been stated that the amount entering into chemical combination may be of the order of 10%.

One of the most interesting properties of wool is its resistance to acid attack; and sulphuric acid solutions even up to 80% concentration, exert little effect provided that the temperature does not exceed 50°C., above this temperature and concentration the attack is very rapid. Numerous trials have been made from time to time to determine the concentration and temperature limits for sulphuric acid during the carbonising process, and some of these have been discussed by Brown (Textile Colorist, 1935, *57*, 49). It is, however, well known that the temperature and time of heating have a greater effect on the dyeing properties of the treated wool than has the concentration of the acid. Although wool will resist the action of dilute mineral acids at temperatures below 90°C., when even 5% sulphuric acid is used at temperatures above 90°C. the wool becomes brown and discoloured.

It is interesting to note that the absorption of acid by wool requires some time and may necessitate a period of about thirty minutes. In the actual carbonising process it is common practice to use concentrations of acid varying from 4·5 to 7·5°Tw., which, on drying, become concentrated and destroy the vegetable contaminant at 90°C. The effect of the acid on the wool fibre is to make it rather harsh, so that numerous attempts have been made to vary the carbonising process in different ways, such as employing a lower concentration of acid, reducing the surface tension of the impregnating solution, and protecting the wool by other means.

For instance the common wetting agents which are stable to acid, such as the naphthalene-sulphonic acid derivatives, have been used, and it has also been proposed that an emulsion of wool fat could be employed as a protective agent.

Where sulphuric acid is employed, it is customary to use solutions of 5 to 6°Tw. and to dry the material at 55°C. or thereabouts. The question of temperature is rather important, for wool is apt to be damaged at temperatures above 60°C. (140°F.). The drying stage of the treatment requires care, but if the goods are thoroughly dried before baking there is little danger of damage; some moisture is necessary for reaction but the amount should be kept to a minimum, and the dried goods should not be allowed to absorb moisture from the atmosphere before heating to the final high temperature.

Harris (Bur. Stand. J. Res., 1934, *12*, 475) has shown that the amount of sulphuric acid absorbed by wool increases with acid concentration up to about 5%; for each concentration of sulphuric acid there appears to be a critical moisture content of the wool above

320   TEXTILE BLEACHING

which the breaking load of the wool decreases and the ammonia-nitrogen decreases as a result of carbonising.

MOISTURE AND CARBONISING

Solution of H_2SO_4	Critical moisture content
2·5%	50%
5%	40%
7·5%	30%

The modern methods of carbonising appear to date from about 1854 and arose in France and England about the same time. The French method seems to have relied on the action of hydrochloric acid applied either as a warm aqueous solution or alternatively in vapour form, the goods being subsequently dried and then washed thoroughly. On the other hand, the English method made use of sulphuric or nitric acid in the liquid state, but many of the early attempts appear to have resulted in considerable damage to the fibre. A common method of protection was to mordant the wool with a mixture of soluble salts of tin and aluminium followed by treatment in a soap solution before impregnating in acid and heating to 98°C.

An alternative method of carbonising is the use of aluminium chloride and this is based on the hydrolysis of the salt at 115 to 120°C., with the liberation of hydrochloric acid; the hydrolysis soon reaches a state of equilibrium which can be destroyed by removal of the hydrochloric acid by increasing the temperature; hydrolysis is forced back in presence of acid and in many works it is customary to add a small quantity of hydrochloric acid to the aluminium chloride solution in order to retard premature decomposition and permit of the use of solutions of 7 to 15°Tw., compared with 15 to 30°Tw., for aluminium chloride alone. The time of treatment is at least 5 minutes and if prolonged beyond 45 minutes the scales of the wool are apt to be attacked.

Dry Carbonising

The so-called "dry" carbonising process consists in treating the wool with hydrochloric acid in vapour form in an enclosed chamber. It is mainly utilised in the carbonising of rags, which are placed in large perforated cylinders capable of dealing with 6 to 12 cwts. at a time. The cylinder is placed inside the carbonising chamber and driven by external gearing. A retort is situated on the opposite side of the chamber and connected to the end of the shaft which carries the cylinder containing the wool.

A furnace serves the double purpose of providing hot air for the drying of the wool, and for heating the retort; a scrubber is also provided to prevent the acid vapours from passing into the outer atmosphere. Hydrochloric acid is allowed to drip into a bowl from which it falls into the hot retort, where it is vaporised and passes through the hollow shaft of the cylinder into the wool; about 1 cwt. of hydrochloric acid (28 to 32°Tw.) is necessary to carbonise 10 cwt. of wool in 3 to 4 hours. During the process, the temperature is maintained at 90 to 94°C., and the perforated cylinder is slowly rotated (about 4 r.p.m.), but before admitting acid vapours it is preferable to raise the temperature to about 60°C. in order to dry the wool. Towards the end of the carbonising process, samples may be withdrawn through the hollow shaft by a long rod with a hook at the end, and examined to ascertain if the cellulosic matter has been "killed." A damper is opened which allows the acid vapours to pass through the scrubber; when the process is complete, the wool is allowed to cool before passing to the "shaker" where the friable hydrocellulose is beaten out, leaving the wool.

Hydrochloric acid is never used for the carbonising of woollen or worsted piece goods.

Wet Carbonising

The wet process with sulphuric acid or aluminium chloride solutions is rather slower and dearer than the dry process, but there is a more thorough cleansing of the wool on account of the rinsing and better shaking.

Solutions of about 9°Tw. sulphuric acid are generally employed, although with thick and compact rags, this may be increased to 20°Tw. In one process, the wool is loaded into wooden cages holding about 3 to 5 cwts., which are then lowered into the liquor where they remain until the wool is thoroughly soaked for perhaps as long as 2 to 3 hours. Excess liquor is removed by hydro-extraction, after which the wool passes on to the drying chambers where the wool is spread on racks and heated at a temperature of 80 to 95°C. Samples are withdrawn from time to time to ensure that the wool is not over-treated. The wool is then taken to the shaker which removes the friable hydrocellulose, after which the material is thoroughly washed with copious supplies of water. A second drying, and often a second shaking follows to ensure the complete removal of the cellulosic material.

The above process does not represent modern practice for the treatment of raw wool although it is still used to some extent in dealing with rags. Modern carbonising is carried out by a continuous process in a bowl corresponding to that used for scouring; the passage of the wool through the acid solution is somewhat slower, however, and may take about 20 minutes.

The concentration of sulphuric acid used for carbonising is determined to some extent by the state of the wool itself; freshly scoured and dried wool may be treated with sulphuric acid of 4 to 6°Tw., but if the wool is still moist from the scouring bowls (the general practice) then it is necessary to increase the concentration accordingly to about 8°Tw. or even higher. The immersion may take place in bowls which are similar to the scouring bowls, but lined with lead; the impregnated wool is then squeezed or hydro-extracted so as to contain about 35 to 40% of liquor, and the concentration of acid at this stage, estimated on the wool, should be approximately 5%.

The wool next enters a special drying and baking oven in which it is preferable to conduct operations in two stages, first a drying process at 50 to 60°C. and secondly a heating process at 95 to 100°C. This two-stage treatment is rather important, for it must be remembered that the acid is becoming concentrated and hot at the same time, and without suitable precautions there is a danger of the wool being hydrolysed by hot concentrated solutions of sulphuric acid. The aim should be to subject the wool to the anhydrous acid at the higher temperature, when there is less risk of damage. The concentrated acid attacks the vegetable matter and turns it into the friable hydrocellulose.

The time of passage through the drying and heating machine is generally about 30 minutes.

Immediately after the heating, the wool is passed through corrugated rollers which crush the burrs, and then into a willow where the hydrocellulose is shaken from the wool; it is sometimes desirable to have two sets of crushing rollers and willows in series.

The crushing and willeying of the carbonised vegetable matter should take place as rapidly as possible in order to avoid absorption of moisture or leaving the wool in contact with the strong acid any longer than may be necessary; similarly, after the mechanical removal of the carbonised vegetable matter, the acid in the wool should be removed by washing with water and neutralising with sodium carbonate solution as soon as possible.

It has been estimated that for 100 lbs. of wool in the air-dry state and containing 20% of vegetable matter, it is necessary to use 5 lbs. of the pure sulphuric acid, applied in a concentration of about 6 to 7°Tw.; 1 lb. of the sulphuric acid is used in the actual carbonisation and about 2·5 lbs. are removed in the wash-water, but it is possible to recover this and regenerate it for use in carbonising further quantities of wool. It appears, therefore, that about 1·5 lbs. of acid is retained by the wool, and this acid should be neutralised; the neutralising bath usually consists of sodium carbonate solution of 1·5 to 3°Tw.

The neutralising part of the plant is similar to the scouring and

SUBSIDIARY CLEANSING PROCESSES

acidifying bowls, but consists of 3 or 4 bowls, in the first of which clean water is supplied, the alkali being provided in the second, and water in the third, sometimes with a little softener or emollient. A little "blue" may also be added if required.

It has been suggested that strong salt solutions may be used instead of water for the removal of the acid, in order to repress swelling.

The carbonising plant is thus arranged with the various machines in series, with brattice feeds from one machine to another, and concluding with the drying machine.

Carbonising with sulphuric acid solutions appears to be the commonest method; it is very convenient and may be applied to both the raw wool or to the fabric. The temperature of carbonisation is relatively low and varies between 80 and 100°C.

It has been suggested to carbonise wool in the greasy state as the grease may exert some protective action; it must be remembered, however, that raw wool contains potassium salts which would neutralise part of the acid, so that if greasy wool is carbonised, then it should first be steeped to remove the suint. Subsequent scouring of carbonised wool, unfortunately, is sometimes made more difficult.

In view of the ease with which the spinning properties of wool may be adversely affected, it is often better to carbonise the material after it has been woven. It is possible to effect economies by using cheap wool containing vegetable matter which is not removed until after the weaving stage. Wool flannels and blankets are generally carbonised in the piece, but in the worsted trade it is preferred to carbonise the wool before manufacture. In continental Europe, carbonising in the piece is generally practised.

BACKWASHING

Although wool is given an initial cleansing or scouring in the loose state, it is often found that after the mechanical operation of carding, the slivers present a dingy or dirty appearance; hence it is customary to give a subsidiary washing treatment to the slivers and this is termed backwashing.

It appears that the initial washing treatment of the raw wool does not remove all the impurities on account of the stapled condition of the wool during processing; after carding, however, the fibres are more evenly distributed and the dirt from the centre of the original fibrous material is more readily accessible. In addition, the wool is apt to acquire certain impurities during the mechanical operations up to the carding stage; this so-called top-dirt would tend to accumulate in the machines used in the subsequent preparation of the worsted tops and interfere with the operation of the machines and the production of tops of good quality. It has been pointed out by Schofield (loc. cit.) that a further reason for backwashing is to obtain tops of good colour, for colour is one of the factors which is considered in assessing wool quality, not only in passing from crossbred to merino, but also in deciding if the wool has been subjected to any excessive chemical treatment in the previous processes of purification.

The fibres in the carded sliver present a better form for cleansing than the inadequately opened raw wool, but their presence in sliver form requires a different machine from that used in scouring raw wool.

The backwashing plant is essentially a small two-bowl scouring machine modified for the treatment of slivers; it is generally equipped

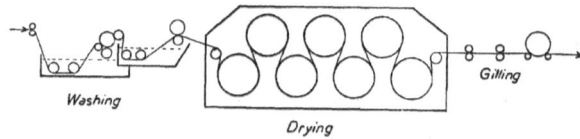

FIG. 88.—Diagram of Backwashing.

with a drying system and sometimes passes direct to a gill-box for oiling. The bowls are situated so that the second is higher than the first, thus allowing the liquor from the second bowl to flow into the first; the second bowl is supplied with a constant flow of warm water. As the amount of cleansing which is required is not very great, the first bowl usually contains soap and warm water, although sometimes a little ammonia is added.

About 16 to 30 slivers are guided into the first bowl by revolving brass rollers, and submerged in the liquor itself are further rollers between which the sliver is passed. The liquor is squeezed out of the slivers at the end of the first bowl before passing into the second which operates in a similar manner to the first. The amount of soap in the first bowl is replenished as the lather diminishes. The second bowl often contains a little blue colouring matter in addition to the warm water; this is used, particularly for work done on commission, to improve the colour of the tops by the complementary colour effect. It seems to be based on the assumption that colour is an index of wool quality; the use of Methyl Violet for this purpose was forbidden in England during the war of 1914–1918, but has been revived.

The drying stage of the backwashing process originally consisted of passing the slivers over steam-heated cylinders but overheating is apt to occur on the outside and underheating on the inside; the wool was sometimes slightly scorched and its spinning quality diminished. More modern drying arrangements utilise perforated

SUBSIDIARY CLEANSING PROCESSES

cylinders around which the slivers pursue a serpentine path; hot air is blown through the perforations drying the wool and maintaining it in an open and lofty state. The commonest type consists of about 6 cylinders, but another arrangement is the so-called multi-cylinder machine in which about 40 small cylinders of 4-inch diameter, are fitted around a large cylinder; the slivers pass around the small cylinders and are dried by heat from the large cylinder.

Hot-air drying without cylinders has been used in two ways, one of which is to feed the slivers on to a wire lattice or conveyor which passes through a heated chamber; the multilayer type causes the slivers to fall from one conveyor to the next (see page 427). Lattices may also be used, and it is possible to blow warm air through the strips as the sliver passes forward.

In most mills, it is necessary to oil the slivers after the back-washing process which has removed any remaining grease and left the fibres somewhat brittle; an addition of about 3% of a suitable oil emulsion is sufficient to protect the wool during the subsequent combing and spinning operations.

YARN SCOURING

Practically all yarns of wool fibres contain some oil or other lubricant which has been added to assist in the spinning operations and will therefore require scouring before dyeing; a high proportion of wool is dyed in the yarn form for the production of coloured-woven fabrics. The amount of oil in the yarn may vary from 2 to 3% in worsted yarns to 20% or more in carpet yarns.

Some types of yarn tend to curl and distort during the scouring process and others are apt to felt; it is therefore necessary to obviate these defects by imparting some permanent set before the scouring operation. The commonest setting process for yarn is to stretch on frames and then immerse in boiling water for 30 to 60 minutes, turning the hanks half-way through the treatment, and cooling them while they are still in the stretched state; with certain machines, it is possible to impart permanent set along with the scouring operation. An alternative setting process utilises steam instead of boiling water.

The setting process is fundamentally similar to the crabbing or blowing of piece goods before scouring, and the discussion of these effects is given on page 327.

There are various methods which can be adopted for the scouring of yarns; in one type of machine, the yarn may be placed on rods or bars carried on a frame which is lowered into the scouring liquor in which the hanks are rotated. Machines of this type are commonly used for yarn-dyeing and they may be used also for imparting permanent set as mentioned above. (See Fig. 85.)

Another method is to make a rope or chain of the hanks and scour them on the dolly machine as used for piece goods and described on page 331.

One of the oldest methods is merely to suspend the hanks on sticks as for dyeing and turn them by hand in the scouring liquor from time to time.

Large-scale scouring of wool in hank form may be carried out by a series of bowls equipped with travelling lattices or aprons; as with wool-scouring bowls, each bowl is fitted with squeeze-rollers. A modern yarn-scouring machine of good design has been made by Petrie and McNaught, and may consist of 3 or 4 bowls, 3 bowls for easily washed material such as knitting yarn, and 4 bowls for more difficult material including carpet yarn. Each bowl is fitted with a powerful squeezing press, and equipped with two endless brattices between which the hanks are carried. The hanks are not tied or fastened, but merely laid end to end on the brattice of the first bowl and carried underneath a shower of washing-liquor. The shower is produced by liquor pumped from the bowl, so that the yarn is saturated before it enters the liquor in the bowl; the yarn travels along the brattice for the length of the bowl and then, held between the two endless brattices, it is carried twice the length of the bowl through the liquor before being delivered to the squeezing rollers.

In the second and subsequent bowls, the shower-system may be dispensed with, as the yarn is already wet. After the first bowl, it is advantageous to fit reciprocating guides to keep the yarn in the middle of the brattices, as the squeezing rollers tend to press the hanks out sideways.

The scouring operation is generally effected with 2 to 3% of soap with or without the addition of soda ash according to the type of wool being scoured. When the liquor in the first bowl becomes dirty, it is discharged into the drain and the inside of the bowl is swilled with water, and the contents of the second bowl are transferred or "blown-back" by an injector into the first bowl; similarly, with the other bowls, the last of which is filled with clear water again. The transferred liquors are then adjusted to the required concentrations, clear water being used in the last bowl. The concentration of soap is lower in the second bowl than in the first.

Where soda is required, the solution may contain as much as 3 to 4% according to the type of yarn; with woollen yarns which may contain 15 to 20% of oleine oil it is possible to scour without the addition of soap, as the soda ash will form a soap with the oleic acid present in the oil. Many worsted yarns are oiled with olive oil, in which case soap alone may be used.

In no circumstances should the temperature of the scouring liquors be allowed to exceed 60°C.

CHAPTER XXVI
FABRIC SCOURING

The scouring of woollens and worsteds, in itself, is a comparatively simple process, but the complete purification of the material necessitates a somewhat wider discussion; owing to the large and varied number of fabrics it is not possible to dogmatise on their treatment, which will be considered under the following headings:

(a) Crabbing
(b) Scouring
(c) Carbonising
(d) Milling.

As will appear later, not all fabrics pass through the above sequence of operations, and with many fabrics some of these processes are operated in a different order.

The crabbing process effects very little purification of the cloth, but is an important preliminary step to the treatment of fabrics in hot aqueous solutions; it is intended to set the fabric in a flat state and obviate cockling and similar distortions which might otherwise occur, particularly with worsteds.

Carbonising may rightly be regarded as a purification process for it is responsible for the removal of the unwanted vegetable matter by destruction with acid.

Milling or fulling, on the other hand, is really a finishing process, but in certain forms of milling a considerable amount of purification takes place at the same time, and it may therefore be considered as a special type of scouring process in these circumstances.

Before proceeding to the wet process, the fabrics are numbered and labelled in one of the obvious ways. They are then subjected to the preliminary operations of perching and mending; perching involves examining the fabric in a good light and marking any defects, many of which may be mended by knotting, replacing broken threads, and "bad places." Stains may also receive attention at this stage.

CRABBING

The processes of crabbing and blowing are applied to wool in the greasy or unscoured state in order to fix the woven texture against the distortions which would otherwise occur during the scouring and milling treatments. Crabbing utilises boiling water and blowing makes use of steam; a full crabbing process originally meant treatment by both hot water and by steam applied to the fabric in roll-form.

Many worsted fabrics, particularly of crossbred wool, become cockled and distorted on wetting because of the release of the latent strains and unless some degree of permanent set is given to these goods they are very difficult to manipulate and to finish in an attractive manner. Many of the more complicated woven structures would become distorted and irregular without being set before scouring, dyeing and finishing.

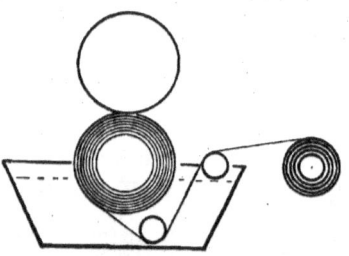

FIG. 89.—Diagram of crabbing machine.

In the simple crabbing process, the fabric is wound on to a roller by passing through a trough of hot water; the roll of cloth is built up on a second roller which may be weighted to a greater or less degree. The application of pressure in conjunction with heat and moisture is responsible for setting the fabric in a smooth and even manner. Immersion in the hot water may occupy from 5 to 15 minutes, after which the fabric may be allowed to cool naturally or it may be cooled by running through cold water.

The simplest form of crabbing may be appreciated by reference to Fig. 89. The fabric, previously wound under tension on a roller, is passed over and under guide-rollers through hot water, from which it goes between two larger rollers and is tightly wound on the lower roller which continues to revolve in the hot water. As the roll of cloth builds up on the lower roller, pressure is applied from the upper roller which operates in slots to allow for the increasing diameter of the lower roller. A system of levers, wheels, weights and chains supplies the pressure to the upper roller.

When the crabbing process is finished, the cloth may be passed to a second trough where the process is repeated; finally the cloth is drawn through a bath of cold water and allowed to cool.

It is often advisable to give a double crabbing treatment whereby the fabric becomes more evenly treated throughout its length and

PLATE XXIII

FIG. 90. Single-bowl crabbing machine.

FIG. 91. Double combined crabbing and steaming machine.

Courtesy of Sellers and Co. (Huddersfield).

To face page 328.

PLATE XXIV

FIG. 92. The Williams-Peace Scouring Machine for pieces in rope-form.

Courtesy of J. Mitchell and Sons.

PLATE XXV

FIG. 93. Open-width washing machine.

Courtesy of William Whiteley and Sons.

PLATE XXVI

Fig. 94. The tub-scouring machine or tom-tom.

Courtesy of Samuel Pegg and Son.

PLATE XXVII

FIG. 95. Fulling stocks (side action).

Courtesy of Sellers and Co. (Huddersfield).

PLATE XXVIII

Fig. 96. Williams-Peace combined milling and scouring machine.

Courtesy of J. Mitchell and Sons.

To face page 329.

FABRIC SCOURING

the possibility of ending is obviated; the outer layer of the roll during the first crabbing becomes the inner layer during the second treatment. Machines are available for double crabbing and also for treble crabbing, but the last roller may be used with cold water.

The crabbing process is rather slow and is often replaced by blowing—a process in which the pieces of cloth in the unscoured state are wound on a perforated roller through which steam is blown under about 50 lbs. pressure for from 1 to 3 minutes, after which the cloth is cooled by drawing a current of cool air through while it is still on the roller. Here again the pressure between the layers of the fabric varies from the outside to the inside of the roll as in crabbing, so it is customary to repeat the process after re-winding the fabric so that the external layers form the interior of the roll.

Machines are available in which the crabbing and blowing processes may be carried out consecutively; they are equipped with both crabbing and blowing rollers.

It may be remarked that it is possible, to some extent, to combine crabbing and scouring in the open-width scouring machine.

In both the crabbing and blowing treatments, all irregularities such as cockling, creasing and so forth should be removed and the fabric present a uniform and even appearance; further, as previously stated, all latent strains from spinning and weaving are removed, and the fabric is set in a smooth and regular state capable of withstanding the later processes of scouring and milling. The cloth generally shrinks a little in the weft direction, its lustre is improved, the weave clarified and the handle becomes thinner and firmer. Where the weave is pronounced, it is usual to interpose a cotton wrapper between the layers of wool to avoid marking or embossing during the crabbing or blowing when the cloth is in a plastic and susceptible condition; the cotton wrappers usually have a surface devoid of structure and produced by raising and then grinding.

In general, an interlayer wrapper is not used when crabbing or wet blowing unless it is absolutely essential; with most goods it is not necessary. The use of an interlayer increases the time and expense of the process, particularly as the interlayer requires frequent cleansing.

Although wet-blowing is a substitute for crabbing, dry blowing is a finishing process and not one of the preliminaries to scouring.

The crabbing solutions may vary from plain water to alkaline solutions; the latter may be preferred as the great majority of pieces are crabbed in the greasy state. When goods are treated with plain water, there is a tendency to "set" the grease, for it is well known that such goods are more difficult to scour. Most of the goods which require crabbing are worsteds with low percentages of oil; such fabrics are comparatively clean and may be put through the "Leeds finish" in which the goods are dyed after crabbing and the scouring operation is omitted. In actual fact, it is only the cheaper fabrics which are so treated, and the better qualities are scoured.

It has been established by Speakman (J.S.D.C., 1936, *52*, 338) that the extent of setting is largely dependent on the pH of the solution in which the cloth is crabbed; solutions of low pH give little setting, but the maximum effect is reached at pH 9·2. The pH of the solution, therefore, should be adjusted to the highest value below 9·2 which does not unduly interfere with the strength of the fabric, and a 2% solution of borax was used for the original work on fibres; many fabrics, however, already contain a little soap or alkali and may not require any addition to the water used for crabbing.

It was found by Astbury and Woods (Phil. Trans. Roy. Soc., 1933, *A232*, 333) that a short period in steam or hot water did not impart permanent set but, on the contrary, led to super-contraction. A long time is essential for true permanent set, which is due to a twofold reaction, first the breakdown of the cross-linkages in the wool, and secondly, a rebuilding in the new positions.

The actual chemical mechanism has been explained by Speakman (J.S.D.C., 1936, *52*, 338) and by Speakman and Stoves (ibid., 1937, *53*, 237); the cystine linkages are broken according to the following scheme:

$$R—S—S—R \longrightarrow R—SH + HO—S—R.$$

The structure relaxes owing to the dissipation of the internal stresses in the fibre, and permanent set is acquired in the relaxed structure by the formation of new linkages through the condensation of the amino-acids and the products of hydrolysis of the original cystine linkages.

$$R—S—OH + H_2N—R \longrightarrow R—S—NH—R.$$

The new linkages between the main peptide chains are formed with the fibres in their new positions and thus impart permanent set.

It is important to realise that the set is only permanent to temperatures not exceeding that at which it has been applied; for example, setting woollen or worsted fabrics at 70°C. will not render them immune from distortion in water at higher temperatures later.

SCOURING

Fabrics are generally woven from unscoured yarn, except in the case of coloured-woven goods, but most piece goods contain the spinning oils and size which have to be removed.

The two chief methods of scouring fabrics are in rope-form or in open-width; the former is more efficient and may also be used to

FABRIC SCOURING

give a slight fulling action. Hosiery is generally scoured in the stocks or in the tom-tom machine.

SCOURING MACHINES

Rope-washing machines are most commonly used for the scouring of wool fabrics. These machines generally consist of a pair of squeeze-rollers below which is situated a small trough to receive the dirty liquor expressed from the fabric; the whole is surrounded by a larger trough to contain the detergent solution. This larger trough

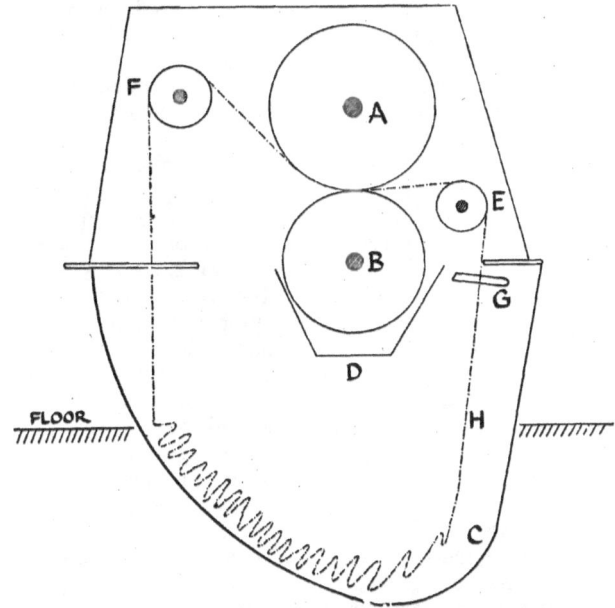

FIG. 97.—Diagram of rope-scouring machine.
(A and B are squeeze-rollers, C the external trough, D the internal trough, E and F guide-rollers, G draft-board, and H is the fabric.)

is fitted with suitable guide-rollers and also a draft-board which separates the various pieces of cloth which are being scoured. The small inner trough is equipped with an outlet system whereby the liquor which it contains may be discharged into the main trough or into the drain; this small box is popularly termed the "sud-box." In general, during emulsification, the liquor is returned to the main trough, but during the rinsing process the expressed liquor is discharged by a separate exit-pipe to the drain.

332 TEXTILE BLEACHING

The draft-board, with its divisions of wood or stainless steel, not only separates the pieces under treatment, but may also be fitted to a stop-motion which operates when any entanglement takes place.

The rope-washing machine, or dolly, is often about 10 feet high, 6 feet wide and 7 feet deep from back to front; some recent machines are wider from back to front, but shallower in order to give a greater capacity per unit length and afford freer circulation. The squeeze-rollers are generally made of birch or beech, the top roller being 26 to 30 inches in diameter and the lower roller 20 inches in diameter; the effective weight of the top roller may be reduced by a screw system, and increased pressure may be applied by weight and levers. The roller system should be resilient enough to yield as the stitchings of the pieces pass through the nip.

The recent tendency towards rubber-covered rollers obviates the difficulty of stitchings; these rubber-covered rollers are generally of smaller diameter—about 12 inches being typical.

A water-pipe is often fixed in front of the dolly so that washing is facilitated; hot water may also be supplied by the same piping if it is fitted with a steam injector, but some pipes are equipped with two-way valves for cold water and for a supply of hot water at a constant temperature. The more modern mills also have pipes for bringing sodium carbonate solution of the required concentration to the machine, instead of carrying the liquor in buckets.

It seems common practice to arrange the dollies in groups of six to ten for purposes of drive, but individual drives may be used. Eight to ten pieces may be treated side by side in one machine.

The dolly is a very versatile machine and may be used for many different fabrics ranging from light dress-goods to heavy low woollens; tandem machines with two pairs of squeeze-rollers have been designed for use with heavy goods of low quality. Another modification of the dolly is somewhat similar to the milling or fulling machine in that it has a small spout and flanged rollers to which heavy pressure may be applied.

The open-width or broad-washing machine is specially suited to the treatment of worsteds and other goods where a clear surface is required.

The general design is similar to that of the rope-scouring machine, but there is an opening device which consists of two rollers covered with copper twigging and arranged to rotate in opposite directions; the relative positions of the rollers may be altered by a hand-wheel to give different amounts of "bind" on the fabric, and eliminate the creases as the rollers rotate. The fabric is lifted from the machine by passing over a guide-roller; after passing through the squeeze-rollers it is lifted over a lagged roller from which it falls to the bottom of the main trough.

FABRIC SCOURING

Before the guide-roller is situated a swivel-guide to keep the fabric in a central position relative to the opening rollers; the swivel-guide consists of a brass frame pivoted in the centre and arranged to tilt as the fabric departs from its central position. Another feature of the open-width machine is the provision of a working roller immediately above the sud-box and below the squeeze-rollers; the working

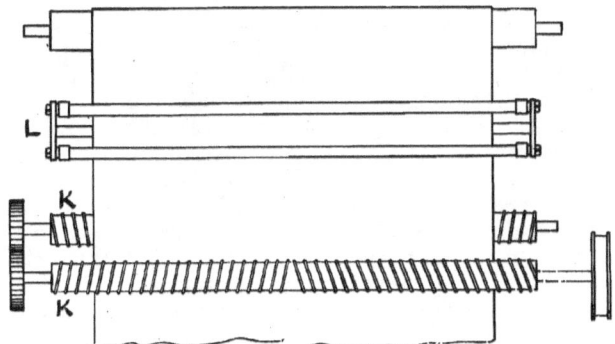

FIG. 98.—Diagram of part of full-width scouring machine.
(K and K are rollers covered with copper twigging, and L is a swivel-guide.)

roller is fluted and exerts a frictional action on the fabric and so promotes rapid scouring.

A new type of open-width washing machine has been developed by Spooner in which the fabric passes between two impinging jets of liquor projected from parallel pipes with specially shaped and directed slots. Another suggested machine for open-width scouring relies on passing the fabric over a suction-slot so that the liquor is drawn through the cloth which then passes on to the usual squeeze-rollers; the whole is enclosed in a large trough, as customary.

In addition to the use of the rope-scouring machine and the open-width machine, it is possible to scour fabrics in the wash-wheel or laundry wheel. This device has been used for garments and also for knit goods and is described on page 350. It has also been found suitable for fine woollen material, particularly crepes; the individual pieces are placed in the separate compartments of the machine and the cylinder rotates or oscillates thus imparting the necessary agitation in presence of the scouring liquor. The time of treatment varies from one to two hours.

THE SCOURING PROCESS

The actual method of scouring fabrics of wool is decided by several factors, including the type of fabric to be scoured and the nature of the finish required; an important consideration, however, is connected with the nature and extent of the oil in the fabric.

As the natural grease from the wool hairs is generally removed during the scouring of the raw material, it is necessary to utilise some form of lubricant in order to facilitate the subsequent mechanical processes of yarn manufacture; oils are employed for this purpose, and although it may appear that mineral oils would be suitable on account of their low cost, yet the difficulties of removal are considerable. At one time, olive oil was used almost exclusively for high-class goods such as worsteds, but the European situation within recent years has been responsible for a great increase in the price of this commodity so that attention has been turned to possible substitutes. Where other vegetable oils of an unsaturated character are used, it is generally necessary to add some form of antioxidant; another type of lubricant has been suggested by Garner in B.P. 487,949, in the form of the synthetic esters of monohydric alcohols and unsaturated fatty acids, or of glycols and unsaturated fatty acids, which not only resist oxidation but are readily removed in scouring. It has also been suggested to hydrogenate certain fatty oils, such as cotton seed, and convert the linoleic glycerides into those of oleic acid, followed by removal of the stearates by freezing (U.S.P. 2,026,735).

Lloyd (Wool Record, 1936, *50*, 985) considers that the water-white mineral oils are eminently suitable for wool oils, if small quantities of oil-soluble polar compounds are added to promote removal by soap and soda. This recalls the work of Speakman and Chamberlain (Trans. Farad. Soc., 1933, *29*, 358), who showed that the chief cause of difficulty in removing mineral oil from wool lay in its high interfacial tension against water, together with the adhesion to the wool which increases with the molecular weight of the oil. In order to overcome difficulties in scouring the oiled wool, the surface tension of the wool must be reduced by polar compounds.

For efficient scouring, 70% of oleic acid or 80% of olive must be present in the mineral oil mixture, but with oil-soluble polar compounds efficient scouring can be realised with much smaller additions; for example, with long-chain aliphatic alcohols such as oleyl alcohol, about 6% is adequate. The polar compound must be retained by the oil during the scouring process, and for this reason the sulphated products are less effective than the alcohols in promoting the removal of the mineral oil from wool.

It may be remarked that Trotman and Horner (J.S.D.C., 1932, *48*; 325) had previously found that the best scouring assistant for mineral oils was sodium silicate, but this is a somewhat drastic agent.

The general trend of research on wetting agents and scouring assistants, however, has been towards the production of wool oils which are self-scouring; products of this type include the so-called ionised oils which are made self-scouring by sulphonation and may also possess bleaching properties. Borosulphates and borophosphates (B.P. 409,598) have been made by treating higher aliphatic alcohols with boric acid and sulphonating or phosphorylating agents.

It is possible to remove mineral oils from wool by means of a neutral detergent instead of soap, provided that the mineral oil contains about 15 to 20% of oleic acid. Speakman and Wang (J.S.D.C., 1940, *56*, 259) have shown that with detergents such as cetyl sodium sulphate, instead of soap and soda, the oleic acid is retained by the mineral oil during scouring so that its effective emulsifying action is preserved. By using a water-white mineral oil as the basis of the blend and selecting a suitable fatty acid, or incorporating an antioxidant, there is no discoloration on exposure to light and air, and the mixture is satisfactory with regard to spontaneous combustion.

Mixtures of cholesterol and long-chain alcohols are also available; these are obtained by the hydrolysis of recovered wool fat. When mixed with mineral oils, they enable the latter to be removed in the ordinary scour.

A well-known lubricant for wool is termed "oleine" and this consists of the liquid portion of the fatty acids obtained by the splitting of fats; according to the method of preparation, one may have saponification or distillation oleines.

For the textile industries, the oleine should be liquid at 20°C. and its setting-point should not exceed 10°C.; the chief sources of the oleines are the animal and vegetable fats, bone fats, palm oil and wool fat. One of the most important tests in connection with the oleines is that of spontaneous inflammability, in Mackay's apparatus, and should this be found dangerous, the inflammability may be depressed by the addition of β-naphthol.

The use of oleines for the oiling of wool not only presents little difficulty in removal during the scouring operation, but is of value in connection with milling processes by the soap which is readily formed in presence of alkali.

In so far as the impurities in the piece goods determine the type of scouring process, it is possible to divide the materials into three main classes:

(a) Worsteds, which contain 2 to 5% of oil on the weight of the wool; olive oil or neutral oil with a little free oleine is commonly used, and mineral oils are not supposed to be employed.

(b) Woollens, which contain 8 to 12% of an oleine preparation, consisting chiefly of crude oleic acid, although small amounts of mineral oil are sometimes present.

(c) Low woollens, which contain 15 to 25% of a low-grade olein, containing substantial amounts of mineral oil.

As will appear later, there are some variations on this broad classification, which, nevertheless, is sufficient to represent the general methods of scouring.

With the neutral oils present in the worsted materials, there is comparatively little free fatty acid, so that the emulsification process must be adopted in which the goods are cleansed by the action of soap, although a little alkali may be added to increase the efficiency of the process and prevent the formation of acid soap.

With woollen fabrics, the oleine is mainly fatty acids in the higher quality products which may be cleansed by saponification of the oil with sodium carbonate solution; soap is formed by interaction of the alkali and the oleine, and this emulsifies the loose dirt.

Low woollens contain substantial amounts of mineral oil with the low grades of oleine and are difficult to scour. Sodium carbonate solution is first used and sufficient soap is added to make a lather.

Fuller's Earth

Before the days of soap, fuller's earth was one of the chief detergents for textile materials; the term probably originated in the custom for milling or fulling to be carried on by the same under the same roof as scouring.

The fuller's earths are fine clays which generally have a slight bluish colour; they are hydrated aluminium silicates which are often contaminated with the oxides of magnesium, iron and calcium. The reason for their detergent action is not very clear, for they may only remove the wool grease by gentle surface friction, or there may be some specific surface adsorption; many inert substances in a fine state of division will exert some scouring effect on greasy wool.

The quantity of goods which are now scoured with fuller's earth is relatively small, but it is still used in the low end of the trade in place of soap as a mild scouring agent; low quality crossbred goods are sometimes treated in the grease with fuller's earth as some of the fatty matter may be left in them and provide a better handle. As the cloth is not cleared to anything like the same extent as with soap, the treatment is almost limited to black, navy, brown and other dark shades. Another application of this mild scouring material is to remove the loose colour from cloth which has been dyed with indigo.

The time occupied by a mild scour with fuller's earth is approximately the same as that required with soap and soda, but in the rare cases where milling is carried out with the earth it is much slower in its action than soap, soap and soda, or soda alone. It is occasionally employed for milling in the grease where a very mild fulling is required, and for the removal of watered effects which may have appeared on the goods during processing.

Trials have also been made with Kaolin and with China clay, sometimes in colloidal suspension; Bentonite has been investigated in the U.S.A.

WOOLLENS

The average woollen fabrics may contain 8 to 10% of oil which, with the higher-quality productions, will consist entirely of an oleine; as the lubricant is composed of fatty acid, the scouring operation may be conducted by saponification. Some of the oleine preparations may contain small amounts of mineral oil, but they are often emulsified by the soap produced from saponification of the fatty acids; the modern tendency with regard to wool oils is to make them self-scouring or, at all events, readily removable.

The saponification scour produces soap by the interaction of the fatty acid in the oleine with the alkali in the scouring liquor; the goods are first treated with sodium carbonate solution of 5 to 6°Tw. (3 to 4%) when soap is formed which emulsifies the dirt with which the fabric is contaminated. Care must be taken to keep the temperature below 35 to 40°C., particularly with coloured-woven goods, and conditions should also be arranged so that the time of scouring is not excessive.

The amount of sodium carbonate is naturally in excess of the theoretical amount for saponification and it is generally the custom to work with a liquor to cloth ratio of 2 : 1, so that 100 lbs. of cloth will require 20 gallons of liquor.

It is preferable to have the sodium carbonate solution prepared in a separate tank and piped to the dolly; the custom of making the solution in the machine suffers from the danger of local concentration as well as the possibility of undissolved particles of the soda ash becoming lodged in the fabric.

When the goods are worked in the dolly, there should be a rapid conversion of the oleine to soap as shown by the formation of a froth which builds up to a lather. For many goods, it is generally wise to drain away this liquor after about 15 minutes; this is done by passing it from the sud-box to the drain. A fresh supply of sodium carbonate solution is then added and scouring continued for a further period of 30 to 45 minutes; where difficulty is encountered in establishing a satisfactory lather, a little soap solution may be added.

At the end of the scour, warm water is gradually added to the goods and the soap solution is diluted over a period of 15 minutes or more; it is believed that this is a most important part of the scouring operation and one which is largely responsible for the actual cleansing of the cloth. The soap solution is removed by warm water (approx. 50°C.) and not by cold water, the pieces running in the dolly until they are quite clean, after which they may be cooled with cold water, removed from the dolly and hydro-extracted. It is important that the emulsion should not be cooled or diluted too quickly or the dirt will be re-deposited on the wool.

The lather in the bottom of the dolly should cover the material completely, but the presence of a copious lather is not in itself a good criterion of efficient scouring, for it may be largely due to agitation of the liquor by the fabric; a good lather must be permanent, and this may be tested by examining a handful, when an impermanent lather will drain away.

Low Woollens

Low woollens are often very difficult to scour because they may contain substantial amounts of mineral oil in addition to the fatty acids. The total oil content of the wool may be from 15 to 25%. A lather may be produced by a treatment of 15 minutes or so, with sodium carbonate solution of 6 to 8°Tw. concentration; if the oil is mainly of the unsaponifiable variety and no lather forms, then about 2 gallons of 5% soap solution should be added. Solvent preparations are very useful in helping to remove the unsaponifiable matter, and about 1 to 4 pints of Astol A may be added to every 10 gallons of the scouring liquor; alternatively, about 0.5 pint of carbon tetrachloride or of Sextol (methylcyclohexanol) may be added to every 10 gallons of the scouring solution.

With the low woollens, which are often dirty, it is generally necessary to withdraw the contaminated liquor after about 15 or 20 minutes scouring, and replace it with further quantities of sodium carbonate solution, which may be of 4 to 5°Tw. concentration, and the necessary amount of 5% soap solution for the production of a good lather. Exceptionally dirty fabrics are often treated with sodium carbonate solution only for the first 15 minutes and then with soap and soda for the second part of the scour.

The general procedure, however, is to produce a lather with soap and soda, replacing the dirty liquors as necessary until the goods are clean, when they are rinsed first with warm water and finally with cold water.

WORSTEDS

As previously indicated, worsteds generally contain about 3% of a neutral oil of high quality; as there is comparatively little free fatty acid, the oil may be removed by emulsification with soap, although it is customary to add some alkali to improve the efficiency of the soap and to neutralise the small amount of fatty acid, as well as to prevent the formation of acid soaps.

Many worsteds are crabbed before scouring, so as to obviate distortion and cockling. There is sometimes a danger of "setting" the grease and dirt at this stage, and the addition of an auxiliary such as Igepon T or Prestabit Oil may be a useful preventive measure.

The scouring process is almost invariably carried out in the dolly, and it is customary to scour about four pieces of 70 yards in length together; 100 lbs. of cloth will need about 20 gallons of the detergent solution, corresponding to a liquor to cloth ratio of 2 : 1, which is much smaller than that employed for the scouring of cotton. With certain fabrics, however, it may be necessary to use a higher ratio and enable the cloth to float in the liquor which opens it somewhat and prevents washer-marks.

Although the scouring of worsteds is an emulsification process, it is usual to start by wetting the goods in soda-ash solution of about 4°Tw.; higher concentrations may cause colour-bleeding as well as damage to the wool itself. As the soda ash is added to the dry goods and does not cause saponification, it remains at about the same concentration, in contrast to the treatment of woollens, during the 10 to 20 minutes in which the cloth is worked in the dolly. There is, of course, some absorption of alkali by the wool; a further reason for the addition of alkali is to obtain the optimum pH of the scouring liquor. The preliminary treatment with soda ash disintegrates any size and loosens the dirt; if the liquor becomes very dirty at this stage it should be run down the drain and replaced by an equal volume of soda ash of about 1 to 2°Tw. Soap solution is then added in sufficient quantity to establish a good lather, and between 1 and 2 gallons of 5% soap solution may be required for every 100 lbs. of cloth; a superior and more permanent lather is obtained when the goods have previously been wet in the soda-ash solution compared with wetting in water alone. When a good lather has been established, continuous rings of foam will appear on the scouring rollers; at this stage, scouring may be continued for some 30 minutes, after which the liquor should be diluted gradually with warm water and part of it allowed to drain away through the sud-box. Rinsing with warm water is important and cold water should not be used until the goods appear clean; the rinsing process may take about 30 to

45 minutes. The addition of a little ammonia during the final rinse is sometimes advantageous as it helps to brighten the colours.

The residual oil in a well-scoured worsted may be estimated by extraction with ether and should not exceed 0·5%; the goods should be thoroughly cleansed but scouring should not be prolonged or some felting may occur. If a rapid reduction in the width of the cloth is required, then the pieces may be arranged to cross at the back of the dolly which should be fitted with a draft-board having an automatic stop-motion to deal with entanglements. The chief reason for the use of the cross-drafting method is to obviate the formation of creases.

Certain fabrics are preferably scoured in the open scouring machine, where the pieces are stitched end to end to make one loop, which passes round and round in open-width; production is low as the scouring is much slower than on the dolly, but with fine goods where creases must be avoided the open-width treatment may be essential.

In general, the progress of the scouring operation may be followed to some extent by the handle of the pieces; the dirty goods have a slippery feel, but the clean cloth possesses a slight scroop.

In the emulsification scour with soap, it is customary to use oleic acid soaps on account of their low titre (14°C.); the soaps of high titre, such as stearate (70°C.) and palmitate (63°C.) are preferred for milling but not for scouring.

CARBONISING

The carbonising of fabrics is essentially similar to the carbonising of the wool before spinning; the tendency towards carbonising in the form of piece goods is due to the preservation of wool quality during spinning, and also to the fact that it is cheaper.

It is not possible to dogmatise on the position of the carbonising process in the sequence of operations; the goods may have been scoured, milled and dyed beforehand. If carbonising takes place after dyeing, then considerable care is necessary to ensure that there is no change in shade of the dyed goods. It has been suggested that a possible advantage of carbonising before dyeing would be to permit the acid to be carried forward to the dyebath, with acid colours, and so effect certain economies of time and material; on the other hand, it has been found that carbonised goods do not always dye evenly.

With regard to milling, goods carbonised before milling take longer to full or mill, but where acid milling is done, the acid from the carbonising process may be allowed to remain in the goods. It is easier to carbonise goods which have not been milled, as the burrs and other impurities have not been "fastened" in the cloth by milling, and are more readily removed.

It is possible to carbonise goods which have not been scoured, but penetration of the acid is better in clean, scoured fabrics; the amount of acid required is less, and the goods are easier to cleanse afterwards.

In general, there is a tendency to carbonise the goods between the scouring and the milling processes. The usual procedure is to immerse in 5°Tw. H_2SO_4 solution, dry and heat at 95°C.; occasionally, an alternative procedure is to immerse in 12°Tw. $AlCl_3$ solution, dry, and heat at 120°C.

The carbonising of cloth may be effected either in rope-form or in open-width.

Most commonly, the goods are in the wet state when they come to be carbonised, but if dry they should be uniformly wetted by immersion in a tank of water followed by squeeze-rollers or a vacuum slot; if the cloth is in rope-form, then it will require treatment on a centrifuge to give an even amount of water.

The wet fabric is then immersed in the soaking tank which contains sulphuric acid of the appropriate concentration, generally about 5 to 6°Tw., but this is apt to vary with the weight of the fabric and the amount of vegetable matter it contains. Lightweight goods of low cellulose content may require only 2°Tw. sulphuric acid.

With goods in open-width, the fabric passes over the usual openers and through the tank containing the acid; the cloth is led over and under a series of rollers in the usual manner, and may even pass between squeeze-rollers in the middle of the tank. At the end of the impregnation trough or tank, the fabric is passed between squeeze-rollers and over a suction slot to remove excess of acid. The speed of the material through the tank may vary from 6 to 30 yards per minute according to its weight, but generally the speed is determined by the necessity of obtaining an even impregnation which, after suction, amounts to about 40% expression and leaves the wool containing not more than 5% of its weight of sulphuric acid.

Modern practice utilises the acid at room temperatures, and the old method of using boiling acid of 2 to 4% concentration is rarely encountered to-day.

Where the carbonising of piece goods with sulphuric acid is carried out continuously on a large scale, it is possible to utilise the tank of acid over a period of months merely adding water and acid as required. After a time, however, it becomes necessary to use a higher concentration of acid than that originally employed; this is due to the buffering action of the salts which are brought into the tank with the wool and left there. Whereas 7°Tw. of fresh sulphuric acid may be adequate at the start of the process, a concentration of 10°Tw. may be necessary after some months, and it then becomes a question of the cost of the stronger acid compared with the cost of cleaning the tank or trough.

The treatment of goods in open-width can be effected in a continuous manner, and after passing through the impregnation tank, the fabric may be allowed to lie in a shell from which it is drawn to the drying and baking section of the continuous range. This section generally consists of a series of chambers, possibly 4 or 6 in number, the first part of which acts as a drying unit and the last as a heating unit. The drying portion of the range is equipped with fans and blowers, in a similar manner to the ordinary drying chambers and stenters. It is important that the drying sections do not exceed a temperature of 60°C. or 140°F. The final heating, on the other hand, may be 95 to 100°C.

For many goods, the sequence of drying and heating may require 30 minutes, about 10 of which are required for the final heating.

During this process, the cellulosic impurities will have been converted into hydrocellulose by the action of the acid; the hydrocellulose is brittle and friable and may be removed by simple mechanical means. The usual procedure is to pass the goods through a rotary milling machine or fulling mill, but without any liquid being present; the crushing action on the dry cloth is at its best while the fabric is still warm, when the hydrocellulose readily falls away. With some open goods of fine quality, this operation is not necessary on account of the ease with which the hydrocellulose falls from the fabric. In the milling or fulling machine, the goods have been treated or dusted in rope-form, but they should be returned to open-width for washing and neutralising on the open-width washing machine. A preliminary rinse for about 20 minutes is sufficient to reduce the concentration of acid in the wool to a minimum, after which the residue is neutralised with 2% sodium carbonate solution followed by a thorough rinsing for 20 minutes or so. Dyed material may be finished with dilute acetic acid solution to brighten the colours in the case of coloured-woven cloths, but with goods dyed in the piece, it is customary to remove the acid by water alone, and so avoid any possible removal of colour by the soda ash. Some coloured goods contain dyes which are sensitive to sulphuric acid, and here the use of aluminium chloride is preferred. Where the latter has been employed for carbonising, it may be necessary to rinse the cloth, after heating and dry-milling, with hot dilute hydrochloric acid, followed by washing with water, rinsing with ammonia solution and again with water.

Magnesium chloride, in 10 to 15% solution, is also used occasionally for the carbonising of white goods; after impregnation and drying, the final heating should take place at a temperature of 130 to 150°C. The magnesium salt sometimes leaves a basic chloride

FABRIC SCOURING 343

or oxychloride in the wool, so that souring may be necessary after carbonising.

Both aluminium and magnesium chlorides, of course, carbonise in virtue of the hydrogen chloride liberated at a high temperature, the ideal reactions being

$$AlCl_3 \longrightarrow Al(OH)_3 + 3 HCl$$
$$MgCl_2 \longrightarrow Mg(OH)_2 + 2 HCl$$

One of the most recent investigations into carbonising is due to Ryberg (Am. Dyes. Rep., 1934, *23*, 230; 1935, *24*, 150). The temperature and humidity under which the acid-treated wool is dried before baking have a marked effect on the affinity for dyes and should be standardised; with sulphuric acid, there is an increased affinity for basic dyes and a decreased affinity for acid dyes after carbonising, but with aluminium chloride, the opposite holds. All acid-treated wool should be dried as rapidly as possible before baking, and should not be left unduly afterwards. Ryberg also established that the washing of freshly carbonised wool with hot 3% sodium sulphate solution removes more residual sulphuric acid than does washing with water, and results in more even dyeing later.

MILLING

The ability to felt is probably the most characteristic feature of wool and has been known from very early times. It seems likely that felting originated in the primitive methods of facilitating the cleansing of fabrics by pounding in the detergent solution. The hand pounding was probably followed by trampling underfoot to produce the required felting, and this process featured among the curious mural decorations of Pompeii. The original term for the process was "fulling" which signifies a treading action and persists in the French *fouler*; the term "walker" was also used and is still seen in the German *walken*.

The original fulling machines were stocks which reproduced the treading and pounding action, but rotary milling machines were devised in 1833 by John Dyer of Trowbridge. The term "milling" is now generally applied in England to all compressive processes which cause felting, but in the U.S.A. the word fulling persists.

The felting properties of wool are due to three factors:

(*a*) a surface scale structure
(*b*) ease of deformation
(*c*) recovery from deformation.

The surface scale structure is such that the tips of the scales all point to the tip of the fibre which can therefore only move amongst its fellows in the direction of its root end. The ease of deformation and powers of recovery are such that the fibre moves in much the same way as an earthworm crawls; fibre travel has been demonstrated on numerous occasions and is responsible for the mass contracting in bulk by a series of self-tightening mechanisms. The pressure of the milling machine or stocks forces the hairs into close contact whereby the scales of adjacent hairs find the necessary mutual frictional purchase. The movement during milling is due to the intermittent nature of the pressure, which causes the hair to move in the direction of its root end, drawing others with it, but the elasticity of the moist hair is such that the fibre contracts when the pressure is released.

Milling processes may be divided into two chief classes:

(*1*) acid milling
(*2*) alkaline milling, including soap milling.

The investigations of Speakman and his co-workers established that the felting of wool is most rapid at about pH 10 on the alkaline side of neutrality (J.T.I., 1933, *24*, 273), but that soap exerts some specific action in promoting milling shrinkage apart from the pH of the solution. Hence it is not possible successfully to replace soap with caustic soda or sodium carbonate where milling shrinkage is required, except for milling in the grease, which is a variation of soap milling.

On the acid side of neutrality, the rate of shrinkage on milling starts to increase at pH 4 and continues to increase with decreasing pH, whereas, as stated above, on the alkaline side of neutrality milling reaches a maximum rate around pH 10.

The effect of temperature is such that rate of shrinkage increases up to 35 to 37°C. and then decreases; this is probably due to difficulty in recovery from extension at the higher temperatures and the same reason accounts the fall in milling shrinkage above pH 10.

The optimum conditions for milling, therefore, are those which break down salt linkages between the peptide chains, and so facilitate extension, but leave the powers of recovery unchanged; high alkalinity and high temperatures would affect the cystine linkage and so reduce the powers of recovery. The work of Speakman, Menkart and Liu (J.T.I., 1944, *35*, 41) is of interest in this connection; there seems to be no critical temperature for acid milling, which is often carried out at 70°C.

As previously stated, the milling process is really a finishing process and logically falls outside the scope of scouring and bleaching. This is definitely the case with acid milling for the wool has to be well scoured and rinsed before the milling; it may be sufficient to remark that the process requires about 3 lbs. of B.O.V. (*d*. 1·72) or

FABRIC SCOURING

2 lbs. of D.O.V. (d. 1·84) per 100 lbs. of cloth, the acid being diluted so as to have a concentration of 0·2 to 0·5% sulphuric acid.

On the other hand, alkaline milling involves some cleansing of the goods at the same time as milling; this is evidenced to a very high degree with milling in the grease, where soap is formed by reaction between the fatty acid in the fabric and the alkali applied to it. This type of milling is often utilised on low woollens, as well as with overcoatings for greatcoats, because a scour would entail a loss in weight due to the disappearance of the shorter fibres, which, however, become bound into the fabric by the milling process. Solutions of sodium carbonate of 5 to 8°Tw. are often used; the process is often rather dirty owing to the soapy oily scum which forms and is not regarded as suitable for fine goods. With heavy uniform cloths, such as military greatcoats, greasy milling is commonly practised, as high shrinkages are relatively quickly developed. Greasy milling saves a process.

Soap milling is probably the commonest type of milling and takes place on fabrics which have already been scoured to some extent; 5 to 10% of soap is required, depending on the extent of milling and estimated on bar soap (70%) and not flake soap (85 to 90%).

The choice of a soap is very important. Whereas for the scouring of raw wool, the prime consideration is to have a solution of low pH but good detergent properties, with "rinsability" as a secondary factor, the same does not apply to the later stages of cleansing where detergent and rinsing powers are both very necessary. In the scouring of piece goods, the soap should operate at a low temperature, and the simpler oleates are very good in respect of rinsability.

Milling soaps should have good solubility and stability, and according to the Dyer (1943, *90*, 201) they may be made from tallow and palm oil, modified with silicate to adjust the cost.

Milling Soaps

	Good	Medium	Cheap
Tallow	47	31	25
Palm oil	11	10	8
Na_2CO_3 (66°Tw.)	6	9	11
NaOH (66°Tw.)	31	26	25
Silicate	—	9	17
Starch	—	—	1
Water	5	15	13
	100	100	100

As the amount of sodium silicate increases, the ease of rinsability decreases.

346   TEXTILE BLEACHING

The use of stearates and palmitates, in preference to oleates, provides a soap of higher melting-point which encourages felting for the soap does not "run" during the milling process, as the heat develops. This is seen, as previously mentioned, from the titre of the various soaps, oleate (14°C.), palmitate (63°C.) and stearate (70°C.). Milling soaps should be hard and heavy bodied; they should not contain free fat or large amounts of free alkali.

It should be realised that, in general, no heat is supplied to the goods during the milling or fulling process from any external source; the heat is generated by friction and is very largely influenced by the speed of the process. A moderate temperature favours rapid milling, but above 38 to 40°C. the rate of milling decreases again. During the fulling process, the temperature is largely controlled by opening or closing the doors of the machine.

There are three chief types of fulling processes: (*a*) grease milling, (*b*) soap milling, and (*c*) acid milling.

Milling in the grease is the slowest method, and is somewhat dirty, the fabric finishing with a poor colour. It is claimed to give the heaviest fabrics, but they are often weaker and less elastic than from other fulling processes.

Soap milling gives the best colour and handle, producing material of the highest quality.

Acid milling is the most rapid method and produces the lightest, strongest and most elastic fabrics; the handle is apt to be harsh, however.

The use of an aqueous suspension of fuller's earth for milling is rarely encountered in modern times.

FULLING AND MILLING MACHINES

There are two chief types of machine for fulling or milling fabrics composed of, or containing, wool. These are as follows:

(*1*) The fulling stocks.
(*2*) The rotary milling or fulling machine.

The fulling stocks are the oldest mechanical method of felting wool; they simulate the original trampling processes. The fabric is placed in a roll or in the form of a ball in the stocks where it is hammered by the fallers; during this process, the pounding causes the fabric to turn, so that the simple process applies the same force to both warp and weft of the fabric. In the rotary fulling machine, on the other hand, different forces may be separately applied to the length and width of the cloth. Another interesting point about the stocks, is that the force applied increases as the bulk diminishes, in the simplest machines.

Two kinds of fulling stocks are available, first, the gravity stocks

FABRIC SCOURING 347

where the fallers are lifted by cams and fall by their own weight; and, secondly, stocks where the length of the stroke is fixed and controlled by cranked gearing or by eccentric cams. With the gravity type, the fulling process adapts itself to the type of cloth to a large extent, as indicated above.

The oldest type of fulling stocks comprised an oak-lined trough, in which the cloth was placed, and at the back of the trough were situated pillars to which the wooden fallers were pivoted. The fallers or specially shaped hammers worked in the trough, as shown in Fig. 99, and had beams projecting through them to make occasional

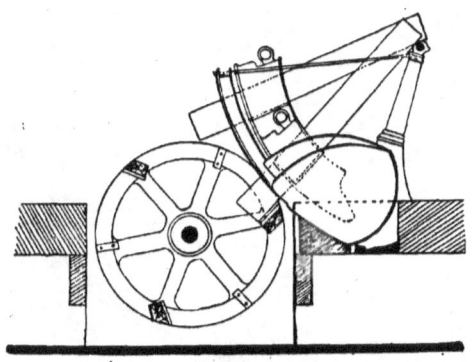

FIG. 99.—Diagram of fulling stocks.

contact with the projections situated on each side of a tappet wheel which was sunk in a pit in the floor immediately in front of the machine. As the wheel revolved, it lifted the stocks in turn and then allowed them to fall, compressing the cloth and also turning it slightly. Each foot struck about forty blows per minute.

A simple type of fulling stocks, made by Sellers and Co., is illustrated in Fig. 95; it is 24 inches wide, and 150 cubic feet in capacity. These stocks operate at high speed, the two feet striking the cloth 240 times per minute; it is also interesting to note that the stock or trough can be moved either towards or away from the feet during the process, thus increasing or decreasing the size of the trough to suit the conditions. A circular cover is also fixed by hinges at the top and this enables the goods to be compressed somewhat with acceleration of the fulling process.

A pendulum-type of fulling mill is made by S. Pegg and Son. It is crank-driven, the beaters being made of hard wood and the container is lined with sheet zinc or with stainless steel. This machine is largely used for knitwear which may be fulled for

5 minutes or more with a lukewarm soap solution. Three sizes of machine are available, capable of dealing with 20, 40 and 55 lbs. of fabric respectively.

The rotary milling or fulling machine consists essentially of a large trough with a curved bottom to enable the cloth to slide easily. As will be seen by reference to Fig. 100, the cloth passes, as an endless band, through an earthenware mouthpiece which guides it between

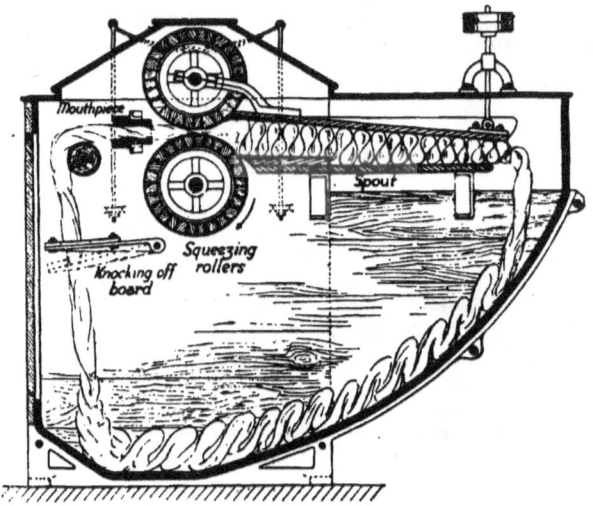

FIG. 100.—Diagram of rotary milling machine.

the squeeze-rollers feeding it into the spout of the machine. (A roller mouthpiece is popular with some machine-makers.) Some of the fulling or milling takes place in the spout, the lid of which is weighted to vary the intensity of the effect. In the spout, the fabric is folded or cuttled, but the oncoming fabric fed by the rollers lifts the lid and the cloth falls down the curved base of the machine. It then passes upwards through the draft-board to the mouthpiece again; the draft-board separates the strands of fabric and is also connected with an automatic stop-motion which comes into operation should entanglements occur. The stop-motion and draft-board is popularly termed the "knocking-off board."

A full account of various milling machines has been given by "B.H." (Textile Manufacturer, 1924, *50*, 384, 520; 1925, *51*, 166, 244, 315 and 351); a more recent account of some milling machines has been contributed by Kilburn Scott (J.T.I., 1933, *24*, 247P).

FABRIC SCOURING

Some of the latest types of rotary milling machines, and also the combined scouring and milling machines, are due to the work of Williams and Peace (B.P. 312, 459; 336,016; and 409,869); the machines in question are supplied by J. Mitchell and Sons. One of the chief features is that the top roller is carried in such a manner that the actual working weight may be adjusted in a few moments from zero to 17 cwts.; this weight is constant and does not vary with the thickness of the cloth. The lid of the trough and also the draft-board are likewise accurately weighted.

Combined scouring and milling machines have been devised, and, as might be expected, they are either adapted scouring machines or adapted milling machines.

It is possible to modify the draft-board of the rope-scouring machine or dolly to accommodate a glass or porcelain throat which exerts some constriction on the fabric as it passes through and therefore produces some milling. An alternative idea is to replace the draft-board by a guide-roller and place near the rollers a set of horizontal throats through which the fabric is drawn in a compressed state.

The rotary milling machine may readily be adapted for scouring and milling by placing a sud-box under the milling rollers. When used for scouring, the throat is wide open and the spout free from pressure.

The machine is first run under very light pressure when most of the dirt and grease is removed and passed out of the machine via the sud-box and drain; the fabric is then thoroughly soaped and the pressure applied to the rollers and spout to effect the milling or fulling action. The pressure may be removed again for the final washing process.

The advantages of the combined scouring and milling machine include a considerable economy in time, labour, soap, water and steam; the finished pieces are better in appearance and handle.

Knitted Material

A special type of combined scouring and milling machine has been devised for the treatment of hosiery; it is termed the dolly, or tom-tom machine, and is fundamentally a tub-scouring machine, similar in its action to the old household "poss-tub." Knit goods are generally scoured as garments or in short lengths, which are much shorter than the woven pieces.

The tom-tom or tub-scouring machine may be made with rectangular or circular tubs; the latter are made to revolve during the process, whereas the former move to and fro. The rectangular type is more popular, and the well-known Pegg model is made in five different sizes accommodating from 30 to 240 lbs. of material. A

350   TEXTILE BLEACHING

thorough scouring action is achieved as the tubs travel backwards and forwards at a uniform speed, bringing every portion of the goods under the beaters in turn. Machines are made with either one or two tubs; the one-tub machine of 30 lbs. capacity is generally used for gloves and other small articles, and contains two beaters, but the larger scouring machines, whether of the one- or two-tub type, contain three or four beaters per tub. Each tub can be started or stopped independently. The beaters are made of hardwood, and are lifted by a set of cams and then allowed to fall by gravity (see Fig. 94).

Scouring generally takes place with soap solutions of concentration amounting to 5% on the weight of the goods, and a little ammonia is often added; the temperature of the liquor is approximately 80°C., and the average time for treatment is about 20 minutes. Obviously a great deal depends on the amount and type of impurity present in determining the conditions of treatment; prolonged or severe action may be used to produce a fulling machine on woollen goods.

It should be realised that although this type of machine is commonly referred to in the Midlands of England as the "dolly," it must not be confused with the rope-scouring machine for woven fabrics (see page 331), which in Yorkshire is also known as the "dolly."

A rotary scouring machine, devised by S. Pegg and Son, is also suitable for washing and bleaching hosiery and garments. The chief use of the machine is for the scouring and washing of goods that require delicate treatment, and for high-grade woollens which should not be felted; the principal parts of the machine are the outer and inner drums, the outer drum, with its closely fitting doors, is stationary. The goods to be scoured are placed in the inner drum, which is caused to rotate first in one direction and then in the other. The liquor is allowed to circulate through the perforations of the staves of the inner drum; a series of beaters is fixed longitudinally to these staves, and this feature is partly responsible for the efficiency of the machine. Two sizes are available accommodating 40 and 80 lbs. of material; the inner cylinders are 28 inches by 48 inches, and 36 inches by 54 inches respectively.

Most knitwear does not contain a high grease content as it comes from the knitting machines and is comparatively easy to scour. As many goods require a slight milling, it is convenient to utilise a combined scouring and milling machine, such as that illustrated in Fig. 101, designed by S. Pegg and Son. The machine is designed primarily as a scouring machine for hosiery, but it is also capable of scouring, milling and rinsing in sequence; after scouring and milling, the liquor is withdrawn to a separate container for strengthening and re-use, and clean water is admitted to the machine itself

FABRIC SCOURING

for rinsing. The method of operation is to pass the goods over the guide-roller G, and between the two rollers W, to the main squeeze-rolls R, after which they pass through the milling box M, down into the soap solution, and up through the "knocking-off" board K, where the two ends of the fabric are joined to make an endless band.

The squeeze-rolls R, have an iron core about 14 inches in diameter, constructed so as to receive a number of oak sectors, turned to a diameter of 20 inches. The pressure on the fabric is adjusted by the spring S, and after passing through the rollers it is cuttled in the

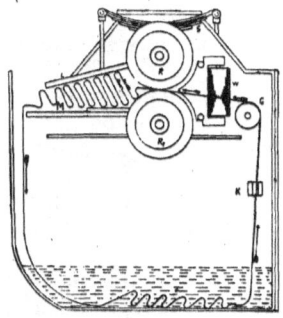

FIG. 101.—Line diagram of combined scouring and milling machine.

milling box M, whose lid L is weighted to exert the necessary amount of pressure to give a light fulling or milling. The fabric, as it accumulates, lifts the lid and drops into the liquor in the bottom of the large enclosing trough below. The "knocking-off" board K has slots of varying widths so that any accumulation of fabric will lift the board and stop the machine.

The nominal speed of the squeeze-rolls is 60 r.p.m., so that the fabric passes through at 300 feet per minute. The sequence of operations is repeated as often as is necessary to give thorough scouring and milling; the average output is 160 lbs. per hour. The size of the machine is 8 feet long, 5 feet 6 inches wide and 7 feet 6 inches high.

THE MILLING PROCESS

In the rotary milling or fulling machine, the fabric may be treated in rope-form and is sewn end to end to form loops or endless chains; these are known as drafts. In the single-draft system, the fabric merely passes once through the machine as a single strand, but in the double-draft system the piece is formed into a double loop before its ends are joined. The double-draft system is generally adopted as it tends to hasten shrinkage by the friction of twice the normal amount of fabric in the throats.

With some goods, it is customary to join the lists or selvedges together forming a bag with the face of the fabric inside; this protects certain weaves from chafing or marking during the milling process.

Fulling or milling is not a standardised process, and the shrinkage and milling effect is controlled by stopping the machine at intervals to examine the pieces and measure their length and width. An opportunity is also taken to open the pieces and so prevent their running in the same folds with the formation of mill riggs.

For milling in the grease, it is often customary first to impregnate the pieces by passing through a "lecker" containing sodium carbonate solution of 5 to 8°Tw. (2·4 to 3·8% Na_2CO_3). The lecker consists essentially of a box or trough for the liquor and a pair of squeeze-rollers to ensure even distribution and remove any excess of liquor; it may also be used to apply hot soap solution to the fabric as a preliminary to soap milling.

Alternatively, the goods may be placed dry in the milling machine, and the lukewarm liquor added from a can thoroughly to saturate the goods before milling starts.

With soap milling, a full scouring treatment beforehand is not essential and it is also possible to leave residual soap in the goods from the scour; it is better, however, to start with a clean fabric, for dirty goods mean using more soap for milling. Where the pieces come from the scouring machine, a thorough hydro-extraction is advisable, for an excess of water will reduce the actual concentration of soap when it is applied, frequent additions of soap may become necessary during the process and the cloth may become too wet to mill properly. It must be appreciated that the correct amount of liquor is of great importance in fulling or milling; if the goods are too dry there will be a loss of fibre by flocking, and if they are too wet they will slip and not felt.

Solutions of 10% soap are generally prepared and about 2·5 gallons of such a solution is adequate to start a piece of 100 lbs. in weight; the temperature may take some time to rise during the process, and it is therefore customary to close the doors of the machine after the first soaping. If signs of dryness are detected, then further small additions of soap solution may be made during the milling operation. Correct soaping is essential for good milling.

The time of milling depends to a very large extent on the type of fabric undergoing treatment as well as the finish required. With overcoatings, for example, two hours' milling may be required, but they should be measured frequently after the first hour. When milling is complete, the goods may be removed to a scouring machine for a final scour in straight drafts to prevent extension in length; warm sodium carbonate solution may be used to remove the soap from the milled cloth, and a little fresh soap added to assist in the cleansing of the fabric. After scouring for 30 minutes or so, the goods are rinsed with warm water over a period of two hours to remove all soap and alkali; the temperature of the wash-waters may be reduced gradually during this period.

Striped flannels, on the other hand, may only require a moderate amount of milling, and from 30 to 60 minutes treatment is often adequate. Fancy goods require careful handling because of the possibilities of the colours bleeding; more dilute solutions of soap are applied than with other goods and the temperature of the process is not allowed to rise to the same extent.

CHAPTER XXVII
BLEACHING WOOL

SCOURED wool is not perfectly white but possesses a cream or slightly yellow colour due to the presence of certain pigments which have not, as yet, been isolated or identified.

The proportion of wool which is bleached is very much lower than that of cotton; practically all men's underwear is sold in the natural colour or even a deeper shade produced by dyes. Most dyed or coloured-woven fabrics do not require a bleaching process during the course of their manufacture, with the exception of white flannel and similar material. Knit goods for ladies' underwear are generally bleached to a good white and considerable quantities of white yarn for knitting are also produced; where wool has to be dyed in the fashionable pastel shades which need to be bright and clear, it is preferable to bleach the material beforehand or the result will be rather yellow and "saddened."

Very substantial amounts of both woollen and worsted fabrics are woven from coloured material to an infinitely greater extent than with the vegetable fibres.

Bleaching, therefore, plays only a small part in the purification of wool, and is not a logical extension of the scouring process. It must also be remarked that the bleacher would not regard the washing or scouring of raw wool to fall within his province.

The two chief methods of bleaching wool are with

(a) sulphur dioxide
(b) hydrogen peroxide.

Hypochlorite solutions cannot be used for the bleaching of wool or mixtures of wool and cellulosic fibres, as they produce a yellow colour which is accentuated on subsequent washing. Two minor methods of bleaching wool consist of a treatment with permanganate solution followed by sulphurous acid or a simple tinting process. Tinting is often carried out during the backwashing process to mask the creamy colour of the tops and is merely a superficial and temporary expedient to improve the selling value of the material; it is also utilised after the sulphur dioxide bleach to improve the colour. The permanganate bleach is not frequently employed but consists in steeping the wool for about an hour in a cold solution of potassium permanganate containing 1 to 3 lbs. per 200 to 300 gallons of water (0·05 to 0·1%) after which the material is squeezed and steeped for a few hours in acidified bisulphite solution, followed by washing.

BLEACHING WOOL

When woollen goods are to be bleached, they should be on the alkaline side of neutrality; "unshrinkable" hosiery, carbonised material and acid-milled wool may therefore need special attention beforehand.

SULPHUR DIOXIDE

The simplest method of bleaching wool with sulphur dioxide is to expose the scoured material in the wet state to the action of sulphur dioxide gas; a variation of this process is to bleach the wool by immersion in a solution of sulphur dioxide or sodium bisulphite.

Stoving

The classical method of bleaching wool by stoving figures amongst the interesting illustrations which are to be found on the walls of Pompeii; the material was spread over a frame beneath which sulphur was burned in a small pot. The present process is not vastly different, for it is customary to suspend the wet material, after scouring and squeezing, over poles or rods in a closed chamber where it is exposed to the fumes of burning sulphur. The absence of metal is necessary, and the "stove" is made of wood, brick or stone; the wool is exposed for from 4 to 12 hours and it is usual to conduct the process at night. About 6 to 7 lbs. of sulphur are required per 100 lbs. of wool.

In the simplest form, a series of parallel wooden poles is evenly spaced about 8 inches apart and at such a height that the operative may hang the wet fabric from them in loops or festoons; the walls of the room are generally boarded, and a small vertical flue projects through the roof to ensure a very moderate circulation of air for otherwise the burning of the sulphur would cease prematurely. The "stove" may vary in size in different works, but it is often about 6 feet wide and 16 feet long.

In one corner of the floor is placed a small iron tray or pot containing the sulphur which is ignited by dropping on to it a red-hot piece of iron; the door of the chamber is closed, and the burning sulphur generates sulphur dioxide which fills the chamber and is absorbed by the wet fabric. Bleaching usually starts in the evening and is allowed to continue overnight.

On account of the fabric being wet, the surrounding atmosphere becomes humid, but during the night the temperature falls and some moisture is apt to condense, dropping on to the wool and causing stains. It is possible to obviate this by placing a fabric shield between the roof and the poles on which the cloth is suspended; there is some evidence of tendering where the condensed drops have fallen occasionally, but it is not clear if this is due to sulphuric acid or not.

Another defect of the simplest method of stoving wool is that the burning sulphur may become so hot that some vaporisation takes place and particles of sulphur finally deposit themselves on the wool; for this reason, an alternative method of stoving is to burn the sulphur in an ante-chamber and circulate it through the main chamber by a fan.

These somewhat crude methods are still widely utilised in spite of their disadvantages. The fumes have to be disposed of and are apt to be objectionable; sulphur trioxide formation is apt to affect dyeing properties.

Although sulphur dioxide is readily available in commercial quantities, little systematic attempt seems to have been made to utilise it in the bleaching of wool. Ward (J.S.D.C., 1939, 55, 445) has described a method in which the sulphur dioxide is obtained from cylinders of the liquid, and the fume problem was dealt with by using the waste lime sludge from the water-softening plant. The main feature of the process lies in controlling the rate of evaporation of the liquid sulphur dioxide. Hydro-extracted material is arranged in the stove in a more economical manner as no space is required for the sulphur pots; air is circulated by a specially designed lead fan, and the sulphur dioxide liquid discharged from its cylinder on to a heated surface. As the cylinder is supported in a cradle suspended from a weighing machine, the calculated amount may be admitted at a suitable rate, after which the fans are run for a further 10 minutes, and then the goods are left for from 2 to 12 hours according to circumstances. The ventilating system is then put into operation by opening the absorbing system in a tower above the stove, running the fans and opening the doors. Ventilation is maintained during the emptying of the stove. Washing completes the treatment.

This process is protected in B.P. 480,448 and has been operated on a large commercial scale to give results which are equal to those of the older system; the cost, too, is very similar on account of the greater efficiency.

No form of sulphur dioxide bleaching gives a perfect white nor is the bleaching effect permanent; the natural colour of the wool gradually returns but this is not a serious disadvantage for underwear.

The stationary method of stoving wool may be the oldest, cheapest and simplest bleaching process, but it is by no means foolproof; the amount of moisture in the wool needs careful control for if it is excessive then it tends to accumulate at the bottom of the material with irregular results. On the other hand, when the cloth is not sufficiently damp, the bleaching action is diminished. On this account, the "running" stove offers an alternative to the "hanging" stove; the cloth is passed through the apparatus by means of a series of rollers and then allowed to remain on batch-rollers overnight. A second passage through the stove completes the process, after which the goods are washed as usual. In some cases the sulphur dioxide is generated in a separate burner and forced into the stove by fans. In general, however, the bleaching effect is less pronounced than from the "hanging" stove.

The avidity with which wool retains sulphur dioxide is such that it is very difficult to rid the bleached material of the unpleasant odour; in spite of very thorough washing it is not possible completely to remove the odour of the sulphur dioxide and it is often necessary to give a second chemical treatment to remove the sulphurous residue. As might be expected, oxidising agents form the chief reagents for this purpose, and dilute solutions of calcium or sodium hypochlorite may be used especially with fabrics intended for printing; the hypochlorites not only remove the sulphurous acid by converting it into sulphuric acid, which is more readily removed by washing, but they increase the affinity of the wool for dyestuffs.

The use of hypochlorites cannot be universally recommended for they are apt to produce a somewhat harsh feel on the goods; the safest reagent to employ is a dilute solution of hydrogen peroxide or sodium peroxide. This was first suggested by Lange (D.R.P. 31,741 of 1884) with the idea of oxidising the H_2SO_3 to H_2SO_4, as the latter could be removed more readily. At this time it was also considered that the presence of the H_2SO_3 was responsible for the restoration of some of the colour of wool. It has been suggested later that the action of the peroxide on the soluble reduction products is to render them soluble and so remove them.

Sulphites

The liquid methods of bleaching with sulphur dioxide usually depend on the use of sulphurous acid, although it has been suggested to employ a two-stage process, first an impregnation with sodium bisulphite solution followed by acidification with sulphuric acid. It is more general to work the goods in a bath containing sodium bisulphite and either sulphuric or hydrochloric acid; the water should be free from iron.

The sodium bisulphite processes may be operated in the following manner. For a single-bath treatment, about 1 gallon of sodium bisulphite solution of 70°Tw. and 1 pint of sulphuric acid are required per 100 lbs. of wool; fabric may be treated for 2 hours or more on the jig, whereas yarns may be saturated and steeped for 2 to 3 hours. In both cases, it is wise to ensure good ventilation on account of the evolution of sulphur dioxide.

An alternative method is to steep the wool for 6 to 10 hours in

2°Tw. sodium bisulphite solution, squeeze, and then immerse in 1°Tw. sulphuric acid solution. Whatever method may be employed should be followed by a thorough rinsing afterwards.

The white obtained by the aqueous method is no more permanent than that from the gaseous method, but a better control may be exercised especially with piece goods. The bleaching action may take a little longer than with stoving, and is not quite as good.

About 1912, Schofield (*The Finishing of Wool Goods*, p. 698), devised a process for utilising an aqueous solution of sulphurous acid, separately prepared but employed in the ordinary rope or dolly scouring machine, the repeated squeezing effect of whose rollers promotes an even result.

King (J.S.D.C., 1930, **46**, 225) found that mixtures of sodium sulphite and bisulphite having the molecular ratio $SO_2/NaOH$ equal to $1/1 \cdot 25$ or $1/1 \cdot 35$ ($2NaHSO_3 \cdot Na_2SO_3$), exert very strong bleaching effects and may be employed in aqueous solution corresponding to 2% SO_2 content; the pH of the solution is about 6·8. For ordinary bleaching in cold solutions about 24 hours is required, but 2 hours is sufficient at 50°C. The process is protected by B.P. 332,389.

Numerous small variations on the sulphur dioxide bleach are obviously possible and many of these have been collected by Cook (J.T.I., 1926, **17**, 371) in a good summary of the literature. The fabric may be treated on the jig for about 2 hours, by running to and fro in the solution containing 2·5% of sulphur dioxide prepared from the above-mentioned mixture of sodium sulphite and bisulphite.

It is well known that sodium hydrosulphite is a much more powerful reducing agent than sulphur dioxide, and attempts have been made to use this compound as a bleaching agent for wool. The results are only moderately successful, giving a pinkish-white which is less acceptable than the white effect resulting from the old-fashioned stoving process.

General

In general, it must be remembered that the bleaching action of sulphur dioxide on wool is very slow and requires many hours for a good effect; double stoving produces a better white, but numerous repetitions of the process or very prolonged treatments do not bring about further improvements. It might be expected that attempts would be made to subject wool to higher concentrations of sulphur dioxide than those used in normal practice; sulphur dioxide is more soluble in ethyl alcohol than in water, but even with this solution, Schofield (loc. cit.) found that the resulting bleach was no better than that obtained by the ordinary method. Schofield considers that the sulphur dioxide bleach has its natural limit.

Some of the practical problems connected with the stoving of wool indicate that a slightly superior effect is obtained when the wool contains a little soap or is slightly alkaline; wool previously treated with sulphuric acid is apt to give inferior results. In neither case is the permanency of the white affected. The concentration of sulphur dioxide in the stove should be as high as possible, but although the rate of bleaching increases with temperature, the quality of the white is not improved.

The wetness of the wool is an important factor in the stoving process and needs a little care; if the wool is too dry, the bleaching is poor, but if the goods are too wet there is a tendency for the water to accumulate in the bottoms of the loops of the fabric and give an irregular bleaching action throughout the piece. The latter defect is generally obviated in the "running stove" method.

The theoretical aspect of the sulphur bleach has attracted some notice, and it is most generally assumed that the action is one of reduction. It is well known that the sulphur dioxide bleaching of wool does not give a permanent effect; the original brilliant white gradually disappears until the creamy yellow appearance is restored as is commonly noticed with white coats, skirts and trousers made from wool. This phenomenon disposes of the suggestion that in the bleaching process, the sulphur dioxide forms a soluble compound with the colouring matter.

Other theories are that the colouring matter is reduced to a colourless form, and that the colour gradually returns on oxidation, or that a colourless bisulphite compound is formed with the wool substance and this slowly decomposes restoring the original colour.

It is well known that wool has a great affinity for sulphur dioxide, and Patterson (J.S.D.C., 1909, *25*, 112) found that stoved yarn on bobbins still retained SO_2 on the inside after twelve years; drying the stoved yarn at 75 to 100°C. will not remove the SO_2. The difficulties of removing the gas are notorious and very often the goods are passed through a weak solution of an oxidising agent to hasten the removal of the sulphurous odour.

The great affinity of wool for sulphur dioxide has been studied by Raynes (J.T.I., 1926, *17*, 379), who found that air-dry wool absorbs a large excess of the gas from a closed vessel and acquires a yellow colour; if the wool is then exposed to the atmosphere, the yellow colour gradually disappears and the wool becomes white. One gram of wool previously dried at 103°C. and then conditioned will absorb about 55 c.c. of SO_2, but wet wool will absorb almost twice this amount; the extra absorption corresponds to that required to saturate the water which is present.

Time-absorption curves give no indication of chemical combination, but if true chemical combination occurs between the colouring matter and the sulphur dioxide this may be obscured by the greater absorption of the gas by the wool substance.

The presence of water appears to be essential for the bleaching action, for if completely dried wool is exposed to dry sulphur dioxide, it acquires a yellow colour, and a great amount of SO_2 is absorbed; when the saturated wool is placed in an evacuated vessel, it loses the absorbed gas and the yellow colour, but the wool is not bleached. Repetition of this experiment with wet wool leaves the material in a fully bleached state. The intense yellow coloration with both wet and dry wool suggests the formation of some additional compound with the wool substance and not the colouring matter which is not affected in absence of moisture.

Another interesting point is that the white of wool which has been bleached with sulphur dioxide is not stable to dilute sulphuric acid, for after acidification, part of the original colour gradually returns and a small amount of sulphur dioxide is liberated. It is known that sulphurous acid will combine readily with substances containing carbonyl groups to give products which are readily decomposed.

Raynes (*supra*) considered that at least two distinct components of wool combine with the sulphur dioxide, and one of these is responsible for the formation of a lemon-yellow product at high concentrations. This may be the same compound accepted as a sign of good bleaching in works' practice; Schofield (loc. cit.) also obtained it when using his alcoholic solutions of sulphur dioxide. The second component is formed only in presence of appreciable quantities of water, and is colourless. Raynes assumed that ketonic groups came into play and showed that wool combined with small amounts of hydroxylamine and semicarbazide. These reagents, however, also combine with aldehyde groups which are present in wool that has been degraded by air and light, according to the work of Race, Rowe, Speakman and Vickerstaff (J.S.D.C., 1938, *54*, 141).

King's bleaching process arose from examination of the effect of various sulphites on the azo dyes, and it was considered that solutions which destroyed these colours might bleach the colouring matter in wool. The particular mixture referred to, has a pH value of 6·8 in solution and is sometimes termed the neutral bleach.

Phillips (J.S.D.C., 1938, *54*, 503) considers that a better theoretical approach to the question of sulphur dioxide bleaching is made possible by consideration of reactions of the disulphide bond. Wool contains about 13% of cystine and this amino-acid reacts with sulphites as established by Clarke (J. Biol. Chem., 1936, *97*, 235), who prepared sodium-S-cysteine-sulphonate. It has also been suggested by Speakman (J.S.D.C., 1936, *52*, 335) that when wool is treated with boiling bisulphite solutions, the cystine disulphide linkages are broken with the formation of thiol, —CH_2SH, and

BLEACHING WOOL

S-cysteinesulphonate, $NaO.SO_2.S.CH_2-$ side-chains. These reactions are considered by Phillips to be of importance in the technical bleaching of wool. Maximum reaction between the wool and sulphite occurs in solutions of pH 5 and most sodium bisulphite solutions are near this figure; with sulphur-stoving, wool in a moist alkaline condition is suspended in an atmosphere of sulphur dioxide which dissolves in the alkaline water to form alkali bisulphites. Stoving therefore becomes a method of obtaining a fairly concentrated

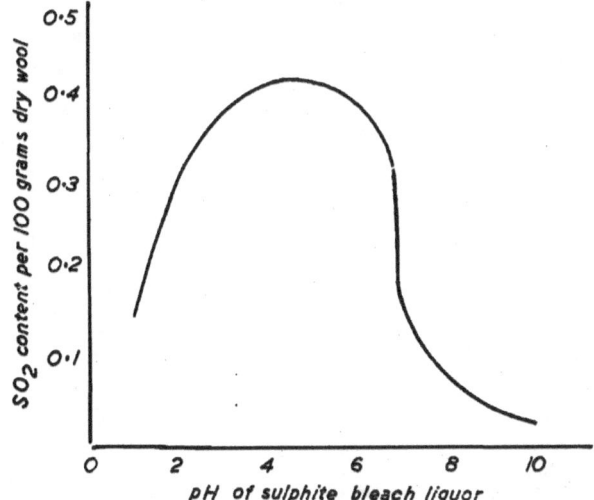

FIG. 102.—Influence of pH of sulphite solution on sulphur dioxide content of wool.

solution of alkali bisulphite in contact with wool. The extent of the reaction is commensurate with that which occurs in 3% sodium bisulphite solution after 17 hours.

The best bleach obtained by Phillips came from solutions of pH 5, and outside the range pH 3 to 7, the bleaching effect was definitely inferior; solutions within this range contain the highest proportion of sodium bisulphite and cause greatest combination with the wool.

In addition to the natural colouring matter in wool, it is also possible that some colour is due to the chromophoric nature of some of the disulphide groups; some organic disulphides are thermochromic, becoming coloured by reason of the strained state of the disulphide linkages on warming. Speakman (J.T.I., 1936, *27*, 231P) has suggested that some of the disulphide linkages in wool are also under strain.

Harris and Smith (Bur. Stand. J. Res., 1936, *17*, 577) have shown that hydrogen peroxide attacks the disulphide group and breaks the linkage; the bleaching action of sulphur dioxide, a reducing agent, and hydrogen peroxide, an oxidising agent, have a common explanation. It should be added that the chief factor which causes the sulphur-sulphur linkage in organic disulphides to be chromophoric is the linkage of both sulphur atoms to large organic radicals.

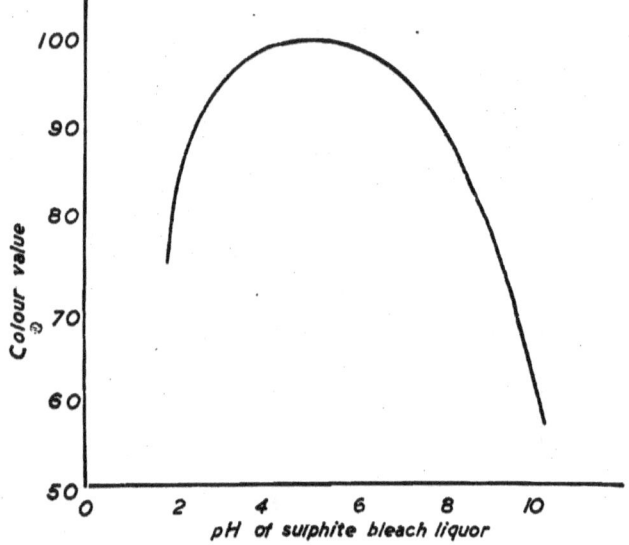

FIG. 103.—Influence of pH of sulphite solution on the whiteness produced during bleaching.

HYDROGEN PEROXIDE

Hydrogen peroxide gives a permanent bleaching effect on wool and is therefore used for the highest-quality material where whiteness is required; the use of peroxide has been facilitated by the commercial production of stable solutions of high concentration as previously explained in discussing the peroxide bleach for cotton.

Two methods may be employed: (*a*) the steeping method in which the goods are immersed in H_2O_2 solution (0·5 to 5 volumes) either at room temperature or at 50° C. for from 1 to 24 hours according to conditions and the type of wool undergoing bleaching; (*b*) an ageing process in which the goods are impregnated in 1·5 to 10 vol. H_2O_2 and then dried at a temperature of 17 to 27°C. In both cases, the bleaching process is followed by thorough washing.

More drastic treatment is required for the very yellow cross-bred wools than for the less-coloured varieties, but in all cases it appears that a really good white can only be obtained with some degradation of the wool itself; actually, hydrogen peroxide gives a pinkish-white which may be improved by a subsequent treatment with sulphur dioxide, either by stoving or bisulphite solution.

It has also been suggested that the pinkish colour of peroxide-bleached wool may be due to over-bleaching.

It is necessary to adopt similar precautions in using hydrogen peroxide on wool as with cotton in so far as the presence of metals is concerned; stainless steel or monel vessels are preferable as wood is gradually disintegrated and further, the solution may penetrate the wood and come in contact with metal bolts, etc., where it is catalytically decomposed, destroying the surrounding wool.

With wool, sodium silicate is preferable to ammonia for neutralising the bleaching bath as it exerts a better stabilising action, although wool itself has an appreciable effect. Weber (J.T.I., 1933, 24, 178P) has described an improved method for wool in which sodium pyrophosphate and sodium oxalate or oxalic acid is used; this agent is available commercially as Stabiliser C. The bleaching bath is made up to 1 or 2 vols. according to the degree of white required and 5·5 lbs. of Stabiliser C is added for every 100 gallons of the bath irrespective of the concentration of peroxide. Bleaching is carried out at 50°C., and not only is the necessary time of treatment less than with other stabilisers, but the white is better and obviates the necessity for a further treatment with sulphur dioxide. This stabiliser not only neutralises the peroxide liquor but exerts a buffering action during the bleaching process.

With the silicate method, the peroxide is brought to pH 7·2 to 7·4 using phenol red as indicator, and in these circumstances a steep for 12 to 15 hours is necessary in 1 vol. H_2O_2 solution at 50°C.; improvements may be made by adding more silicate, using phenol phthalein as indicator, or by employing a higher concentration of peroxide. With piece goods, it is sometimes convenient to work the material on a jig during the bleaching process, as this gives a more even result which may be necessary for material to be dyed subsequently; similarly, hosiery fabric may be treated on a wince machine. In these two cases, the time of treatment may be reduced on account of the more even treatment, and the greater possibility of maintaining a uniform temperature.

Heavy piece goods such as worsteds and gaberdines may need 6 to 12 vol. H_2O_2 for the bleaching bath and after thorough impregnation on the jig, may be batched and allowed to bleach overnight. Blankets are generally bleached in 4 vol. H_2O_2 and again are thoroughly impregnated, batched and allowed to stand for 24 hours or so; if the blankets are dried on frames or stenters, then the drying and bleaching may be arranged to take place simultaneously.

Where the ageing methods at room temperatures are used, it may be advisable to employ a solution at pH 9 to 10 on account of the lower temperature.

The strength of the bleaching bath may be followed by titration with potassium permanganate or by the iodine method; the peroxide bleach permits of the baths being replenished from time to time.

Although it is possible to bleach wool at 40 to 50°C. in 1 to 2 vol. hydrogen peroxide solution at pH 9 in about 2 hours, the conditions of time and temperature are often varied to suit the convenience of the bleacher; hence a great deal of wool, previously scoured with soap, is allowed to bleach overnight. In these circumstances, the temperature of the liquor gradually decreases; peroxide liquor of 1·5 vol. concentration is often used and the goods are entered at about 50°C., but the temperature falls to about 35 or 30°C. in the twelve hours. It will be noted that the so-called stagnant process is adopted, whereby the wool is immersed in the liquor and then neither the liquor nor the material is moved during the treatment; this is somewhat unusual in textile processing. Some relative motion is generally essential if a uniform action is to be obtained, but in this case, the wool is packed somewhat tightly into the liquor so that there is almost complete absorption which ensures an even effect; with a large excess of liquor the tendency would be for the outer surfaces of the woollen material to be bleached more than the rest of the goods, unless there was some movement of the liquor.

The effect of hydrogen peroxide in oxidising wool has been examined by Smith and Harris (Am. Dyes. Rep., 1936, 25, 180). Samples of wool were treated for 3 hours at 50°C. with hydrogen peroxide solutions of different concentrations varying from 0·1 to 10 volume. There was no significant change in wet breaking strain, resilience or total sulphur; cystine and nitrogen contents decreased however.

HYDROGEN PEROXIDE AT 50°C. FOR 3 HOURS

Concn.	Cystine	Nitrogen
0·0 vol.	11·6%	14·23%
0·1 vol.	11·6%	14·19%
0·5 vol.	11·6%	14·14%
1·0 vol.	10·8%	14·17%
2·0 vol.	10·7%	14·18%
4·0 vol.	10·5%	14·17%
6·0 vol.	9·6%	13·99%
8·0 vol.	9·3%	14·04%
10·0 vol.	8·4%	14·05%

A fairly sharp drop may be seen between 4 and 6 vol. concentration and this was accentuated when the treated wool was immersed in $N/10$ NaOH solution at 65°C. for an hour.

Effect of $N/10$ NaOH at 65°C. for 1 Hour

Concn.	Loss in weight	Cystine	Nitrogen	Wet strength
0·0 vol.	12·8%	2·2%	14·03%	0·30 kg.
0·1 vol.	18·3%	2·2%	14·07%	—
0·5 vol.	16·8%	2·2%	13·97%	0·19 kg.
1·0 vol.	17·7%	2·2%	13·88%	0·17 kg.
2·0 vol.	20·0%	2·0%	13·91%	0·13 kg.
4·0 vol.	20·9%	2·2%	14·01%	0·08 kg.
6·0 vol.	41·7%	1·4%	13·83%	—
8·0 vol.	46·1%	1·4%	13·63%	—
10·0 vol.	55·6%	1·3%	13·67%	—

The samples treated with peroxide solution of concentrations greater than 4 vol. became so gelatinous during the alkali treatment that they dried to horny masses to which ordinary physical tests could not be applied.

A second series of experiments was carried out in 2 vol. peroxide solutions for 3 hours at temperatures varying from 23 to 80°C.

Effect of Temperature on Peroxide Solutions

Temperature	Cystine content	Alkali-solubility
23°C.	11·2%	9·7%
35°C.	11·3%	13·6%
50°C.	10·8%	19·4%
65°C.	9·8%	31·6%
80°C.	7·4%	100·0%

It will be seen that there is severe attack on the wool when the temperature of the peroxide solution exceeds 50°C.

The loss of weight in alkali increases with the time of treatment but in an even manner, there being no sharp rise. At 50°C. after 8 hours the loss is 20% and this increases to 40% in 16 hours.

The solutions employed in these two series of experiments were slightly acid, and the results indicate that even under acid conditions it is possible to produce serious damage with peroxide. The results show, however, that it is preferable not to exceed temperatures of 50°C. or concentrations of 4 vol. hydrogen peroxide.

As previously mentioned, however, alkaline conditions are always employed in large-scale bleaching with peroxide.

The effect of pH was examined for solutions of 2 vol. concentration, which were applied to wool for 3 hours at 50°C. There is no appreciable effect of pH below 7, but between pH 7 and 10 the alkali-solubility increases from 20% to 40%; above pH 10, the alkali concentration in the peroxide bath becomes sufficiently high to dissolve wool, so that subsequent alkali-treatment shows less apparent effect.

A commercial bleaching process was examined by Smith and Harris (ibid., 542), the wool being treated in peroxide at pH 9·8 at 50°C.

Effect of Commercial Bleaching

	Treatment	Alkali-solubility	Cystine	Swelling
	None	11·1%	12·8%	12·4%
1st	2·71 vol. for 80 mins.	17·2%	11·7%	14·7%
2nd	2·62 vol. for 90 mins.	20·3%	11·3%	18·6%
3rd	1·70 vol. for 60 mins.	3·14%	10·8%	21·2%

The bleaching treatments were additive, with acid treatment and washing between each bleaching stage. The swelling was measured in $N/10$ NaOH, according to the method of von Bergen (Proc. Am. Soc. Testing Materials, 1935, *11*, 35; Am. Dyes. Rep., 1936, *25*, 542).

From these data it seems that a simple bleaching treatment is capable of producing a relatively large increase in the alkali-solubility of the wool, but without serious changes in the cystine content or in the ease of swelling. Repetition of the bleaching process, brings about serious deterioration of the wool which may be seen, not only from the above data, but by examination of the actual product which, on wetting, exhibits the characteristic slipperiness of damaged wool.

Oxidation

The chemical reactions which take place during the bleaching of wool with hydrogen peroxide have received attention from Smith and Harris (*supra*), and they are of the opinion that in addition to the destructive action on the natural yellow colouring matter, there is also some attack on the cystine or disulphide linkage of wool. In the extreme, oxidation of the cystine linkage could produce sulphuric acid, but the bleaching action is so controlled that oxidation does not proceed to this extent. The amount of oxidation may be roughly determined by testing with lead acetate.

Both cystine and cysteine are well-known compounds; the latter is derived from the former by reduction. Now the formula for cystine is $HOOC.NH_2CH.CH_2.S.S.CH_2.CHNH_2.COOH$, whereas that for cysteine is $SH.CH_2.CHNH_2.COOH$. The sulphur in the grouping —$CH_2.S.S.CH_2$—reacts with lead acetate to give lead sulphide, but oxidised cystine does not give this test except under drastic conditions, nor does wool bleached with peroxide. The reaction is greatly influenced by the presence of acid which hinders the formation of the black colour. Within the pH range of 4 to 6, the bleached wools were much lighter in colour than the ordinary wools, so Harris and Smith (loc. cit.) made use of this phenomenon to determine the

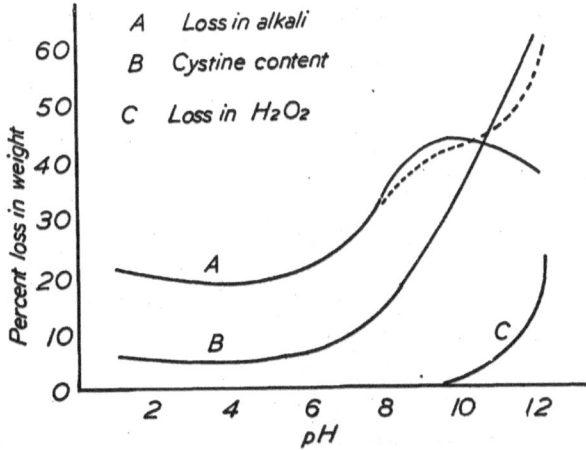

FIG. 104.—Effect on wool of 2 vol. H_2O_2 solutions of different pH for 3 hours at 50° C. (The dotted curve shows the losses due to both peroxide and alkali.)

extent of the oxidation of wool during the peroxide bleach; the colour of the wool was measured by the reflectance of a wavelength of some 650 millimicrons.

The bleaching process given to the wool was one hour in 2 vol. peroxide at pH 9·5 and 55°C.; the alkali treatment was with 0·5% sodium carbonate solution for 15 minutes at 55°C. From the curves of Fig. 105 it appears that within the pH range of 1 to 6, both the untreated wool and the alkali-treated wool behave similarly to lead acetate, but that bleached wool possesses a characteristic behaviour. At pH 5, the blackening is much less with the bleached wool, and this appears to be a suitable pH for determining the effect of bleaching.

From investigation of the action of oxidising agents on cystine and cysteine, apart from wool, other workers have produced evidence

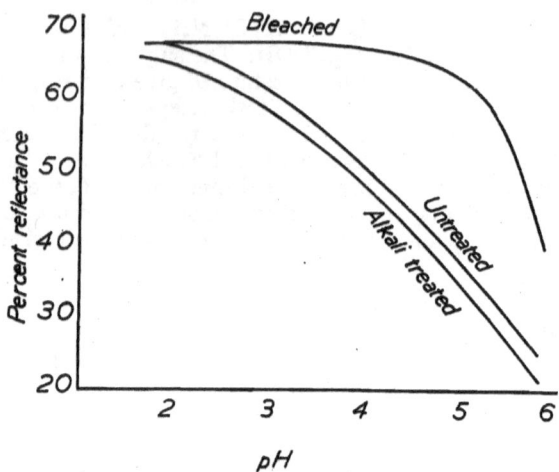

FIG. 105.—Relation between pH of lead acetate solution and the reflectance.

which throws some light on the probable behaviour of the cystine linkage in wool. Oxidation may proceed in the following manner:

$$R-CH_2-S-S-CH_2-R'$$
$$\downarrow$$
$$R-CH_2-SO-S-CH_2-R' \quad (1)$$
$$\downarrow$$
$$R-CH_2-SO_2-S-CH_2-R' \quad (2)$$
$$\downarrow$$
$$R-CH_2-SO-SO-CH_2-R' \quad (3)$$
$$\downarrow$$
$$R-CH_2-SO_2-SO-CH_2-R' \quad (4)$$
$$\downarrow$$
$$R-CH_2-SO_2-SO_2-CH_2-R' \quad (5)$$

The lead acetate test was applied at the boil and under conditions of pH 5·0 and 12·5 to cystine and to cystine disulphide.

$$HOOC-(NH_2)CH-CH_2-S-S-CH_2-CH(NH_2)-COOH \quad \text{cystine}$$
$$HOOC-(NH_2)CH-CH_2-SO-SO-CH_2-CH(NH_2)-COOH$$
$$\text{cystine disulphoxide}$$

No darkening of either substance took place under acid conditions, but at pH 12·5 the cystine darkened at the end of 5 minutes whereas the cystine disulphoxide only darkened slightly after prolonged boiling. Hence it appears that when the cystine linkage in wool has been changed to type (*3*) by bleaching with hydrogen peroxide, it is in a state which resists reaction with lead acetate. It will be noted that cystine itself only darkens with lead acetate solution under alkaline conditions; this does not apply to wool, although the darkening occurs more readily when the wool is alkaline. This may be due to the following sequence of reactions:

$$R-CH_2-S-S-CH_2-R'$$
$$\downarrow \text{alkali}$$
$$R-CH_2-SH \quad HOS-CH_2-R'$$
$$\downarrow$$
$$SH_2 + R'CHO$$
$$\downarrow \text{lead acetate}$$
$$PbS.$$

Ryberg (Am. Dyes. Rep., 1940, *29*, 588) has extended this test by replacing the optical examination of the material by a determination of the insoluble lead-sulphur compound, as described on page 479.

Owing to the grid-structure of wool, it is not possible to exercise the fluidity control of bleaching which has proved so successful in the case of the vegetable fibres and also with silk. Wool may be damaged by agents which break the cross-linkages without affecting the main peptide chains, and the cleavage of these linkages appears to be a necessary preliminary to dissolution; hence, questions of hydrolysis of the main peptide chains apart, it is not possible to obtain solutions of the wool substance without considerable degradation. The low viscosity of these solutions, from which the wool cannot be regenerated, points to hydrolysis of the peptide chains also.

Elöd, Nowotny and Zahn (Textilber., 1942, *23*, 313) have examined the action of hydrogen peroxide on wool. Treatment with 2·4% H_2O_2 solution at 80°C. for 8 hours transforms all the cystine in wool, the sulphur content is lowered, and free H_2SO_4 appears in the liquor. As increasing amounts of nitrogenous compounds of low molecular weight are formed and dissolved, it appears that some breakdown of the main polypeptide chains occurs. The free amino-groups and the salt links are also reduced in number. The oxidised wool also becomes increasingly acid, as shown by a shifting of the isoelectric region, an increased affinity for alkali, and a greatly increased solubility in alkali.

Although processes have been protected for bleaching wool with peroxide under acid conditions, it is very doubtful if they are used on any commercial productions.

The fact that alkaline peroxide is invariably employed may account for the fact that there is some impoverishment of the wool at the same time as the destruction of the natural colour. The attack on the wool substance is quite moderate and is roughly similar to that which takes place during the bleaching of vegetable fibres under careful control with hypochlorite solutions. The bleached product is satisfactory in its mechanical properties even though these are affected according to the degree of bleaching.

Considerable quantities of wool are "chlorinated" in order to produce non-felting properties; the general procedure is to use about 3% of chlorine on the weight of the goods, for an hour at pH 4·5. Careful control is necessary, for over-chlorinated wool is apt to have a yellow colour.

Trotman and Trotman (J.S.D.C., 1926, *42*, 154) have investigated the bleaching of chlorinated wool and found there was an increase in damage proportional to the pH of the peroxide solution; there was no damage with unchlorinated fibres provided that the alkalinity of the solution did not exceed pH 10. The loss in weight during bleaching in 2 vol. H_2O_2 for 4 hours at 55°C. was greater with chlorinated wool; in general, the best colour and least damage is seen at pH 10, over which there is a very great increase in the damage to chlorinated wool.

CHAPTER XXVIII
SILK DEGUMMING

Raw silk possesses neither the lustre nor the softness commonly associated with silk; these properties do not appear until the raw material has been scoured. The scouring process is generally referred to as degumming, boiling off, and occasionally as stripping; in point of fact, the term "boiling-off" is a misnomer for the silk is not boiled.

The raw silk is composed in the main of two materials, fibroin which constitutes the fibre proper, and sericin of which the silk gum or silk glue is formed; the chief function of the scouring process is to remove the silk gum or sericin and it is based on the fact that the sericin is soluble in hot solutions of soap.

When examined under the microscope, the appearance of the degummed silk is very different from that of the two filaments of the raw silk, as discussed on page 56.

IMPURITIES

Whereas sericin is the chief natural impurity associated with the silk fibre or filament, there is also a small amount of silk wax and coloured pigment; the nature of the pigment varies with the type of silk, and that from the Italian silkworms is usually bright yellow. Many forms of silk contain acquired impurities which are mainly the oils with which the silk has been treated during the throwing process.

SOAKING OILS

As previously mentioned, the continuous thread or bave from the cocoon of the silkworm consists of two filaments or brins which are cemented together by the sericin or silk gum. The bave may be about 2,000 metres in length of which about one half can be reeled, after which the remainder is broken into short fibres and spun like cotton. At the filatures, the cocoons are softened in warm water and six or more baves are laid side by side as they are unwound; the threads adhere again as the gum hardens to give the commercial raw silk product.

The quality and strength of the raw silk threads is still inadequate for knitting or weaving so they must be subjected to the process of throwing which consists of rewinding on suitable holders and doubling or twisting the threads together. Before throwing, however.

the raw silk is generally given a preliminary soaking treatment which consists in a treatment with an emulsion of soap and oil in order to render the silk more supple and facilitate weaving and knitting. It is essential that the oils used in this process should not only possess good lubricating properties, but have little or no tendency to oxidise; they should also be readily removable in the degumming process.

The usual method of soaking silk is to employ an emulsion of oil and soap, containing about 1% of each component, in which the silk is soaked for 16 to 24 hours at a temperature of 32 to 33°C. One of the chief effects of the soaking process is the absorption of water but there is also some absorption of the oil. Alkali carbonates are often added to the emulsion in order to neutralise the free fatty acid produced by hydrolysis of the soap and absorption of the alkali by the sericin. The oils commonly employed are olive oil and neatsfoot oil, but partially saponified oils are sometimes used; blends of sulphonated oils and mineral oils have been utilised, and a large number of proprietary preparations is marketed.

Hart and Searell (Ind. Eng. Chem., 1930, *22*, 980) have examined the behaviour of raw silk when soaked in soap and oil emulsions; the sericin combines with some of the alkali from the hydrolysis of the soap, but the silk itself appears to absorb the free fatty acid and also the soap. Hence, during the soaking process, the concentration of the fatty acid, and more especially the soap, is reduced in the soaking solution.

The reaction seems to take place quite quickly as the acidity of the soap solution is the same after soaking for 15 minutes and for 10 days. The amount of sericin which dissolves in the solution is rarely 1% of the weight of the silk; the alkali which combines with the sericin was equal to 21% of the total alkali in the bath, but this does not represent the total amount of alkali which will combine with the sericin as the liberation of the free fatty acid represses hydrolysis of the soap. The presence of additional alkali in the soaking bath brings about further combination with the sericin.

As pointed out by Snell and Kimball (Am. Dyes. Rep., 1931, *20*, 79) in a discussion on the soaking of silk, Hart and Searell (*supra*) used 2·25% of soap which is higher than that employed in commercial practice, so that the loss of sericin would be less than 1%. The average concentration of soap is usually nearer 1%, and it has been established that a minimum soap concentration gives maximum softening.

The stability of the emulsion is also affected by the amount of soap, smaller concentrations giving better dispersions; this may account for the improved softening effect. It may be that whereas the soap reacts with the sericin and is adsorbed by the silk, yet its prime function is that of an emulsifying agent for the oil and not as a softening agent for the silk.

The oil does not appear to react in any way with the silk, but is deposited on it; about 20% of the neutral fat in the emulsion is deposited on the silk.

Although soap gives a better effect than sulphonated oils as dispersing agents, yet the latter are often preferred on account of a superior dispersion and a more stable emulsion; the use of triethanolamine is useful in this respect.

Examination of the effect of alkali has shown that the plasticity of the silk increases up to pH 10, so that the presence of a limited amount of alkali is useful for increased plasticity as well as the neutralising of free fatty acid from the soap. The alkalinity of the bath is always reduced during the process, and if initially in excess of pH 9·7 there is a tendency for the silk to become harsh. In addition, the presence of an excess of alkali renders the emulsion unstable.

Many throwsters limit the amount of free fatty acid in the soaking oil to less than 0·1% of the weight of the silk, to obviate rusting the needles of the knitting machines; much more than this amount of acid seems to be liberated during the soaking process, so that the free acid of the soaking oil can only exert a comparatively slight effect. A popular soap is that from olive oil or from oleic acid; the tendency to oxidation may be reduced by suitable anti-oxidants. The oil in the soaking mixture may be olive oil, neatsfoot oil, sulphonated castor oil or coconut oil.

The soaking of the silk must not be allowed to proceed too long, but the silks with hard gums will naturally require a longer time than the soft-gum silks.

Johnson (Text. World, 1936, *86*, 85) has suggested the use of an aqueous solution of a soluble oil, a wetting agent, a plasticiser, a mildew-preventive and potassium carbonate.

A method adopted in France is to use about 1 to 3 lbs. of potassium carbonate and 15 to 30 lbs. of oil per 100 lbs. of silk; the soluble oils give better penetration than can be realised with emulsions.

Some of the more recent developments in the application of soaking oils depend on the use of the oil in a negatively charged emulsion to which a discharging agent may be added slowly to exhaust the bath; suitable discharging agents are ammonium sulphate or fluosilicate. Sodium lactate has also been used in presence of sulphonated oils (B.P. 477,066), whilst the general application of ammonium salts to emulsions of soaking oils helps to maintain the correct pH (B.P. 485,398).

In general, there are three methods whereby the soaking oils may be applied: (*a*) immersion in open standing baths, (*b*) circulation, and (*c*) forced circulation.

SERICIN

The removal of the sericin or silk gum from the fibroin or filament proper, is based on solubility in hot aqueous solutions; the solubility increases considerably with acidity or alkalinity of the solution.

The first serious publication on the nature of sericin seems to be *La Dissolution du Vernis de Soie* (Rigaut de Saint Quentin, 1784), when the possible existence of various fractions of sericin was discussed. Cramer (J. prakt. Chem., 1865, *96*, 76) extended the scope of previous investigations and obtained several proteins by fractionation of the sericin; he suggested that these fractions may be decomposition products of the natural material. Subsequent workers came to the conclusion that the sericin of raw silk exists in more than one modification, although the evidence would not be acceptable to modern organic chemistry; different methods were utilised by various investigators and widely different results obtained. Anderlini (Chem. Zentr., 1898, I, 795) separated three fractions after extracting raw silk with water at 50 to 60°C., whereas Bondi (Z. physiol. Chem., 1901, *34*, 481) reported two fractions from sericin obtained by boiling raw silk in water. Shelton and Johnson (J.A.C.S., 1925, *47*, 412) treated raw silk in an autoclave at 115°C., and were able to isolate two fractions, one of which appeared to be soluble in cold water whereas the other fraction was insoluble. The findings of numerous later investigators seemed to substantiate this work, and attempts were made to characterise the two fractions. Mosher (Am. Dyes. Rep., 1932, *21*, 341) considered the two fractions to be very similar in chemical properties, but to differ in physical properties, particularly solubility. Sericin A is soluble in hot water, and is readily attacked by enzymes, sericin B is sparingly soluble in hot water, and resists proteolysis, and a third fraction (Canad. Text. J., 1934, *51*, 31), which Mosher termed sericin C, is insoluble in the common solvents and is unaffected by enzymes. The insoluble fraction may be made soluble by boiling in aqueous solution, and the conversion is accelerated by increasing the temperature, acidity or alkalinity of the solutions. Many of the results may be interpreted to show that sericin does not necessarily exist as a mixture of proteins in the raw silk but that the various fractions may be decomposition products.

Investigators of the Japanese school consider that the various fractions differ chemically, and their results are mainly based on determinations of tyrosine contents and nitrogen distribution. Some investigators have found more nitrogen in the insoluble fraction, whereas others found the soluble portion of the extract to be richer in nitrogen.

Rutherford and Harris (Am. Dyes. Rep., 1940, *29*, 213) have recently examined the question of the various fractions of sericin; the naturally-occurring waxes were first removed from the white Japanese silk by extraction with alcohol and ether at room temperatures for 24 hours. The silk was then placed in a beaker of water at 100°C. in an autoclave and steam admitted until the temperature increased to 114–115°C. for the duration of the treatment. The fraction which was precipitated at pH 4 was separated from the remainder by a centrifuge, washed with alcohol and ether, and dried; this gives sericin B. The remaining solution was evaporated to a small volume and the protein in the concentrate precipitated by the addition of sufficient 95% alcohol to give a solution of 75% alcohol at about 0°C.; the fraction soluble at pH 4 but insoluble in 75% alcohol is sericin A. Rutherford and Harris also found a small fraction which was soluble in 75% alcohol and this was termed sericin D. It will be remembered that Mosher (Am. Dyes. Rep., 1932, *21*, 341) reported the existence of another fraction, sericin C, which was described as extremely insoluble.

Rutherford and Harris found that 80% of the total sericin is removed in 20 minutes and complete removal occurs in an hour. The yields of A and B vary with the time of treatment; after 5 minutes treatment, 97% of the sericin removed was in the B form, but after 3 hours treatment only 18% of the sericin removed was recovered in this form. The low yield of 2·7% sericin A in 5 minutes increased to 77·4% in 3 hours.

Sericin A and B are thus substances whose composition varies with the time of treatment; the bulk of the sericin is removed in the B form, and the increasing yield of the A form with time of treatment must result from decomposition of sericin B after its removal from the silk. Hence it is concluded that the sericin in raw silk does not exist as a mixture of fractions, but that the fractions are artifacts which result from the hydrolytic decomposition of the natural sericin during its removal.

The increase in the yield of the very soluble fraction D with increased time of autoclaving indicates extreme decomposition of the sericin; the chief interest in this fraction lies in the relatively high carbohydrate and amino-sugar contents—about three times as much as the original sericin.

It is also interesting to note that the conversion of B to A is accompanied by an increase in amino-nitrogen content, sufficient to account for an appreciable amount of hydrolysis.

WAX

The fact that raw silk contains about 0·5% of a waxy substance soluble in alcohol, was mentioned by Roard (Ann. chim., 1808, *65*, 63) in his investigation of the composition of silk; the work was extended by Mulder (Ann. Phys. u. Chem., 1936, *113*, 610), who stated that the ether-alcohol extracts of raw silk consist of a mixture of wax, fat and resin, but the yellow silk also contains pigment.

Later investigators estimated the ether-alcohol soluble portion of raw silk to be between 1% and 1·5%, that of cocoon silk lying 2·8 and 3·2%, but here there are also water-soluble substances in addition to the wax.

Oku (Bull. Agric. Chem., Japan, 1928, *4*, 123) considered that silk wax resembled beeswax, and contained alcohols such as ceryl alcohol, fatty acids such as melissic acid, together with some hydrocarbons. Bergmann (Text. Res., 1938, *8*, 221) extended these observations with the following results:

ETHER EXTRACT OF RAW SILK

Primary alcohols	35%
Hydrocarbons	25%
Higher fatty acids	11·5%
Lower fatty acids	20%

Apart from the ill-defined liquid acids, the ether-extract of raw silk mainly consists of normal primary alcohols, acids and hydrocarbons. Waxes of this type are known to be resistant to chemical attack, and Mulder (*supra*) has pointed out that silk wax is not readily affected by warm concentrated nitric or hydrochloric acid.

It does not appear to be certain whether the silk wax is synthesised by the silkworm or if it is derived from the mulberry leaf which is the diet of the worm. The waxes consist mainly of hydrocarbons C_{25} and C_{31}, together with esters of alcohols and acids C_{26} and C_{32}, and the shiny surface layer of the mulberry leaf contains the same components.

Bergmann (Text. Res., 1939, *9*, 175, has found that the silk fibres which form the outside layers of the cocoon contain the highest concentration of wax, and it may be that the importance of the wax lies in conferring impermeability upon the underlying organic matter. Estimates of the wax-like material soluble in ether, obtained from white Chinese silk, give an average figure of 0·54%.

Some oily and fatty matter may also be extracted, and Bergmann prefers to describe the whole extract as the "ether-soluble number" which is the number of milligrams of ether-soluble material present in 10 grams of the raw silk. The average number for white Japanese silk was 55, but yellow silks, whether of Italian or Japanese origin, gave numbers between 90 and 100; the ether-extracts of the coloured silks contain some pigment in addition to wax.

There appears to be no relation between the ether-soluble number and the various physical properties of the silk. Evaluation of the wetting-time by the Draves test (see page 144) before and after removal of the wax demonstrated its influence on the wetting properties of the silk; it may be remarked, however, that the wettability of the silk is determined by the distribution of the wax rather than by the absolute amount, although in general, the white silks wet faster than the yellow.

PIGMENTS

It is well known that different varieties of silkworms produce coloured cocoons; the domesticated *Bombyx mori* can be divided into types which spin white, yellow or yellow-green silk, and the majority of wild silkworms produce material which may range from light brown to black-brown. From the above facts, it is possible to divide the natural colouring matters of silk into yellow, green and brown pigments.

The natural pigments of silk have been studied in some detail by Bergmann (Text. Res., 1939, *9*, 397).

The yellow pigments were the first to attract the attention of chemists because of wide distribution. The pigment does not make its appearance in the gland until a few days before spinning, and as silkworms which spin yellow silk possess yellow blood, it was assumed that there was some relation between the pigments of blood and silk. As a result of many and varied experiments, two points of view developed, one of which held that the colour came from the leaves of the mulberry tree, whereas the other maintained that the colour was produced by the silkworm. The yellow pigment was first isolated in the pure crystalline state by Oku (Bull. Agric. Chem. Soc., Japan, 1929, *5*, 81; 1932, *8*, 7 and 89; 1933, *9*, 91), who identified it as xanthophyll (lutein); a small amount of carotin in various oxidised forms was also found. The xanthophyll content of normal yellow silk is 0·022 to 0·024% and the carotin content is 0·0006 to 0·0008%; the intensity of the colour appears to depend on the concentration and combination of the carotinoid pigments. Oku also isolated xanthophyll from mulberry leaves and proved the identity of leaf and silk pigments, from which it appears that the yellow pigment of silk is derived from the mulberry leaves.

Oku appears to be of the opinion that the pigment is combined in a physico-chemical manner with the sericin. Prolonged extraction with solvents for carotinoids is not sufficient to remove all of the pigment and some form of bleaching is necessary for complete removal of the colouring matter.

The yellow pigments are generally unevenly distributed in the various layers of the cocoons; the first silk secreted by the worm is highly pigmented so that there is a gradual decline in pigmentation from the outer to the inner layers of the cocoon.

Green pigments are produced in certain Japanese types of *Bombyx mori*. Jucci and Manunta (Boll. soc. ital. biol. sper., 1932, *7*, 162) concluded that the pigment was closely related to the flavones, whilst Oku (Bull. Agric. Chem. Soc., Japan, 1934, *10*, 158) isolated a green pigment which gave all the typical colour reactions of the flavones and was a monoglucoside; its absorption spectrum resembles that of quercetin. Oku also found that mulberry leaves contain another flavonol glucoside which is identical with isoquercetin, so that it may be that the isoquercetin is absorbed by the silkworm and changed biochemically into the similar pigment in the cocoon.

Brown pigments are widely distributed among the silks of the wild silkworms; a striking difference from the yellow and green pigments, however, is shown in the fact that the brown colour does not appear until a day or two after the completion of the cocoon. The change of colour is due to an excretory liquid with which the worm wets the cocoon, and if the anus of the worm is sealed during spinning, then the silk remains white. White silkworms frequently utilise excrements in making the cocoon, but the *Bombyx mori* undergo a thorough purgation before starting to spin their cocoons.

The presence of an excretory fluid is not essential for the formation of colour, and it may be replaced by water or glycerol, as shown by Dewitz (Zool. Anz., 1904, *27*, 161). The fresh white cocoons of some wild silkworms turn brown when placed in water containing air or oxidising agents, and this colour change may be impeded by the presence of hydrogen or reducing agents. If treated with boiling water, the cocoons remain colourless but the extract becomes brown. Observations of this type indicate the presence of a soluble chromogen which becomes brown on enzymatic oxidation. Przibram and Schmalfuss (Biochem. Z., 1927, *187*, 467) concluded that the chromogen was related to o-dihydroxy benzene; they actually isolated 3 : 4-dihydroxy-phenyl-alanine from certain cocoons and showed it was responsible for the development of brown colour.

Many attempts have naturally been made to obtain silks of various colours by feeding the worms with certain dyes. Indigo only soils the silk, and worms which were washed before starting to spin gave white cocoons. Blanc (Compt. rend., 1890, *111*, 280) and Dubois (Lab. d'Etudes de la Soie Lyon, 1890, *5*, 347) carried out feeding experiments with about thirty different aniline dyes but only obtained white silk. There is evidence, however, that certain dyes such as neutral red and orcein may be absorbed and passed on to the silk, as shown by Vaney and Pelosse (Compt. rend., 1922, *174*, 1566).

The production of coloured silks does not seem to have any practical outcome; experiments with dyes by feeding or injection have been described by Campbell (J. Econ. Entomol., 1932, *25*, 905) for the identification of silkworms in biological assays.

DEGUMMING

The common method of removing the gum from raw silk is to treat the fibres in a moderately concentrated solution of soap. The degumming operation is sometimes termed "boiling off," but in actual fact, the silk is not boiled, and the temperature is, in general, maintained at about 95° C., for the act of ebullition is apt to disturb the fine filaments and bring about entanglements and damage; some chemical damage may also result from the higher temperature.

If the whole of the gum has to be removed, it is necessary to prolong the process, and also to use fresh solutions of soap. The concentration of the soap in the bath and also the number of baths are determined by the amount of sericin which has to be removed.

The amount of silk gum, or sericin, may vary from 22 to 30%, so that its complete removal will result in a corresponding loss in weight. For some manufacturing purposes, however, this is undesirable, and qualities of silk have been classified as

(a) Ecru, from which only 2 to 5% of sericin has been removed;
(b) Souple, from which 8 to 15% of gum has been removed;
(c) Boiled-off silk, which has been completely degummed.

It will be noted that the "boiling-off" process in silk corresponds to the "boiling-out" or scouring of cotton.

Ecru silk is generally simply scoured in a weak soap solution in order to remove fatty and waxy constituents without appreciable loss of cericin; it resembles raw silk in feel and appearance and is generally employed in warps where the friction of weaving is resisted by the relatively harsh fibres.

Souple silk is much softer and fuller, as more of the sericin has been removed; it is therefore suitable for wefts.

Most silk is degummed in hank or skein form, but for certain fabrics where weaving in the gum is undertaken, the degumming operation takes place in the form of piece goods; in both cases, however, the method is very similar with the exception of the machinery and manipulation. Similar considerations apply to the large quantities of silk degummed in the form of hosiery.

As the only agents which are used in actual practice for the degumming of silk are substances which give mildly alkaline solutions, it is believed that the basis of the process is due to the formation of soluble salts by reaction between the carboxyl groups of the sericin and the alkali; even with soap solutions, the action is due to the alkali liberated from the soap by hydrolysis.

The ideal degumming agent, therefore, would be an alkaline reagent which attacked the sericin without affecting the fibroin which is also built up of amino-acids and hence susceptible to alkaline hydrolysis. For this reason, the degumming process must be confined to rather narrow limits of alkalinity, time and temperature. It must not be assumed that all the sericin reacts with the alkali to form a salt, for some of it passes into colloidal solution.

The fact that a considerable amount of salt formation does in fact take place may be shown by the fall in pH value of the degumming bath during the process; the extent of the decrease depends on the alkalinity of the bath and the quantity of degumming agent.

The chief degumming agent is soap, which is formed from a strong alkali and a weak fatty acid; hydrolysis takes place in water with the formation of free fatty acid and sodium hydroxide, until equilibrium is reached. As the sericin dissolves, some of the alkali is used up in forming the amino-acid salt, so there is a fall in the pH of the liquor. The sericin salt also acts as a buffer.

In the actual degumming process, a great deal depends on the type of silk and the nature of the yarns; in addition to the sericin, there are also certain acquired impurities such as soaking oils and the sighting colours used for the identification of various yarns. The ingredients of the various soaking oils, nowadays, include soaps and allied products which facilitate rather than hinder the removal of sericin. Attention should be drawn to the fact that yarns of high twist resist the action of the degumming bath to a greater extent than yarns of low twist; the penetration is less, and so the extraction of sericin is not so efficient. Any unevenness in degumming is apt to be revealed by irregularity in the subsequent dyeing. It also seems that the final 5 to 10% of gum is very difficult to remove.

Soaps

Soap is the basis of most degumming baths; it is preferably used in soft water and sometimes small amounts of alkali are added. There has been a recent tendency to give the silk a preliminary treatment through a solution of a wetting agent at 65°C. to soften the sericin and render it more susceptible to removal during degumming.

The best soaps for degumming silk are made from olive oil, which is now somewhat expensive, so that the poorer oleic acid soaps have been used to a greater extent than heretofore; tallow soaps are not so easy to use on account of the difficulty of removal from the silk after degumming. Soaps from cotton-seed oil or coconut oil have sometimes been suggested, but they are apt to leave an unpleasant odour on the material; the addition of thiourea, according to B.P. 499,371, retards the oxidation of many of these unsaturated compounds. Where the neutral olive oil soap can be used, together with small amounts of alkaline reagents such as sodium carbonate, phosphate or silicate, the pH of the liquor should not fall outside the range of 9·2 to 10·5.

The concentration of soap is generally not less than 20% of the weight of the material and often higher amounts are employed corresponding to 30 to 50% of the weight of the silk. As the pH of 1% soap solution is often around 10, care should be taken to

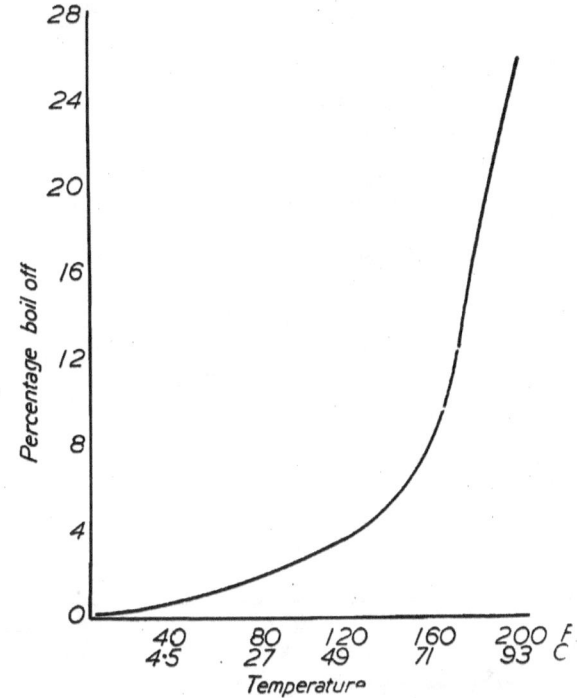

FIG. 106.—Effect of temperature in degumming silk.

ensure that the concentration of soap in the liquor is not too low or the pH of the solution will be below the optimum and degumming will be retarded.

Unless the concentration of soap is adequate to remove and to retain the impurities in suspension, there is always a danger of the scum and gummy matter being deposited on the silk again; this danger may be obviated by the addition of a sulphated alcohol which acts as a dispersing agent.

The effect on the degumming process of temperature, time and the amount of soap is shown in Figs. 106, 107 and 108, taken from publications of the Textile Research Laboratories of Proctor and Gamble (Paterson, New Jersey).

The temperature of the degumming bath generally lies between 90 and 95°C.; higher temperatures are apt to bring about some yellowing of the silk. A further difficulty of high temperatures is

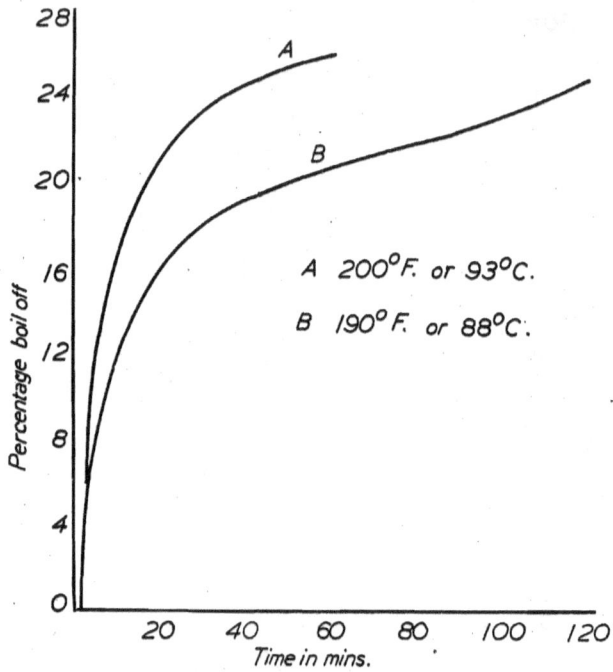

FIG. 107.—Rate of degumming raw silk.

that ebullition leads to disturbances of the material accompanied by chafing and lousiness. One of the difficulties encountered in degumming is that the reagents not only remove the sericin and free the parallel filaments of fibroin, but they also penetrate the filaments, so that with an uneven structure there is a danger of separating the fibrillae; further processing may break them, bringing about the exfoliation known as "lousiness." The term is probably due to the fact that the broken filaments are apt to be a much lighter shade than the rest of the material after dyeing. Protective colloids, such as protein decomposition products, are often added to scouring solutions to minimise the effect of alkali and also of prolonged

SILK DEGUMMING

boiling; some of the sulphated fatty alcohols may also be used as their lubricating properties tend to reduce lousiness.

The foam process for degumming (see page 392) is also of interest in connection with lousiness. Investigation of the Schmid method showed that in 1,000 metres of silk degummed by the usual process lousiness occurred 1,010 times compared with 60 times for the foam treatment.

Returning to the use of soap solutions, it will be remembered that

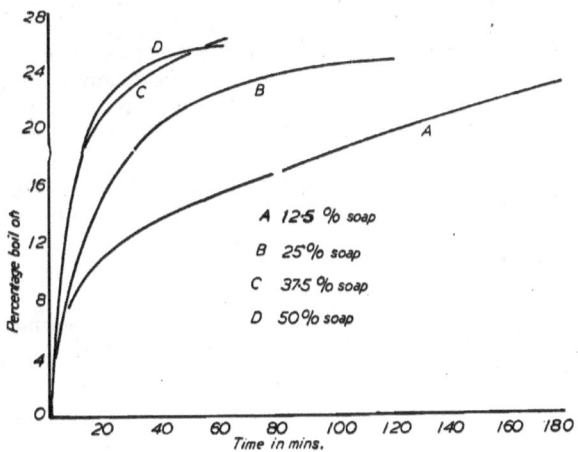

Fig. 108.—Influence of time and concentration of soap in degumming raw silk.

the function of a detergent is not merely to remove the impurities by emulsification and dispersion, but also to prevent their redeposition. This fact accounts for the use of successive degumming baths whereby the first in the series is discarded as it becomes saturated. The alkaline solution of sericin possesses protective-colloid properties and is often used in the preparation of dye-baths. It is also of value in dispersing lime soaps. The "boil-off" liquor can be treated with lime water to precipitate the fatty acid as a calcium soap from which the free fatty acid may be recovered; the lime may be replaced by aluminium sulphate.

As will appear later, where successive baths are employed for degumming, on a counterflow basis, the later baths are the cleaner, so the silk is progressively purified (see page 391).

Soap appears to be the best degumming agent because it supplies the alkali required for the reaction with sericin at a rate which is sufficiently slow to avoid any appreciable damage to the fibroin;

an additional feature of soap, as previously stated, lies in its powers of suspending the impurities in the liquor. Olive oil soaps and those from oleic acid have several advantages; they have a low rate of hydrolysis and a high solubility which facilitates removal from the fibre after the degumming process.

According to Tsunokaye (see J.S.D.C., 1928, *44*, 142), the degumming powers of a soap are proportional to its degree of hydrolysis. The work of Stockhausen (Seide, 1932, *37*, 387; Rev. Gén. Mat. Col., 1933, *37*, 95) has shown that the effective degumming agent in soap solutions is the alkali liberated by hydrolysis, so that the soap solution functions as a continuous source of alkali. The synthetic detergents which do not furnish alkali have no degumming action, and may even reduce the efficacy of soap by repressing hydrolysis; it was also shown that the fat-solvent soaps were not as good as the ordinary types.

In view of the fact that the soaps function as degumming agents in view of the liberation of alkali by hydrolysis, it is of interest to compare the different soaps.

HYDROLYSIS OF SOAP

Soap	pH of 0·1% solution
Na stearate	9·6
Na palmitate	9·6
Na myristate	8·1
Na laurate	7·5
Na oleate	8·15

From these data it seems that sodium laurate is the least efficient, but there is no great difference between the others. However, the efficiency of actual degumming is only one of the factors to be considered in the choice of a soap.

Alkali

During the degumming operation, the alkali is consumed by the sericin and hence an acid soap is liberated which is adsorbed by the fibre protecting it from hydrolysis to some extent; in time, however, the adsorbed acid soap will retard the rate of degumming.

Although alkali is frequently used to maintain or restore the pH of degumming baths, its use alone is rarely practised in spite of suggestions to the contrary; alkalis leave the silk with a thin, harsh and yellow appearance, and also tend to reduce the lustre, strength and elasticity of the silk itself. Whereas there appears to be no theoretical reason why the use of a suitable buffer mixture, such as sodium carbonate and bicarbonate, should not provide a solution whose pH falls within the required limits, yet, as previously stated, it is also necessary to keep in dispersion the metallic soaps, soaking oils and impurities; hence soap possesses advantages over buffered alkaline solutions.

Where printed silk fabrics are considered, the complete absence of gum is desirable, and this may be assured by the addition of a little sodium carbonate to the soap solution; a similar addition is valuable for degumming hard-twisted crepe yarn where the prolonged treatment in pure soap solution is apt to have an adverse effect. The addition of sodium carbonate provides a manifold increase in the rate of degumming, but care should be taken to ensure that the concentration does not exceed 0·25% in a 1% soap solution.

Alkaline baths whose pH exceeds 11 are definitely dangerous on account of the attack on the fibroin. Alkali may be used with soap, however, possibly on account of the buffering action of soap and sericin; as previously stated, alkali is often used to restore the pH of used baths. It has been suggested that the pH of the solution is a direct measure of the rate of degumming, but there is some difference of opinion on the matter; there seems to be little doubt, however, that pH affords a useful control of degumming.

The use of small amounts of sodium carbonate or phosphate entails less risk than the addition of sodium hydroxide should it be decided to add alkali to the degumming bath. Toyoda (J.S.C.I., Japan, 1932, *35*, 43) has stated that the use of alkali in conjunction with soap is better than the use of soap alone; he has also produced evidence that in the degumming process it is possible to leave silk for an hour in 0·001 N NaOH or in 0·0067 N Na_2CO_3 solution, together with soap without adverse effects. Toyoda (ibid., 1933, *36*, 368) has also shown that the sericin may be removed from silk without damage to the fibroin by boiling in 0·001 N NaOH or 0·007 N Na_2CO_3 solution with the addition of a little soap. It was later claimed (ibid., 1934, *37*, 150) that the use of sodium hydroxide or potassium carbonate with the soap gave better strength and extension data on silk than did the use of soap alone. In this connection, it should be remembered that the work of Stockhausen (Seide, 1932, *37*, 387; Rev. Gén. Mat. Col., 1933, *37*, 95) showed that alkali in concentrations over 0·02 N at moderately high temperatures definitely reduced the strength and extension of silk. Hart (Melliand Text. Monthly, 1933, *5*, 68) has stated that the action of soap in the degumming process may be increased by the addition of small amounts of sodium sulphate.

Assistants

Comparatively little information has been published on the use of surface-active agents in degumming, although it has been known for some time that an alkaline solution of such a product could be used in the degumming and dyeing of silk.

Morgan and Seyferth (Am. Dyes. Rep., 1940, *29*, 616) have described experiments in the degumming of silk hosiery with various concentrations of soap, alkali, soap and alkali, the alkyl aryl sodium sulphonate known as Nacconol alone and with alkali; the alkalis used were sodium silicate and sodium sesquicarbonate. The use of alkali alone is not feasible, for difficulties arise with emulsification and a heavy scum is apt to form. When alkali and soap are used together, the presence of alkali is apt to have an adverse effect on the handle of the silk; as previously indicated, soap alone is a very good degumming agent.

The use of the surface-active agents (Nacconols) alone gave a poor degumming, probably on account of the lack of alkalinity, but when used in conjunction with alkali a very good effect was obtained.

Roberts (Textile World, 1931, *79*, 2684) has given a useful list of degumming agents which have been suggested at various times. Naturally, soap is the most important of these, and low titre soaps are preferred.

Soaps of high titre are those in which the melting-point of the fatty acid is above 35°C. Soaps of this type form a gel at concentrations of the order of 2 ounces per gallon (1·25%) and generally give a good lather.

Soaps of low titre are those in which the melting-point of the fatty acid is 25°C. or less; they are soluble in lukewarm water and form copious suds at temperatures below 60°C. They are readily removed from textile materials by rinsing with water.

The liquid soaps are usually potassium soaps of low titre; they are often slightly alkaline and may be useful for degumming hosiery of high-twist yarn.

Nearly all silk hosiery is degummed with neutral soaps of low titre or with proprietary preparations of various types; the chief products may be classified as follows:

(a) Solvent preparations in which organic solvents are used in conjunction with sulphonated oils.

(b) Cresylic acid products, where the cresylic acid (a mixture of ortho, meta, and para cresol) is used in conjunction with sulphonated oleic acid or sulphonated castor oil.

(c) Sulphonated oils which are brought to the required pH by the addition of caustic soda or sodium carbonate.

(d) Silicated products, in which sulphonated castor oil is combined with sodium silicate to give the correct alkalinity.

The solvent preparations are useful for the removal of oil stains and grease spots; local application is the usual procedure for this purpose. If employed for degumming they are advantageous in retaining the impurities in solution, but the solvent should not be

SILK DEGUMMING

volatile at the boiling-point of the degumming bath; otherwise, it may be preferable to soak the silk in the solvent preparation before degumming with soap.

The use of sulphonated oils without silicate, as applied to hosiery, necessitates very strict control as the usual composition of sulphonated castor oil and excess sodium hydroxide is not buffered. Sulphite waste liquor is sometimes added. This type of oil is not used with piece goods.

Sulphonated oils and silicates are the oldest form of liquid degumming agent; the chief application is for hosiery. The silicate exerts a buffer action maintaining the pH between 10·5 and 9·5; without silicate, the sulphonated oil and caustic soda soon fall from pH 12 to pH 8.

General

Returning to the consideration of degumming with soap, it is general to base the amount required on the weight of silk; 20 to 25% is common. Hard-twisted yarn and a close fabric construction, however, will necessitate a higher concentration and even 50% has been suggested for certain types of hosiery. The ratio of liquor to silk may vary from 10 : 1 to 50 : 1.

Mosher (Am. Silk J., 1930, *49*, 53) advises that the pH of the degumming bath should be maintained between 9·5 and 10·5; below pH 9·5 the rate of sericin removal is too slow and the silk is exposed to damage from the prolonged treatment, whereas above pH 10·5 there is a great danger of chemical attack. Below pH 8·5, degumming is both slow and incomplete.

Further work by Mosher (Am. Dyes. Rep., 1930, *19*, 395) dealt with the effects of pH and temperature in some detail. There is relatively little degumming below 82°C.; other work has shown that the rate of degumming at 94°C. is three times that at 84°C.

With regard to the effect of pH, Mosher found that there were two regions in which the sericin removal was at a maximum, above pH 9·5 and below pH 2·5. From Fig. 109 it is clear that the degumming solutions should lie between pH 9·5 and 10·5, or between pH 1·7 and 2·5. It is well known that the best conditions for sericin removal are not necessarily those for degumming in practice, particularly as acid solutions have no emulsifying powers.

Hougen (Text. Res., 1935, *5*, 191) has shown that the concentration of soap in solution should not exceed 0·053 N or the time for degumming will be excessive.

Some interesting work on the degumming of silk has been carried out by Hougen and his co-workers (ibid., 1935, *5*, 92, 134 and 191). When alkalis or soap act on sericin there is a progressive degradation of the sericin particles in solution after they have left the silk. After defining as the "binding weight," the number of grams of sericin combining with one gram-equivalent of soap or alkali, when a point of equivalence is attained in the solution, it was shown that the binding weight decreases progressively as the alkali concentration increases. The binding weight of the sericin was also shown to decrease with increasing pH, but no simple relation exists.

The degumming of silk is assumed to be a chemical reaction because of the reduction of the active alkali content during the process, the liberation of free acid from soap and the liberation of carbon dioxide from sodium carbonate; it has also been shown that

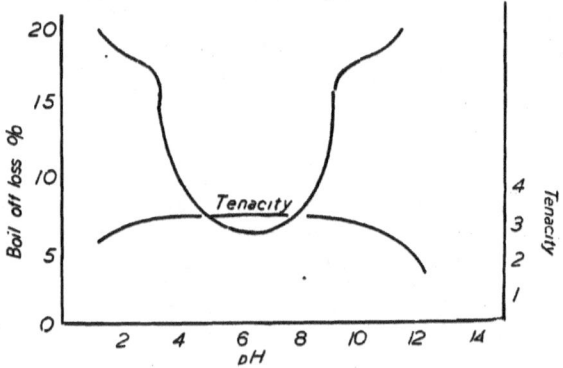

FIG. 109.—Effect of pH on tenacity and removal of gum.

an increase of 10°C. during degumming more than doubles the rate of sericin removal. Hougen produced an equation to represent the mechanism of degumming, in which he made use of the degumming coefficient denoted by the symbol k; this was found to depend on various factors such as temperature, denier, agitation, liquor ratio, and the nature of the soap and alkali. For alkalis, k is nearly independent of concentration, but for soaps k decreases with increasing concentration. Values of k are given for various alkalis and for soap solutions; under the given experimental conditions, the maximum rate of degumming was obtained in 0·029 N soap solution.

The order of effectiveness of some common alkalis is shown in the following table, on the basis of equivalent weights.

DEGUMMING COEFFICIENTS

Sodium hydroxide	984
Sodium carbonate	827
Sodium silicate	427
Trisodium phosphate	349
Borax	189
Sodium bicarbonate	173

SILK DEGUMMING   389

With soaps, the degumming coefficient varies from 208 to 516, as the normality of the solution decreases from 0·032 to 0·004.

The increasing effectiveness of soap with greater dilution is in accordance with the greater hydrolysis at lower concentrations; it is also of interest to consider that soap loses its colloid properties and approaches the behaviour of a true solution as the dilution is increased.

THE DEGUMMING PROCESS

Silk may be degummed in the form of hanks, or as hosiery or as woven piece goods. The greater part of silk is treated in hank-form.

With ecru silk, the loss in weight during the very mild process is limited to about 2 to 5%; indeed, the treatment in a weak soap solution is directed to the removal of fatty and waxy matter without serious loss of sericin. Ecru silk is unsuitable for bleaching, or for dyeing in many shades, but it can be used in the undyed state or for black material.

With souple silk, the loss in weight during the degumming process is restricted to about 5 or 10%; the raw material is treated for 1 to 2 hours at 30 to 35°C. in 10% soap solution followed by rinsing. Souple or soupled silk is used mostly for weft yarns. It may be dyed in dark shades without further treatment, but during the dyeing process care should be taken not to exceed the temperature of 60°C. or more of the silk gum may be removed.

The boiled-off or fully scoured silk has had practically all of the sericin removed in the degumming process and has, therefore, suffered a decrease in weight of about 20 to 25%. The complete degumming of silk generally requires two baths, the first of which may need about 30% of soap on the weight of the silk, whereas the second will only require 10 to 15%. The amount of water in the bath corresponds to a liquor to silk ratio of at least 10 to 1, and may be as high as 50 : 1. The time of treatment may vary from 1 to 3 hours and the temperature from 90 to 95°C.

Hanks

Where hanks of silk have to be degummed, they are generally allowed to hang freely in the bath, and about one gallon of solution is required for every pound of silk. During the process, the fibres swell and become sticky, but the gum soon starts to dissolve and reveal the smooth, soft and lustrous silk fibres.

Various methods are available for the treatment of hanks or skeins of silk, one of the simplest being to suspend the silk on smooth wooden rods in a rectangular vat of the soap solution and turn the hanks by hand from time to time. More modern vats may be made of stainless steel instead of wood to avoid the filaments being caught on the sides of the vat. Special machines are available in which the hanks are mounted on porcelain arms and rotated in the soapy liquor; the well-known Klauder-Weldon machine is often used for the treatment of silk in hank form.

Where the silk is treated in two stages, it is sometimes customary to pack the hanks in cloth bags or pockets, to avoid entanglement of the filaments as they become freed from the gum; the first stage of degumming may take place in hank form on poles, after which the hanks are packed into their pockets of fabric and placed in a vat for the final treatment. It is sometimes customary to stretch the hanks between these two processes, for degumming tends to contract and crinkle the filaments, the handle and lustre of which may be improved by stretching; it may be remarked, at this stage, that hanks of silk after degumming are often shaken out by twisting and untwisting to improve the appearance and handle.

Where the hanks are placed in bags for the final degumming, the ratio of liquor to silk may be reduced to about 5 : 1. It should be noted that machines in which the silk is maintained in a stretched state during the degumming process avoid any entanglement of the fibre and also save the time otherwise spent in putting the hanks into bags for the final treatment.

No matter what method of degumming is adopted, it is important to rinse the material with warm water first, for if cold water is used, there is a strong possibility that some of the impurities may be deposited on the filaments again. In some cases, it is customary to use slightly alkaline rinsing water to prevent any precipitation of the soap, but more modern technique utilises compounds such as Calgon or sodium hexametaphosphate.

After degumming and before rinsing, the silk should not be allowed to stand for any appreciable time as there is a danger of the soapy residues drying into the filaments, when they become difficult to remove and are apt to cause difficulties in dyeing, printing and finishing.

Amongst the many factors which contribute to an effective degumming liquor, the most important is to have enough soap to remove not only the sericin but also the soaking oils, and to maintain them in solution or suspension. The products of the reaction between the soap and the sericin are apt to slow down the degumming process but this may be obviated to some extent by degumming in two or more baths instead of a single solution.

The advantages of an excess of soap are well known; for instance, if one regards 20% of soap in the dry flake form as an average quantity for most qualities of silk, the use of 40% soap not only considerably increases the rate of degumming, but also ensures a

SILK DEGUMMING

more complete removal of gum, oil and colour. These advantages may be obtained without a substantial increase in the consumption of soap, by the use of successive baths.

Where a single bath is used for degumming, it is possible to start the process with 40% of soap, one half of which is used in the degumming treatment, and then add fresh soap to replenish the bath for every fresh lot of silk; the usual proportions employed at one time for degumming four lots of silk in the same bath amounted to 40% for the first lot, 33% for the second lot, 27% for the third lot, and 20% for the last. It will be seen, therefore, that for 2,000 lbs. of silk, it is possible to use 400 lbs. of soap (i.e. 20%) by dividing the silk into four lots of 500 lbs. The first 500 lbs. will be treated with 200 lbs. of soap (i.e. 40%), of which one half will be used by reaction with the sericin; the remaining 200 lbs. of soap may be distributed between the three successive lots of silk in equal amounts, to give the above-mentioned percentages. Hence, although the total consumption of soap is 20% of the weight of the silk, the first three lots are degummed in percentages of soap in excess of that figure. The variation in the excess of soap from start to finish is compensated for by the time of degumming, the last lot taking two or three times as long as the first; this is apt to cause a slight yellowing and possibly some weakness.

It will be remembered that the ability of soap to degum raw silk depends on the hydrolysis of soap in aqueous solution with the formation of some free alkali which attacks and dissolves the gum; hence alkali is used by combination with the sericin and the fatty acid from the soap is apt to build up in the bath. The increase in the concentration of free fatty acid lowers the efficiency of the degumming liquor, for before one half of the soap has been used in degumming, there is enough free fatty acid to prevent the liberation of free alkali from the soap; further, the fatty acid requires soap to keep it in the dispersed form and so reduces its effectiveness towards the soaking oils. The decreased powers of emulsification and suspension are also seen in the behaviour of the soap solution towards the pigments of the silk, and successive lots become darker in colour. The reduced powers of suspension are exhibited by the increasing tendency of the soaking oils to form a scum or cream on the surface of the degumming bath, which is apt to stain the silk as it is lifted out of the box.

The use of a double bath process obviates many of these defects; it has been described in some detail by a publication from the Textile Research Laboratories of Proctor and Gamble in Paterson, New Jersey (The Dyer, 1929, *61*, 225).

The double-bath method starts with a partially exhausted soap solution and finishes in a fresh liquor before washing in water; three boxes operate on the counter-current principle, two being filled with soap solutions and the third with water. For example, for 2,000 lbs. of silk which will require 400 lbs. of soap (i.e. 20%), about 260 to 300 lbs. of soap will be used in the first box and the remainder in the second box. The first lot of silk (500 lbs.) is boiled until 80% of the gum is removed, which may take 45 minutes, and is then transferred to the second box, where the degumming is rapidly completed. The soap solution may be used for four lots of silk with about 500 lbs. in each lot; the first bath may be reinforced by the addition of 20 lbs. of soap after each lot. At the end of the 2,000 lbs. of silk, the first bath is replaced by the liquor in the second bath on the counter-current principle.

The advantages of this method are seen in the uniformity of the degumming and colour; the second bath is not overloaded with gum and oil, and the excess of soap it contains prevents the silk from retaining the gum. The last of the four lots of silk is degummed as efficiently as the first, whereas with the single-bath process, there is a gradual decline over the four lots. An additional advantage is that the oil is retained by the first liquor and not carried forward.

It is also claimed that there is a saving of 10% of the soap required by degumming in the above manner, for the double-bath process only requires 18% of soap where the single bath needs 20%. It is also possible to standardise the time, temperature and other factors with the two-bath process, in a manner which is not possible for single baths.

A method of utilising three baths for the complete degumming of raw silk has been given by Matthews (Bleaching; Chemical Catalog Co., New York, 1921). On the basis of 10 lbs. of raw silk, the first bath should contain 2·5 to 3·5 lbs. of soap, the second bath about 1·5 lbs. and the third bath 1 lb.; when the last bath has been used once, 0·5 lb. of soap is added and the liquor used as the second bath, and in the same way the second bath may receive an addition of 1 lb. of soap when it is then used as the first bath. This method of operation gives better results and is more economical than using one bath until it becomes unfit for further degumming. Soap baths may be used about three times in this manner, but it is essential that the final bath should be fresh in order thoroughly to cleanse the silk filaments.

The degumming of silk may be accomplished by using the foam from a soap solution, instead of immersing the silk in the solution itself; this process was due to Schmid of Basle (D.R.P. 179,229).

The hanks of silk are suspended on rotating arms in the upper part of a rectangular trough in such a manner that they do not come into contact with the liquor but only with the foam rising from it. For instance, a solution of soap in the lower part of the vessel is heated by a closed steam-pipe so that the liquor is converted into a foaming lather which rises in the vessel and covers the hanks of silk; the strength, elasticity and dyeing properties of the silk are stated to be the same as with the older process. Complete freedom from gum may be assured, and the lustre and handle of the material is better than with the ordinary treatment.

It is possible to add dyestuffs during the foam-degumming process, as the sericin becomes so softened that the dyestuff can penetrate

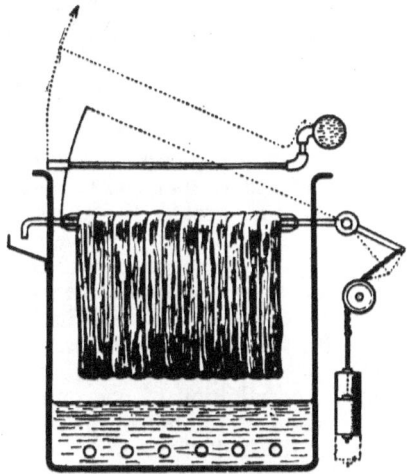

FIG. 110.—The foam method of degumming raw silk (Schmid).

the silk itself; after the silk has been subjected to the action of the foam for the necessary period of time, the hanks are removed and washed with hot water, which removes the softened gum from the silk.

It has also been suggested to treat hanks of raw silk in 30% soap solution at 65°C. until thoroughly impregnated, and then steam for 20 to 30 minutes at 7 lbs. pressure, followed by washing. It is claimed that the softened gum melts, and runs off the silk during the steaming treatment, but it seems doubtful if such a process is of commercial interest.

Hosiery

Where silk hosiery is being treated, it is possible to combine the degumming and dyeing processes to some extent; some useful notes on the methods have been given by Gaede (Am. Dyes. Rep., 1934,

23, 117). Three processes are generally available, the two-bath process, the one-bath process, and the split-bath process.

In the two-bath method, the silk is degummed, rinsed, and then dyed in a fresh bath. About 15% of 88% olive oil soap is used or 8 to 14% of a boiled-off oil; the bath is prepared and the goods entered at the boil and treated for an hour. Degumming water is then run into the bath which becomes flooded so that scum or suspended dirt is washed over the top; any draining method would filter the impurities on to the hose. Two rinses with lukewarm water are generally sufficient to remove any soap or oil. It is claimed that the two-bath method gives superior results because all the gum and oils have been removed before dyeing, and it is better to dye fabric from which the impurities have been removed; it is also stated that less dyestuff is required and the shades are clearer and faster.

In the one-bath process, the degumming agent is added to the bath and the goods are entered at about 60°C. and worked for 5 or 10 minutes in order to become wetted; the dye is then added and the temperature raised to the boil. It is possible to add dyes at the start, but it is better to wet the goods first. It is claimed that the one-bath process gives goods of better handle and quality because the silk gum acts as a protective colloid, and the time of treatment is much shorter than with the two-bath process.

The split-bath process claims all the advantages of the one-bath method. It consists in the addition of one half of the necessary amount of degumming agent at the outset, in which the goods are worked for 15 minutes at the boil; the bath is then made up to the full strength and the dyestuffs added. By the use of this method, the goods receive a preliminary treatment which removes the throwing oils and tinting colours, and also softens the gum.

Hosiery is often treated inside a reciprocating drum, but the degree of agitation obviously depends on the nature of the material. Some authorities maintain that the disturbance produced by a gentle boil is adequate, and that mechanical tumbling is apt to produce chafe-marks. The use of slowly reciprocating drums is common for many types of hosiery, but they may also be treated in tubs or vats. About three or four dozen hose are placed in muslin bags and boiled gently in 10% soap solution calculated on the weight of the hose. The ratio of liquor to silk may be about 10 : 1. The tub or vat in which the hose are placed should be filled almost to overflowing, and boiling should be allowed to proceed gently for an hour, at the end of which time the dirty liquor may be removed by overflowing for about 10 minutes. The vat is then drained and the goods thoroughly rinsed by the overflow method; finally, the hose are withdrawn and excess water removed by hydro-extraction.

Woven Fabrics

The chemistry of the degumming process for silk fabrics is the same as for hanks of yarn and for hosiery. The important question is that of manipulation; silk fabrics are generally rather fragile and cannot be treated in rope-form in the ordinary way. Indeed, Roberts (Textile World, 1930, 78, 1679) has suggested the use of a stagnant bath for sheer delicate fabrics of silk.

Silk fabrics are not easy to treat on the jig unless they are woven with strong selvedges which are apt to complicate processing on account of rolling and also of festooning across the weft. Where selvedges are substantially different from the body of the fabric, differential shrinkage and extension is apt to cause creasing and other irregularities.

With many of the delicate fabrics, the simplest method of treatment is on the star-frame, and this may also be employed for bleaching and dyeing.

It is also possible to treat fabrics in rope-form by forming them into a large loop or hank for degumming; alternatively, they may be treated in open-width by forming into book-fold when they are suspended from one selvedge by loops of cotton cord and allowed to remain in the vat of degumming liquor for the requisite period of time. Heavier fabrics may be treated on the jig.

Woven goods consisting of wool and silk need careful attention, for if given the usual boil-off treatment there would be substantial damage to the wool. Hence it is customary to use lower temperatures of the order of 40°C. and a dilute solution of soap and sodium carbonate.

With fabrics composed of real silk and cellulose acetate, care must be taken to avoid hydrolysis of the acetate during the degumming of the silk. A method suggested by British Celanese is to pad the goods in a solution of Turkey red oil of 40 to 50% concentration containing 10 grams per litre of sodium silicate, sodium carbonate or sodium borate; the solution is maintained at 75°C. After impregnation, the goods are rolled on a batch and allowed to lie for several hours, after which they are finally degummed in 5% soap solution containing 0·3 grams per litre of the above alkali at a temperature of 75°C.; the degumming should take place in two stages. It is stated that the cellulose acetate is not affected if the temperature does not exceed 75°C. and the alkalinity does not exceed pH 10·5.

An alternative method suggested by the I.G. is to treat the material with acid solutions in presence of a sulphonated oil and a fatty product which is stable in presence of acid. For example, the fabric may be treated for 3 hours in a bath containing 15 c.c. of hydrochloric acid and 25 c.c. of sulphonated oil per litre. The ratio of liquor to fabric should be 40 : 1. The goods are then well rinsed and soaped in a solution of olive oil soap containing 10 grams per litre at 60°C.

Another method is to use a bath containing 5 c.c. of acetic acid per litre and sulphonated oil up to 3 c.c. per litre at 80°C. for 2 to 3 hours; the liquor to fabric ratio should be 250 : 1. The goods are then rinsed and treated for 1 to 2 hours at 65°C. in a solution containing 3 c.c. of sulphonated oil and 10 grams of olive oil soap per litre.

Silk Waste

Silk waste which is used in the manufacture of spun silk, must be degummed in order to render it suitable for the mechanical operations of spinning. The waste is derived from the silk reeling and also from the throwing process.

The general procedure is to pack the waste silk into net bags with about 0·25-inch mesh; the bags generally accommodate about 1 lb. of silk which should be packed in an open and bulky condition. The loose packing is important on account of the swelling of the silk during the degumming process. The actual degumming generally takes place in large wooden vats about 4 to 5 feet in diameter and 4 to 6 feet in depth; a perforated false bottom holds the silk above the steam coil which is used to heat the liquor.

The vat is filled to about one-third capacity with water, to which has been added about 10 to 15 lbs. of soap which will be adequate for the treatment of 100 lbs. of silk waste. The water is heated to dissolve the soap and the silk waste is entered when the temperature is about 70 to 80°C.; care is taken to ensure that the bags of waste are thoroughly submerged, and the steam is turned on and treatment continued for about 2 hours, during which the bags may be turned from time to time. The dirty liquor is then withdrawn, the waste silk centrifuged, and the process repeated.

An alternative method which is sometimes used in Europe is termed the "schappe" process. The waste silk is evenly spread in a tank and covered with water at 60°C.; fermentation for a week is sufficient to attack the sericin, after which the waste silk is washed in a stamping machine under the action of wooden hammers or stampers where a spray of water removes the gum which has been loosened by fermentation.

Tussah

Tussah silk and all wild silks are more difficult to scour than the ordinary varieties, on account of the fact that they contain more mineral matter and also the resistant nature of the gum. In general, however, the same methods are employed as with cultivated silks, but in a more drastic manner; the time of treatment and the number of treatments are adjusted accordingly, and, further, a more alkaline liquor is employed because the sericin of wild silks is not completely removed by soap alone.

One method of treatment is first to process the silk with a solution containing 8 to 10% of sodium carbonate estimated on the weight of the silk, followed by a bath of 10 to 15% soap solution.

It may be remarked that the "boil-off liquor" from tussah silk contains a large amount of brown colouring matter and cannot be used for addition to dye-baths.

DEGUMMING WITH ENZYMES

Although sericin is generally removed from silk by the action of soap solutions, with or without the addition of soda or some of the modern synthetic detergents and auxiliaries, alternative methods have been suggested, particularly for waste silk.

Enzymes such as degummase and thermodegummase show an optimum effect at pH 6·0 and 40°C., and at pH 7·5 and 55°C. respectively. Katagiri and Nakahama (J. Agric. Chem. Soc. Japan, 1941, *17*, 165) have shown that these reagents also attack peptides, peptones, gelatin edestin and caseinogen as well as sericin.

As an example of degumming with proteolytic enzymes, the raw silk may be treated in a solution containing 3 g. per litre of a mixture consisting of 20 parts of papain, 40 parts of sodium thiosulphate and 40 parts of sodium hydrosulphite. Munch (Textilber., English Ed., 1938, *19*, 37) has investigated the action of papain because of the possible injury to cellulose acetate or wool in the degumming of mixed materials with soap. In a series of experiments with various activators for papain, sodium thiosulphate and hydrosulphite were found to be most suitable; a mixture is best for silk-acetate mixtures, but for silk-wool fabrics it is better to use thiosulphate alone in a weak acetic acid liquor. After the degumming with papain the goods should be soaped for 30 minutes in a solution containing 5 g. per litre of olive oil soap at 65°C. to complete the degumming. The results of breaking load measurements show that the tenacity of the silk degummed with papain is less than that treated with soap, so that the treatment should not exceed two hours' duration; the wool and acetate rayon are unaffected by the papain in presence of thiosulphate.

Enzyme products, such as papain, bromelin, etc., may also be activated by hydrogen cyanide or by hydrogen sulphide; the two may be used in conjunction according to B.P. 320,327. The presence of an electrolyte in the solution is useful when degumming raw silk which has been hardened or dyed.

The Wallerstein Company in U.S.P. 1,855,431 has suggested the removal of sericin at temperatures below 50°C. with proteolytic enzymes in neutral, slightly acid, or slightly alkaline solutions, together with an activator such as sodium sulphoxylate; the treatment may last for 1 or 2 hours, or it may be for a short time followed by exposure to moist air and washing afterwards.

Fermentative methods of degumming have been examined by Katagiri and Nakahama (J. Agric. Chem. Soc., Japan, 1937, *13*, 1003), who found that no noticeable changes occurred when raw silk was kept in peptone water with toluene, but without toluene 45% of the sericin and 23% of the crude fat were decomposed by fermentation. Eighteen strains of aerobic bacteria were isolated from the degumming liquids, and two were found to be very effective; they were similar to *B. cereus*, the active agent of which is a soluble enzyme "degummase." It attacks sericin A and B at the same rate, but does not affect fibroin; its maximum activity is at pH 7 to 8, at 40°C. Degummase is sensitive to acid and is inactivated by heating for 2 hours at 65°C.

Some interesting data on the use of enzymes have been given by Mosher (Am. Silk J., 1931, *50*, No. 8, 55), who pointed out two important facts: first, that a short preliminary treatment with hot neutral or alkaline liquors renders the sericin very susceptible to proteolysis, and secondly, that the residual sericin after a treatment with pepsin or trypsin is very soluble in warm soap solutions.

A comparison was made of pre-treatments under different conditions, hydrochloric acid at pH 1·09, distilled water, and olive oil soap at pH 9·89; the ratio of liquor to silk was 50 : 1, and the conditions of treatment 5 minutes at 100°C. Following upon the pre-treatment, the silk was subjected to the action of trypsin at pH 8 and 40°C.

RAW SILK DIGESTION BY TRYPSIN

Pre-treatment	Enzyme treatment (0·25%)	Raw silk removed as sericin	Raw silk remaining as sericin
0·1% HCl, 5 mins., 100°C.,	60 mins.	17·55%	3·33%
,, ,, ,,	2 hours	18·23%	2·93%
,, ,, ,,	4 hours	18·69%	2·79%
,, ,, ,,	none	7·96%	12·92%
Distilled water,	1 hour	14·06%	6·54%
,, ,,	2 hours	15·57%	5·13%
,, ,,	4 hours	16·66%	4·20%
,, ,,	none	3·31%	17·21%
0·25% soap,	1 hour	18·29%	2·77%
,, ,,	2 hours	18·47%	2·66%
,, ,,	4 hours	18·64%	2·58%
,, ,,	none	9·01%	11·41%

SILK DEGUMMING

Influence of Soap on Pre-enzyme Digestion

Preliminary (4 hours 40°C.)	Subsequent	Raw silk removed as sericin	Raw silk remaining as sericin
0·25% trypsin.	0·25% soap; 30 mins, 70°C.	18·08	2·51
,,	,, 1 hour, 70°C.	18·31	2·08
,,	,, 2 hours, 70°C.	18·84	1·91
Soap alone. 0·25% for 2 hours at 70°C.		10·34	10·06
,,	,, 24 hours at 70°C.	10·94	9·32
0·25% trypsin.	0·25% soap, 30 mins., 80°C.	18·88	1·51
,,	,, 1 hour, 80°C.	18·42	1·99
,,	,, 2 hours, 80°C.	18·71	1·72
,,	,, 5 mins., 80°C.	19·01	1·64
Soap alone (0·25%; 2 hours at 80°C.)		12·03	8·31
,, (0·25%; 24 hours at 80°C.)		19·44	1·29
0·25% trypsin.	0·25% soap; 2 hours, 80°C.)	19·01	1·64
,,	1% soap; 2 hours, 80°C.	19·18	1·49

With most degumming treatments there is a small fraction of sericin which is very difficult to strip, and whereas this is not a serious matter with many goods, yet it is sufficient to render crepes harsh and stiff; another disadvantage of the residual gum is that the pigments are not completely eliminated. If the prepared fabrics are treated with dilute soap solutions at 60 to 80°C., then the colour is slowly eliminated and also the remaining sericin; a preliminary fermentative operation, therefore, enables the difficulty of the residual sericin to be overcome by mild soaping.

It must be remembered that the degumming process is not directed solely to the removal of sericin, but that oils, waxes and pigments should also be eliminated with the minimum of degradation of the fibroin. With 1% soap solution for 2 hours at 100°C. there appears to be about 1% hydrolysis of the fibroin, but some of the enzymes are more active in their attack on the fibroin, as shown by the following tests on Japanese degummed silk, with a ratio of liquor to silk of 50 : 1.

Influence of Enzyme or Soap on Fibroin

Hydrolytic agent	Treatment	Hydrolysis
1% olive oil soap	2 hours at 100°C.	1·07%
0·25% pepsin, 0·1% HCl	4 hours at 40°C.	0·81%
0·25% trypsin, 0·25% NaHCO$_3$	4 hours at 40°C.	1·18%
0·5% papain, 0·25% HCN	2 hours at 75°C.	3·33%
2% Korofor, 0·25% NaHCO$_3$	2 hours at 54°C.	1·28%

Some further experiments by Mosher on raw silk gave the following data:

INFLUENCE OF DEGUMMING ON STRENGTH

Degumming	Time and Temp.	Tenacity
1% olive oil soap	2 hours at 100°C.	3·87%
0·25% trypsin, bicarbonate	4 hours at 40°C.	3·79%
0·25% papain, hydrocyanic acid	2 hours at 75°C.	2·49%
2% Korofor, bicarbonate	2 hours at 54°C.	3·72%
0·25% pepsin, hydrochloric acid	4 hours at 40°C.	—
followed by 0·25% soap	1 hour at 80°C.	3·92%

Although the loss of strength by enzymatic degumming is not very great, yet it is apt to be reflected in an increased susceptibility to chafing.

It appears, therefore, that in no case do the enzymes effect complete elimination of the sericin, but the amount removed may be increased by a preliminary treatment in hot acid, neutral or alkaline liquor. The enzyme itself does not affect the oils, wax or pigments.

Sericin Fixation

Whereas it is general practice to remove the sericin from the raw silk, it is sometimes required to be fixed more permanently to the fibre; one of the commonest means of effecting this is by the ubiquitous formaldehyde. The effectiveness of the treatment is probably due to combination between the formaldehyde and the amino-groups in the protein. According to Nakahama and Sakaguchi (J. Agric. Chem. Soc., Japan, 1941, *17*, 192), the best method is to treat the raw silk with 4% formaldehyde solution at pH 7 for 3 hours at 50°C.

Chromium salts have also been used for the fixation of sericin, and potassium oxalatochromiate at pH 4·7 is very effective. Oku and Hirose (Bull. Agric. Chem. Soc., Japan, 1937, *13*, 1257) have examined the fixation of sericin by chromium salts; the absorption of 1% of Cr_2O_3 facilitates fixation of the sericin. The strength and extension of the silk are not impaired by the presence of chromium, but the surface of the fibre is apt to feel rather "woolly"; white silk requires rather more Cr_2O_3 than yellow silks for the fixation of the sericin. The absorption of the chromium may be effected from either the basic sulphate (cationic Cr) or the basic oxalatochromiate (anionic Cr).

Various suggestions have been put forward for measuring the efficiency of the degumming process, and most of them depend on a staining method which reveals the presence of the silk gum. Denham and Dickinson (J.S.D.C., 1935, *51*, 93) mention the use of Benzopurpurin 4BS, Purpurin 10BS, Diamine Green G, and Acid Magenta N. Neocarmine, which is a mixture of dyestuffs, has also been put forward as has Shirlastain, which was produced by the I.C.I. in conjunction with the B.C.I.R.A.

CHAPTER XXIX
SILK BLEACHING

FROM the account of the chemical constitution of the animal fibres on page 57, it would appear that the similarity between wool and silk is such that reagents which damage the one would also damage the other; hence, in considering the substances available for bleaching, the use of compounds liberating chlorine is inadvisable, and one is left with sulphur dioxide and hydrogen peroxide as the chief bleaching agents for silk.

Sulphur dioxide bleaching has been discussed on page 355, and the use of hydrogen peroxide is described on pages 250 and 362.

There is, however, one important difference between wool and silk where bleaching is concerned, and this is due to the silk gum which is not always removed completely. A great proportion of Italian silk possesses a yellow colour which is due to a pigment contained in the gum and not in the filament; hence when the gum is completely removed, the silk acquires a near-white colour without the intervention of a bleaching process. The removal of gum alone is often adequate for silk which has to be dyed.

In some types of silk, however, more or less of the gum is retained and this creates special problems.

Ecru silk, for example, has merely been treated in tepid soap and water, and is generally regarded as unsuitable for bleaching or dyeing without a further degumming treatment, except when used in black material.

Souple silk has the gum softened, but without removing the whole of it; the loss in weight may vary from 5 to 10%. Hence, in bleaching, the gum must be considered as well as the fibroin. It is often required to retain the gum which facilitates subsequent weighting.

With silks containing their gum, therefore, it was customary at one time to give a pre-treatment with aqua regia of about 3°Tw. before the soupling process which preceded the bleaching process proper. Similar considerations apply to the treatment of ecru silk where required. It must be emphasised that the use of aqua regia does not apply to boiled-off silk, but only to those yellow-gum silks which are to be stoved with sulphur dioxide.

The usual method, according to Matthews (*Bleaching*, Chemical Catalog Co., New York, 1921) was first to soften the silk by working in a warm bath containing about 10% soap solution, followed by rinsing to remove all traces of soap. The aqua regia was made by mixing 5 parts of commercial hydrochloric acid with 1 part of nitric acid; the mixture was allowed to stand for 4 or 5 days before use. The mixture was then diluted to the required concentration and the silk steeped in the bath until the fibres acquired a greenish colour, when they were rinsed with water, and the bleaching completed by stoving in sulphur dioxide as for wool. Care had to be taken that the silk did not remain too long in the aqua regia solution or it acquired a permanent yellow coloration.

As an alternative to aqua regia, the use of chamber crystals from sulphuric acid manufacture was also adopted; a similar solution may be made by acidifying a 10% solution of sodium nitrite with sulphuric acid to give a strongly acid solution. It has also been suggested to use a bath containing 1% sodium nitrate and 5% sulphuric acid.

Ecru silk yarn is sometimes required in the bleached state for use as warps when it may be semi-bleached by stoving twice; if a good white was required, yellow silk ecru yarn may be stoved twice, treated with aqua regia, washed, rinsed, and stoved again two or three times. As with wool, the material must be wet when it enters the stove for treatment with sulphur dioxide.

Ecru or hard silk is often found in the warp of fabrics which need to be scoured in the piece; the method of scouring or degumming is similar to that which is given to the yarn. As the ecru silk still contains most of its silk gum, it also contains most of the natural colouring matter, which it will be remembered largely exists in the gum; where the fabric is required in the white state it must be bleached, and even for the bright shades of colour the fabric should be bleached before dyeing.

An old method of treating material of this type is to treat the silk in 10% soap solution at 35°C., followed by "stoving" in sulphur dioxide as for wool, and then bleaching in aqua regia, followed by washing and stoving again. For a good white on ecru silk it may be necessary to stove three times, in addition to the treatment with aqua regia; for instance, Knecht (*Manual of Dyeing*, Knecht, Rawson and Loewenthal, Griffin, London, 1916, p. 157) has suggested washing with 10% soap solution at room temperatures, rinsing, stoving twice, bleaching in aqua regia, rinsing, treating with dilute sodium carbonate followed by soap, and then stoving twice again, before the process is completed by rinsing.

The old process for the treatment of souple silk was to treat the scoured material for about 15 minutes in aqua regia of 3 to 4°Tw. As soon as a greenish-grey colour appeared on the goods they were removed and washed, for too prolonged an immersion causes yellowing. The wet goods then had excess liquor removed, and were stoved in the wet state for 6 to 8 hours; if a good white was required, then the stoving in sulphur dioxide was repeated several times. The actual process of soupling is a softening treatment to remove the harshness caused by the above methods; it consists in working the silk for 60 to 90 minutes in 0·3 to 0·4% cream of tartar solution (i.e. potassium hydrogen tartrate). Finally the material was centrifuged and dried.

These methods for treating either ecru or souple silk are rarely practised to-day except under special circumstances. Where retention of the sericin is required, it is possible to treat the silk with a solution of formaldehyde (0·25 to 0·5% CH_2O) for several hours, followed by rinsing in ammonia solution. It is sometimes possible to give the effect of silk containing gum by treating the degummed material with a solution of gelatin, about 2%, at 30 to 40°C., followed by immersion in another solution containing about 20 c.c. of 40% formaldehyde solution per litre.

The use of aqua regia for yellow-gum silk belonged to the period when sulphur dioxide was the only common bleaching agent for silk; the SO_2 is applied in gaseous form so that there is no possibility of the removal of sericin during the process.

STOVING

Bleaching with sulphur dioxide, or stoving, is rapidly being replaced by bleaching with hydrogen peroxide; as with wool, the sulphur dioxide bleach is not permanent. The costs of manipulation enter into the question of which method of bleaching to employ, and it is here that the main argument is found for the preference of stoving silk in hank form, but not for hosiery or woven piece goods.

The wet material may be bleached by stoving from 3 to 8 times according to requirements, and is then often tinted to produce the required shade. Before tinting, however, the residual sulphur dioxide should be removed by treating with hydrogen peroxide solution as for wool.

In view of what has already appeared on page 355, with regard to the use of sulphur dioxide, it is not considered necessary to repeat the account of stoving.

Solutions of sulphurous acid or of sodium sulphite have been utilised to some extent in the treatment of wool, but are rarely employed for silk. Solutions of sodium hydrosulphite, $Na_2S_2O_4$, however, have been used occasionally; the silk may be treated at room temperatures with 2°Tw. $Na_2S_2O_4$ solution for about 6 hours, squeezed, rinsed, acidified and rinsed again. An alternative method is to treat the silk for several hours in 2 to 3°Tw. sodium hydrosulphite solution containing 1% of acetic acid. Sodium hydrosulphite solutions give better results with silk than with wool.

Pokorny (Rev. Gén. Mat. Col., 1925, *29*, 288) observed that the reducing powers of sodium hydrosulphite were greatly increased in the presence of alcohol and suggested that this activity might be utilised in the bleaching of tussah silk.

HYDROGEN PEROXIDE

The properties of hydrogen peroxide solutions have already been described on pages 250 and 362. Hydrogen peroxide may be regarded as a universal bleaching agent, for in addition to its use on cotton, linen, rayon, wool and silk, it may also be applied to furs, feathers, skins, bones, fats and oils.

The strength of the aqueous solution is represented by the number of volumes of oxygen it is capable of providing; hence, 1 litre of 1 vol. H_2O_2 solution will give 1 litre of oxygen, whereas 1 litre of 10 vol. H_2O_2 will give 10 litres of oxygen. The concentration of 10 vol. is equivalent to 3% by weight of H_2O_2, whereas 20 vol. corresponds to 5% concentration by weight.

Hydrogen peroxide is almost universally employed for the bleaching of silk and the most common concentration is 2 vol.; this contains 2·858 grams per litre of available oxygen.

It will be remembered that for the treatment of scoured cotton it is customary to use concentrations of 0·5 to 1 vol. H_2O_2 at pH 11 to 11·5 and at a temperature of 85°C. Where hydrogen peroxide is used for the bleaching of wool, however, it is necessary to employ lower temperatures and less alkaline solutions to avoid damage to the fibre; the concentration of the peroxide is therefore raised to about 2 vol. for use at 40 to 50°C. with solutions of pH 7·5 to 8, but with more prolonged bleaching at lower temperatures, solutions of pH 9 may be used. It is not wise to exceed temperatures of 50°C. or an alkalinity of pH 9 with wool, although some heavy goods are bleached in 2 hours in 2 vol. H_2O_2 solutions of pH 9 at 50°C. (Phenol red is a useful indicator for pH 6·8 to 8·4, and phenol phthalein for pH 8·3 to 10.)

Cotton is greatly valued as a textile material because of its ability to withstand repeated launderings; its resistance to boiling alkaline solutions is of a high order. Wool, on the other hand, is susceptible both to boiling water and to aqueous solutions of alkali. Silk occupies an intermediate position and will withstand treatment with water at higher temperatures than will wool, and also with solutions of greater alkalinity; nevertheless silk is alkali-sensitive and must not be treated with impunity. In general, therefore, it is possible to use bleaching baths of peroxide at higher temperatures and of slightly greater alkalinity with silk than with wool. Although sodium hydroxide may be employed for the peroxide bleaching of cotton, its use should be avoided with silk, and the necessary alkalinity

SILK BLEACHING

provided by ammonia or by sodium silicate. In some publications, confusion has been caused by the custom of expressing the alkalinity of the bath in terms of grams of sodium hydroxide per litre; the value is somewhat fictitious where no sodium hydroxide has been used. The alkalinity is often estimated by titration with $0.1\ N$ HCl solution with phenol red as indicator; phenol phthalein is not satisfactory. Many bleaching liquors for peroxide are defined as having an alkalinity of 0·3 to 0·8 grams NaOH per litre, although, as previously stated, they contain no sodium hydroxide, and are made alkaline with ammonia or sodium silicate. When expressed in these terms, the alkalinity should not exceed 0·8 gram NaOH per litre.

It is generally preferred to express the alkalinity in terms of pH values, but some works are not equipped with an apparatus based on the glass electrode and find colorimetric indicators unsuitable for their purpose.

This may be a suitable opportunity of remarking that, useful and important as pH control may be, yet it is possible to over-estimate its value, unless utilised with intelligence. Erroneous determinations of the alkali content of bleaching solutions may occur as the result of the absorption of alkali by the fibres; it is also possible to be misled by buffered and unbuffered solutions of the same pH value, but which will behave very differently in use.

Where phenol phthalein is used to indicate the alkalinity of the peroxide bleach-liquor, it is useful to remember that it covers the range pH 8·3 to 10. As previously mentioned, the alkalinity is not provided by sodium hydroxide but by ammonia or by sodium silicate; the alkalinity, expressed as NaOH, may range from 0·3 to 0·8 gram NaOH per litre, and this would correspond to pH 12–12·3 (0·004 grams NaOH per litre corresponds to 0·0001 N, i.e. pH 10).

Returning to the discussion of hydrogen peroxide as a bleaching agent for silk, it is of interest to recall that it was originally suggested, in 1875 (B.P. 4172), that barium peroxide should be used for bleaching silk; this appears to be the first reference to the peroxide bleach for any purpose. In 1878, however (B.P. 1414), hydrogen peroxide was suggested for bleaching silk, the goods being steeped in 0·5-vol. H_2O_2 for 24 hours at room temperatures.

It will be remembered that there are two main methods of bleaching with hydrogen peroxide:

(a) the steeping process in which the goods are immersed in an aqueous solution of H_2O_2 for a definite time,

(b) the ageing process whereby the goods are saturated with H_2O_2, and then removed and dried with the H_2O_2 in the material.

The latter is utilised with cotton or wool to some extent, and forms the basis for continuous treatments; it is not generally employed for silk, although its application was patented by Kershaw (B.P. 162,198) in 1921. The goods were impregnated in 1·2 vol. hydrogen peroxide solution brought to pH 10 to 11 with ammonia, and then aged for 16 hours; the silk was subsequently washed and given a light bleach with sodium hydrosulphite solution.

Most silk, however, is bleached by the steeping process, and with many piece goods the wince machine is a convenient apparatus as it is relatively easy to maintain the required temperature. The manipulation of the fabric, whether woven or knitted, is facilitated by using the wince machine; it is easy to feed the fabric into the bath and also to perform the rinsing operations after bleaching, as well as to withdraw the goods when the sequence of processes is over. The jig may also be used for heavy goods where the dangers of extension are less acute. As with the bleaching of cotton or wool, it is important to take the usual precautions in respect of the presence of metals. It is general to allow 1·5 gallons of liquor per lb. of silk, i.e. a liquor to fabric ratio of 15 : 1.

Where a particularly good white is required, it may be necessary to stove the silk after the peroxide treatment; before stoving in sulphur dioxide, however, the goods should be impregnated for 30 minutes in a solution of soap containing 6 to 8 grams per litre.

The use of hydrogen peroxide for bleaching silk has many advantages over stoving; the white is permanent and there are no unpleasant residual odours of sulphur dioxide in the goods. Three fundamental differences between stoving and the peroxide treatment are (*1*) reduction as compared with oxidation, (*2*) an acid bleach as against an alkaline bleach, and (*3*) the use of gas in contrast to liquid. Factors (*2*) and (*3*) should enable hydrogen peroxide to be used as a combined scour and bleach; this has also been suggested for cotton and for wool. The combined process is possible but is apt to be impracticable, in general, on a large scale; it is better to scour the goods first, although some silk hosiery is degummed and bleached in one solution. The use of peroxide to bleach impurities which are removed in a normal scour is somewhat expensive. With yellow gum silk, the gum is not bleached (or removed) with the stoving treatment, but the hydrogen peroxide bleaches and removes the yellow gum, so that treatment with aqua regia is not necessary.

When hydrogen peroxide is used for bleaching silk, it is essential that the usual precautions are followed; these are control of time, temperature, alkalinity and concentration. The strength of the peroxide solution may be determined with permanganate in the ordinary manner, and the active oxygen content should be maintained between 1·6 and 2·5 grams per litre; the lower value is suitable for bleaching weighted silk and the higher figure for the treatment of unweighted fabrics. (1 vol. H_2O_2 contains 1·429 grams of available oxygen per litre.)

The alkalinity of the bath needs careful supervision and it should be maintained at about pH 10; above this figure there is a risk of damage to the silk. Stabilisers such as sodium silicate may be added, but ammonia is preferred to caustic soda for producing the required alkalinity.

Lightweight fabrics may be treated at 70 to 75°C. for 5 to 6 hours in a bath containing 2·5 grams per litre of active oxygen; about 0·8 gram of ammonia and 1·5 grams of sodium silicate per litre are also added. The bleaching operation is often conducted in open vats, the fabric having been reeled into large hanks which are supported on sticks; these are turned for the first two hours and also towards the end of the treatment. Bleaching is followed by rinsing with warm water at 40°C. and then with cold water. The wet material may then be stoved if a particularly good white is required; the goods are soaped at 80 to 90°C., centrifuged and then suspended in the wet state in the stoving chamber overnight. The damp, soapy state seems necessary for successful stoving.

With heavier fabrics, the concentration of active oxygen should lie between 1·8 and 2·5 grams per litre, but if the goods are allowed to lie overnight, then the lower concentration should not be exceeded; the higher figure is adequate for a bleaching time of about 6 hours at 70°C. It is not generally customary to stove the heavier fabrics.

When unweighted silk is bleached, it is important to ensure that the fabric is properly acidified before the weighting process; for this reason the bleaching bath is often made alkaline with ammonia only and not with sodium silicate which would create difficulties on acidifying and weighting.

There are naturally many variations in the treatment of silk with hydrogen peroxide solutions. The short treatment for 100 kg. of degummed silk is to add 3 litres of 100 vol. hydrogen peroxide to 2,500 litres of water and then add the necessary amount of sodium silicate; the bath is then heated to 35 or 40°C. when the silk is entered and the temperature raised to 85°C. over a period of one hour. The goods are then rinsed in the usual manner.

A more prolonged treatment may be given in 4 vol. peroxide for 48 to 60 hours at room temperatures.

Many bleachers consider that it is better to bleach silk after the weighting process, for it has been found that the weighting of bleached silk is accompanied by irregularities and local tendering. F. Weber (Textilber., 1939, *20*, 209) has provided some data showing the effect of the peroxide bleach and of weighting on the strength of five fabrics; the weighting followed the peroxide bleach.

EFFECT OF BLEACHING AND WEIGHTING
(relative to the degummed fabric)

Fabric	Degummed	Bleached	Weighted	Silicate-treated
A	100%	92%	65%	60%
B	100%	85%	60%	60%
C	100%	96%	78%	58%
D	100%	82%	59%	53%
E	100%	87%	63%	44%

It will be seen that the silk is damaged to a greater extent by the weighting process than by bleaching.

Where weighted silks are bleached, it is important that they should have had a final treatment with sodium silicate or they may lose 5 to 10% of their weight during the bleaching process.

With the heavier weighted silks, the bleaching bath may be 1·5 to 2 vol. H_2O_2, made alkaline with sodium silicate to about pH 9 to 10. The goods may be bleached in 6 hours at 60 to 70°C. or allowed to lie overnight after a preliminary working in the liquor.

Weber (*supra*) has given some interesting data on the consumption of oxygen during the bleaching process.

OXYGEN CONSUMPTION BY SILK VELVET
(4 hours at 75°C.)

Time of bleaching	Available oxygen
0 hours	2 grams per litre
2 hours	1·87 ,, ,,
3 hours	1·75 ,, ,,
4 hours	1·60 ,, ,,

Some further data refer to the bleaching of flat goods. A standing bath was prepared of 2,500 litres of water containing 36 litres of hydrogen peroxide and 7 litres of sodium silicate; this was used for the bleaching of 11·5 kg. of weighted silk.

OXYGEN CONSUMPTION ON BLEACHING

Time	Oxygen	Temp.
4 p.m. (1st day)	1·95 g./l.	70°C.
7 a.m. (2nd day)	1·44 g./l.	52°C.

After bleaching overnight the silk was withdrawn and the liquor allowed to stand.

SILK BLEACHING

OXYGEN CONSUMPTION ON BLEACHING—*contd.*

Time	Oxygen	Temp.
8 a.m. (2nd day)	1·38 g./l.	52°C.
9 a.m. ,,	1·36 g./l.	52°C.
10 a.m. ,,	1·34 g./l.	51°C.
11 a.m. ,,	1·32 g./l.	50°C.
2 p.m. ,,	1·28 g./l.	48°C.
5 p.m. ,,	1·23 g./l.	48°C.

At this stage, the bath was warmed to 70°C. and 8·6 Kg. of silk entered for bleaching overnight.

Time	Oxygen	Temp.
7 a.m. (3rd day)	0·60 g./l.	53°C.

The silk was then withdrawn and the bath allowed to stand.

Time	Oxygen	Temp.
3 p.m.	0·48 g./l.	48°C.

The above table is of interest in showing the course of the available oxygen content of a peroxide bleaching bath when used overnight and allowed to stand during the day.

Some details of the bleaching of silk with hydrogen peroxide have been given by I. E. Weber (J.T.I., 1933, *24*, 178P); before the bleaching operation it is usual to remove the sericin by degumming either with an olive oil soap or a degumming oil such as sodium silicate and sulphonated castor oil.

Pure silk stockings may be degummed in 10% of soap and then bleached; for 200 dozen stockings, it is necessary to have 300 gallons of 1 vol. hydrogen peroxide to which has been added 20 lbs. of sodium silicate. The goods may be bleached over a period of two hours, starting at 75°C. and finishing at 90°C.

It is also possible to degum and bleach in one bath by using 300 gallons of 1 vol. hydrogen peroxide solution, containing 25 lbs. of olive oil soap and 2 lbs. of sodium silicate. The stockings are preferably placed in net bags and treated in Monel metal rotary machines for an hour, starting at 75°C. and finishing at 90°C. Instead of olive oil soap, sulphonated castor oil may be used in which case, for 600 lbs. of stockings, 300 gallons of 1 vol. hydrogen peroxide should be used to which has been added 20 lbs. of sodium silicate and 6 lbs. of sulphonated castor oil. The time and temperature of treatment is similar to those given above.

Pure silk crepe fabrics may be degummed and bleached in winch machines, 300 gallons of liquor being necessary for 100 lbs. of silk. The silk is first degummed with olive oil soap using about 30% on the weight of the goods; degumming may require 2 or 3 hours at

85 to 90°C. depending on the material and the finish. The fabric is then washed thoroughly and bleached with 1·5 vol. peroxide on the same winch machine; sodium silicate may be added until the liquor is alkaline to phenol phthalein (pH 8·5 to 10). Bleaching may require 2 hours at 70 to 75°C.

According to Scott (Am. Dyes. Rep., 1938, *27*, 710), it has been suggested to bleach silk with a solution containing 1 molecular proportion of hydrogen peroxide and 3 molecular proportions of urea.

It should be appreciated that it is possible to overbleach silk with hydrogen peroxide solutions, when the material acquires a yellow colour again.

PERMANGANATE

Solutions of potassium permanganate have been suggested for the bleaching of silk, but are not commonly utilised. The reagent is comparatively expensive and the same objection holds with sodium perborate.

Potassium permanganate is sometimes used, however, for clearing silk before dyeing as distinct from bleaching to a good white. The degummed silk may be treated at room temperature in 2·5% $KMnO_4$ solution, followed by immersing for 45 minutes in a bath of 100 litres of cold water, containing 5 litres of sodium sulphite solution (64°Tw.) and 375 grams of sulphuric acid (168°Tw.). Hurst (J.S.D.C., 1905, *21*, 99) proposed treating the silk in a solution containing 5 grams per litre of potassium permanganate for 4 hours at 50°C., after which the goods were squeezed, without rinsing, and then passed into a bath of 50 litres of water containing 1 litre of sodium bisulphite solution (64°Tw.) and 250 c.c. of hydrochloric acid (34°Tw.) at 40 to 50°C. The silk is rinsed and then treated with dilute hydrogen peroxide solution.

Arnold (Färber Z., 1912, *23*, 513) has described the bleaching of degummed silk in a solution containing 3 grams per litre of $KMnO_4$ for 20 minutes at room temperatures, followed by squeezing and passing into 100 litres of water containing 4 litres of sodium bisulphite solution (64°Tw.) and 400 grams of sulphuric acid (168°Tw.); the silk is allowed to rest in the bisulphite for 30 to 45 minutes, after which it is rinsed and soured, before the final rinsing, hydroextraction and drying.

The use of permanganate may sometimes be advisable with Tussah silk.

PERBORATE

Sodium perborate may be employed for the bleaching of silk in the following manner. For a bath of 100 litres of water, 1 kg. of sodium perborate is adequate; the perborate may be neutralised with about 340 grams of sulphuric acid (168°Tw.) and then brought to the required alkalinity (pH 10) with sodium silicate.

Schappe silk is sometimes bleached with sodium perborate, using 5 to 10% on the weight of the silk; the solution is first neutralised with sulphuric acid and then brought to the required alkalinity with sodium silicate.

Before leaving the discussion of the bleaching processes for the commoner types of silk, it may be of interest briefly to refer to certain treatments which are often applied to silk in the yarn form.

Hanks of silk, after degumming, bleaching and dyeing, are often submitted to certain subsidiary processes in order to improve their appearance.

Stretching may take place before or after dyeing, and consists in placing the hanks on pegs or poles where they are sharply jerked to straighten and stretch them; originally, this was done by hand but machines are available for the purpose.

Glossing consists in stretching and twisting the hanks; they are allowed to remain in the twisted state for several hours.

Lustring is a similar process, but steaming plays an important part.

TUSSAH

Tussah silk is often a dark brown colour and it is not easy to obtain a satisfactory bleaching effect without some damage to the fibre. In spite of the numerous bleaching agents which have been suggested from time to time, it appears that hydrogen peroxide is the only product capable of giving the required effect; the best results appear to be realised when additions of sodium silicate are made to the bath. Tussah silk cannot be bleached by stoving.

As with the other animal fibres, a compromise must be reached between a good white on the one hand, and a satisfactory retention of tensile strength, on the other. It does not appear possible to obtain a good bleaching effect except at the expense of some 15% of the tensile strength of the silk.

With a ratio of liquor to yarn of 30 : 1, a typical modern method of bleaching with peroxide is to utilise 2 vol. peroxide solution containing 6 grams per litre of sodium silicate; this corresponds to 11 grams per litre of water-glass of 135°Tw. Six hours' treatment at 70°C. may be considered adequate, followed by 2 grams of hydrosulphite per litre at 100°C.

If the temperature is raised above 75°C. there seems to be further diminution in the tensile strength of the bleached material; similarly, an increase in the concentration of sodium silicate may improve the efficiency of the bleach, but at the expense of the tensile strength. It has even been suggested to boil tussah silk in 10 vol. H_2O_2 solution made alkaline with silicate.

Shantung material is often bleached in alkaline peroxide solutions containing an active oxygen content of 3 g./l. and 0·8 g./l. of silicate; tussah silk may require a solution of higher available oxygen content (4 g./l.). A subsequent treatment with hyposulphite is advisable and repetitions are helpful for a good white.

The fact that tussah silk may be treated more severely than other varieties is probably responsible for the suggestion to bleach it for 24 hours at room temperatures in a dilute solution of ammonium hypochlorite, prepared from bleaching powder by treatment with ammonium carbonate or ammonium sulphate.

Bleaching with a solution of potassium permanganate containing magnesium sulphate has also been recommended followed by a treatment with sulphurous acid. One method of treatment with potassium permanganate is to prepare a solution containing 25 g./l. of $KMnO_4$ and 20 g./l. of H_2SO_4 (106°Tw.), in which the silk is immersed for 15 to 60 minutes, after which excess liquor is removed, and the goods are passed into 3 to 4% sodium bisulphite solution acidified with hydrochloric acid.

Another suggestion for the bleaching of tussah silk is to use sodium or ammonium persulphate solution at pH 10 and at a temperature of 40 to 60°C.

CASEIN AND NYLON

Manufactured fibres, whether regenerated from natural protein substances or synthesised to simulate the peptide type of molecular chain structure, are supplied in the pure state by the manufacturer and do not require bleaching. Fabrics containing these fibres may also contain acquired impurities which need removing by a mild scouring process.

Casein Fibre

Casein fibres, such as Lanital, Tiolan and Aralac, are only used in admixture with other materials; the high degree of swelling and the loss of strength on wetting are serious disadvantages. The fibre is also very susceptible to alkaline solutions and readily absorbs alkali; hence scouring under alkaline conditions should be avoided and use be made of sulphated fatty alcohols which can be employed under acid or neutral conditions. In the preparation of fabrics, therefore, care should be taken to avoid creases, and the goods should be manipulated with the minimum of tension. Severe treatment of the swollen material which causes softening will also give hard and brittle product on subsequent drying; in order to avoid deterioration it is advisable to add acetic acid to the final rinsing waters.

Drying should be effected at the lowest possible practicable temperature.

The casein fibres do not felt like wool, but it may be milled with soap in the presence of wool; carbonising reduces the resistance to alkali still further, so that alkaline milling should precede carbonising but the amounts of alkali and acid should be reduced to a minimum.

The bleaching of the casein fibres is not easy, and the best results are generally obtained with hydrogen peroxide or with sodium bisulphite. A useful series of articles on the processing of Aralac has been provided by Millson (Text. Col., 1943, *65*, 151, 198, 206, 245, 280 and 294).

Nylon

Nylon appears to be very resistant to many chemical reagents *not* including hydrogen peroxide and sodium hypochlorite; one of its most interesting properties, however, is the ability to acquire considerable permanent set in hot water or steam. Advantage is taken of this property to "pre-board" stockings so that the original shape will be retained even after repeated washings; once the nylon is set, it requires conditions of temperature and moisture which are greater than those of the original set for re-setting in another shape.

Up to the present the chief application of nylon has been in stockings, so that the conditions of set are rather important. There does not appear to be complete agreement as to the best temperature, for although 130°C. has much to recommend it, there is sometimes a danger that the size and knitting oils are likely to become fixed to the fibre and cause difficulties in the subsequent dyeing operations; there is no doubt, however, that a superior set is obtained at 130°C. than at 110°C. There is a general tendency to utilise temperatures between 115 and 120°C.

The scouring and dyeing of nylon should not take place at a temperature in excess of 95°C. in order to avoid interference with the set. In general, it seems wise to employ a temperature about 20 to 30 degrees below that at which the set was imparted.

A suitable scouring bath consists of 2 to 4% of a sulphated fatty alcohol, 1·5 to 3% of soap, and 1 to 2% of trisodium phosphate; the alkalinity of the liquor is often about pH 12 and it may be utilised at 85°C. for 30 to 45 minutes. An alternative scouring bath consists of 3 g. of soap and 1 c.c. of ammonia (d. 0·88) per litre; this may be used at 60°C. for 20 to 30 minutes.

As there are no natural impurities to remove from nylon, the scouring is directed to the removal of acquired impurities such as size and oils; other contaminants such as grease may be difficult to remove except by the usual spotting agents. It may be helpful to pre-treat the fabric by padding in hot concentrated soap solutions and allowing the goods to stand in a roll for 3 or ·4 hours before scouring. Tensions should be relaxed during scouring.

Treatment on a jig has the advantage of keeping the fabric in a flat state during processing, as does the scouring in a continuous machine, although the latter does not maintain an even tension. Scouring in skein or book-fold fashion permits full shrinkage to develop and exerts very little tension. Treatment in rope-form on the wince machine is not advised because of the extra trouble involved in pre-setting the fabric to obviate creases and running marks or in post-setting it to remove them.

It is not usually necessary to bleach nylon which is generally quite white, but the finer changes in cast may be produced by tinting. According to a trade pamphlet produced by Du Pont, the usual bleach with chlorine or peroxide is not to be considered; either the nylon is actually discoloured instead of being bleached, or it may be considerably weakened. Where mixtures of nylon and other fibres have to be considered, it is preferred to use bleached yarns with the nylon.

PART FIVE
THE DRYING OF TEXTILES

CHAPTER XXX
DRYING TEXTILES

THE drying of textiles usually takes place in two stages: (a) the mechanical removal of excess water, and (b) drying by evaporation.

Excess of water may be removed by mangling under more or less heavy pressure, or by hydro-extraction, or suction.

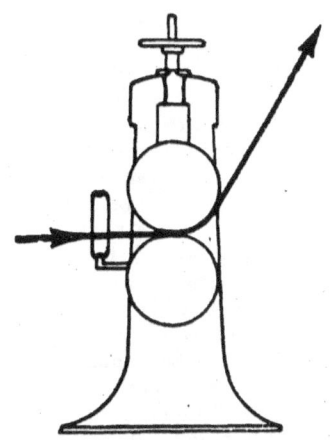

FIG. 111.—Diagram of mangle for cloth in rope-form.

Where the fabric is in rope-form, a small two-bowl mangle or squeezer is commonly employed. This is about 15 to 18 inches in width and of robust construction; two ropes of cotton fabric generally pass through the nip of the machine before passing to the scutcher, which is an apparatus for opening the cloth to full width before drying. The top bowl of the mangle may be made of wood, compressed cotton, compressed coco-nut fibre or of rubber; the bottom bowl, which is driven, is usually made of brass. Pressure may be applied by compound levers and weights or by springs. Fine worsted fabrics are not mangled.

REMOVAL OF EXCESS WATER

Where the fabric is in open width, larger mangles are used and an account of these is given below. Mangling cotton goods in the wet state is sometimes utilised as a form of wet calendering to form a foundation for later finishes or even to close the interstices of the fabric as a finishing operation in itself. It is obviously important in all open-width treatments that the cloth should be free from creases, and this applies particularly to mangling; otherwise the fabric will be permanently marked, or even cut, by the heavy pressure exerted. The type of apparatus used for maintaining the cloth at width is called an expander, and an account of these is given on page 417.

MANGLES

The modern mangle is available in different forms which do not show great variation in principle; the bowls may be of various materials including wood, compressed cotton, brass, vulcanite, or hard or soft rubber. Bowls of wood and cotton are generally built on to steel centres, whereas other bowls have cast-iron centres with steel journal ends; most bowls are made with a certain amount of camber, so that when pressure is applied the expression is even along the entire length and not different in the centre from the ends. The two-bowl mangle is a common type but a three-bowl mangle is more versatile; two metal bowls are never used together as they would damage the fabric—soft bowls, however, are frequently run together. The more usual practice is to have the bottom or driving bowl of metal and the upper bowl covered with rubber of medium hardness in the case of the two-bowl machines; with three-bowl mangles, the centre and driving bowl may be metal, but the other arrangement in which the centre bowl is rubber-covered and the top and bottom bowls are of metal has much in its favour—the metal bowls are the driving bowls.

Pressure is generally applied to mangles by a system of weights and compound levers; a heavy screw-pressure applied to each top bowl-bearing enables the load to be applied or relieved, the load being caused by weights at the ends of levers and varied with the weights. This system enables pressures of 6 tons to be applied, i.e. 3 tons on each end of the bowl.

Hydraulic pressure may also be applied to mangles giving a load up to 10 tons; a small hydraulic pump drives a weighted accumulator which feeds a pair of cylinders under the bottom bearings of the mangle. The pressure is varied by the weights on the accumulator and is indicated on a gauge; a control valve enables the pressure to be applied or withdrawn instantly.

PLATE XXIX

Fig. 112. The Foxwell Guider.

Fig. 113. Scrimp Rail (Farmer Norton).

Fig. 114. Revolving Expander (Farmer Norton).

To face page 416.

PLATE XXX

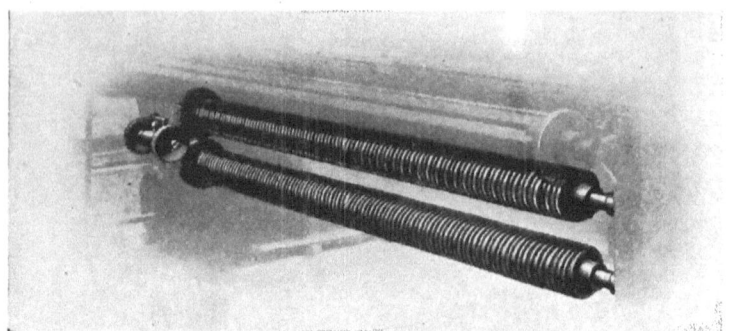

Fig. 115. Scroll Opener (Farmer Norton).

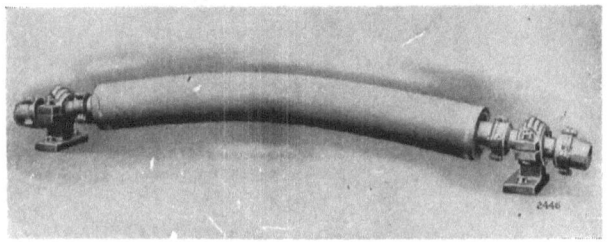

Fig. 116. Curved Rubber Expander (Mather & Platt).

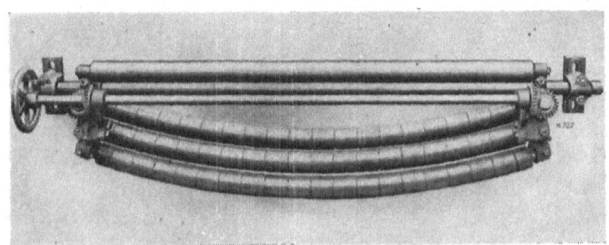

Fig. 117. Mycock Expander (Mather & Platt).

To face page 417.

Pneumatic mangles are of comparatively recent introduction; the hydraulic cylinders are replaced by large pneumatic cylinders and compressed air is supplied from the air-line of the works or from a small compressor. The pneumatic system is cleaner than the hydraulic system.

Expanders and Guiders

Fabric guiders or expanders are standard fittings to be used in conjunction with mangles and other machines in order to keep the cloth free from wrinkles and creases. One of the best known is the Foxwell guider which operates automatically. A driven rubber roller presses against a flat plate and the fabric passes between them; the inclination of the axis of the roller to the weft of the cloth draws the fabric to the side of the machine. The plates are set to the width of the cloth and two stops are adjusted to the position in which the fabric should enter the machine. Any outward displacement of either stop releases the pressure and frees the cloth which is then drawn to the other side by the opposite roller; the cloth is thus maintained between the stops. The rollers may be operated by compressed air, electromagnetically, or by the force from a freely rotating roller in contact with the moving cloth. In the Foxwell-Dungler guider, the rollers are held together by spring-pressure, and the frame carrying them is mounted on a swivel-bearing; the angle of the roller-axis varies automatically with the position of the fabric and regulates the travel of the selvedges.

Expanders which free the fabric from creases and stretch it slightly in the weft direction are available in many forms, one of the simplest being the scrimp rail which is a grooved bar, the ridges of which diverge from the centre; slight warp tension causes the fabric to "bind" on the rail and so slightly stretch the weft, freeing it from creases. Another type of scrimp rail is composed of oval metal discs, with fibre washers between them, and threaded on to a square-section steel bar.

In the Mycock expander, a number of grooved bobbins are interlocked and mounted on a curved shaft; the interlocking is sufficiently loose for the bobbins to rotate as a whole when the cloth passes over them and is thus expanded. It is possible to raise alternate bars, increase the tension on the fabric and hence the lateral stretch; the usual types have 3 or 5 bars.

The simple curved rubber expander is formed by a rubber sleeve which covers a spring mounted on a curved bar; in a modern type, flexibility is obtained by supporting the sleeve with bobbins with plain rims running in ball-bearings—between each spring is fitted a spiral spring and a number of distance pieces.

Scroll opening rollers are very efficient in opening and straightening fabric; they are driven in pairs in a counter-direction to the cloth which passes between them. A device somewhat similar in principle, is formed by conical opening rollers which are set diagonally; the cones are pivoted at a central point, and each is fitted with a brake contrivance which causes one cone to stop should the cloth run to the side where it is situated. This action also causes the two cones to swing sideways from the pivot and make the cloth move back to its central position.

Revolving and spreading expanders consist of a series of floats which move laterally from the centre of the expander to its extremities; they are mounted to form a hollow cylinder, with cam-wheels at the ends to bring about the lateral movement as the expander is rotated by the cloth which passes over it. The floats may be made of grooved brass rails or of felt-covered metal for use on delicate fabrics.

HYDRO-EXTRACTORS

Hydro-extraction may be carried out in the common basket form of centrifuge which consists essentially of a perforated cage of copper or galvanised iron mounted on a central spindle and contained in a steel or cast-iron casing. The wet fabric is placed in the cage, which is then caused to rotate at a high speed, developing centrifugal

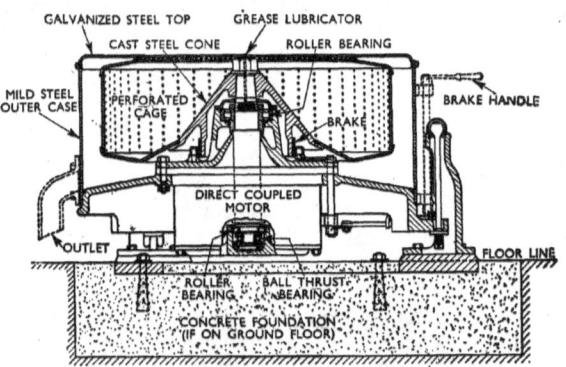

FIG. 118.—Diagram of centrifuge (Broadbent).

pressure which forces the fabric to the inner wall of the cage; the excess of water passes through the perforations and is drained away. Centrifuges of this type are available in various sizes from 36 to 72 inches in diameter and developing speeds of from 1,000 to 600 revolutions per minute respectively; the drive may be by belt, friction or directly by electric motor. Depending on the size of the machine

DRYING TEXTILES

and the type of fabric, the centrifugal action may require from 2 to 10 minutes to remove the excess of water; about 50 or 60% of water remains in the goods, estimated on the weight of the dry material.

With some goods, particularly smooth fabrics and rayons, there is a danger of the perforations of the cage marking the outer layers of the fabric where they are forced against the walls by centrifugal action; this may be obviated to some extent by lining the inside with a loose wrapper of cheap cotton cloth. Centrifugal machines

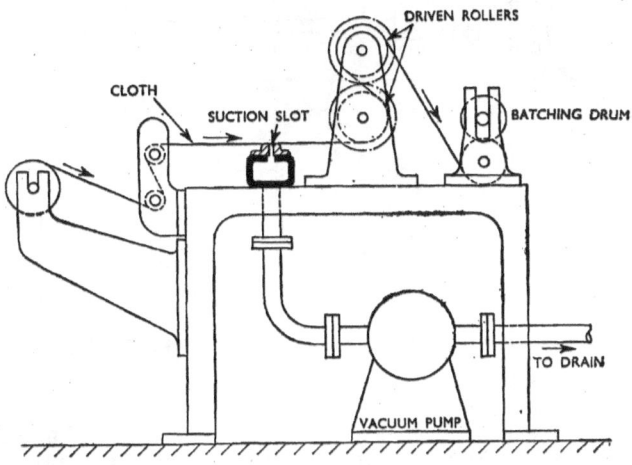

FIG. 119.—Diagram of machine for hydro-extraction by suction (Dalglish).

are also available where the cloth is wound at full width on a perforated cylinder which is rotated at high speed.

The above centrifugal machines extract water from the fabric in a discontinuous manner, but continuous extractors may be used where the cloth is treated at full width and the water removed by a vacuum pump. These machines consist fundamentally of a cylinder with a narrow slot across the top, and the cloth passes over the slot, water being sucked out of it by a vacuum pump which is connected with the cylinder. The length of the slot is adjusted according to the width of the cloth, and water is extracted very evenly with the minimum of disturbance to the face of the fabric. This machine has been found to be quite satisfactory with heavy fabrics, but with lighter materials it is advisable to place a heavy felt over the fabric as it passes the slot, and so create a better vacuum. The suction extractor is thus very suitable for cloths which have to be maintained at full width but may not be subjected to the pressure of a mangle.

TEXTILE BLEACHING

When the three mechanical methods of removing excess water are compared, it is found that the mangle is the cheapest in respect of power costs, and the suction extractor is probably the most expensive. The centrifuge is very useful on account of its versatility as it is capable of dealing with a large variety of fabrics; the output, however, is low.

It is generally advisable to remove as much water as possible by mechanical means, not only on account of cheapness, but also

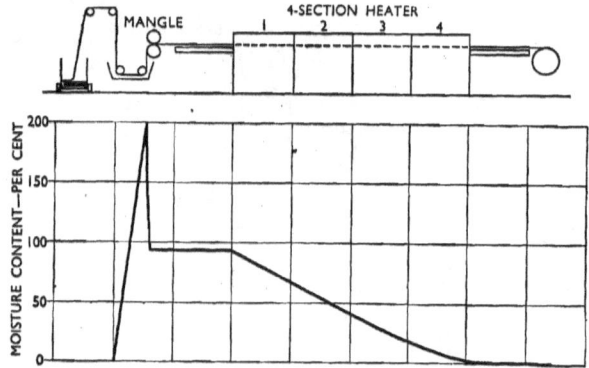

FIG. 120.—Diagram of typical process of drying cloth.

because as these processes reach their limit of efficiency, the residual water is evenly distributed and this is a great aid to the subsequent drying operations. According to Laurie (Engineer, 1943, *175*, 187), a mangle may remove moisture from heavy cloth at the rate of 3 tons per hour at the cost of about 10 kW.-hr., whereas to remove the same quantity by heat would require the equivalent of about 4,000 kW.-hr. The limits of mechanical "drying" seem to be about 50% of the weight of the cloth, but more commonly the residual water amounts to between 80 and 100% of the weight of the fabric. The removal of this residue is the real problem in the drying of textile materials.

DRYING

The evaporation of water requires heat, and whereas this could be supplied by the atmosphere at ordinary temperatures, conditions would not be standardised and the drying would require a long time; hence it is generally the custom to utilise higher temperatures. Although no exact data are available to limit the safe temperatures of heating textile fibres, the usual range employed in actual practice is well below these danger limits. This does not mean that textile materials may be dried without reasonable precautions, for the handle or finish of many fabrics is readily affected by over-drying or by baking.

Studies of the rate of drying of many textile materials indicate that there are four stages to be considered; first, a constant rate of drying whilst the fabric is wet until it contains about 30% of moisture, secondly, a slower rate as the water comes from the interior of the fabric, leaving the surface dry, but retaining moisture of condition, thirdly, the removal of the water vapour from the fibres, and lastly, the removal of combined water. The aim of efficient textile drying methods is to complete the process at the second stage, so that the moisture content of the fabric is nearly normal.

The two general methods of drying textiles are by constant temperatures as in cylinder drying, or by hot-air drying in which the heat content of the air in the immediate vicinity remains constant but the temperature falls as moisture is absorbed, leaving the fabric at a constant temperature as long as water is actually present.

The three available methods of heat transfer are by conduction, convection and radiation. Conduction is commonly employed in cylinder drying machines, or "cans," where the fabric is heated by direct contact with a metallic surface; these machines are rarely used for the drying of woollen goods, and even have certain disadvantages with cottons, as will appear later. The most frequent method of drying textile materials is by convection currents as utilised in the numerous hot-air drying machines where the air is circulated and the fabric is in movement. Even in the old type of drying machines with fixed steam-pipes and no fans to circulate the air, about 70% of the heat transfer was by convection, and the remainder by radiation. Radiation methods have been almost abandoned, except for a few special purposes; however, recent developments, particularly in the U.S.A., indicate a growing interest in infra-red radiations as a means of drying, but the actual examples in the literature appear to combine convection and radiation. Many old-fashioned stenters, however, are still in daily use and some of them rely on radiation from steam-pipes; wool tenters are examples of this, in many cases.

With cylinder drying, the metallic surfaces are kept at a constant temperature by steam, but with air drying it is necessary to provide sufficient heat to evaporate the water and also enough air to remove the moisture coming from the fabric. It has not yet been conclusively proved that a certain temperature is best for hot-air drying but many finishers consider that the goods are adversely affected if the temperature exceeds 70 to 80°C., and with rayons, temperatures of 60°C. are often preferred. It may be argued that as low a temperature as possible should be aimed at from considerations of the behaviour of colloids. With air temperatures of 130°C., however, the temperature of the wet fabric remains low, on account of the heat taken up by the evaporation of moisture, and it is only when the fabric reaches its normal moisture content that its temperature rises appreciably. On the other hand, it has been stated that the rate of removal of moisture affects the handle of the cloth, apart from its temperature, so that a high air-temperature may be detrimental. Presumably the argument is that sufficient time is not allowed for the moisture to diffuse from the interior of the fibre to the surface, with the result that a superficial over-drying or baking may result.

When the textile material is wet, the evaporation of water depends more on the difference between dry and wet bulb temperatures than on the actual temperature. Although great attention has been paid to the dry-bulb temperature of most drying machines, comparatively little attention has been given to the wet-bulb temperature and due regard to this point should obviate over-drying. It does not appear to be sufficiently appreciated that the wet-bulb reading gives the temperature of the wet fabric, and that with accurate control of humidity it is possible to obviate the ill effects of high dry-bulb temperatures. Economy in drying often overshadows the quality of drying, but with many types of silk, rayon and wool the effects of over-drying are difficult to remove even with sulphonated oils and other softening preparations. The operation of the so-called Palmer machine, for example, recognises this principle in a practical manner, for the steam entrapped in the woollen blanket prevents over-drying even though the cylinder may be at a high temperature; the effective action of the decatiser is another case in point.

The chief factors influencing drying are temperature, humidity and air movement; without the last-mentioned there would be little drying for the atmosphere would become saturated. In the counter-flow system of drying, the wet fabric at the entering end encounters hot humid air which has already passed through the rest of the drier, and the high moisture content of the fabric prevents it from overheating, but as the cloth becomes dry the cooling effect of evaporation is lost and the temperature should be lowered. When evaporation starts it should be slow, for a sudden change is almost as harmful as excessive heat in its effect on surface drying and distortions; humid air obviates sudden action or shock to the textile fibres and ensures penetration.

Drying under conditions of controlled humidity is commonly practised with many materials, such as wood, in order to obviate strains and distortions which otherwise occur; the textile industry has been slow to follow suit. It may be remarked that the ideal drying technique, from certain points of view, is to ensure that the rate of evaporation from the surface does not greatly exceed that of diffusion to the surface.

The drying process does not proceed at a uniform rate, for the first 75% of the moisture may be removed in 20 or 25% of the total drying time, but when the moisture content falls to somewhere in the neighbourhood of 30 to 15%, according to the type of material, the temperature of the cloth will start to rise and approach the dry-bulb temperature of the surrounding air. At this stage, the temperature must be lowered to a point at which there is no danger of injury to the material.

Considerations of this type are responsible for many drying machines being divided into separate compartments or zones, with the air circulated by fans. In order to reduce the drying period as much as possible, high temperatures and high air velocities are employed and the air is generally recirculated; the partially saturated air takes up moisture very slowly when it has reached about 60% of its capacity, and it is wise to provide such a rate of flow through the machine that the air is only about 50% saturated as it leaves the outlet.

When the heat is supplied by radiation or by contact with a hot surface, the temperature of the cloth is raised above the wet-bulb temperature; the internal vapour pressure exceeds that at the surface, and the rate of diffusion is increased, with a decrease in time of drying. Although the drying process is faster, the absence of a protective surface layer of moisture is a disadvantage.

With some of the old-fashioned stenters, heat was radiated from steam-coils, but a uniformity of temperature was very rare and shading was common on many dyed goods. This has been obviated by the more modern methods of air-circulation. Further, no air is blown on to the chains, so as to avoid oil spots, and the air is so directed as to balance the fabric and avoid stretching it with the production of "slack."

Reference has already been made to the protective nature of a surface layer of moisture laden air, but in the early stages of drying this is a disadvantage as it insulates the fabric from the main volume of air and hinders free diffusion. When the velocity of the air is increased, the film becomes thinner and permits more rapid drying; the effect of air currents parallel to the fabric is less than that of currents at an angle which may exert a "scrubbing" action. In the later stages of drying, however, diffusion becomes an important factor, so that the question of air velocity is secondary to that of the supply of heat; the air humidity will affect the extent of drying, but not its rate.

It must be remembered that moisture is an excellent plasticiser for textile fibres, which are impaired by its absence. As long as vaporisation does not greatly exceed diffusion, and some moisture remains in the fibres, the protective film of moisture will prevent the more serious aspects of over-drying.

Some interesting data on drying technique were given at a conference of the U.S. Institute for Textile Research (Text. Res., 1936, 6, 359); the principles of drying have been discussed by Cowen (J.S.D.C., 1939, 55, 290).

Most drying machines for fabrics operate at a constant speed for any particular cloth, and no attempt is made to cope with variations in the textile material; hence, the wettest part of the cloth is dried and the remainder of the fabric is often over-dried. This is supported by the well-known fact that for many finishing processes, it is necessary to condition the cloth after drying, i.e. to recondition the fabric. It is customary to allow a very considerable margin of safety in drying fabrics in order to obviate damp spots and irregularities which would appear in the subsequent dyeing or printing processes; most cloth is over-dried and this results in a deterioration in quality and a reduced output.

Within the past few years, developments have been made to measure the amount of moisture in the fabric as it passes through the drying machine and to operate the speed of the machine accordingly. These principles have been applied to the drying of cotton warps after sizing, and are in course of application to fabrics. The detection of the amount of moisture may be effected by measuring the electric resistance or the dielectric constant of the cloth.

With the resistance method, the electrodes may be in the form of wheels or spring plates which press on the drying fabric at fixed intervals; a potential is established between the electrodes and current flows through the fabric between the electrodes, because the conductivity of the textile material varies with its moisture content. By means of a suitable control unit, variations in the current are caused to control the speed of drying. Methods of this type are being developed by Laurie and Dalglish (B.P. 563,480 and 567,259).

It is also possible to estimate the moisture content of the fabric on the capacity principle; an air condenser may be formed by two metal plates situated above and below the cloth which passes between them. The capacity of the condenser is determined by the area of the plates, the gap, and the dielectric constant of the moist fabric. As the dielectric constant of water is about 80 and that of cotton is about 2·5, there is ample opportunity to measure the moisture in the fabric and use this to control the rate of drying.

The precise control of drying and the rate of drying are important factors in textile processing, for many textile properties are greatly affected by minor changes in moisture content.

In bleachworks, the wet cotton is often left lying for very considerable periods of time during which partial local drying occurs on the exposed portions; this may result in uneven dyeing subsequently because of oxidation of the type mentioned by Bone (J.S.D.C., 1934, *50*, 307), who showed that the evaporation of water through cellulose may produce effects of some importance. Strips of cotton were suspended vertically with the lower end dipping into water; tests on the boundary region, as distinct from the wet and dry areas, showed marked reducing properties, increased methylene blue absorption, increased cuprammonia fluidity, and finally complete tendering of the cotton. The modification of the boundary line is accompanied by a thin deposit of a brown water-soluble material. The changes in the cellulose, as detected by dyeing and fluidity tests, occurred after 4 days' exposure in the dark, but oxycellulose is also produced by exposure to light. The modification of the cellulose seems to depend more on the quantity of water evaporated through the strip rather than on the temperature; although this type of modification of cellulose should be borne in mind, it is doubtful if cloth is often left to lie in the wet state sufficiently long to cause tendering, but uneven dyeing is not uncommon.

CHAPTER XXXI
DRYING MACHINES

THE textile materials are often subjected to wet processes before they reach the fabric state; this is particularly the case with wool. Hence arrangements must be made to dry the fibrous material in its various forms of loose fibre, yarn and fabric.

LOOSE FIBRES

Of all textile materials which are scoured, washed and dried in the loose fibrous form, by far the greater part is comprised by wool; hence drying machines for loose fibres are mainly designed for the taeatment of wool, but are generally suitable for other fibrous material also.

One of the oldest types of drier was the table type which is rarely seen to-day, although some of them are still in use. The most popular type was that with a flat top and sloping sides, on a foundation of wire netting upon which the loose wool was laid by hand to a thickness of 3 to 4 inches. A number of steam-pipes were arranged beneath the table, and the warm air circulated through the wool by means of a fan at the end of the machine; alternatively, the pipes were arranged above the table, and the direction of the flow of warm air reversed.

Modern wool driers operate on a continuous basis and are of two main types: (a) single conveyor, and (b) multiple conveyor.

With the single conveyor machine, the wool or other fibrous material falls on to the conveyor at the feeding end of the machine and is carried forward horizontally through the various heated sections of the drying machine. Conveyors may be made on the brattice principle or constructed in the fo m of wire aprons; they vary in length from 30 to 45 feet, but are generally only about 5 to 10 feet wide. As the material passes forward, giving up its moisture, the temperature of the air is gradually lowered according to the moisture content, and the last section of the drier often functions as a cooling section, so that the material leaves the delivery end of the machine in a conditioned state. The "Turblex" drier of Messrs. Tomlinsons is well known, and this machine not only operates on the counter-flow principle in so far as heating is concerned, but the air currents are arranged so as to lift the drying material slightly during its passage through the machine and thus produce uniformity throughout the mass. This method is illustrated in Figs. 121 and 122.

DRYING MACHINES

In some other machines of the single-conveyor type, the air blast is directed downwards and there is no lifting of the mass of fibres. Practically all the modern machines operate on the multifan principle instead of one large fan; this gives better control of the drying and uses the heat more efficiently. The Petrie and McNaught machine, for example, is constructed in sections of 6 feet in length with a fan to each section.

The air at the delivery end is warmed by passing through a battery

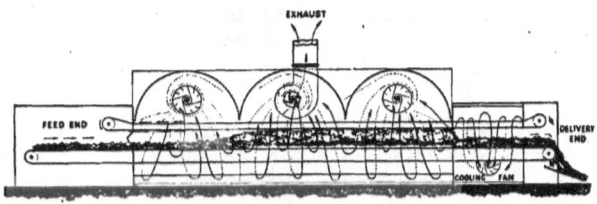

FIG. 121.—Longitudinal section of "Turblex" Drying Machine (Tomlinson) for loose fibres.

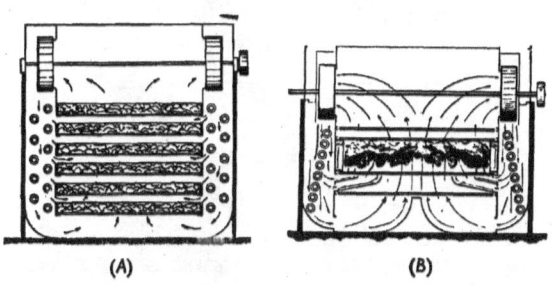

FIG. 122.—Cross-sections of Turblex Drying Machines (Tomlinson). (*A*) six-conveyor, and (*B*) single conveyor.

of heaters, blown through the wool, and then heated again and passed on to the next section; a repetition of this process ensures that the wet fibres at the entering end of the machine encounter air at the highest temperature and at a fairly high humidity. Sometimes the last compartments are arranged to circulate the air several times.

The wet material is generally supplied to an automatic feeder which spreads it evenly on to the brattice which carries it through the machine.

The advantages of the single conveyor drier are its simple mechanical construction and operation, ease of cleaning, and the great control of temperature and humidity throughout the sections.

The multi-conveyor type of drier, on the other hand, requires less space, and the drying surface is obtained by height rather than length. The number of conveyors may vary from 2 to 6, but a few machines are made with a greater number; the conveyors or aprons are staggered so that when the material has traversed the full length of the top conveyor, it drops on to the second and lower conveyor and returns the full length of the chamber, falling on to the next conveyor and so on. It will be realised that with this arrangement, the material is turned over during its passage through the machine and this assists in the even nature of the drying; it is also an advantage with the long wools and with matted material of all types.

HANKS

The "pole" type of drying machine for yarn in hank form, as manufactured by Messrs. Petrie and McNaught, has several points of interest. The yarn is mounted on poles which in their passage through the machine form a series of inverted V's, beneath each of which runs a beater fan mounted transversely across the machine; it is stated that this arrangement brings about a full and open state of the hanks together with a good handle. The air is enabled to pass backwards and forwards through and between the hanks at all

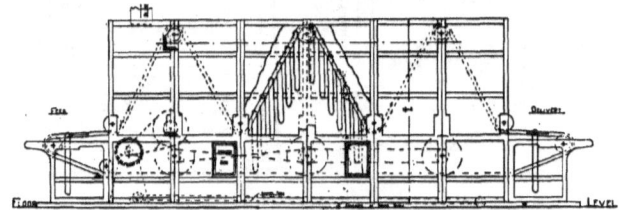

FIG. 123.—Yarn Drying Machine (Petrie and McNaught).

points, and the beater fans are so designed not only to give continuous circulation but to set up pulsations which separate the individual fibres or filaments in the yarn; the extent of pulsation is varied for different materials and is controlled by variation of the speed of the fans.

At the point of each V, each pole is given a complete revolution to assist in even drying; the poles are of large diameter and this, too, ensures a more certain turning of the yarn.

Heat is provided by gilled radiator-pipes mounted on the floor of the machine, and the air is circulated through the pipes and through the yarn; the direction of air-travel is opposed to that of the yarn, so that cold air is drawn in at the delivery end of the machine, thus cooling and partly conditioning the yarn as it leaves. Inside the

DRYING MACHINES

machine, the air is immediately heated by contact with the radiator-pipes and continually reheated as moisture is taken from the yarn, until when it reaches the feeding end of the machine, it has reached its highest temperature where it is in contact with the incoming wet yarn; the air is exhausted at this point, charged with moisture to the highest amount which enables reasonably efficient drying to be

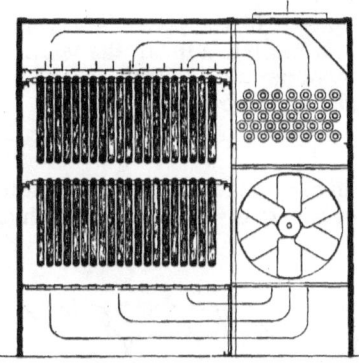

FIG. 124.—Hank Drier (Tomlinson) cross-section.

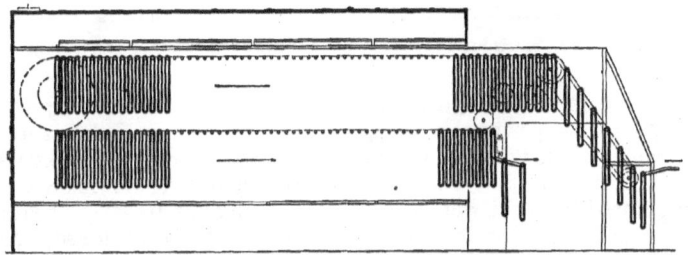

FIG. 125.—Hank Drier (Tomlinson) longitudinal section.

undertaken. It will be seen, therefore, that the incoming wet yarn meets air of high humidity as well as at a high temperature, and so obviates many of the defects often associated with rapid drying at high temperatures and low humidities.

The "Turblex" automatic hank-drying machine of Messrs. Tomlinsons has several points of interest; the entrance of the wet yarn and the delivery of the dried and conditioned yarn take place at the same end of the drying tunnel, so that the feeding and delivery points are both within easy reach of the operative. By means of the special system of circulation, the yarn is not completely dried but is allowed to retain its natural moisture content. This machine is illustrated in the diagrams of Figs. 124 and 125.

FABRICS

The chief machines for the drying of textile fabrics may be considered under the following headings:

(*a*) Cylinders.
(*b*) Hot-air drying machines.

A special development of the latter is seen in the stenter which not only dries the fabric but adjusts the width at the same time; special over-feed stenters are available whereby subsequent warp-shrinkage on laundering is obviated.

CYLINDERS

The simplest form of drying machine is undoubtedly a set of steam-heated revolving cylinders or drums around which the cloth passes

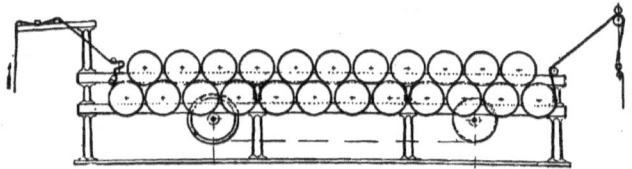

FIG. 126.—Horizontal Range of Drying Cylinders (Mather and Platt).

giving up its moisture, as it moves forward. Steam, under pressure of 5 to 40 lbs. per square inch, is passed into the cylinders through the hollow framework of the machine, a special bearing known as a "doll-head," and the end nozzle of the cylinder. As the water is evaporated from the wet cloth, the steam gives up its latent heat and water condenses inside the cylinder and must be removed; this is effected by an ingenious system of collectors or "buckets" which deliver the condensed water to the opposite nozzle where it drains through the hollow framework on the water-side of the machine, complete with steam traps and other devices.

The cylinders are usually made of copper, tinned copper, or tinned iron to suit various requirements, but for certain purposes stainless steel cylinders may be preferred. A set of drums may be arranged vertically, horizontally or in a combined vertical and horizontal unit; the number of drums in the machine is from six to thirty or more. The horizontal units are generally arranged with the drums in two tiers, but the vertical units may have several stacks of two vertical rows of drums and give a large production in a small floor space where sufficient height is available. The combined vertical and horizontal machine is a development of the vertical machine by bridging the spaces between the stacks with additional cylinders.

The working steam pressure of these cylinders is usually 15 or 30 lbs. per square inch.

Although the use of the "steam-cans" is the cheapest method of drying cloth on a large scale it has certain disadvantages, the chief of which is that the goods are stretched lengthwise and there is no control over the width which is apt to be irregular; the warp tension and firm contact with the smooth surface of the drums impart a harsh papery finish which is not generally acceptable, particularly for silks and rayons. Although more cloth is dried over cylinders than by

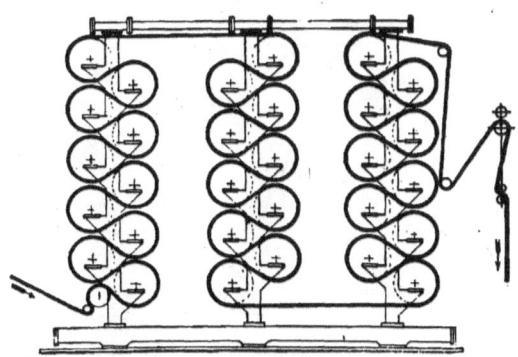

FIG. 127.—Vertical Range of Drying Cylinders (Mather and Platt).

any other method, they are never used with some fabrics, such as crepes, and are rarely used for the final drying process in general. It is possible, however, to effect a partial or preliminary drying on cylinders, and then complete the drying on stenters where control of width and a softer finish are possible. To a certain extent, the harsh finish produced by tin-drying may be obviated by wrapping the surface of the cylinders with a thin cloth to prevent contact with the smooth metallic surface; the question of excessive warp tension still remains, with its effect on handle and subsequent difficulties in respect of laundry shrinkage.

With some fabrics, such as cotton sheetings, the question of warp shrinkage is not serious, nor is the handle such a critical matter as with dress goods; cloths of this type may be brought to the required width on a short stretching frame and then dried on steam-heated cylinders.

Electrically heated cylinders have been introduced by the Lang Bridge Company in 1928; the cylinders comprise a stationary inner cylinder, mounted on a hollow shaft and carrying robust heating elements, and a thin outer shell mounted on ball-bearings and moving freely around the stationary cylinder. The elements heat the small annular space, about 0·25 inch wide, between the inner and outer shells, the former being insulated to keep the heat in the cavity.

The power required to drive these cylinders is much less than with the steam-heated drums and the difficulties associated with leaky doll-heads is entirely eliminated; there is no internal pressure, with attendant risk, in the cylinder, and no steam to condense and necessitate buckets and other devices for its removal. It is possible

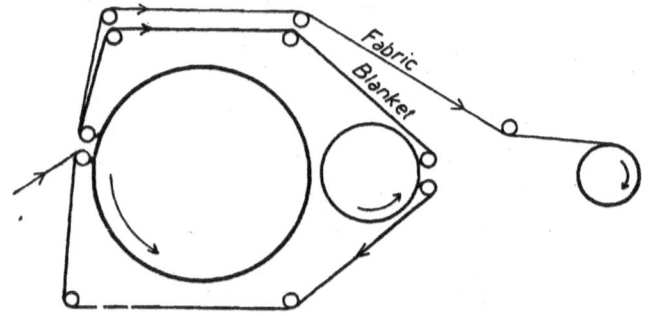

FIG. 128.—Diagram of blanket drying machine or Palmer.

to control the temperature of each cylinder separately and also to have high temperatures which could only be realised by steam at high pressure. It is claimed that the usual cylinder 48 inches in width and 22 inches in diameter only requires 3 kilowatts per hour for full heat, but many of the later cylinders in the range require less than the full amount. With larger cylinders the cost favours the use of electricity, for the steam used increases as the square of the diameter as the inside has to be filled with steam, but with electricity, only the surface has to be heated which means that the current consumption increases in direct proportion with the diameter. For example, a cylinder of 4 feet diameter requires four times as much steam as a cylinder of 2 feet in diameter, whereas the electric cylinder only requires twice the current.

There are two main difficulties in connection with the ordinary methods of cylinder drying; first, there is no control of width, and secondly, some warp tension is necessary to realise the required "bind" of the fabric on the cylinder to remove the creases and distortions—this gives an unwelcome finish to many fabrics.

DRYING MACHINES

A special method of cylinder drying, however, obviates the second drawback and enables the cylinder to be used as a valuable drying and finishing machine; this is the blanket-drying machine or felt calender, sometimes called the Palmer machine. The main cylinder is generally 6 to 8 feet in diameter, but larger types are available; around the greater part of the cylinder passes an endless felt blanket. The fabric is fed between the blanket and the cylinder, and is kept against its surface by the pressure of the blanket during the drying period. In this manner there is no undue tension or warp extension; indeed, a special modification of this machine brings about a contraction in the warp direction. A full handle is given to the fabric, together with a very rich appearance.

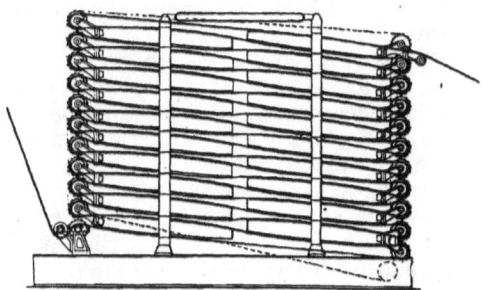

FIG. 129.—The Cell Drier (S. Walker & Sons).

In many of these machines, it is customary to bring the cloth to the required width by a special spreader called a "Palmer" immediately in front of the cylinder; this spreader appears to have passed on its name to the drying machine itself. Occasionally the spreader is replaced by a short stenter to stretch the fabric weft-way, and sometimes a number of smaller cylinders are used to evaporate part of the water in the fabric before it reaches the large cylinder.

Cell-driers have been put forward as alternatives to cylinders. In this machine a series of steam-heated cells, made of a special alloy, are arranged one above the other in two tiers, both of which are slightly inclined to the horizontal. The rectangular cells are 5 inches deep, 4 feet 6 inches wide, and vary in length according to the width of the cloth it is required to dry.

The source of heat is steam at 25 to 30 lbs. pressure, and the fabric may actually make contact with the smooth surfaces of the cells or it may pass between them at a slight distance from the surface; the passage of the fabric may be assisted by rollers.

The cell-drier is a very compact machine and is stated to be very efficient, each cell being equivalent in drying capacity to two ordinary steam-heated cylinders. The cell-drier is manufactured by S. Walker and Sons.

This type of drying machine operates largely by radiation.

HOT AIR DRYING MACHINES

Many machines depend on the use of a current of hot air to absorb and remove the water from textile fabrics; probably the earliest attempts were in some form of the old hanging room, in which the wet fabric was suspended from poles and allowed to dry in the warm air. Air movements were gradually increased with improvements in output, and the air stream was usually heated by passing over steam-heated metallic surfaces on account of the convenience and cheapness of steam as a source of heat.

Although this type of air-drying gave a greatly improved finish to the fabric, yet, compared with direct drying by contact with hot cylinders, a considerable amount of heat must be used in raising the air stream to its working temperature, and only a part of the heat can be utilised in evaporation of the moisture in the cloth.

From special drying-rooms, hanging driers were developed comprising an insulated metal casing inside which a series of poles moved forward slowly on a pair of conveyor chains; the wet fabric was fed into the entering end by rollers with a mechanical device to cause the cloth to fall over the moving poles in loops, the length of which could be varied. Hot air was circulated through the casing of the machine, either vertically or horizontally, but only comparatively low velocities could be employed or the hanging loops would become disarranged or even entangled. This type of Festoon machine was improved in many respects; some of the later types adopt a counter-flow system of air circulation in which the wet cloth at the entry encounters the hottest air, but the dry fabric at the delivery end of the machine meets cooler air.

Increased output was obtained in the hot-flue type of drier, built of panels and framework to form a heated enclosure through which the cloth was drawn over a series of rollers at the top and bottom of the machine. Many of the top rollers in these machines are driven and the bottom rollers act as guide-rollers; the rollers are placed only short distances apart to prevent creasing of the cloth, including turning of the selvedges. A slipping clutch on the driven rollers allows for adjustment of the fabric throughout its passage. In the later types of hot-flue driers, the drying is effected in two stages, first by heat radiated from steam-chests between the laps of cloth and also by hot air, and, secondly, by a final stage of hot air supplied by a fan and multitubular heater, and discharged through vents between the loops of the supported fabric.

DRYING MACHINES

One of the best-known loop drying machines is that made by Messrs. Tomlinsons; the fresh air enters the machine at the delivery end and leaves it, in a highly saturated condition, at the point of entry; hence the drying operates on the counter-flow principle, and the heat applied to the fabric increases in direct proportion to the moisture content. The wettest material, least sensitive to heat, is subjected to the highest temperature, the partially dried material is subjected to gradually decreasing temperatures, and, finally, the

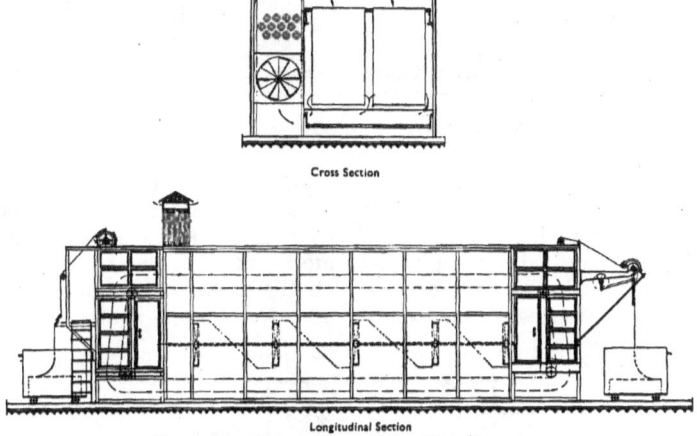

FIG. 130.—Fabric Loop Drier (Tomlinson).

almost dry material, which is most sensitive to heat, is subjected to the lowest temperatures.

The air is drawn by the fans through the heating batteries and made to impinge on the loops of fabric, after which it is again sucked through the heating batteries, and so on, so that each fan subjects the air coming under its influence to a new heating action. Thus the temperature of the drying air increases during its passage through the machine, and its humidity also increases.

Inside the tunnel, the conveying mechanism for the poles carrying the loops of fabric through the machine operates in such a way that the endless conveyor chain moves by the length of one pitch at the exact moment when the loop in process of formation has reached a predetermined length; the length of the loops may be altered in order to adapt it to the type of goods being dried.

The heating coils are generally separated from the drying chamber proper by a vertical partition open at top and bottom, to permit re-circulation of the air over the steam coils and through the festoons of fabric. The air is generally re-circulated a number of times, but a small amount of moisture-laden air is continually drawn off from the machine and a corresponding amount of fresh air admitted. These machines were highly developed in the U.S.A. in the form of Proctor driers and Hurricane driers.

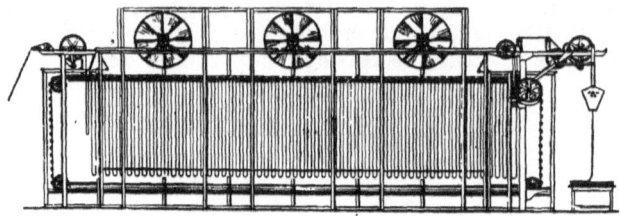

FIG. 131.—The Hurricane Loop Drier.

It became clear that a greater output could be reached if some support could be given to the fabric so that higher air velocities could be utilised; numerous drying machines of the creeper type have been devised, and a typical model utilises several layers of travelling lattice or netting on which the cloth could rest while a transverse air stream blows on it. The well-known brattice driers

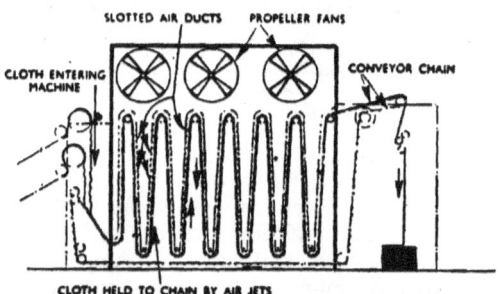

FIG. 132.—Diagram of Air-lay drying machine (Dalglish).

are good examples of this type of machine, and although a very good finish is given to the fabric, the support given to the cloth is not sufficient to permit a high air velocity and consequently a large output.

The Multipass Airlay Drier, however, has been developed to ensure a high output and gentle manipulation. The fabric rests on the flat cork faces of poles which are carried up and down, almost vertically, by a pair of conveyor chains. Fine jets of warm air are blown from

DRYING MACHINES

slots on to one side of the cloth and thus hold it against the cork faces as well as drying it. In this manner, a great proportion of the air comes into intimate contact with fabric and gives a more efficient heat transfer.

Drum-drying may be carried out in special machines of the Buti or Weisbach type. These machines consist essentially of an outer casing in which is situated a drum of large diameter and covered by a blanket of felt supported on a perforated steel shell. The cloth

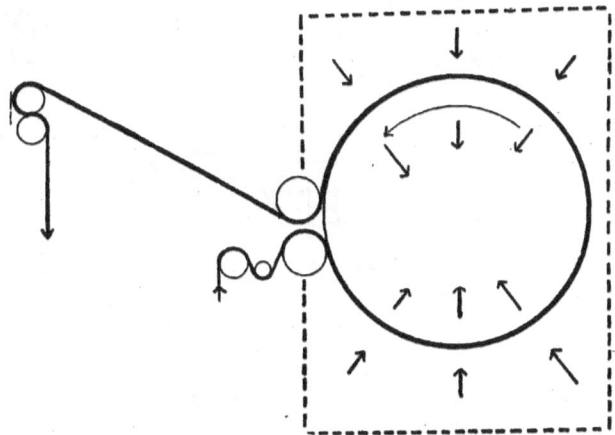

FIG. 133.—Diagram of Buti drying machine.

to be dried is fed into the casing and falls against the blanket on the drum; hot air is blown into the casing and extracted from the inside of the revolving drum after passing through the cloth and the blanket. The air pressure keeps the fabric against the blanket which supports it against any practicable air pressure, but at the same time allows shrinkage to take place if the cloth has been fed at a greater speed than that of the drum itself. This machine is stated to give excellent results in the drying and finishing of light and delicate fabrics, particularly crepes.

With both the Airlay drier and the Buti machine, it is possible to make use of the over-feed principle and so bring about warp-shrinkage during drying and thus remove the potential laundry shrinkage.

Stenters

One of the commonest pieces of textile finishing machinery is the stenter, which is also termed tenter; occasionally it is referred to as the stenter-frame, and in the U.S.A. the word "frame" alone is used.

TEXTILE BLEACHING

The main function of the stenter is undoubtedly to stretch the fabric, and some machines are devoted to this purpose alone, but the great majority combine stretching with drying.

The oldest form of stenter was the fixed hand-frame, which comprised two parallel rails on which were mounted rows of pins to hold the fabric; when the cloth had been impaled on the pins by hand, the rails were caused to diverge by adjustable cross-rails and the cloth dried in the stretched state. The stenter, therefore, dried the fabric in a stretched state, free from creases, and permitted some control over the dimensions of the material. With the older system, which is still used for lace and some knit goods, the frames were often placed in tiers in heated rooms, in such circumstances that as

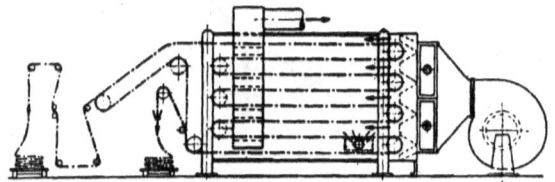

FIG. 134.—Diagram of Multilayer stenter.

the last frame was in course of preparation, the first frame was dry; hence the drying was more or less continuous, but the length of cloth which could be dried at any one time was limited by the length of the frame.

The hand-frame is of value for the drying of certain cloths, such as crepe georgettes, where some control of length is as important as the control of width.

Hand-frames may deal with fabrics up to 400 inches in width, whereas most continuous stenters are limited to about 70 inches for woven goods, although some special stenters for knit goods will operate up to 100 inches.

The first continuous stenter was also a multilayer drying machine, and in the wool and worsted industry, multilayer stenters or tenters continue to be used, but the single layer or Scotch stenter is generally employed for cotton, silk and rayon, although some two-tier or return stenters are in operation for the drying of silk and rayon fabrics.

In tenter or frame drying, the general method appears to rely on surface drying, as distinct from heating, followed by the removal of internal water by diffusion. The aim of the modern stenters with air-circulating systems is to penetrate the fabric and remove the moisture uniformly throughout.

The first tentering machine was made in 1854 by Whiteley of Huddersfield.

DRYING MACHINES 439

The steam coils were placed at different levels to form a number of storeys, and various vertical sections formed a number of bays.

With the continuous stenter machine, the pin-rails of the hand-frame are replaced by moving pin-chains in suitable guides; the fabric was usually fed to the stenter by hand, and the first few yards of rails were caused to diverge until the required width was obtained,

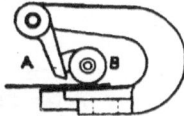 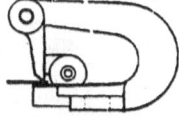 

FIG. 135.—Diagram of automatic clip (Mather and Platt). (*A* is the tongue and *B* the pawl).

FIG. 136.—Diagram of a link in the pin-chain.

after which the chains ran parallel. Steam-heating in one form or another appears to be the chief method of warming the air which dried the fabric, although a carriage of glowing charcoal is still used in certain works which specialise in the hand-finishing of pile fabrics such as velvets. Other uncommon methods of heating include gas burners, and more recently, interest has been aroused in infra-red heating methods.

An alternative method of gripping the selvedges of fabrics during the drying process is by spring clips, and the invention of an automatic or self-feeding clip gave a great impetus to the use of automatic clip stenters which are now so widely used for cotton and linen fabrics. This type of machine is generally made with flat chain-rails, the chains returning in a horizontal plane; the chain-rails are made of cast iron, machined to accommodate the chains, and are mounted on cross-rails. The chain-rails, carrying the chains of clips, are

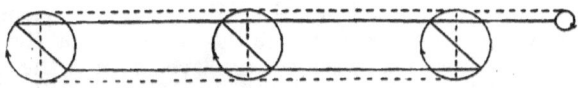

FIG. 137.—Diagram of oscillating motion for jigging.

adjusted to width by screws which are operated together by a shaft which passes along the stenter, but the two sections of rail at the entering end are separately adjustable and are set narrower than the rest, but diverge as they join the longer rails, and so gradually bring the fabric to its full width.

Some clip-stenters of the single layer type are arranged as "jigging" stenters, that is, the cross-rails are caused to oscillate about their vertical centre support and carry the chain-rails with them; this gives the cloth a diagonal or "to-and-fro" movement which helps to straighten weft threads, break any temporary adhesions at the points of intersection of warp and weft, and give a softer effect sometimes termed the elastic finish. The jigging motion is generally provided by a crank disc and connecting rod; the stroke is often variable up to 24 inches, and speed up to 30 strokes per minute. Pin-stenters are not provided with a jigging motion.

The automatic clip-stenters as used for cotton goods may vary in length from 60 to 120 feet, but a popular type is the 90-feet stenter, which may have an output ranging up to 140 yards per minute.

Although the modern automatic clip-stenter is an efficient and economical machine to run, it is not suitable for many of the finer qualities of silk and rayon; further, it does not permit of shrinkage in the warp direction which is so desirable in many shirtings. This may require subsequent correction by some method of controlled compressive shrinkage as in the Sanforised products.

The pin-stenter, however, in its modern form, is suited to the manipulation of very delicate goods, and may also be fitted with over-feed gear which enables the fabric to be placed slackly on the pins by feeding at a known rate in excess of that of the pin chains; the fabric may then shrink in warp direction during drying. Some modern machines have a special tapering main-frame to allow for weft-shrinkage of silks and rayons during drying.

The chains of most pin-stenters return in a vertical plane, in contrast to the horizontal return of the clip machine; this lends itself to lighter construction and simpler driving mechanism. The adjustment of width is similar to that of the clip-stenter.

Wool tenters are generally of the multilayer type on account of the greater amount of heat necessary for drying woollen and worsted materials which are nearly always heavier than cotton and linen goods. A modern type of stenter may comprise as many as ten layers, the fabric being fed into the top layer of the machine up an incline; the hottest air is generally in the top layers and the temperature gradually decreases to the bottom layer, so that the cloth may be conditioned by the incoming fresh air. These multilayer tenters may be built with from eight to ten layers and hold from 55 to 150 yards of fabric at one time.

All modern stenters, whether clip or pin types, have some automatic method of feeding the cloth to the clips or pins. There are two main methods of carrying out this feeding; first, by moving the cloth selvedge into line with the chains; and, secondly, the alternative method of moving the rails to follow the position of the selvedge. The first method is commonly used with clip stenters where no high degree of accuracy is necessary; a small pair of rollers is situated at each selvedge to pull the fabric to the side of the machine, but at a certain point the selvedge meets a feeler mechanism which

PLATE XXXI

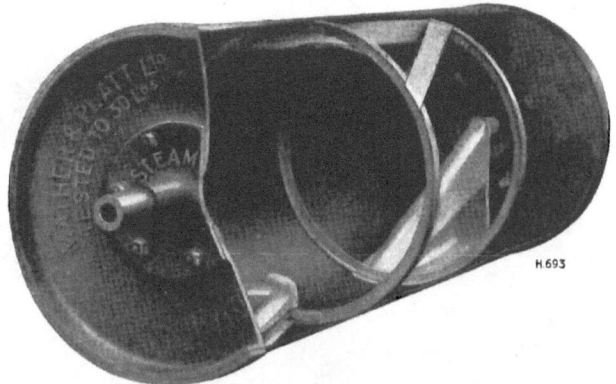

Fig. 138. Drying Cylinder showing buckets (Mather & Platt).

Fig. 139. The Charlesworth-Whiteley Tentering Machine (10 layer, 4 bay).

Courtesy of Wm. Whiteley & Sons.

To face page 440.

PLATE XXXII

Fig. 140. Mather and Platt clip-stenter showing trunking and air-ducts.

Fig. 141. Mills and Platts Pin Stenter.

Courtesy of Mather & Platt.

DRYING MACHINES

releases the grip of the rollers. Typical mechanisms are actuated by mechanical means (Durrant), by electricity (Wood), or by compressed air (Foxwell). A special Foxwell guider is also available for pin-stenters, but most automatic guiders for these machines utilise some method of rail-guiding rather than cloth-guiding. A popular type of rail-guider employs a reversible electric motor on each rail, driving a screw through reduction gearing; the nut on the screw imparts the motion to the rail. A light cloth feeler operates between two electric contacts which in turn close the coil circuit of one or

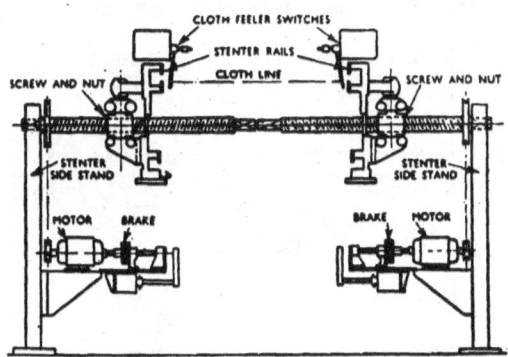

FIG. 142.—Diagram of rail-guider with two motors.

other of a pair of reversing contactor switches controlling the motor. The inertia of the moving parts demands a lightness which is often accompanied by a lack of robustness; hence a reciprocating type of rail-guider has been developed in which the rail is actuated by a piston in a cylinder, the motive power being compressed air or a liquid. The all-pneumatic type has been manufactured by Mather and Platt, and the combined pneumatic and hydraulic type by John Dalglish and Sons. The guiding and control of modern stenters has been described by Laurie (Silk Journal and Rayon World, 1939, October, 23; J.S.D.C., 1940, 56, 289).

Nearly all modern stenters remove the moisture from the cloth by hot air. Multilayer stenters originally relied on heat from steam-pipes which were placed between the tiers or layers of fabric, with the cloth passing over and under each layer. The next step was to circulate hot air through the machine, but the latest methods combine radiation and circulation. Although the fabric still travels around the steam coils, the drying chamber is divided into compartments by partial floors which are fitted with fans. Air is circulated from the two bottom layers into the next two, and distributed

TEXTILE BLEACHING

warp-way; in turn, the next fan and distributing boxes convey the hot air to succeeding layers, the top layer being fitted with a vertical baffle which causes the heated air to penetrate the cloth.

With single-layer stenters, as used for cotton and rayon, the standard method of heating for many years was by streams of hot air directed on to the cloth by branches or "puffer-pipes" which projected from a trunking below, and often above, the frame. Pin-stenters, on the other hand, usually relied on radiation and convection from steam-pipes under the fabric. As the rate of removal of moisture from the fabric increases with the velocity of the air

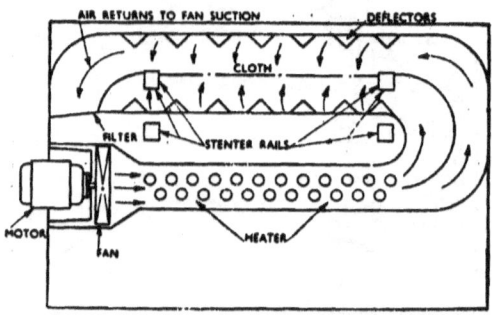

FIG. 143.—Krantz system of air-circulation.

stream, attempts were made to cause air streams to impinge on and along the surface of the fabric from both sides, but at high velocity the old puffer-pipe gave a local and unbalanced effect. The return of the pin-stenter for delicate goods made it essential to balance the high velocity air-streams to avoid distortions. The Krantz and Jahr cross-current systems of air-heating relied on the circulation of air through a heater, across the fabric, both above and below, at right angles to the long axis of the stenter, and back to the fans. The air velocity at the cloth surface was increased by baffles, and the whole machine carefully enclosed by insulated panels. The actual heating of the air is done by gilled-tube radiators, with steam in the tubes; air is blown over the tubes by a fan or by several fans, and it is possible to have the stenter chamber served by fans blowing air at different temperatures, usually highest at the entering end.

The necessary hot air may be produced by a multitubular heater, supplied with low-pressure steam, and one blowing fan, or by one or more battery-type heaters, supplied with high-pressure steam, each with its own blowing fan. On leaving the heater or heaters, the air is distributed by sheet-iron trunking and directed on to the fabric by swivel nozzles.

The efficiency of the heating apparatus depends to a large extent on the design of the enclosure for the stenter. At one time, it was customary to have a number of stenters side by side in a room, but the later developments resulted in the enclosure of each separate stenter in a chamber of brick with cavity walls, although fibre boards on a metal frame and lined with asbestos sheet are almost as efficient. It is also possible to have a chamber which encloses the area around

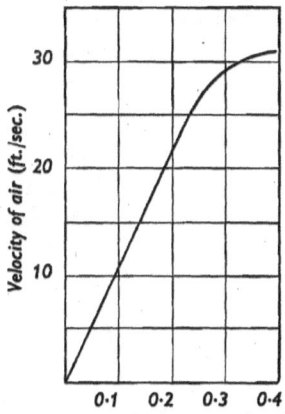

Fig. 144.—Moisture removal from 16-oz. cloth by air at 210° F.

the chain-rails and the fabric, instead of a room or passage through which one may walk.

An important contribution to drying technique has been made by Spooner whose inventions have been protected in B.P. 374,152; 388,167, 396,106; and 409,094. One of the great difficulties encountered when drying textile materials by heated air is that the fabric becomes surrounded by a sheath of warm but stagnant air which is apt to form an effective barrier against the absorption of more heat. The heating medium should therefore be made to pass quickly over the surface of the cloth, continually replacing the atmosphere in actual contact with the fabric. It has been established that the rate of drying is proportional to the speed of the air over the surface up to a speed of 32 feet per second, and above this rate evaporation takes place more slowly; the speed of evaporation is also roughly proportional to the temperature employed, as shown by Spooner (J.S.D.C., 1939, *55*, 303).

The two controlling factors, therefore, are the temperature and velocity of the air currents; both Krantz and Jahr utilised crosscurrents of air kept at comparatively high velocity by baffles, but Spooner subdivided the projected air into thin streams, which enable it to establish intimate contact with the cloth and realise a velocity of impact of 32 feet per second. Nozzles of a truncated venturi type convert the static pressure into velocity and bring the air-stream into direct contact with the fabric; the air-streams impinging on the cloth create sufficient turbulence to destroy the stagnant boundary layer and permit the warm dry air to replace it. The air is

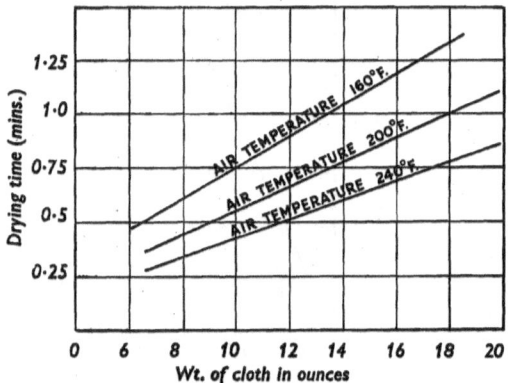

FIG. 145.—Time required to reduce moisture in fabric from 50% to 10%.

caused to impinge transversely on the fabric at high velocity as it passes through the drying machine and in most cases the air is projected evenly on to both sides of the cloth.

The nozzles of the Spooner drying machine perform three separate functions; first, they distribute the static pressure from the pressure chambers evenly over the entire surface of the cloth; secondly, they give the required velocity to the air streams and destroy the stagnant boundary layer; and thirdly, they form with their neighbours a channel which produces a turbulence which is very desirable for effective drying. The temperature of the air streams is such that the material will not be damaged should a stoppage occur.

For stenter-drying, the pressure chambers are situated above and below the fabric with the nozzles or slits stretching the whole width of the cloth. After the air is projected on to the material, it is removed sideways and the main volume is passed through a heater and re-circulated by a fan of special design with two runners, the larger for circulating the air and the smaller for introducing fresh air. An interesting feature of the high evaporative effect produced

DRYING MACHINES 445

is that the fabric itself is kept relatively cool during evaporation of the water with a beneficial effect on the handle and finish.

The Spooner drying sections can be built into almost any shape and used for various purposes. For instance, in the vertical type of drying machine where the pieces are carried upwards by rollers driven at the top, there is practically no tension and the fabric may be passed vertically between banks of opposing nozzles which, being in equilibrium, obviate any ballooning of the cloth.

Nearly all modern stenters are electrically driven and fitted with push-button control.

In many cases, it is essential that the fabric should enter the stenter with its weft straight and at right angles to the warp of

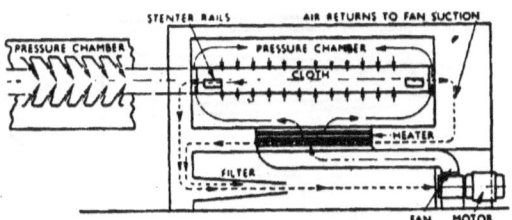

FIG. 146.—Diagram of Spooner method of air circulation.

the cloth; this can be done by mounting the roll of fabric on a swivelling stand and making the necessary adjustments by hand. There are, however, two mechanical devices for straightening the fabric, one of which is to use a series of canting rollers so controlled as to lengthen the path of that selvedge which requires retarding to bring it at right angles to the warp; the second method is to mount a differential gear on the vertical driving shafts of the stenter and retard or advance the speed of one of the chains as necessary.

A somewhat elaborate system operated by photo-electric cells has been devised by the G.E.C.

Some stenters are fitted with steam-boxes at the entering end, so that fabrics which have been dried by festoon driers, for example, may be damped just sufficiently to enable the stenter to bring them to the required width. Again, some stenters operate in conjunction with drying cylinders, which may be placed at the entering end to assist in the drying of the fabric which is completed in the stenter; alternatively, some clip stenters are followed by cylinders to dry the selvedges.

At the delivery end of the machine, the fabric is usually plaited on to tables or into wagons; the short type of stenter used for stretching without heating may be fitted with a drum batcher, but the fabrics on which it operates should be able to withstand tension and somewhat rough handling.

Other methods of batching, of course, may be used even with delicate fabrics.

KNIT GOODS

Although knit goods are available in tubular form and also in the flat state, most of the latter are converted into a tubular form by stitching the selvedges together before wet processing. When ready for drying and finishing, the excess of water may be removed on the centrifuge, after which the goods are often dried by hot air in a brattice machine.

The brattice drying machine generally takes the form of a

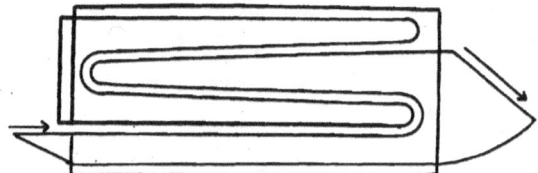

FIG. 147.—Diagram of brattice drying machine.

rectangular chamber through which warm air is circulated; the brattices are preferably made of stainless perforated strips which form long horizontal loops on which the fabric is carried, falling on to the lower lattice at the end of each loop, and thus by a circuitous route traversing the length of the chamber several times as it falls from tier to tier until it finally emerges in the dry state.

The dried material is actually "inside-out" and must be turned for the dry-finishing process; this is usually done by threading the fabric on to a metal tube of large diameter and then withdrawing it outwards through the centre. There are no continuous methods for turning tubular fabrics.

An alternative to the brattice drier is the tubular drier where the fabric is turned with the face side out, and then piled around a vertical tube through which hot air is blown upwards. Above the tube is a special stretching device and a two-bowl calender.

The drying and finishing of tubular fabrics presents certain difficulties, one of which is that they must generally be manipulated as single pieces with consequent loss in production; by using a cartridge-loading mechanism, however, it is possible to approximate to continuous treatment. A modern drying and finishing unit, made by Samuel Pegg and Son, comprises the cartridge-loading device, the drying tube, drying head, stretcher and, finally, a calender.

DRYING MACHINES

The two cartridges are mounted opposite one another on a turntable, with one cartridge under the drying tube; the latter is a continuation of the tubular centre of the cartridge, and the warm air is supplied from below. The drying tube consists of a steel tube with slots and perforations designed to direct the warm air in an upward direction and also at right angles to the fabric as it is drawn upwards to the nip of a calender which forms a type of seal forcing the warm air to pass through the ascending fabric and dry it. The

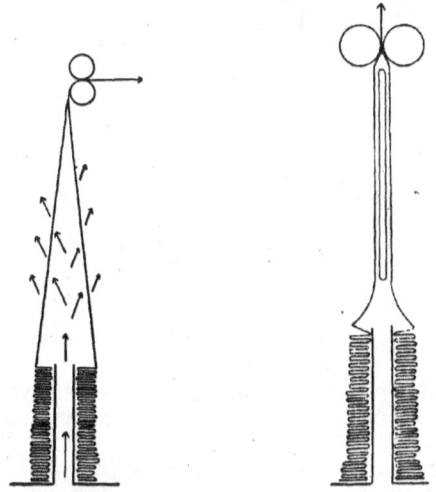

FIG. 148.—Diagrams of drying machines for tubular fabrics.

drying head, which is detachable, comprises a steel frame on which are mounted five aluminium rings and an adjustable stretcher; the rings are supplied in various diameters so that a wide range of fabrics can be handled. The bottom ring is a tensioning ring which controls the length of the material, while the width is adjusted by the four upper rings. Drying the fabric in a circular form evenly distends the loops and eliminates bowing and distortion.

Stretchers may be flat or tubular, and a common flat type has anti-friction guiding wheels along its outer members, over which the tubular fabric passes; another type is a tubular stretcher with spaced metal rings on the outside of the perforated tube. The "drag" of the fabric is distributed over the whole surface of the tube so that bowing of the wales does not arise; the flat metal frame stretcher is usually attached to the top of the drying tube, but may be suspended vertically and kept in position by fabric feeding wheels which press lightly on the outside of the tubular fabric.

A special type of pin-stenter has been devised for drying and finishing locknit material in open width; special spiral rollers open the curled selvedges which are then impaled on the pins by rotating brushes. Feeding devices on each side of the stenter act in accordance with the varying width of the fabric, and the length is also controlled by varying the rate of feed on to the stenter, enabling the fabric to be extended or contracted.

Another type of stenter for knit goods consists of a short frame to bring the fabric to the required width, after which it passes over a series of steam-heated cylinders of small diameter.

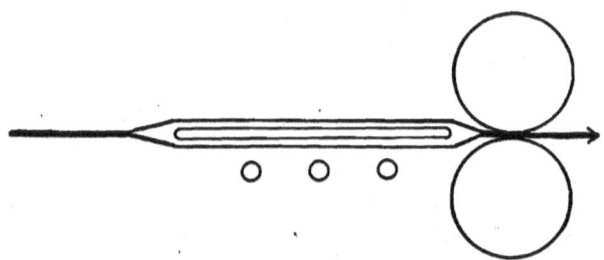

FIG. 149.—Diagram of steaming and drying tubular-knit fabrics.

The handle of goods dried on the stenter is somewhat firmer than from other drying machines, but this may be adjusted by running through soft rubber rollers.

Tubular knit materials are not finished when dried as described above, but still require a final calendering during which the width is also adjusted. Moderately heavy calenders are used for woollen material, but lighter calenders for silk and rayon. It is customary to cover the steel bowls, which are about 8 inches in diameter, with an endless blanket, but this may be removed when it is desired to finish the fabric in direct contact with the metal surface. Endless blankets also guide the fabric round the bowls so that it passes between the blanket and the heated bowl being kept in contact for about 240° of the circumference. Stretchers maintain the width of the material, and each edge of the flat tube of the fabric passes between the stretcher and a small rotating rubber wheel, which drives the fabric forward; bowing of the wales may be controlled by the pressure between the fabric and the rubber wheel.

The fabric is steamed just before it passes through the calender and the moisture enables it to be set to the required width; jets of steam may be blown into the fabric from the inside where the fixed type of stretcher is employed, or, alternatively, with the floating stretcher, the fabric may pass through a steam box.

PART SIX
TESTS FOR DAMAGE

CHAPTER XXXII
DAMAGE IN THE CELLULOSIC FIBRES

THE measurement of damage in textile materials which have been subjected to various processes is associated with the growth of scientific control on the one hand, and the growing demand for guaranteed goods on the other.

The term "damage" in this connection is obviously wider than the appearance of gross mechanical defects, and includes certain modifications of the material which may be acceptable to the consumer within certain limits. Various properties are believed to be associated with the performance of the fibres in actual use, and it is customary to rely on methods which have been devised for their accurate measurement. Some of these tests are concerned with incipient or even latent damage, and are now regarded as very important, whereas other tests relate to damage which has reached completion and may not increase during washing and wear.

Tests may perhaps be conveniently divided into physical and chemical methods.

The physical tests include the well-known measurements of tensile strength which should include both breaking load and extension at break; bursting tests are also of some value on fabrics. Within recent years there has been a very welcome tendency to perform these tests under standardised conditions of relative humidity and temperature; and also to test the material in the wet state as well as in the dry state. Measurements on fabrics give results which are complicated by the large number of factors associated with fabric strength so that interpretation of the results is difficult; single-thread testing is often more satisfactory as the results are less difficult to interpret. Tensile strength and extension tests on single fibres necessitate special apparatus and technique, and it is essential to perform a sufficiently large number of tests for the results to be statistically accurate; single-fibre testing is not generally convenient, except in research organisations.

In addition to the tensile tests, attention has been focused on measurements of resistance to wear; numerous machines are available but there is still considerable difference of opinion as to their usefulness. Some tests rely on rubbing the sample against a standard abrasive surface until the threads are worn through, whereas others measure the loss in strength after a given number of rubs.

Other physical tests are not of the mechanical variety, but rely on examination of the material under the microscope; such tests may be qualitative or quantitative. Qualitative tests depend on the detection of changes in the morphological component of the fibres, whereas the quantitative tests necessitate the counting of large numbers of fibres graded, for example, as sound, modified or damaged.

The chemical tests are often of the simple staining variety and may be associated with a counting method when examined under the microscope; in general, these tests rely on the difference in affinity of textile fibres for various dyes or stains according to the degree of modification.

A second set of staining tests may be classified as depending on the reaction of various groups in the fibres with inorganic or organic compounds; the reducing properties of modified cellulose illustrate this class.

Chemical tests are also available which rely on changes in composition of the parent material, as, for example, the measurement of sulphur or nitrogen in wool. With the vegetable fibres, however, such tests are of minor importance.

Greater weight may be attached to the physico-chemical tests, such as solubility in alkali or even swelling in dilute alkali; one drawback to this type of test is that most materials have been washed with alkali at some stage in their processing. The cellulose fibres and also silk are noteworthy in that they are built up of long-chain molecules without primary valency bonds between the chains; they can, therefore, unlike wool, be dissolved in certain solvents without degradation, to give solutions whose viscosity (or fluidity) may be determined. Tests of this type are of very great value and have been of major importance in the scientific control of textile processing. Not only are they very precise, but they have the advantage that the results give a direct measure of the extent of damage so that comparison with undegraded material is not necessary; they may also be carried out on small samples, unlike measurements of breaking load.

A considerable amount of information on textile testing may be found in the various textile year-books and handbooks, but there are also books, such as the following:

Testing of Yarns and Fabrics by Curtis (Pitman, London, 1938).
Textile Testing, by Lomax (Longmans, Green & Co., London, 1937).
Textile Analysis, by Trotman and Trotman (Griffin, London, 1932).
Textiles on Test, by Williams (Chapman and Hall, London, 1931).
Textile Testing, by Skinkle (Macmillan, London, 1940).

Mention may be made of a useful bibliography of *Physical Tests*, by Harvey (Am. Dyes. Rep., 1935, 24, 703).

Methods of determining damage by the usual physical tests are so well known and have been described so frequently that it is not proposed to recapitulate them here. However, it should be realised that tests of strength are not always satisfactory in judging the efficiency of a process or the soundness of the material. For example, the strength of cotton yarns depends not only on the strength of the fibres, but also on the twist of the yarn and the lubrication of the fibres; during the scouring process, the natural lubricant is removed, and this may affect the strength of the yarn according to the degree of twist. With high-twist yarns the effect is small, but with yarns of lower twist the strength may be increased by scouring on account of the improved binding of the fibres; poorer cottons also show an improvement because they are short and coarse. During the scouring process there may also be some chemical attack on the fibres themselves, which may, or may not, be compensated by yarn structure, but would be shown in hard-twist yarn. The structure of the fabric may also increase the binding of the fibres by interlacing of the fine yarns, and this is generally more prominent in the warp; hence many cloths may show an increase in the weft direction and a decrease in the warp direction after scouring.

In addition to mechanical damage and attack by mildew and bacteria, cotton is susceptible to attack by acids and by oxidising agents; as previously mentioned, cotton is remarkable among textile fibres in its resistance to hot dilute alkaline solutions, and is therefore recognised as the best "washing-fabric."

The soundness or otherwise of the material is not always revealed by simple mechanical tests; as will appear later, it is possible for cellulose to be damaged by acid oxidising agents and appear quite strong, but the strength disappears on laundering. Latent damage, however, may be detected by chemical means.

MICROSCOPIC TESTS

Swelling in NaOH and CS_2

The original microscopical test for damage to the cotton hair was developed by Fleming and Thaysen (Biochem. J., 1920, 14, 25; 1921, 15, 407) and was based on Balls' experiments with the viscose process of Cross and Bevan. A small sample of cotton is shaken for 2 to 3 minutes with a mixture of equal volumes of carbon disulphide and 15% (by weight) of sodium hydroxide. The cellulose expands inwardly to compress the central canal, and outwards to the limit imposed by the cuticle; after 30 minutes the stress is sufficient to burst the cuticle through which the cellulose protrudes. Undamaged fibres show the characteristic bead-like swellings, as reproduced on Plate II., whereas damaged cotton hairs are uniformly swollen.

The numbers of damaged and undamaged fibres may be counted under the microscope and the percentage of damaged fibres calculated.

The test has been criticised by Denham (J.T.I., 1922, *13*, 240) and by Burns (ibid., 1925, *16*, 185), but Bright (ibid., 1926, *17*, 396) considers that there was some misunderstanding over the method of making the count; according to Thaysen and Fleming, that portion of a cotton hair in the field of the microscope at any one time was reckoned as one unit, so that a single hair may figure as several units in the count.

The Congo Red Test

During work on the fine structure of the cotton hair, it was observed by Clegg that hairs which had been swollen in sodium hydroxide solution and then stained with Congo Red did not become uniformly coloured, but stained more intensely where the cuticle had been damaged. The original description of the various effects by Bright (J.T.I., 1926, *17*, 396) has recently been considerably amplified by Clegg (ibid., 1940, *31*, 49).

The test is based on the different rates of diffusion of direct dyes into the exposed secondary cellulose and the cuticle (i.e. primary cell wall and cuticle proper) of the cotton hair after swelling in NaOH solution; the secondary cellulose stains more deeply. Subsidiary factors in the test are the spiral splitting of the cuticle and the behaviour of the cotton hair when swollen in NaOH solution, particularly with respect to the constricting action of the cuticle.

The sample of cotton under examination is first immersed for 3 mins. in a solution of NaOH of definite strength to be discussed later; it is then washed in water, placed in a concentrated solution of Congo Red for 10 mins., washed, and immersed in 18% NaOH solution. The first immersion in NaOH solution is intended to promote a controlled amount of swelling, and for this purpose different concentrations of NaOH are used; the choice of concentration is important. For example, if it is suspected that the fibres have been mechanically damaged, and the cuticle ruptured, then 9% NaOH solution (by weight) should be used, for this will close the lumen without stretching the cuticle. Where it is suspected that the cuticle has been weakened, but not actually broken, as in attack by heat, light, mildew or chemical reagents, then a concentration of 11% NaOH should be used in the preliminary swelling, as this will cause a spiral splitting of a weakened but unruptured cuticle. With slight damage which may be produced by mild processing, the first immersion should take place in 18% NaOH solution.

During the staining step, the exposed secondary cellulose (if any) takes up the stain rapidly, so that whereas the cuticle may only be pink, the exposed secondary cellulose becomes bright red; the final immersion in 18% NaOH solution produces a swelling in the differentially stained material and therefore accentuates the colour contrast.

DAMAGED CELLULOSIC FIBRES

It may be remarked that considerable experience with material, which has been subjected to known treatments, is essential before interpretating observations on material of unknown history; neither printed descriptions nor photomicrographs afford a satisfactory substitute for actual observation under the microscope.

With surface damage where the fibre has been bruised or cut by mechanical means, the swollen and stained secondary cellulose will protrude through the cuticle.

Where the cotton hair has been damaged by heat or by over-bleaching or by acid attack, there is a breakdown of the cuticle shown by a splitting into spiral bands which are stained red; with strong acid tendering, however, there is a characteristic bright red blotchiness.

Biological tendering may bear a general resemblance to chemical damage, but the disintegrated appearance of the hairs and the tendency to break along the quick spirals are distinguishing features.

Sodium Zincate

The reagent for this test, which is due to Lewis (J.T.I., 1933, 24, 122), is prepared in the following manner: zinc chloride is dissolved in water and brought almost to boiling-point when dilute ammonia (1 : 3) is added gradually to precipitate the hydroxide. The zinc hydroxide is removed by sedimentation and washed until free from chloride; it is then filtered and dissolved in hot NaOH of 60°Tw. concentration to give a saturated solution which is allowed to cool and finally filtered through glass-wool.

Small portions of cotton fibres are placed on a microscope slide and covered with a cover-slip; a few drops of the zincate solution are run under the cover-slip where they are allowed to remain for a few minutes before removing excess liquor with filter paper. A few drops of water are then run on to one side of the cover-slip and drawn through by filter paper at the other side to wash the fibres.

With perfect cotton hairs, or those subjected to mechanical damage, the fibres are quite definite in outline, but the swollen cellulose projects at the ends to give a characteristic dumb-bell shape. In scoured and bleached material, the hairs are rather less definite in outline and the dumb-bell effect is less pronounced, but with chemically damaged goods, particularly where holes have appeared in the fabric, the fibres are greatly swollen, the ends and edges are blurred and partial dissolution may occur.

The Segmentation Test

It was found by Searle (J.T.I., 1924, 15, 371) that when tendered flax fibres are mounted in water or paraffin wax, they reveal breaks transverse to the fibre length; jagged edges appear giving the appearance of cracks at different levels. There seems to be no tendency to break in a longitudinal direction.

The suspected fibres are mounted and subjected to slight pressure through the cover glass; a normal fibre shows a spiral structure, but a tendered fibre exhibits transverse fissures, and increased pressure may cause separation into fragments.

A further test is to swell the mounted fibres in 15% NaOH solution, when normal fibres may show transverse segmentation, but damaged fibres will break at the planes of segmentation into parallelepipeds.

CHEMICAL TESTS FOR DAMAGE

The degradation of cellulose, whether native or regenerated, is accompanied by a reduction in the length of the molecular chain due to hydrolysis of the 1 : 4 linkage between the glucose units; owing to the larger number of smaller molecules there will be an increase in the number of terminal aldehyde groups which may be revealed by an increased reducing ability. Hence with the formation of hydrocellulose by acid-tendering, there is an increase in reducing power which may be seen in the action on alkaline copper salts, as in the determination of Copper Number.

Degradation by oxidation is usually accompanied by hydrolysis, but there may also be an oxidation of the hydroxyl groups to aldehyde groups followed by a further oxidation to carboxyl groups. If oxidation is carried out in acid solution, the product will contain a large number of aldehyde groups and a small number of carboxyl groups, but if oxidation is effected in alkaline solution the product will contain a small number of aldehyde groups and a large number of carboxyl groups.

The chemical tests for damage depend on determinations of reducing power and acidity.

The physico-chemical methods, however, depend on measurements of chain length either by viscosity (or fluidity) determinations or by the solubility number.

It should be appreciated that with oxidation in acid solution, the molecular chain is not immediately broken but remains in an alkali-sensitive state; treatment with alkali, either by boiling or in the alkaline solution of cuprammonia used for fluidity work, brings about a cleavage of the chain, and results in low strength and high fluidities.

The oxidation of cellulose by alkaline oxidising agents appears to start at the more exposed carbon atoms in the 6 position and the cleavage of the glucose linkages takes place at the same time, probably being influenced by the reaction of oxidation. The alcoholic group is converted to aldehyde and then to carboxyl.

Kalb and von Falkenhausen (Ber., 1927, *60*, 2514) dissolved cellulose in cuprammonia and oxidised it with potassium permanganate to give poly-glucuronic acid; the alcohol groups in the

6 position of cellulose were thus transformed to the carboxyl groups in the poly-glucuronic acid, from which glucuronic acid itself was obtained by acid hydrolysis.

COPPER NUMBER

The reduction of Fehling's solution has long been recognised as a characteristic test for oxycellulose and C. G. Schwalbe (Ber., 1907, *40*, 1347) proposed the adoption of standard conditions in order to obtain comparable results. This led to the "Copper Number" test, which was later modified by Clibbens and Geake (J.T.I., 1924, *15*, 27) of the British Cotton Industry Research Association, who substituted for Fehling's solution the mixture of copper sulphate and sodium carbonate—bicarbonate as advocated by Braidy (Rev. Gén. Mat. Col., 1921, *15*, 35) and estimated the cuprous oxide by the volumetric method used by Knecht and Thompson (J.S.D.C., 1920, *36*, 255).

The following solutions are used in the determination:

(a) pure copper sulphate, $CuSO_4, 5H_2O$ . . . . . 100 g.
    water . . . . to 1 litre
(b) sodium bicarbonate . . . . . . 50 g.
    crystallised sodium carbonate . . . . . 350 g.
    water . . . . to 1 litre

Immediately before use 5 c.c. of solution (a) are run from a burette into 65 c.c. of solution (b), the mixture is raised to the boil and poured over 2·5 g. of the material to be examined, contained in a conical flask of capacity only very slightly greater than 100 c.c. By means of a glass rod the cotton is distributed through the liquid and any air bubbles are allowed to escape, after which the flask is closed with a pear-shaped glass bulb and immersed in a rapidly boiling constant-level water bath. The flask should be deeply immersed in the water, and care should be taken to cover the top of the bath sufficiently to prevent cooling of the reaction mixture by currents of cold air; several determinations may, of course, be carried out simultaneously in a suitable bath. The flask is allowed to remain in the boiling bath for exactly three hours; the contents are then filtered with suction, and the cotton, impregnated with cuprous oxide, is washed first with dilute sodium carbonate and then with hot water. The cuprous oxide is dissolved by treating the cotton on the filter with the following solution:

Iron alum . . . . 100 g.
Concentrated sulphuric acid . . . . . . 140 c.c.
Water . . . . to 1 litre

Two portions of this solution, of volume 15 c.c. and 10 c.c., respectively, are usually sufficient for this purpose, though a further treatment with 10 c.c. may occasionally be necessary in the case of highly reducing products. The cotton is then washed with 2 N sulphuric acid, and the combined filtrates and washings are titrated with standard potassium permanganate solution of concentration approximately $N/25$, corresponding to about 2·5 mg. of reduced copper per cubic centimetre. The end-point of the titration is sharp and stable, which is not the case when Fehling's solution is used as the copper-containing reagent, owing probably to the fact that cotton absorbs tartrates from this solution, and that these, like the absorbed bivalent copper are only removed by acid washing. It is certainly the case that when Fehling's solution is used in this determination the pink colour produced by a slight excess of permanganate soon disappears when the titrated liquid is allowed to stand, and this may well be due to reduction of permanganate by the absorbed tartrates which have been extracted from the cotton by the acidic iron alum solution.

A micro-method of estimating copper numbers has been described by Heyes (J.S.C.I., 1928, *47*, 90).

<small>A sample of cotton, 0·25 g., is placed in a covered test tube (4 ins. by 0·75 in.); 9·5 c.c. of solution (*b*) is added to 0·5 c.c. of solution (*a*), and the mixture heated to boiling-point, when it is poured on to the sample. The tube is then immersed in boiling water for three hours, but should be stirred after 10 mins. to remove carbon dioxide. The mixture should be filtered on a Gooch crucible with an asbestos mat, or on a fritted glass filter, and washed three times with distilled water, after which 5 c.c. of the ferric alum solution should be added to the tube and poured over the sample in several portions. The residue is then washed with several applications of 2 c.c. of distilled water. Finally, the ferric alum and the washings are titrated with standard permanganate by means of a microburet reading to 0·005 c.c. A control test may be made with 5 c.c. of the iron alum solution.</small>

Clibbens and his associates found that the average copper number for raw cotton and grey cloth was 0·9 and bleached cotton gave a value of 0·2. The copper number of bleached material should not exceed 0·3, except in the case of the regenerated cellulose, where values of 0·8 to 1·2 are recorded for viscose rayon.

Further work revealed the point that copper number measurements were of little general value as a guide to the loss in strength of chemically modified cellulose. A rise of one unit may be accompanied by almost complete loss of strength, as in the case of the action of alkaline hypochlorite solution, by 50% decrease in strength, as in acid tendering, or by a very small decrease in strength, as in the action of dichromate on cotton. Further, the copper number reveals no clue as to the source of the damage, and even ceases to be a characteristic of chemically modified cotton which has been subsequently treated with a hot alkaline solution. Fluidity measurements, on the other hand, are a valuable control of the extent of tendering.

Other Reducing Actions

The reduction of solutions of other metallic salts has also been described, e.g. Nessler's solution (Dietz—Chem. Zeit., 1907, *31*, 833); silver nitrate (see later) and ferric salts (Knecht—J.S.D.C., 1920, *36*, 251). The colour of Schiff's reagent is restored by most forms of oxycellulose.

A rapid method of determining the reducing power of modified cellulose has been described by Foster, Kaji and Venkataraman (J.S.C.I., 1938, *57*, 310). The material is extracted with NaOH, acidified, treated with excess of ceric sulphate solution and the latter is then determined by titration with ferrous ammonium sulphate. The conditions of extraction and ceric sulphate oxidation are adjusted to give numbers for reducing power in accordance with the Braidy values. The new method gives lower values for copper numbers above 2 with oxycellulose but not with hydrocellulose.

Another type of reducing action was noted by Ermen (J.S.D.C.,

1912, *28*, 132), who found that oxycellulose behaves in a similar manner to hydrosulphite in the reduction of indanthrene yellow when it is dyed from a hot alkaline suspension. Scholl (Ber., 1907, *40*, 1692; 1911, *44*, 1312) had also remarked on the reduction of vat dyes by oxycellulose.

The Silver Number test of Götze (Textilber., 1927, *8*, 624) has been reviewed by Rinse (Ind. Eng. Chem., 1928, *20*, 1228), who considered it to have certain advantages over the Copper Number.

About 0·5 g. of the bone-dry sample is placed in 50 c.c. of a boiling liquor containing 1% of silver nitrate and 0·7% of hydrated sodium acetate in water; the mixture is boiled gently for about 24 hours in a flask fitted with a long glass tube, and any water which escapes should be replaced. The contents of the flask are filtered through a porous glass filter and washed with silver nitrate solution to avoid colloidal dissolution of the silver which has been precipitated on the cotton; the silver is then removed with hot dilute 25% nitric acid solution followed by three washings with distilled water which is adequate to bring all the silver nitrate into the flask. The liquor is titrated with potassium thiocyanate (0·02 to 0·05 N) with 2 c.c. of saturated ferric alum solution as indicator.

According to Rinse, the time of reaction is longer than for the Copper Number, but the method is easier and more accurate.

The Silver test of Harrison (J.S.D.C., 1912, *28*, 359) is a useful qualitative test for damage.

The reagent is prepared by adding 1% silver nitrate solution, with stirring, to 4% sodium thiosulphate solution and then adding 4% NaOH solution until the liquor is clear. The suspected material may be boiled in the solution or impregnated and steamed; dark patches varying from grey to black indicate local oxycellulose or hydrocellulose.

Most oxycelluloses form coloured compounds with phenyl hydrazine and its derivatives; this phenomenon is evidence of the presence of CO groups.

During the examination of dyed materials, Muller (Helv. Chim. Acta, 1939, *22*, 208) found that treatment with hydrosulphite solutions increased the response of oxycellulose (but not hydrocellulose) to tests based on reducing powers; it seems that an addition compound is formed with the $Na_2S_2O_4$.

The original reaction with phenyl hydrazine and modified cellulose was first established by Witz, and has since been examined by numerous workers. Muller, however, found that hydrocellulose only reacts with phenyl hydrazine and β-naphthyl hydrazine, whereas the oxycellulose will react with many aryl hydrazines; hence it appears that the aldehyde groups of hydrocellulose are less reactive than those of the oxycellulose.

Tests with the aryl hydrazones are better performed by utilising their sodium sulphonates; the reaction product with the oxycellulose may be coupled with diazonium compounds to give azo-dyes.

For testing with the sulphonates, 1% solution for one hour at the boil is suggested; if the hydrosulphite test is applied, then 10 mins. with 0·2% sodium hydrosulphite at 80 to 90°C. is adequate.

PRESENCE OF -COOH

A second set of typical reactions is regarded as characteristic of the presence of the carboxyl group. Schwalbe and his co-workers (Zellstoff u. Papier, 1921, *1*, 100; 1922, *2*, 75) suggested two methods for the measurement of the acidity of oxycellulose, (*a*) estimation of the amount of barium absorbed from barium hydroxide solution, and (*b*) estimation by titration with $N/100$ NaOH, the figures being corrected for ash alkalinity.

The evolution of carbon dioxide during acid hydrolysis is regarded as characteristic of oxycellulose, for Heuser and Stockigt (Cellulosechem., 1922, *3*, 61) have employed the reaction to measure the carboxyl content of oxycellulose.

Typical results are shown below.

ACIDITY OF OXYCELLULOSE

Substance	%COOH
Oxycellulose (H_2O_2)	0·29
($KMnO_4$)	0·65; 1·04
(CrO_3)	1·32
($KClO_3$)	0·66
(HNO_3)	0·97; 0·83
Cotton cellulose	0·03
Hydrocellulose (Girard)	0·04

The quantitative estimation of carboxyl groups has been described by Neale and Stringfellow (Trans. Farad. Soc., 1937, *33*, 881).

The material must first be washed free from cations in order that the whole of the carboxylic acid may be originally present as such; this is done by leaving the oxycellulose in contact with cold 2 N HCl for 30 mins. and then washing repeatedly with distilled water until the washings are neutral to brom-cresol-purple.

The sample is cut into small pieces and 1 g. (dry weight) is placed in a hard-glass, 250 c.c. stoppered and conical flask. 20 c.c. of NaCl solution (50 g./litre) are then added and also 20 c.c. of $N/50$ NaOH solution. The flask is allowed to stand, stoppered, for 30 mins. in the cold, and then four drops of brom-cresol-purple are added and titration is carried out with $N/50$ HCl, boiling and passing CO_2-free air into the flask towards the end of the titration only.

Methylene Blue

The presence of carboxyl groups in oxycellulose may also be estimated by the absorption of Methylene Blue, but the test is apt to be complicated by the fact that increased absorption of Methylene Blue may also be due to the presence of mineral acid which has dried into the cellulose, and formed hydrocellulose. If the material, however, is previously boiled in dilute NaOH solution, the absorption of Methylene Blue by the oxycellulose will be unaffected, but it will be decreased in the case of the hydrocellulose.

The test has been described in detail by Clibbens and Geake (J.T.I., 1926, *17*, 127), but requires 18 hours for its accomplishment;

the absorption requires either a colorimeter, or a centrifuge if the titration method is followed, Nevertheless, the test may be of value in differentiating between oxycelluloses from acid oxidation (low values of Methylene Blue absorption) and those from alkaline oxidation (high Methylene Blue absorption).

In general, the conditions which produce a high Methylene Blue absorption do not greatly increase the copper number, and vice versa; hence the results of the two tests may be of greater value together than separately. It should also be stated that the test is independent of previous treatment with alkali.

For greater reliability, the test should be made with buffered solutions of Methylene Blue, and if solutions buffered at both pH 7 and pH 2·7 are used, the data will indicate the type of chemical agent which has degraded the cellulose. With oxidation, there is greater absorption at pH 7, whereas hydrocellulose formed by "drying-in" gives greater absorption at pH 2·7.

The commercially pure Methylene Blue hydrochloride should be recrystallised two or three times from twelve times its weight of water and then dissolved to give $M/250$ solutions in the following manner:

The neutral solution consists of

Potassium dihydrogen phosphate KH_2PO_4	6·81 g.
Normal sodium hydroxide, $N.NaOH$	29·63 c.c.
Pure Methylene Blue hydrochloride, $C_{16}H_{18}N_3SCl$	1·279 g.
Water	to 1 litre

The solution at pH 2·7 comprises

Pure Methylene Blue hydrochloride	1·279 g.
$N/5$ Acetic acid solution	to 1 litre

The absorption measurement is made by adding 2·5 g. of air-dry cotton to 15 c.c. of the buffered Methylene Blue solution contained in a narrow tube with a constriction in the middle; after standing overnight, the solution is separated by centrifuging. The tube is inverted into one in which it slides easily, and the two are placed in the centrifuge, when the liquid will be collected in the bottom of the outer tube, whereas the cotton remains above the constriction in the inner tube.

The Methylene Blue in the solution may be determined colorimetrically by diluting the standard to the same colour, or by titration (10 c.c.) with Naphthol Yellow S which brings about mutual precipitation with the Methylene Blue (2 moles of Methylene Blue to 1 mole—358 g.—of Naphthol Yellow S). The Naphthol Yellow S solution is prepared in a concentration of 1·5 to 2 millimoles per litre and standardised against the Methylene Blue. The addition of Naphthol Yellow S to the Methylene Blue produces a brown precipitate, the solution becoming less blue and finally changing to yellow; the end-point may be determined by spotting on filter paper, or by centrifuging the mixture and observing the supernatant liquor after each addition of Naphthol Yellow S (see Birtwell, Clibbens, and Ridge; J.T.I., 1923, *14*, 297).

The Methylene Blue absorption is calculated in millimoles per 100 g. of the dry sample; the absorptions of different scoured cottons vary slightly but average 1 millimole from solutions at pH 7. According to Ridge, Parsons and Corner (ibid., 1931, *22*, 137), the Methylene Blue absorptions of regenerated cellulose rayons range from 1·5 to 2·5.

Cotton damaged by acid shows much the same absorption at both pH values and is greater than for undamaged cotton; with normal cotton and cotton damaged by oxidation the absorption is less at pH 2·7 than at pH 7, but greater absorption at both pH values is seen with the product of alkaline oxidation.

Ash Alkalinity

The sample should first be steeped for an hour in $N/10$ H_2SO_4 solution and washed twelve times in distilled water with centrifuging after every wash; the material is then steeped in water for three hours, washed and dried. In this manner, the alkaline potential ash is removed from undamaged cellulose, but not from oxycellulose containing carboxyl groups. The washed material is weighed, ignited, and the alkalinity of the ash determined in the usual manner; the results are generally expressed in milli-equivalents of alkali per 100 g. of dry cotton.

There is a close correlation between ash alkalinity and Methylene Blue absorption.

An interesting test due to Haller (Helv. Chim. Acta, 1931, *14*, 378; Textilber., 1931, *12*, 257) is stated to differentiate between oxycellulose and hydrocellulose and to operate even when the fabric has been treated with alkali subsequent to attack. The test appears to be due to the increased affinity for metal salts possessed by the acidic oxycelluloses.

The material is soaked for two hours, with occasional shaking, in 1% stannous chloride solution, after which it is washed and immersed in a solution of gold chloride so dilute as to be only faintly yellow (1 or 2 drops of a saturated solution per litre). A purple colour develops with the acidic oxycelluloses.

GENERAL REACTIONS

Amongst the more general reactions of oxycellulose, one of the more important characteristic features is that a larger proportion of furfuraldehyde is formed on distillation with HCl (*d.* 1·06) than is obtained from cellulose, the observed values usually being of the order of 1·5 to 3%, although variations of from 0·8 to 8·2% are recorded in the summary of the literature by Clifford and Fargher (J.T.I., 1922, *13*, 189). The furfuraldehyde is generally estimated as the phloroglucide. Cross, Bevan and Beadle (Ber., 1893, *26*, 2527) appear to have been the first to study this reaction. Heuser and Stockigt (loc. cit.) have shown that the proportion of phloroglucide soluble in alcohol was large in the case of hydrocellulose, lower in the case of cellulose and much lower with oxycellulose.

The increased affinity of oxycellulose for basic dyes was first observed by Witz (loc. cit.), and has been largely used as a test. Knecht (J.S.D.C., 1921, *37*, 76) found that oxycelluloses showing a high affinity for basic dyes resist dyeing with the direct azo colours.

A large number of colour reactions are also recorded in the literature. Whilst the production of a yellow colour on boiling with dilute sodium hydroxide indicates either oxy- or hydrocellulose, only the former gives the colour on steaming according to Freiberger (Farber Zeit., 1917, *28*, 221; 235; 249).

Jandrier (Compt. rend., 1899, *128*, 1407) has recorded a number of colour reactions characteristic of oxycellulose, prepared by the action of potassium chlorate. The material is tested, in presence of

DAMAGED CELLULOSIC FIBRES 461

sulphuric acid, with various phenolic compounds. Berl and Klaye (Zeitschr. Schiess u. Sprengstoffe, 1907, *2*, 381) described a blue coloration with iodine and zinc chloride whilst the coloration with methyl orange was suggested as a test by Schwalbe and Becker (loc. cit.).

Witz (Bull. Soc. Chim., 1886, *45*, 309) observed that oxycellulose exerted an attraction for a number of inorganic salts, particularly those of vanadium, which were withdrawn from very dilute aqueous solutions. This was demonstrated by printing the material containing oxycellulose, with an aniline black mixture.

Everest and Hall (J.S.D.C., 1921, *37*, 227) showed that whilst cotton cloths are normally unaffected by immersion in a solution of tetrazotised benzidine, which has been made alkaline by means of sodium carbonate, a yellowish brown colour, which is fast to washing, is developed if oxycellulose is present.

The colorimetric tests have been reviewed and extended by Thomas (J.S.C.I., 1933, *52*, 79).

FLUIDITY MEASUREMENTS

The viscosity of a solution of cellulose in cuprammonium hydrate is extensively used as a routine test in textile laboratories for estimating the effect of technical processes on the strength of cotton goods. The work originated in connection with cellulose intended for the manufacture of explosives, and its historical development may be traced from the pioneer work of Ost (Z. angew. Chem., 1911, *24*, 1892), Gibson and his collaborators (J.C.S., 1920, *117*, 473 and 479) to that of Joyner (ibid., 1922, *121*, 1511 and 2395).

Farrow and Neale of the British Cotton Industry Research Association, established a method (J.T.I., 1924, *15*, 157) of determining the extent of chemical attack on cotton during the processes of manufacture, which was based broadly on the technique developed by chemists in the explosives industry. The reagent recommended contains about 15 g. of copper and 240 g. of ammonia per litre, and sufficient cotton is dissolved to give a 2% solution. An equation is given connecting viscosity and concentration over the range 0–3% which may be applied to other solutions. The falling sphere method was used to measure the viscosity. It was found that the viscosity of a 2% solution of a carefully bleached cotton cloth should lie between 300 and 10 c.g.s. units (log. η 2·5 to 1·0), and that the tensile strength of cotton hairs and yarn which had been chemically damaged by treatments such as oxidation or acid attack decreased with decreasing viscosity of the solution in cuprammonium hydrate.

Clibbens and Geake (J.T.I., 1928, *19*, 77) continued the work of their colleagues and reduced the technique to its simplest form, and embodied all the precautions found necessary as the result of considerable experimental work. The method is recommended by the Fabrics Research Committee of the Department of Scientific and Industrial Research. (The Viscosity of Cellulose Solutions, 1932.) See also J.T.I., 1936, 27, 285.

As the most important feature of the measurements is the correlation with reduced tensile strength, it was found that neither the viscosity nor its logarithm was very suitable, but that the reciprocal of the viscosity or "fluidity" is more satisfactory, since when this is plotted against the extent of "tendering" the resultant curve approximates to a straight line. The increased extent of chemical attack is therefore accompanied by increased fluidity. The concentration of cellulose in the solution was lowered to 0·5% in order to avoid a multiplication of apparatus. The absolute viscosity of a solution of cotton containing 0·5 g. per 100 c.c. is generally about one poise, but more drastic conditions of preparation may reduce this to 0·2 without serious adverse effects on the quality of the material.

The cuprammonium solution contains 15 g. of copper, 240 g. of ammonia and less than 0·5 g. of nitrous acid per litre. It is prepared in a wide-mouthed earthenware bottle of about five litres capacity and closed by a cork carrying a centrifugal stirrer and air inlet tube, both of iron. The inlet tube ends in an upturned jet which conducts the air into the trumpet-shaped mouth of the stirrer. A mixture containing 2·6 litres of concentrated ammonia solution (sp. gr. 0·88), 0·4 litres of water and 3 g. of cane sugar is stirred at 400 r.p.m., with 180 g. of copper which has passed through a 60-mesh sieve, and air is blown in at the rate of 10 litres per hour through a wash bottle containing an ammonia solution of density 0·9. The approximate copper content of the solution can be followed by colorimetric comparison of a sample with a standard of the correct concentration, both diluted tenfold, and when it attains a value distinctly in excess of 15 g. per litre, the stirring and air supply are stopped. The preparation requires about five hours. If air is replaced by oxygen, the time may be reduced to one hour, but the solution generally contains more nitrous acid than is desirable. After standing for half an hour, the solution is syphoned off, and analysed for copper and ammonia, after the residual copper has been removed. The solution is adjusted to the desired concentration by dilution with the necessary amount of water and ammonia. The nitrous acid content is also checked by analysis and the preparation rejected if this proves to be above 0·5 g. per litre. The nitrometer method is used for the determination of nitrous acid, although this is apt to yield a maximum rather than an accurate value. For the analysis, 15 c.c. of concentrated H_2SO_4 are drawn into the nitrometer and carefully covered with water (1–2 c.c.) before gradually admitting 5 c.c. of the cuprammonium hydrate. If during the latter process, the mercury reservoir is held at its lowest position, and each small volume of cuprammonium solution is neutralised before the admission of succeeding portions, the reaction is not inconveniently vigorous. The cuprammonium hydrate solution is stored in a blackened bottle provided with a tubule and tap at the bottom, and connected at the top through a vessel containing alkaline pyrogallol with a gas holder (e.g. a Kipp's apparatus) filled with nitrogen.

The measurement of viscosity is based on the rate of flow and not on the "falling sphere" technique.

PREPARATION OF THE COTTON SOLUTION AND DETERMINATION OF ITS RATE OF FLOW

A viscometer of the form shown at A in Fig. 150 is used both for dissolving the cotton in cuprammonium and for determining the fluidity of the solution. The wide body of the instrument has an internal diameter of 1 cm. and a length of approximately 26 cm., whilst the lower capillary E is 2·5 cm. long, 0·088 cm. in internal and 0·6 cm. in external diameter. The wide portion is etched with two rings, D and B, at heights of 6·2 and 24·2 cm. vertically above the flat end of the lower capillary, and the upper end of the instrument is closed with a rubber stopper carrying a second capillary F, the dimensions of which are unimportant. Each viscometer has an intermediate timing mark C. The measurement is made with a solution containing 0·5 g. of dry cotton in 100 c.c. of solution, and for each instrument a record is kept of the weight of material required to yield a solution of this concentration when dissolved in the volume of cuprammonium necessary to fill the viscometer.

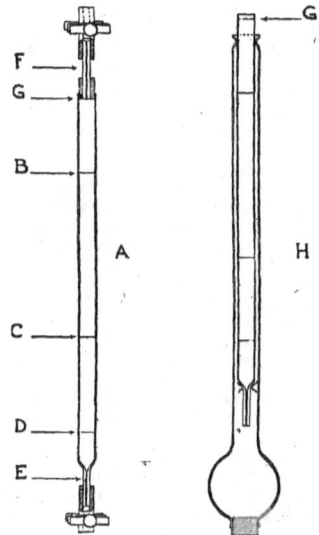

FIG. 150.—Apparatus for measuring the fluidity of cotton and regenerated cellulose rayons.

The cotton for measurement is preferably finely divided, and to secure this, yarn or loose cotton is cut across with scissors into lengths not greatly exceeding one-sixteenth of an inch, whilst cloth in narrow widths is similarly cut in a diagonal direction to break down both warp and weft threads. After the lower capillary has been closed with a short length of pressure tubing and clip, the viscometer with everything in position is filled three-quarters full with cuprammonium, a few drops are run out of the bottom, and the pre-determined weight of cotton added and mixed rapidly with the solvent by means of a thin glass rod. Mercury (0·7 c.c.) is added, the viscometer is then completely filled with cuprammonium and the stopper inserted, so that excess of liquid, displacing all air, overflows through the top capillary and rubber tube. The latter is immediately closed with the clip, the stopper wired into position, and the viscometer, wrapped in black cloth, bound to the spokes of a bicycle wheel rotating at such a rate that during each half revolution the mercury falls from end to end of the liquid column. The agitation produced in this way by overnight running is sufficient to ensure complete and homogeneous solution of the cotton, a wheel rotating at the maximum rate of four revolutions per minute being suitable for the most viscous solutions encountered.

For the specification given above the volume is approximately 20 c.c. so that roughly 0·1 g. of cotton is used in each determination. In view of this small weight

TEXTILE BLEACHING

care should be taken to ensure good sampling. Sufficient accuracy is attained for most purposes if unmercerised cotton, normal or modified, is assumed to contain 6% of moisture and allowance is made for this in calculating the weight of material to be used in each viscometer.

In order to measure the rate of flow of the cotton solution, the viscometer is removed from the wheel, the lower clip and rubber tube withdrawn, and the instrument hung in a wider tube in a thermostat at 20°C. After this temperature has been acquired the viscometer is placed in a glass jacket of the form shown at H in Fig. 150, which is supported vertically in the thermostat. The shape of the jacket is such that the viscometer rests on three glass points at the bottom, and is a sliding fit at the top in the constricted neck. The upper clip is then opened, the stopper removed, and the solution allowed to flow freely through the lower capillary. The time in seconds necessary for the liquid meniscus to fall from the upper to the lower ring is noted. The thermostat may be dispensed with in technical control work if the air temperature is not far from 20°C.

STANDARDISATION OF THE INSTRUMENT AND EXPRESSION OF RESULT

Owing to the fact that the rate of flow is dependent on the fourth power of the capillary diameter, it is not possible to make the viscometer so accurately to specification as to dispense with the necessity for separate standardisation. The instrument is calibrated by means of a mixture of pure glycerine and water, the specific gravity of which is 1·1681 in air at 20°C., compared with water at 20°C. (approximately 64·4% glycerine by weight). The viscometer is filled with this liquid, allowed to attain a temperature of 20°C., and measurement made of the time in seconds required for the meniscus to fall from the upper to the lower ring. For the purpose of routine works control, it is then sufficient to express the viscosity of any cotton solution as the ratio of the time taken by the cotton to that taken by the glycerine solution. The particular concentration of glycerine used for standardisation has, in fact, been chosen because its time of flow is equal to that of a 0·5% solution of cotton, which has suffered the maximum permissible chemical attack in bleaching and finishing. Thus the value one for the ratio referred to above represents a certain minimum standard which should be maintained in technical processes.

As will be explained later, it is more convenient for other purposes to express the results of rate of flow measurements in the form of absolute fluidities (reciprocals of the viscosities stated in poises). In order to do this it is only necessary to divide the "constant of the instrument" obtained as explained in the next paragraph, by the time of flow of the cotton solution.

At 20°C. the specific gravity (d) of the standard glycerine solution is 1·1681, and its fluidity (F) in absolute units is 6·83. If its time of flow in the standard viscometer is t seconds, the constant of the instrument,

$$C' = 1·075 \, (dFt)$$

and can be readily calculated when F, d and t are known. A variation of 0·001 in the specific gravity corresponds with a change of fluidity of about 0·19 absolute units. The constant of instruments of the specified dimensions is approximately 2,000.

When the cotton is highly tendered, the solution consequently very fluid and hence rapidly flowing, it is necessary to apply a correction for the kinetic energy of the liquid. This is conveniently done by subtracting from the time actually observed in the standard viscometer an amount which can be read from a table and which increases with decreasing time of flow. The calculations involved in compiling the table are explained (ibid., *19*, 86). For technical routine purposes the correction can be neglected, as the following illustrations show. A fluidity of 10 corresponds with about 10% loss of strength, and the time of flow in the standard viscometer is approximately 220 seconds; the correction is 2 seconds. A fluidity of 20, corresponding with roughly 20% tendering, gives a time of flow of about 110 seconds and a correction of 5 seconds. The correction evidently becomes rapidly more serious as the fluidity (tendering) increases, but when the tendering is as high as 30 or 40%, an accurate measure is of minor importance in technical control.

The composition of the solvent as recommended by the Fabrics Co-ordinating Research Committee required less ammonia than required when following the method of Clibbens and Geake (200 g. compared with 240 g. per litre).

A further paper on the fluidity of cotton in cuprammonium solution was published by Clibbens and Little (J.T.I., 1936, 27, 285), in which full details of the calibration and calculation of constants are given, together with an account of the applications of fluidity measurement as the result of accumulated experience. When requirements necessitate fluidity determinations on very small samples of material, a rolling sphere viscometer has been designed for use with 0·5% solutions and requiring only 0·01 g. of cellulose. The instrument, calibration, use and limitations are fully described in the above paper.

The complete fluidity scale for cellulose solutions extends from about 2 down to 70, which is the fluidity of the pure solvent. When the fluidity of the material is 40, however, there is almost complete disintegration of the cotton hair, so that for practical purposes the useful range is from 2 to 40. In the case of regenerated cellulose, also measured in 0·5% solution, the fluidity is about 40, and chemical damage may increase this to 60, at which point the filaments are too weak to handle. The fluidity of rayons is, therefore, usually measured in 2% solution when the range of fluidities becomes about 7·5 to 35. These figures are approximately equal to those for cotton in 0·5% solution, and the advantage is therefore gained of the practical measurements being confined to the same range. The fluidity figures for rayon may be recalculated to the basis of 0·5%. The equations of Farrow and Neale (J.T.I., 15, 157) may be used, or, alternatively, the nomograms of Womersley (ibid., 1935, 26, 165) are available.

Tankard and Graham (J.T.I., 1930, 21, 260) point out that the early objections to the rate of flow method as opposed to the "falling sphere" were based on the point that the cellulose solution was prepared in a separate vessel from that in which the determination was made; elaborate precautions were therefore necessary. Clibbens and Geake overcame these objections by preparing the solution in the viscometer itself, but the method of closure does not ensure that the volume of cuprammonium is the same for any other tube, and further, it is only possible to make one determination on each solution. Tankard and Graham also point out that the rate of flow method has other objections such as clogging of the capillary, calibration of each tube being necessary, and the fact that corrections must be applied for surface tension effects and kinetic energy, the latter correction varying with the viscosity of the solution under examination. The rate of flow method was first employed by Ost and improved by Clibbens and Geake; the falling sphere method was utilised by Gibson and his collaborators, by Joyner and by Farrow and Neale.

Tankard and Graham's method is to determine the rate of fall of steel spheres, $\frac{1}{32}$ inch diameter, in the cuprammonium solution of cellulose prepared in the actual tube in which the measurement is to be made. The use of the $\frac{1}{16}$ inch diameter spheres instead of the $\frac{1}{16}$ inch previously used is stated to have definite advantages such as the accurate determination of viscosity over a range going as low as $\bar{1}\cdot 4$ log. viscosity with 2% solutions; it is not necessary to calibrate each tube as Ladenburg's correction for wall effect is applicable, and as small variations in the diameter of the tube do not appreciably affect these calculations, it is possible to plot a graph of observed time of fall against log. viscosity of solution, which applies to all tubes having a diameter of approximately 1·0 cm.

This method requires only a small amount of cotton owing to the fact that the solution is prepared in the actual tube and duplicate readings can be made on each solution. A constant volume of cuprammonium solution for any one tube is assured by means of a new type of stopper.

Strauss and Levy (Paper Trade J., 1942, *114*, TAPPI, 31) prefer cupri-ethylenediamine as a solvent for cellulose in connection with fluidity determinations. The solvent power appears to be sensitive to the molar ratio of amine to copper, and the cellulose dissolves when the ratio is 2; the ethylenediamine solution is completely saturated with cupric hydroxide. The solution, which is 0·5 M in copper, is easy to prepare and standardise; it dissolves cellulose as readily as cuprammonia, and the solutions are stated to suffer no fall in viscosity on exposure to the air for 3 or 4 hours. The specific viscosities are also slightly higher. It appears, therefore, that this new technique may be simpler and quicker, as well as obviating the need for special precautions.

The fluidity of cellulose nitrates, prepared from the suspected cellulose, has been used as a routine test for damage. The method has been developed by Berl and Rueff (Cellulosechem., 1931, *12*, 53) and also by Berl (Ind. Eng. Chem., Anal. Ed., 1941, *13*, 322).

It appears that there is very little degradation when cellulose in nitrated in a mixture containing 50 to 60% nitric acid, 25 to 35% phosphoric acid and 5 to 15% of phosphorus pentoxide; the nitration takes 5 mins. The excess liquor is then removed, and the undissolved nitrate is immersed in 50% alcohol at −10 to −30°C. Stabilisation takes place by boiling for periods of 5 mins. in three changes of 96% alcohol, followed by drying in peroxide-free ether and then at 60°C. The complete process only requires about 45 mins. The measurement of viscosity or fluidity may be effected with 0·25% solutions of the cellulose nitrate in acetone or in butyl acetate; there is no possibility of oxidation.

Davidson (J.T.I., 1938, *29*, 195) has also utilised the cellulose nitrate fluidity method of assessing degradation of cellulose.

He employed a mixture of 48% nitric acid, 50% phosphoric acid and 2% phosphorus pentoxide in which the cotton was steeped for 4 hours at 0°C. The excess liquor was then removed and the nitrate plunged into a large volume of cold water and washed repeatedly for about 24 hours; no stabilising process was adopted.

Measurements of fluidities at 0°C. and 20°C. gave similar results, as did samples of hydrocellulose nitrated at intervals over a period of eighteen months.

Another method has been suggested by Okada and Hayakawa (Cellulosechem., 1931, *12*, 153).

The cellulose is treated with a mixture of 63% H_2SO_4, 27·5% HNO_3 and 9·5% H_2O for two hours at 0°C. or less. The product is washed with water, dried over 40–50% H_2SO_4 and 0·5 g. dissolved in 50 c.c. of acetone. The relative viscosity against acetone is determined with an Ostwald viscometer at 20°C. It is stated that this method does not involve such skilful manipulation as the cuprammonium method, whilst the degradation involved is of the same order as that produced by the dissolving of cellulose in the cuprammonium hydroxide solution.

The newer quaternary benzyl ammonium hydroxides have also been employed for viscosity determinations. Russell and Woodberry (Ind. Eng. Chem., Anal. Ed., 1940, *12*, 151) have compared fluidity measurements in Triton F (dimethyldibenzylammonium hydroxide) and cuprammonia. Measurements have also been made by Brownsett and Clibbens (J.T.I., 1941, *32*, 57), who found that the relation between the fluidities in the two solvents varies according to the method of modification of the cotton, but if the modified cellulose is first treated with dilute NaOH the relation is independent of the manner of modification.

For hydrocelluloses the relative or specific viscosity (0·5% solution) is lowest in cuprammonium solution and is followed in order by the values for solution in NaOH, Triton and cupriethylenediamine, the last giving the highest results.

The viscosity of cellulose in phosphoric acid solution has received attention by Ekenstam (Ber., 1936, *69*, 549), and by Stamm and Cohen (J. Phys. Chem., 1938, *42*, 921), but cuprammonia is still widely favoured for viscosity determinations, although the recent quaternary benzyl ammonium hydroxides appear to have definite advantages, as discussed on page 476, and may become the standard solvents for fluidity determinations with cellulose provided the bases can be obtained in commercial quantities and in a sufficiently pure state.

Comparatively little work has been done on the effect of mercerisation under different conditions, but it seems that the general tendency is slightly to lower the viscosity in cuprammonia, according to Farrow and Neale (loc. cit.) and also Clibbens and Little (loc. cit.).

As a broad generalisation it would appear that in terms of fluidities on the basis of 0·5% solution the most careful preparations of cotton should lie between 1 and 5, whilst the normally bleached material may result in figures between 5 and 10. Fluidities of 20 to 30 indicate over-bleaching or other chemical attack. Most rayons give figures of 40 or more, but some may lie between 30 and 40. Ridge, Parsons and Corner (J.T.I., 1931, *22*, 117) give some interesting fluidity figures for rayons based on a 2% solution, and for wood pulp and cotton linters based on a 0·5% solution. The same authors show that the effect of mercerising on a few samples of cotton material is negligible in so far as fluidity is concerned. The fluidity of regenerated cellulose rayons in 2% solution generally lies between 10 and 12, and should not exceed 12. Although it is possible to determine the fluidity of samples of linen, the presence of substances which are insoluble in cuprammonia necessitates filtering with elaborate precautions, to prevent loss of ammonia and oxidation of the cellulose solution by air. Hence the determination of degradation is better effected by estimating that proportion of the material which dissolves in cold concentrated solutions of NaOH; undegraded native cellulose does not dissolve.

SOLUBILITY NUMBER

The Solubility Number is the number of grams of cellulose dissolved out of 100 grams of the linen material by treatment with NaOH under the above conditions; the solubility number of well-processed linen should not exceed 10. Cotton may also be examined by the same method but the preliminary boil in 2% NaOH solution may be omitted; values of 3 are common for cotton.

Before determining the Solubility Number it is necessary thoroughly to disintegrate the material. This is effected, in the case of cloth, by cutting narrow diagonal widths into shavings not more than one-sixteenth of an inch wide and rubbing these between the hands; 2·5 g. of the modified cotton are steeped in a stoppered bottle with 25 c.c. of 10 N caustic soda for 15 minutes, 100 c.c. of water are then added, diluting the alkali to 2 N, and the mixture shaken occasionally during one hour. The temperature is maintained at 15°C. to within one degree throughout the process. The solution is separated from the residue by inverse filtration through a fritted glass filter, a preliminary separation being effected by centrifuging when filtration is difficult. The filter is of the "inversion" type, 4 cm. in diameter, and made of Jena glass with a glass filter disc of the coarsest grade (Schott. u. Gen., No. 11a, G3/2-3); it is dipped into the bottle containing the mixture, and the solution drawn off into a second vessel. To 10 c.c. of the filtered solution, approximately neutralised with a small volume of sulphuric acid, 25 c.c. of N potassium dichromate are added—or less if the solubility of the cotton is small—followed by 10 c.c. of concentrated H_2SO_4, the total volume being adjusted to roughly 55 c.c. with water. The mixture is boiled under reflux for one hour, cooled, made up to 100 c.c. with water, and 20 c.c. portions titrated with $N/10$ ferrous ammonium sulphate, potassium ferricyanide being used as external indicator. The ferrous ammonium sulphate is standardised immediately before use with $N/10$ potassium dichromate solution. From the volume of dichromate consumed, the weight of cellulosic material in the alkaline solution is calculated on the assumption that its composition is represented by the formula $C_6H_{10}O_5$ and that complete oxidation to water and carbon dioxide occurs. (1,000 c.c. of N. dichromate equivalent to 6·75 g. ot cellulose.)

This method has been adapted by Nodder (J.T.I., 1931, 22, 416) to form a *microchemical* test.

The material to be examined is first boiled for six hours under reflux in a 2% solution of caustic soda. It is then washed free from alkali and carefully dried. The preliminary boil is an essential part of the method, its object being to remove non-cellulosic impurities and cellulose degradation products soluble in boiling dilute caustic soda. The material is then cut into thin shreds. The air-dry powder (exactly 0·1 g.) is weighed into a test tube provided with a ground glass stopper, a convenient size of tube being 5 ins. by ⅝ in. The powder is tapped down into the bottom of the tube and caustic soda (10 N, 1 c.c.) is added, the drops being distributed as evenly

DAMAGED CELLULOSIC FIBRES

as possible. Uniform wetting may be promoted by tapping the tube smartly with the finger-tips.

If wetting is difficult, stirring with a thin glass rod is desirable. As soon as possible the tube is immersed in a water bath maintained at 15°C. (Vacuum flasks have been found very satisfactory for water baths.) After 15 minutes distilled water (4 c.c.) is added in order to dilute the caustic soda to $2\,N$ strength. The tube is replaced in the water bath for a further period of one hour and shaken occasionally during that time. For filtering the solution from the undissolved residue, a small fritted glass filter (Schott. u. Gen., 30a, G3) is used, this being cut down to within 2 mm. of the disc. The filter is attached to the lower end of a 2 c.c. pipette by means of a suitable piece of thick-walled rubber tubing and immersed in the contents of the tube. The pipette is then filled by gentle suction from the water pump. The filtered solution (2 c.c.) is run into a 100 c.c. flask, and to it is added 10 cc. of a half-normal solution of potassium dichromate containing 230 c.c. of H_2SO_4 per litre. The flask is closed with a glass pear stopper and immersed for one hour in a vigorously boiling water bath, being shaken occasionally. The evaporation from the flask is not so great under these conditions to necessitate the addition of water. After one hour, the contents of the flask are cooled and titrated with decinormal ferrous ammonium sulphate, using potassium ferricyanide as external indicator in the usual way. It is rather easy to overshoot the end-point, but if this occurs a back-titration with decinormal potassium bichromate may be made. The half-normal acid bichromate solution (10 c.c.) is likewise titrated. If "a" is the number of c.c. of decinormal ferrous ammonium sulphate required by 10 c.c. of the half-normal acid dichromate and "b" is the number of c.c. of decinormal ferrous ammonium sulphate required by the contents of the flask, then

$$\text{Solubility Number} = 1\cdot 688\ (a-b).$$

This method is very useful for routine work. The solubility number as defined above is the percentage of cellulose (calculated, not on the original weight of the sample, but on the weight of the air-dry state after the preliminary boil in dilute caustic soda), which is dissolved by the caustic soda treatment at 15°C. under the conditions stated.

ALKALI-BOIL

The action of boiling dilute NaOH solution may also be utilised as a test for damage in cotton cellulose.

The original method of Birtwell, Clibbens and Geake (J.T.I., 1926, *17*, 145) was to boil the sample in $N/4$ NaOH solution (i.e. 1%) for four hours, but many chemists now only boil the cotton for one hour. About 2 g. of cotton is dried and weighed; it is then boiled for 60 minutes in 200 c.c. of $N/4$ NaOH solution with refluxing. The liquor is decanted into a Gooch crucible, and the residue washed with NaOH solution, water, and finally with $N/10\ H_2SO_4$. The residue is then washed with water until the washings are neutral to litmus, when the crucible and contents are dried to constant weight at 110°C.

Undegraded cotton cellulose generally shows a loss in weight which may amount to 1·5%, but greater values indicate damage. The percentage loss in weight suffered by modified cellulose is about six times the copper number when the latter does not exceed 2·5, irrespective of the manner of modification. Above this value, hydrocelluloses experience a greater loss, and oxycelluloses a smaller loss, the divergence increasing with increased modification.

THE KAUFFMANN TEST

This method consists in boiling the suspected cellulosic material with aqueous caustic soda and determining the quantity of permanganate required to oxidise the dissolved matter.

The Kauffman "boil-off" number is the number of c.c. of $N/10$ $KMnO_4$ solution required to oxidise the organic matter dissolved from 1 g. of cotton by boiling with 3% NaOH solution; the sample is boiled for 30 minutes with 150 c.c. of 7·4° Tw. NaOH solution with frequent additions of distilled water to conserve the volume. The liquor, together with the water in which the sample is thoroughly washed, is transferred to a 500 c.c. measuring flask and diluted to the mark; 100 c.c. of this solution is acidified with 25 c.c. of 10% H_2SO_4 mixed with 10 c.c. of $N/10$ $KMnO_4$ and boiled for 10 minutes. On cooling to 70°C., excess of oxalic acid is introduced and then permanganate is added until a faint pink colour is obtained. The total volume of permanganate, less the equivalent of the oxalic acid, gives the quantity of permanganate used.

The process should be repeated until the quantity of $KMnO_4$ required is constant; the constant amount of permanganate required by undamaged cotton is thus ascertained and subtracted from the results, as it indicates pure cellulose.

Kauffmann (Textilber., 1923, *4*, 333; 1933, *14*, 138) finds that well-bleached goods generally give numbers under 10; oxycellulose may give boil-off numbers between 30 and 100.

GENERAL

As previously indicated, the characteristic properties of the oxycelluloses depend on the conditions of oxidation. Where cellulose is oxidised by neutral or acid solutions, a reducing type of oxycellulose is formed, which exhibits a high copper number and a low value for Methylene Blue absorption. If, however, the cellulose is oxidised by alkaline solutions, then the product shows a low copper number and a high figure for Methylene Blue absorption.

These two types represent extremes and their properties are shown in the following table:

COMPARISON OF THE PROPERTIES OF THE TWO EXTREME TYPES OF OXYCELLULOSE.

Tests	"Methylene Blue" type	Reducing type
Methylene blue absorption	High	Low
Copper number	Low	High
Loss in weight on alkali boil	Low	High
Properties of residue	as original oxycellulose	as unmodified cotton (excepting viscosity)
Ash Alkalinity	High	Low
Yellow colour with alkali	Negative	Positive
Indanthrene yellow test	Negative	Positive
Resistance to direct dyes	Negative	Positive
Reduction of $AgNO_3$	Negative	Positive

It will be noted that the "high-reducing" type also shows a great loss in weight when boiled in dilute alkali. It was found by Jeanmaire (Bull. Soc. Ind. Mulhouse, 1873, *43*, 334) that cotton goods which had been treated with dichromate solutions and were apparently quite strong, nevertheless became "tendered" when immersed in hot alkali. Witz (Bull. Soc. Ind. Rouen, 1883, *11*, 169) suggested utilising this phenomenon to distinguish between hydrocellulose and oxycellulose, as the former showed a much smaller decrease in strength after treatment with alkali than the latter.

It is not always easy to distinguish between hydrocellulose and the reducing oxycelluloses, but the non-reducing oxycellulose offers less difficulty.

It should be realised that the oxycelluloses of high reducing power which also exhibit an excessive loss in weight when boiled in alkali, leave a residue which is almost identical with the original cellulose in respect of chemical properties; the residue, however, will give solutions of high fluidity. Oxycelluloses with a low copper number only show a slight loss in weight on boiling in alkali, but the high Methylene Blue absorption is only slightly diminished, in spite of the fact that the copper number reverts to that of the original material.

Cotton which has been modified with acid shows very little additional loss in strength after boiling in alkali; the following data by Clibbens and Ridge (J.R.I., 1928, *19*, 390) are interesting in this connection.

LOSS IN STRENGTH (%) AFTER ALKALI-BOILING

Loss in strength after modification	Boiled Yarn after Modification and Re-boiling			
	Modified with acid	Modified with alkaline hypochlorite	Modified with neutral hypochlorite	Modified with dichromate and H_2SO_4
5	5	8	16	20
10	12	16	35	40
15	17	21	48	55
25	24	26	61	67

It will be noted that there is a strong contrast between cotton modified with alkaline hypochlorite and that modified with neutral hypochlorite; the loss is still greater with cotton modified by dichromate and acid. (It should be realised that the concentration of H_2SO_4 used did not, alone, give any tendering.)

Similar results may be obtained by comparing fluidity and strength; neither is greatly affected by boiling material modified with acid or with alkaline hypochlorite.

There is a definite relation between the fluidity of hydrocellulose and the copper number which is irrespective of the conditions of treatment with acid. This relation may be expressed by a simple equation, given by Birtwell, Clibbens and Ridge (J.T.I., 1925, *16*, 13), by means of which it is possible to differentiate between hydrocellulose and the oxycelluloses made in presence of hypochlorous acid which resemble hydrocellulose. For the hydrocelluloses, the copper number-viscosity relation is represented by $N_{cu}V^2 = 2.6$, where N_{cu} denotes copper number and V is obtained by adding 1·82 to the recorded value of the logarithm of the absolute viscosity.

The change in fluidity after treating modified cellulose with alkali is one of the best methods for deciding whether modification is due to oxidation or acid; the other properties do not afford such a sharp distinction, and in certain circumstances may be misleading.

Comparisons have also been made between the copper numbers of modified cottons and the reducing properties of their alkali-soluble fractions; according to Birtwell, Clibbens and Ridge (J.T.I., 1928, *19*, 349), the reducing value of the extract from hydrocellulose is about 4, whereas cotton oxidised with neutral hypochlorite, or with dichromate and sulphuric acid, gave values of about 8, compared with 12 for the hypochlorous acid oxycelluloses, and 16 to 20 for the oxycellulose from dichromate and oxalic acid. The alkali extraction was by $10N - 2N$.NaOH solution.

It appears that the oxycelluloses differ from the hydrocelluloses in yielding extracts of higher reducing value, but it is necessary to exclude from this generalisation any oxycelluloses made from alkaline hypobromite, and also all modified cellulose which has been boiled with alkali. Nevertheless, these differences may be of value in differentiating between hydrocellulose and the reducing oxycelluloses.

With certain reservations, therefore, hydrocellulose and oxycellulose may be distinguished by

(a) Methylene Blue absorption at pH 2·7 and pH 7·0.
(b) Fluidity in cuprammonia after boiling in 1% NaOH.
(c) Fluidity in cuprammonia after treatment with $10N - 2N$.NaOH.
(d) Reducing value of the extract from (c).

Although the fluidity measurements are of the greatest value in assessing damage to the parent cellulose, care is necessary in interpreting the results. It must be remembered that for material attacked by acids or by alkaline hypochlorite, neither the strength nor the fluidity is much affected by boiling in alkali, but where the cotton has been modified by other oxidising agents, both fluidity and tendering are increased by an alkaline boil, although the most marked rise in fluidity on boiling is shown by the neutral hypochlorite oxycelluloses and the greatest additional loss in strength by the dichromate oxycelluloses. The fluidity of cotton attacked by dichromate indicates not only the actual tendering but also a potential tendering which may only appear after boiling with alkali; this is a strong argument for fluidity control of quality, for otherwise material which might lose half its strength on laundering would be passed as satisfactory on a simple estimate of tensile strength.

It cannot be maintained, however, that a given rise in fluidity in cuprammonia always corresponds with the same loss in strength if all causes of chemical modification are taken into account, in addition to the possibility of an alkaline treatment subsequent to modification. Nevertheless, any chemical process which is accompanied by a considerable rise in the fluidity of the cotton will reduce its strength immediately, or after a mild alkali treatment.

The copper number determination is of little real value as a guide to the loss in strength.

Some of the foregoing material is summarised in the following table:

CHEMICAL DAMAGE TO CELLULOSE

	Fluidity		Copper No.	Carboxyl content
	Before alkali-boil	After alkali-boil		
Acid	Varies with attack	Not greatly altered	Varies with attack	Low—scarcely affected
Alkaline oxidation	Varies with attack	Small alteration	Low, relative to attack	High, relative to attack
Acid oxidation	Low, relative to attack	Great alteration	High, relative to attack	Low, relative to attack

From the above comments, it will be seen that alkaline oxidation tends to produce more carboxyl groups than aldehyde groups and acid oxidation gives a low carboxyl content and a high aldehyde formation.

The greatly increased loss in strength after an alkali-boil, as exhibited by oxycelluloses from neutral hypochlorite, acidified dichromate, etc., is evidence of an alkali-sensitive linkage in the cellulose chain-molecule; this probably accounts for the anomalous behaviour of some modified celluloses when cuprammonia-fluidity measurements are made without a previous alkali-boil. With the high-reducing oxycelluloses, the low fluidity is accounted for by the alkaline nature of the cuprammonia itself.

The difficulties associated with the alkaline nature of cuprammonia may be avoided by observing the acetone-fluidity of cellulose nitrates formed from modified cellulose. Davidson (J.T.I., 1938, *29*, 195) showed that within each series of modified cottons there is a definite relation between cellulose fluidity and the cellulose nitrate fluidity, and this relation varies with the method of modification, but that after boiling the modified cellulose with 1% NaOH solution under pressure, the relation between cellulose fluidity and cellulose nitrate fluidity is expressed by a single curve, irrespective of the method of modification, and this corresponds to the curve for hydrocelluloses before alkali-boiling.

The use of cellulose nitrate fluidities enables measurements of chain-length to be made without affecting the alkali-sensitive linkages.

The relation between the solubility and fluidity of modified cotton should be the same for different methods of modification as both properties are functions of the frequency distribution of chain length, provided two conditions are fulfilled: (*a*) if the distribution of actual and latent breaks in the chain molecule changes in the same way with different reagents, and (*b*) if the caustic soda in the solubility measurement, and the cuprammonia in the fluidity measurement, are equally effective in breaking the alkali-sensitive linkage.

Condition (*a*) appears to be fulfilled by modified cottons which do not contain alkali-sensitive linkages (hydrocellulose and alkaline hypochlorite oxycellulose) and by all types after adequate treatment with alkali. The abnormal fluidity solubility relation shown by neutral and acid hypochlorite oxycelluloses suggests that condition (*b*) has not been met; hence, the NaOH in the solubility measurement does not act in the same manner as the alkaline cuprammonia solution.

Brownsett and Davidson (J.T.I., 1941, *32*, 25) showed that the neutral and acid hypochlorite oxycelluloses gave an abnormal rise in fluidity when treated with dilute NaOH from which it appears that cuprammonia is not able to break all the alkali-sensitive linkages which are broken by NaOH.

It seems probable that the alkali-sensitive oxycelluloses with their high reducing powers are formed by an opening of the glucose ring structure without a great reduction in chain-length.

$$\begin{array}{cc}
\overset{|}{CH} & \overset{|}{CH} \\
\diagup \diagdown & \diagup \diagdown \\
O \quad HOCH & O \quad CHO \\
| \quad\quad | & | \\
HOCH_2CH \quad HCOH & HOCH_2CH \quad CHO \\
\diagdown \diagup & \diagdown \diagup \\
CH & CH \\
| & |
\end{array}$$

Additional fragmentation may be brought about by cold dilute NaOH, which is not a solvent for cellulose, or by cuprammonia,

COMPARISON OF HYDROCELLULOSE AND OXYCELLULOSE

Tests	Hydrocellulose	Oxycellulose Reducing	Oxycellulose Acidic
Copper Number	Medium	High	Low
Colour with alkali	Slight yellow	Yellow	None
Reducing action on $Ag(NH_3)_2OH$	Negative	Positive	Negative
Reaction with phenylhydrazine	Positive	Positive	Negative
Reaction with arylhydrazine	Negative	Positive	Negative
Methylene blue absorption	Very Low	Low	High
$SnCl_2$–$AuCl_3$	Negative	Negative	Positive
Loss in weight on alkali-boil	Slight	High	Low
Properties of residue	As unmodified cotton; fluidity as original hydrocellulose	As unmodified cotton, except for fluidity	As original oxycellulose
Change in cellulose nitrate fluidity after alkali-boil	Slight	Great	Moderate

which is a solvent, but produces less additional degradation than NaOH.

Brownsett and Clibbens (J.T.I., 1941, *32*, 57) examined the action of the strong base, dimethyldibenzylammonium hydroxide or Triton F, which is also a solvent for cellulose, and established that it was capable of determining total modification, both actual and latent, and therefore possesses a definite advantage over cuprammonia. With the latter, it is necessary to give a preliminary treatment with NaOH beforehand if the total damage has to be estimated, but the degrading action of Triton F on alkali-sensitive material is similar to that of NaOH.

It is not yet certain, however, whether the Tritons will be available in commercial quantities and of a sufficiently high standard of purity to replace cuprammonia for routine fluidity work.

CHAPTER XXXIII
DAMAGE IN THE PROTEIN FIBRES

THE examination of damage in the animal fibres has not reached such a high pitch of development as with the vegetable fibres. The constitution and structure of wool does not permit of the same type of physico-chemical testing that has been employed with such success for cotton; hence many types of test have been developed but none is completely satisfactory. Silk, however, has recently been examined on a fluidity basis, and it may be expected that detailed information about various types of damage will be available in the near future.

The problem of wool is further complicated by the fact that it is susceptible to a wide range of influences which may bring about modification. For instance, in addition to mechanical damage, wool may be degraded by light probably through oxidation. Other agents which modify wool are bacteria, water, steam, heat, acid, alkali, oxidising agents (including hypochlorite solutions) and reducing agents.

WOOL

The various methods of estimating damage in wool have been given as part of a bibliography of the chemistry of wool by Smith and Harris (Am. Dyes. Rep., 1938, *27*, 183), but a more restricted account of wool damage is due to Ryberg (ibid., 1940, *29*, 588). The most informative review of the measurement of damage in wool materials is probably due to Whewell and Austerlitz (J.S.D.C., 1943, *59*, 45).

Microscopic Examination

Various suggestions have been put forward for detecting damage in wool by examination under the microscope. With the Allworden reaction, for example, the fibres are treated with chlorine water or bromine water; where the wool is undamaged, the characteristic globules appear in 5 to 30 minutes, but if the material has been damaged by light or by alkali, the hairs swell but the globules do not appear projecting from the scales.

Von Bergen (Textilber., 1925, *6*, 745) has suggested estimates of the swelling of wool in $N/10$ NaOH solution. The swelling is expressed as a percentage of the diameter in water; undamaged wool swells 10 to 13%, but if the swelling is greater than 15% then it may be concluded that the wool has been damaged by oxidation, chlorination, light or acid.

Another swelling test is due to Krais, Market and Viertel (Forschungsheft 14, Deutsches Forschungsinstitut fur Textilindustrie in Dresden, 1933), who used a reagent consisting of 20 g. of KOH dissolved in 50 c.c. of concentrated ammonia solution. Normal wool swells in this reagent in 8 to 10 minutes, but alkali-damaged wool takes longer; wool damaged by oxidation, acid or chlorination swells more quickly. The test depends to some extent on the quality of the wool, as coarse wool takes longer to swell than fine wool.

STAINING TESTS

Staining with various dyes may be used as a test for damaged wool and is based on the assumption that the damaged material has a greater affinity than the sound wool. The results are qualitative but, by examination under the microscope and the adoption of a counting technique, it is possible to obtain a quantitative estimate of damage. Cotton Red 10B and Cotton Blue are useful for measurements of alkaline modification but are not so sensitive with wool that has been modified by acid. The use of Indigo Carmine was favoured by Herzog (Textilber., 1931, *12*, 768), but is not very satisfactory for revealing damage by acid; a mixture of Indigo Carmine and picric acid is of interest as the sound portions of the fibre are stained yellow and the damaged parts green.

The Kiton Red test has been applied by the Wool Industries Research Association (Hosiery Trade J., 1941, *48*, 28; J.S.D.C., 1942, *58*, 245) mainly for estimating the damage due to chlorination. The method is a modification of that adopted by Trotman, Bell and Saunderson (J.S.C.I., 1934, *53*, 267); 1 g. of yarn is shaken for 60 minutes with 100 c.c. of a solution containing 0·615 g. of Kiton Red G and 100 c.c. of 0·1 *N*.HCl in 2 litres of distilled water. The stained wool is rinsed until free from loose dye and then dried. The stained sample is then wound spirally around a wooden pencil and a longitudinal cut made along it by two razor blades placed about 0·3 mm. apart. The fibre samples are examined under the microscope, and about 500 are classified as white, light-pink, pink, pink-red and deep-red; these classes are termed 0, 0·25, 0·5, 0·75 and 1·0 for convenience. The percentage of fragments falling into each group is calculated and the result expressed as a "chimney-diagram." The height of the chimney is proportional to the percentage of fibre fragments falling within that group.

Methylene Blue has also been utilised by numerous investigators, particularly in Germany, and is capable of qualitative and quantitative application; it is important to standardise the concentration of the reagent, the pH of the bath, and the time and temperature of staining. Quantitative results may be obtained by a counting method, colorimetrically, or by titration with Naphthol Yellow S (see page 459). As with many other dyes, the results are more reliable in estimating damage by alkali or by mechanical means, than in assessing damage by acid solutions.

Mixtures of dyes under various names have also been suggested for detecting damage in wool; for example, Neocarmine W (Textilber., 1927, *8*, 367; 1931, *12*, 763; 1932, *13*, 29) and Shirlastain (Silk and Rayon, 1935, *9*, 746; 1936, *10*, 35) have both been employed.

An interesting test for estimating small amounts of mechanical damage in wool has been developed by Whewell and Woods (J.S.D.C., 1944, *60*, 148); although the staining technique is employed, the assessment of damage depends on area and not on depth of staining.

A sample of wool (about 0·1 g.) is purified by extraction in alcohol and ether, and then immersed in 100 c.c. of NaOCl solution containing 0·12% chlorine and having a pH of 10, for 15 minutes at 20°C.; the wool is agitated during treatment. The sample is then removed and washed in five changes of 200 c.c. of tap-water for periods of 1 minute in each case, after which it is immediately transferred to 100 c.c. of Methylene Blue solution (0·4 g. of stain per litre) for 5 minutes, and finally washed in running water for 5 minutes. According to the stained area seen under the microscope, the wool may be graded as 0 to 5; the former indicates no sign of attack, whilst the latter indicates heavy general damage with some obscurity of the scale structure, and complete staining.

Stannous Chloride

The modification of wool by alkali and also by exposure may be detected by treatment with a hot aqueous solution of stannous chloride, according to Becke (Farben Zeit., 1912, *23*, 45, 66; 1919, *30*, 101, 116, 128). A brown coloration is produced by reaction between the sulphur-containing protein degradation product and the tin salt; the depth of staining is stated to be related to the degree of damage.

Silver Nitrate

The depth of staining with an ammoniacal solution of silver nitrate has also been stated to afford a sensitive test for damage in wool (Leip. Monat. Textilind., 1927, *42*, 160).

Lead Acetate

The lead acetate test has been discussed on page 366; it is mainly due to Smith and Harris (Am. Dyes. Rep., 1936, *25*, 183, 383). Damage by steam, water or alkali may be estimated by treating wool with lead acetate solution at pH 4·4, when the damaged material exhibits a darkening; if the wool has been damaged by carbonising, peroxide-bleaching, or by chlorination then it shows less colour after treatment with lead acetate solution than untreated wool. Ryberg (Am. Dyes. Rep., 1940, *29*, 588) has suggested treating the sample with 100 volumes of 0·05% lead acetate solution at pH 4·4 for 40 minutes with refluxing; an aliquot portion of the liquor is then titrated with standard ammonium molybdate solution using tannin as an external indicator. A control sample should be included in the test, and the lead acetate solution should be standardised against the molybdate solution; damage is proportional to the loss of lead from the original liquor or to change in concentration.

Thiocyanate Test

Wool which has been oxidised by peroxide or by hypochlorite solutions or by light may be detected by a coloration with potassium thiocyanate under certain conditions described by Rutherford and Harris (Am. Dyes. Rep., 1938, *27*, 179).

The following solutions are required:

(1) Equal parts by volume of distilled water and acetone, acidified with 1% by volume of $6N.H_2SO_4$.
(2) 1 g. of potassium thiocyanate in 30 c.c. of solution (1).
(3) 1 g. of ferrous ammonium sulphate and 15 c.c. of $6N.H_2SO_4$ in 50 c.c. of solution (1). Solution (3) should only be prepared immediately prior to use.

The test is applied by cutting 1 g. of wool into small portions which are placed in a 125-c.c. conical flask; in another flask is placed a similar sample of undamaged wool, whereas a third flask is for the reagents alone. Each flask is given 40 c.c. of solution (1), and 5 c.c. of solution (2); the solutions in the flasks are boiled to expel air, when 5 c.c. of solution (3) are added and the flasks stoppered. If a pink coloration occurs in the flask in which there is no wool, distilled water should be added until the colour disappears, and an equal volume of distilled water should be added to the other flasks.

The presence of oxidised wool is shown by a pink colour, which is absent from the flask containing unoxidised wool.

The test is sometimes complicated by a coloration before the addition of solution *3*; this is due to the presence of iron in the wool, which should be washed with dilute H_2SO_4.

The Biuret Reaction

This test was first described by Becke (Farber-Z., 1912, *23*, 45).

The colour scale is made by preparing standard solutions of the wool substance by dissolving 1 g. of wool in 50 c.c. of $N.NaOH$ heated on the water bath; 200 c.c. of distilled water is then added together with 50 c.c. of $N.HCl$ to neutralise the NaOH. The solution is boiled for 15 minutes to remove the H_2S, and if some of the wool substance is precipitated, it should be dissolved by the addition of $N.NaOH$. The solution is cooled and adjusted to 500 cc.

The solution is poured into each of ten similar tubes to give a graded series containing from 1 to 10 mg. of dissolved wool substance, and the volume is adjusted to 5 c.c. in each tube, to which is added 5 c.c. of $N.NaOH$ and 1 c.c. of $N.CuSO_4$ solution. The solution to be tested is prepared in the same manner and after the precipitates have settled, colour comparisons are made over a period of an hour.

The Pauly Test

This particular type of staining test depends on the formation of a dye with the fibre; wool is treated with diazosulphanilic acid which couples with tyrosine and histidine to form a coloured azo compound. These two amino-acids are present in the cortex of the wool hair but not in the cuticle; hence the production of colour indicates that the scale structure has been ruptured. The test was originally intended by Pauly and Binz (Zeit. fur Farben-u. Textilind., 1904, *3*, 373) as a qualitative measure of damage, but it has been modified by Burgess and Rimington (J. Roy. Microscop. Soc., 1929, *49*, 341) and by Rimington (J.T.I., 1930, *21*, 237) with a view to quantitative estimates of damage.

The stock solutions are 10% sodium sulphanilate, 9% sodium carbonate and 8% sodium nitrite; immediately before use, 10 c.c. of the sulphanilate solution is mixed with 5 c.c. of the nitrite solution and 2 c.c. of concentrated hydrochloric acid are added. The wool is wetted with 15 cc. of the sodium carbonate solution and then the prepared solution is added; after 10 minutes the wool is removed and rinsed. An orange-red colour indicates damage. For a quantitative determination, 0·1 g. of wool is treated as above, and the coloured fibres are transferred to a test tube where they are heated with 4 c.c. of 10% NaOH solution for 5 minutes on a water bath. The dissolved wool is then transferred to a graduated flask and made up to 25 c.c.; the coloured solution is compared with 0·1% solution of New Acid Brown S which represents 100 units of damage. Edwards (J.T.I., 1933, *24*, 1) has suggested that better colour-matching results if 292·5 c.c. of 0·1% of Polar Brilliant Red is added to 1 litre of 0·1% of New Acid Brown S (the latter is now Naphthalene Leather Brown BS).

A recent variation of this test has been developed by Gross, von Roll and Schreiber (Textilber., 1939, *20*, 357) and by Patat (ibid., 1939, *20*, 277), where a more stable diazonium salt is used in place of diazosulphanilic acid. The sample of wool may be treated with the diazonium salt prepared from 5-nitro-2-amino-anisole or Fast Scarlet Salt R; the quantity absorbed may be measured by titration with β-naphthol or dimethylaniline, but it is still necessary to standardise the diazonium solution and keep it at 0°C.

ALKALI-SOLUBILITY

The fact that most forms of damaged wool are more soluble in alkali than the original material has been used by many investigators as a measure of damage. In particular, Smith and Harris (Am. Dyes. Rep., 1936, *25*, 542) developed the method for determining the extent of the oxidation of wool.

The material is dried and treated at 65°C. for 60 minutes in $N/10$ NaOH solution; the ratio of liquor to wool is 100 : 1. The sample is then filtered on a Buchner funnel or Gooch crucible, washed with 2 litres of water, dried and weighed; the percentage loss in weight is calculated, based on the dry weight of the wool. The loss is about 12 to 13% with undamaged material; 18% and above indicate definite damage. Ryberg (Am. Dyes. Rep., 1940, *29*, 588) considered that 65°C. was not generally convenient for works' laboratories, and modified the test to utilise $N/20$ sodium carbonate solution at the boil for 30 minutes.

Damage is shown to be present if the loss in weight exceeds 12%. Ryberg considers that the types of damage may be grouped into three classes by this test; alkali damage is revealed by the fact that the alkali solubility of the product is lower than that of untreated wool; steam, water and dry heat do not alter the alkali-solubility, but acid, light, peroxide or chlorination will increase the alkali-solubility.

The above methods depend on direct weighing of the material but indirect measurements have been suggested depending on estimates of some component of the keratin extracted by the alkali; nitrogen tests have been devised by Krais and Schleber (Leip. Monat. Textilind., 1929, *44*, 211), Folgner and Schneider (ibid., 1934, *49*, 181) and Sauer (Angew. Chem., 1916, *29*, 424).

A disadvantage of this test is that the wool is generally washed with an alkaline liquor in normal processing, and this removes the soluble matter.

SULPHUR CONTENT

For many years past, the sulphur content of wool has been used as a guide to the extent of damage which the fibre may have undergone; the method is open to criticism on various grounds, for the amount of sulphur in wool may vary from 3·1 to 4%, and it is also necessary to have a sample of the original material for comparison.

The actual analysis may be effected by three methods, the oxygen bomb method, the Carius method, and the Benedict-Denis method; the last-named does not required any special apparatus.

The test sample of wool, about 0·2 g., is first warmed with a little NaOH solution until it just dissolves; a few drops of bromine are added and the solution is neutralised with nitric acid after a few minutes. In the meantime, a solution has been made consisting of 25 g. of crystalline copper nitrate, 25 g. of sodium chloride and 10 g. of ammonium nitrate in 100 c.c. of water; 10 c.c. of this solution are added to the solution of wool, and the mixture is evaporated to dryness. The residue is then heated to a dull red heat for 10 minutes, cooled, dissolved in dilute hydrochloric acid and filtered. The sulphate in the filtrate may be determined in the usual manner by the use of barium chloride.

The detection of sulphur in wool has been described by Barritt and King (J.T.I., 1926, *17*, 386; 1929, *20*, 151, 159; 1933, *24*, 119).

DETERMINATION OF THIOL AND DISULPHIDE

The determination of thiol and disulphide compounds, with special reference to cysteine and cystine has been described by Shinohara (J. Biol. Chem., 1937, *120*, 743). Cysteine reacts with phospho-18-tungstic acid, under certain conditions, the reduced tungsten producing a blue colour which may be estimated colorimetrically.

When wool reacts with sodium bisulphite, the following reactions occur:

$$R-S-S-R' + NaHSO_3 \longrightarrow RSNa + HO_3S-S-R'$$

The reaction between cysteine and phospho-18-tungstic acid takes place in the following manner:

$$R-SH + HS-R + (H_2O)_3P_2O_5(WO_3)_{18}$$
$$\downarrow$$
$$R-S-S-R + H_2O + (H_2O)_3P_2O_5(WO_3)_{17}WO.$$

It will be seen, therefore, that only one-half of the cystine is converted into cysteine; the maximum intensity of colour reached by cysteine, or by cystine in presence of bisulphite, is twice that reached by cystine of the same molar concentration without bisulphite.

It is thus possible to determine total sulphur, disulphide sulphur and thiol sulphur by a somewhat elaborate series of tests which may be of value in detecting the particular type of damage which brings about cleavage of the disulphide bond. These tests have been described by Phillips and Elsworth (Biochem. J., 1938, *32*, 837; 1941, *35*, 135).

NITROGEN CONTENT

The determination of the nitrogen content of wool, like that of the sulphur content, is useful as a measure of damage provided that some of the untreated wool is also available for purposes of comparison; the nitrogen content of different wools varies from 16·16% to 17·07%. A useful contribution to this aspect of wool has been made by Barritt (J.S.C.I., 1928, *47*, 69).

> The method generally adopted is based on the Kjeldahl technique. For example, about 0·8 g. of wool is placed in a Kjeldahl flask with 20 c.c. of H_2SO_4, 10 g. of anhydrous K_2SO_4, 0·2 g. of $CuSO_4$, 4 g. of powdered pumice, and a little broken glass; the mixture is digested for an hour, and then dissolved in distilled water and cooled. An excess of sodium hydroxide solution is added cautiously and the flask connected with the Kjeldahl distillation apparatus and distillation performed for 30 minutes, the distillate being collected in 25 c.c. of $N/2\ H_2SO_4$. The excess of sulphuric acid is then titrated with $N/10$.NaOH using Methyl Red as indicator.

The above method is preferred by Skinkle (Textile Testing, Macmillan, London, 1940).

The regain of the original wool is determined on a separate sample.

An estimation of soluble nitrogen has been suggested by Sauer (Z. angew Chem., 1916, *29*, 424), and developed by Krais and Schleber (Leip. Monat. Textilind., 1929, *44*, 165) and by Folgner and Schneider (ibid., 1934, *49*, 181); the test seems to be definite for alkali-damage, but is somewhat lengthy, involving two Kjeldahl determinations with an intermediate treatment for 3 days.

> The total nitrogen content is first determined on one sample of wool, and then about 0·1 g. of a second sample of the same material is treated for 3 days with a mixture of 8 c.c. of water, 10 c.c. of 1% H_2O_2 and 2 c.c. of $N/2$ KOH. The sample is then filtered and washed, the combined filtrate and washings are made acid, and a Kjeldahl determination made on the acid solution.

With undamaged wool, the soluble nitrogen amounts to 4 or 5%, but with damaged material it may be 30% or more.

Where wool has not been exposed to alkali treatment after damage, an estimate of ammonia-nitrogen may be of value. This has been described by Smith and Harris (Am. Dyes. Rep., 1936, *25*, 383).

About 5 g. of wool is placed in a 1-litre distillation flask with 300 c.c. of water and 5 g. of magnesium oxide, free from carbonates; 150 c.c. are distilled into 25 c.c. of $N/100$ H_2SO_4, and the excess of acid is titrated with $N/100$ NaOH with Methyl Red as indicator. The results are calculated as milligrams of ammonia-nitrogen per gram of dry wool.

Undamaged wool gives values of 0·15 to 0·20, carbonised wool will give 0·20 to 0·70, but damaged wool will give still higher figures.

GENERAL

It will be noted that the methods for estimating damage in wool are not so satisfactory as the tests which have been devised for cotton and rayon; indeed, many of the suggested methods are of doubtful value.

The review of testing methods by Whewell and Austerlitz (J.S.D.C., 1943, *59*, 45) was accompanied by the examination of a series of modified wools. It was found that the measurement of tensile strength was insensitive except where the modification or damage was serious, but estimates of bursting strength were more reliable; wearing tests, particularly on wet material, revealed damage at an early stage, when the Ring-Wear machine of Lester was used (J.T.I., 1937, *26*, 95P).

Microscopic examination, although tedious, seems to give reliable results, and the same remark holds for the Kiton Red G test, but the differences between slightly damaged wool and severely damaged wool are not very great; with the Pauly test, Whewell found that the counting method was inferior to the colorimetric comparison which, at its best, is most useful as a qualitative guide to soundness. Staining with Methylene Blue is only of value in revealing comparatively small amounts of damage, but Benzopurpurin is not satisfactory; Indigo Carmine provides the most useful test reagent of all the stains, for it gives the greatest difference between untreated samples and severely damaged material. Whewell's results with the alkali-solubility test show that although it may be useful where serious damage is encountered, the figures are erratic. It appears from this review that wool should be examined, first, for bursting strength and wet resistance to wear, and then under the microscope in conjunction with a staining method, preferably with Indigo Carmine

SILK

The Zimmermann Test

The reagent for this test consists of a solution of *o*-phthalaldehyde acidified with hydrochloric acid; this solution gives no coloration with undamaged degummed silk, but there is a violet colour with damaged silk or with silk gum.

Determinations of Nitrogen

Nitrogen determinations may be carried out by the well-known Kjeldahl method; undamaged silk contains 18 to 19% of nitrogen and tussah silk rather less.

Ammonia-nitrogen determinations may also be effected in the usual manner described for wool; normal silk will give values of 0·07 to 0·12 milligrams per gram of silk, but the damaged material often gives values in excess of 0·2 mg. per gram of silk.

Fluidity

The determination of the fluidity of cellulosic materials in cuprammonia by Clibbens and his co-workers (J.T.I., 1928, *19*, 77; 1936, *27*, 285), as discussed on pages 213 and 461, has proved of great value in assessing chemical damage and affords a remarkable method of control over the scouring and bleaching of cotton and rayon. Fundamentally the method is an estimate of the chain-length of the cellulose, and it is unfortunate that similar means of assessing damage are not available with wool; as silk also consists of chain-molecules without chemical cross-linkages it might be expected that viscosity or fluidity determinations could be made in suitable solvents.

The first method for differentiating between mechanical and chemical damage in degummed silk was due to Trotman and Bell (J.S.C.I., 1935, *54*, 141) and Trotman (ibid., 1936, *55*, 325); it depended on the use of concentrated solutions of zinc chloride as a solvent for silk. Although the rate of flow of solutions of 2·5% concentration changed slowly, it appeared that considerable degradation occurred during dissolution in the zinc chloride. This was revealed by the work of Tweedie (Canad. J. Res., 1938, *16*, 134) which showed that the fluidity of the solution provided an insensitive index of chemical tendering; the fact that the relative viscosity of a 2·5% solution of untendered silk was in the neighbourhood of 2, suggests considerable degradation of the original product with its high molecular weight.

Garrett and Howitt (J.T.I., 1941, *32*, 1) examined the known solvents for silk, such as (*a*) concentrated solutions of certain metallic halides, nitrates and thiocyanates, (*b*) strong acids and bases, and (*c*) basic solutions of complex derivatives of copper or nickel. Cupriethylenediamine was found to have great merit, for although it degraded the dissolved silk at a significant rate, the degradation could be obviated by neutralisation of the reagent after dissolution of the silk. Acetic acid is used for neutralisation.

Viscosity measurements on 5% solutions of silk, covering the range of untendered silk down to complete loss of strength, gave figures from 25 to 2 centipoises; alternatively, the fluidity of untendered silk is about 4 reciprocal poises and that of chemically damaged silk of no tensile strength is approximately 50.

The application of the fluidity measurement has been investigated by Cadwallader, Howitt and Smith (J.T.I., 1941, *32*, 13).

Raw filature silks gave fluidities ranging from 6 to 8, after a careful degumming in 1% soap solution (liquor to silk 200 : 1) for 1 hour at the boil. Nett silk, after soaking and throwing, is often stored for months before weaving, in contrast to knitting yarns; storage of thrown yarns in the soaked state may result in chemical tendering, and the complete degumming of high twist yarn may also be accompanied by tendering, probably due to the prolongation which is necessary to remove the gum from the interspaces of the packed filaments—six hours compared with one hour.

Samples stored for three years and which had not been soaked gave a fluidity of 6·8, compared with 10·5 for soaked material which had been stored also. Degumming of stored soaked yarns of low twist gave fluidities of 11·3 to 13·7, whereas yarns of higher twist, which required longer degumming, gave figures of 18·9 to 22·1.

The effect of the time of boiling in 1% soap solution is shown in the following table:

BOILING AND FLUIDITY

Time (hours)	Strength retained	Fluidity
0	100%	13·5
1	98·7%	14·1
2	92·2%	14·7
3	90·6%	16·4
4	88·0%	17·1
6	84·8%	18·7
8	83·4%	20·7

It will be seen that an increase in fluidity from 13·5 to 20·7 is accompanied by approximately 17% reduction in strength.

In view of the fact that the chief reagents used in the treatment of silk may be classified as acid or alkaline as regards their possible tendering action, the relationship between tensile strength and fluidity of silk which has been treated with acid or alkali was examined. The alkali was found to tender at a greater rate than the acid when decinormal solutions were used.

The general conclusions may be seen from Fig. 151; it will be noticed that the acid hydrolysis was effected at 90°C. but the alkaline treatment at 80°C., in order to give comparative times of treatment. The relation between strength and fluidity is different for acid-tendered and alkali-tendered silks; it is known that the hydrolysis of silk by dilute acid follows a different route from that of dilute alkali. The difference, however, does not seem sufficiently great to preclude the deduction of strength from fluidity irrespective of the acid or alkaline nature of the attack.

A loss of 5% in strength increases the fluidity from 4 to 12 and the fluidity may increase to 20 before the loss in strength is serious;

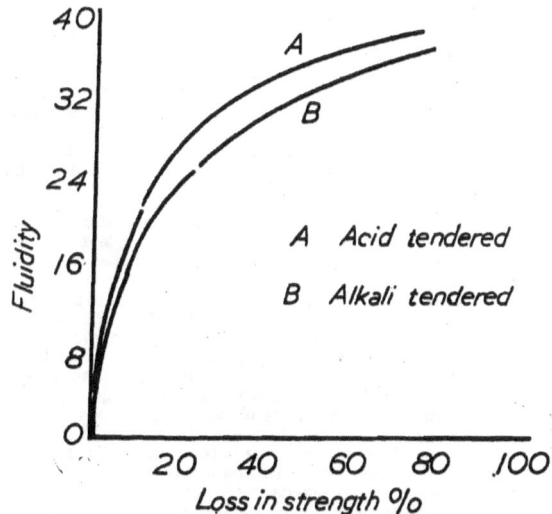

FIG. 151.—Tensile strength and fluidity of degraded silk.

about 50% reduction in tensile strength is shown by a fluidity of 30 to 35.

The general method is to prepare a reagent containing 8% of ethylenediamine and 6% of cupric hydroxide in which silk is dissolved to give a 10% solution (weight/volume) calculated as dry weight. This is neutralised with an equal volume of 1·25 N acetic acid to give a slightly alkaline solution (approx. pH 9); although neutral solutions may be obtained with an equal volume of 1·33 N acetic acid, care is required to avoid local high concentration of acid during the addition with separation of the silk as a colloidal aggregate.

THE REAGENT

The solvent is prepared by dissolving cupric hydroxide in ethylenediamine; the hydroxide should be of good quality and free from sulphate, chloride, nitrate and from caustic alkali. Ethylenediamine is supplied in the form of a concentrated solution (70% weight/volume) in water; this is diluted to 8% with distilled water and 1 litre of this dilute solution is shaken with 70 g. of the cupric hydroxide for an hour, after which the insoluble residue is removed by centrifuging or filtering.

The reagent is then adjusted to give a solution containing 6% of $Cu(OH)_2$ and 8% of ethylenediamine (weight/volume). The standardised reagent may be stored for several months if not exposed to light.

The Viscometer

The viscometer is the small capacity U-tube type. The dimensions of the apparatus shown in Fig. 152 are as follows:

Internal diameter of capillary JK	0·088	± 0·001 cm.
External diameter of capillary JK	0·6	± 0·2 cm.
Length of capillary JK	10·0	± 0·2 cm.
Internal diameter of cylindrical parts of the bulbs BE and FM	1·0	± 0·02 cm.
Length of tube EF	1·5	± 0·3 cm.
Length of tube MJ	1·0	± 0·2 cm.
Length of tube KL	about 7 cm.	
Internal diameter of tubes AB, EF, MJ and KL	0·325	± 0·025 cm.
Length of tube AB	7·5	± 0·5 cm.
Length of tube LN	about 24 cm.	
Internal diameter of tube LN	1·0	± 0·05 cm.
Volume of each bulb BE and FM	1·2	± 0·05 c.c.
Volume between etched rings X' and X"	1·3	± 0·05 c.c.
Clearance between tube NL and bulbs	about 1 cm.	
Over-all length of viscometer	about 26 cm.	

The mark X''' indicates the level to which the silk solution is drawn up by applying suction at A; for 3·0 c.c. of solution, this corresponds with a level in the other limb near the bottom of the wider tube NL. A 3·0 c.c. pipette is used for filling purposes and should be calibrated with accuracy.

The viscometer is held in position by clamping the wide tube arm to two V-shaped blocks by rubber bands; the two blocks fit on to a rod which is held vertically by suitable stands and clamps. In this manner, three or four viscometers may be fitted into a glass tank or glass accumulator cell approximately 9·5 by 11 by 16·5 inches.

The viscometer may be calibrated by measuring the time of flow at 20°C. (t) of a liquid of known fluidity (F) and density (d), flowing sufficiently slowly to keep the kinetic energy correction low; this gives a viscometer constant C in kinematic units (reciprocal stokes). Since the results are to be expressed in dynamic units (reciprocal poises), and since the viscometer is only intended for use with silk solutions of constant composition, the density can be included in the constant term and is given the value 1·039 at 20°C. A constant C' can then be calculated, expressed in seconds per poise, and applicable only to silk solutions, which facilitates subsequent calculations. The fluidity F of a silk solution is given by

$$F = C'/t$$

The kinetic energy correction is generally negligible as fluidities greater than 40 will rarely be encountered, and for this value the correction is only about 1%.

Fluidity Measurement

The silk is first conditioned at 65% R.H. and 65°F., and 193 mg. is introduced into a vessel of about 10 c.c. capacity; flat-bottomed specimen tubes 2·25 cm. in diameter and 3·25 cm. in depth are very suitable. The calibrated pipette is used to deliver 1·75 c.c. of the cupri-ethylenediamine reagent into the tube, and the mixture is stirred vigorously for 3 minutes and the stirring continued during the addition of 1·75 c.c. of 1·25 N acetic acid. The 3-minute period applies from the start of dissolution to the beginning of neutralisation, and should be timed with a stop-watch. The acid is delivered from a micro-burette, the jet of which is constricted to give a delivery of 1·75 c.c. in one minute, during which time careful stirring is applied to avoid local concentration of acid; the stirring may be effected by a small glass propeller driven by a motor.

The neutralised solution is then transferred to a small glass centrifuge tube with a well at the base (see Fig. 152) so that any insoluble matter becomes compact and is not withdrawn into the pipette later. The solution is centrifuged for 5 minutes at 1,000 to 1,200 r.p.m., after which the liquor is withdrawn by the calibrated 3 c.c. pipette and transferred to the viscometer taking care to avoid the introduction of air-bubbles. The viscometer is maintained in a water bath at a temperature of

20° C. and a period of 10 minutes is allowed for the solution to reach this temperature, after which the liquid is drawn by suction at A to the mark X''', and the time observed for the level to fall from X' to X'''; several determinations should be made.

The fluidity of undamaged silk is approximately 4 reciprocal poises, whereas when the silk is chemically damaged until there

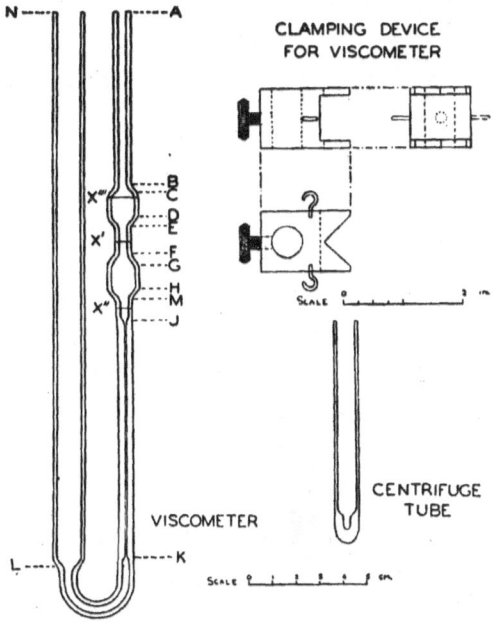

FIG. 152.—Viscometer for fluidity measurements of silk.

is almost complete loss in strength, the fluidity is about 50. In viscosity units, this corresponds to a range of 25 to 2 centipoises.

NYLON

The viscosity of nylon also affords a useful test of chemical damage which produces a reduction in chain-length; m-cresol may be used as the solvent and full details of the method have been published by Boulton and Jackson (J.S.D.C., 1943, *59*, 21).

Solutions of nylon in m-cresol are stable to light and to air, and may be kept for 7 days without appreciable change in viscosity. Solutions of 8 g. of nylon in 100 c.c. of m-cresol were found to be the most convenient for routine viscosity measurements; the type of viscometer utilised is very similar to the viscometers of the B.C.I.R.A. for work with cellulose.

VISCOMETER CHARACTERISTICS

	Viscosity	2 to 5 poises
t	Time of flow	200 to 500 secs.
H	Distance between rings b and d for viscometer body of 0·7 to 0·8 cm. internal diameter	10 cm.
l	Length of capillary	7·5 cm.
r	Radius of capillary	0·10 cm.
v	Volume of efflux	5 c.c. (Total volume used in filling instrument, 10 c.c.)
d	Density of solution	1·037

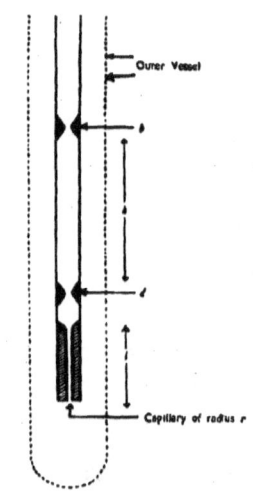

FIG. 153.—Viscometer for fluidity measurements of nylon.

The above table is given as a guide for making a convenient instrument, for the viscosity measurement is independent of the instrument measurement. The viscometer is standardised by measuring the time of flow at 25°C. of castor oil or of a mineral oil of similar viscosity; the viscosity of the calibrating oil is first accurately determined in a No. 3 Ostwald viscometer of known constants.

$$K = \frac{v}{t}$$

where K is the instrument constant in centistokes per sec., v is the viscosity of the standard oil in centistokes and t is the time of flow in secs. This constant K is stated in terms of kinematic viscosity; in order to determine the viscosity of an unknown solution in centipoises, it is necessary to find the time of flow, and multiply by the constant K, and by the density of the solution at 25°C.

DAMAGED PROTEIN FIBRES

$$\eta = Ktd$$

where η is viscosity in centipoises and d is the density at 25°C. Where the instrument is only used for 8% nylon in m-cresol of density 1·037, then a constant C may be used $(C=Kd)$ and the viscosity determined directly from the relation $\eta = Ct$. For the above type of instrument, C is approximately 1·5.

The nylon may be dissolved in a separate vessel or in the outer jacket of the viscometer if suitably stoppered for shaking; dissolution of 0·8 g. of nylon in 10 c.c. of m-cresol will take 2 to 4 hours with gentle agitation, as, for instance, on a slowly revolving wheel.

The viscometer itself is clamped vertically and filled by suction from the top, using a long rubber tube, the meniscus being raised above the mark b and the capillary being raised clear of the remaining liquid. When the time t for the meniscus to flow from b to d has been determined, the operation is repeated and duplicate times should agree within 0·5%. The mean time is multiplied by the instrument constant C to give the viscosity in centipoises. The fluidity may be calculated from the equation

$$F = \frac{10{,}000}{Ct}.$$

It is of interest to compare the fluidity of nylon with that of other materials as shown in the following table:

Normal Range of Fluidities

Cotton in Cuprammonium Hydroxide			Viscose Rayon in Cuprammonium Hydroxide			Silk in Cupriethylene Diamine			Nylon in m-Cresol		
Concentration = 0·5% η solvent = 0·0150 poise			Concentration = 2·0% η solvent = 0·0150 poise			Concentration = 10·0% η solvent = 0·0133 poise			Concentration = 8·0% η solvent = 0·143 poise		
η relative	poises	F reciprocal poises	η relative	poises	F reciprocal poises	η relative	poises	F reciprocal poises	η relative	poises	F reciprocal poises × 100
33 to 1·65	0·5 to 0·025	2 to 40	6·7 to 1·7	0·1 to 0·025	10 to 40	19 to 1·9	0·25 to 0·025	4 to 40	35 to 17·5	5·0 to 2·5	20 to 40

A diminution of 50% in the tensile strength of cellulose corresponds to a drop in relative viscosity from 33 to 1·7, in silk from 19 to 1·9, but with nylon from 35 to 17·5. The drop for nylon is over a smaller range for the same approximate degree of damage than for the natural materials.

It will be seen that the range of fluidities for nylon is 20 for undamaged material and 40 or so for extensively damaged nylon, where the unit is the reciprocal poises multiplied by 100.

Comparison of photodegradation and attack by mineral acid shows that there is a steady rise in fluidity with increasing loss in tensile strength, but the rise in fluidity for a given loss of strength by acid attack is about twice that for the same loss of strength through photodegradation.

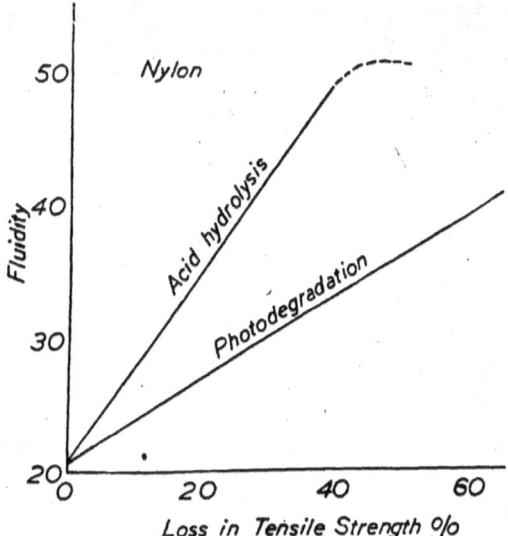

Fig. 154.—Tensile strength and fluidity of degraded nylon.

Although nylon is not strictly a protein fibre, yet its resemblance to the proteins is sufficiently strong to warrant its discussion with them. It will be noted, however, that a decrease of 50% in tensile strength of nylon takes place over a much smaller fluidity range than with natural fibres; this suggests some fundamental difference in molecular arrangement.

In conclusion, it may be remarked that the measurement of chain length is a most important development in the control of the chemical processing of textile fibres, and it is unfortunate that no similar test is available for wool.

APPENDIX

DENSITY TABLES

Baumé	Twaddell	Sp. Gr.	H_2SO_4 %	HCl %	HNO_3 %	NaOH %	KOH %
0	0	1·00	0·09	0·16	0·10	0	0
1·4	2	1·01	1·57	2·14	1·9	0·86	1·18
2·7	4	1·02	3·03	4·13	3·7	1·69	2·28
4·1	6	1·03	4·49	6·15	5·5	2·60	3·36
5·4	8	1·04	5·96	8·16	7·26	3·50	4·44
6·7	10	1·05	7·36	10·17	8·99	4·34	5·53
8·0	12	1·06	8·77	12·19	10·68	5·20	6·6
9·4	14	1·07	10·19	14·17	12·33	6·13	7·68
10·6	16	1·08	11·60	16·15	13·95	7·05	8·76
11·9	18	1·09	12·99	18·11	15·53	7·95	9·82
14·2	20	1·10	14·35	21·92	18·67	8·78	10·87
15·4	24	1·12	17·01	23·82	20·33	10·56	12·96
17·7	28	1·14	19·61	27·66	23·31	12·49	15·04
19·8	32	1·16	22·19	31·52	26·36	14·19	17·13
22·0	36	1·18	24·76	35·59	29·38	16·00	19·15
24·0	40	1·20	27·32	39·11	32·36	17·81	21·17
26·0	44	1·22	29·84		35·28	19·65	23·15
27·9	48	1·24	32·28		38·29	21·47	25·13
29·7	52	1·26	34·57		41·34	23·23	27·07
31·5	56	1·28	36·87		44·41	25·04	29·0
33·3	60	1·30	39·19		47·49	26·85	30·91
35·0	64	1·32	41·50		50·71	28·83	32·78
36·6	68	1·34	43·74		54·07	30·74	34·63
38·2	72	1·36	45·88		57·57	32·79	36·46
39·8	76	1·38	48·0		61·27	34·71	38·28
41·2	80	1·40	50·11		65·30	36·67	40·09
42·7	84	1·42	52·15		69·80	38·67	41·89
44·1	88	1·44	54·07		74·68	40·68	43·63
45·4	92	1·46	55·97		79·98	42·75	48·37
46·8	96	1·48	57·83		86·05	44·33	47·09
48·1	100	1·50	59·70		94·09	46·94	48·78
49·4	104	1·52	61·59		99·67	49·05	50·48
50·6	108	1·54	63·43				52·15
51·8	112	1·56	65·08				
53·0	116	1·58	66·71				
54·1	120	1·60	68·51				
55·2	124	1·62	70·32				
56·3	128	1·64	71·99				
57·4	132	1·66	73·64				
58·4	136	1·68	75·42				
59·5	140	1·70	77·17				
60·4	144	1·72	78·92				
61·4	148	1·74	80·68				
62·3	152	1·76	82·44				
63·2	156	1·78	84·50				
64·2	160	1·80	86·90				
65·0	164	1·82	90·50				
65·9	168	1·84	95·60				
		1·841	97·0				
		1·8415	97·7				
		1·841	98·2				
		1·840	99·2				
		1·839	99·7				

The acid and alkali percentages are by weight—for grams per 100 c.cs., multiply the percentage figure by the specific gravity.

SPECIFIC GRAVITY OF SOLUTIONS OF SODIUM CARBONATE AT 15° C.

Density in Degrees Twaddell	% Soda Ash by weight
1	0·47
2	0·95
3	1·42
4	1·90
5	2·38
6	2·85
7	3·33
8	3·80
9	4·28
10	4·76
11	5·23
12	5·71
13	6·17
14	6·64
15	7·10
20	9·43
25	11·76
30	14·09

SPECIFIC GRAVITY OF SOLUTIONS OF CALCIUM HYPOCHLORITE

Density	Available Chlorine
0·125° Tw.	0·35 g. per litre
0·25 ,,	0·70 ,, ,,
0·36 ,,	1·00 ,, ,,
0·50 ,,	1·40 ,, ,,
0·72 ,,	2·00 ,, ,,
0·75 ,,	2·05 ,, ,,
1·00 ,,	2·71 ,, ,,
1·08 ,,	3·00 ,, ,,
1·40 ,,	4·00 ,, ,,
1·80 ,,	5·00 ,, ,,
2·00 ,,	5·88 ,, ,,

APPENDIX

HYDROGEN PEROXIDE

(From *Das Wasserstoffperoxyd und die Pervenbindungen*, by Machu; Springer, Vienna, 1937)

Weight %	Volume %	Vols. of Oxygen
0	0	0
1·0	1	3·3
1·5	1·5	5
2	2	6·6
3	3	10
4·55	4·55	15
5	5·1	17
6	6·15	20
7·5	7·7	25
10	10·35	34
15	15·8	52
20	21·45	70
25	27·3	90
27·2	30	100
30	33·33	110
35	39·7	132
40	46·25	153
45	53·1	175
50	63·15	208
60	75·05	248
70	90·85	300
80	107·65	355
90	125·4	415
100	144·4	475

As 1 litre of oxygen at N.T.P. weighs 1·429 g., the available oxygen in 1-vol. hydrogen peroxide is 1·429 g./l.

SOME pH VALUES

	$N/10$	$N/100$	$N/1000$
Hydrochloric acid	1·03	2·01	3·0
Sulphuric acid	1·21	2·08	3·01
Oxalic acid	1·52	2·26	3·21
Formic acid	2·35	2·87	3·45
Acetic acid	2·87	3·36	3·89

	N	$N/10$	$N/100$
Sodium hydroxide	14·0	13·0	12·0
Potassium hydroxide	14·0	13·0	12·0
Sodium carbonate	—	11·6	—
Ammonia	11·6	11·1	10·6
Sodium bicarbonate	—	8·4	—

SOME COMMON INDICATORS

	pH	pH	Colour change	
Thymol Blue	1·2 to	2·8	Red	Yellow
Bromo-phenol Blue	2·8 ,,	4·6	Yellow	Violet
Methyl Orange	3·1 ,,	4·4	Red	Yellow
Bromo-cresol Green	3·6 ,,	5·2	Yellow	Blue
Methyl Red	4·2 ,,	6·3	Red	Yellow
Litmus	5·0 ,,	7·0	Red	Blue
Bromo-cresol Purple	5·2 ,,	6·8	Yellow	Violet
Bromo-thymol Blue	6·0 ,,	7·6	Yellow	Blue
Phenol Red	6·8 ,,	8·4	Yellow	Red
Cresol Red	7·2 ,,	8·8	Yellow	Violet-red
Thymol Blue	8·0 ,,	9·6	Yellow	Blue
Phenolphthalein	8·3 ,,	10·5	None	Red
Alizarine Blue	9·0 ,,	11·0	Brown	Green
Alizarine Blue	11·0 ,,	13·0	Green	Violet

CENTIGRADE AND FAHRENHEIT

C.	F.	C.	F.	C.	F.
200	392	120	248	40	104
195	383	115	239	35	95
190	374	110	230	30	86
185	365	105	221	25	77
180	356	100	212	20	68
175	347	95	203	15	59
170	338	90	194	10	50
165	329	85	185	5	41
160	320	80	176	0	32
155	311	75	167	—5	23
150	302	70	158	—10	14
145	293	65	149	—15	5
140	284	60	140	—20	—4
135	275	55	131	—25	—13
130	266	50	122	—30	—22
125	257	45	113	—35	—31

APPENDIX

STEAM PRESSURES AND TEMPERATURES

Pressure lbs./sq. in.	Temperature °C.	Pressure lbs./sq. in.	Temperature °C.
0	100	30	134
5	109	32	136
10	115	34	138
15	121	36	139
20	125	38	140
22	127	40	141
24	129	44	144
26	131	48	147
28	133	50	148

IMPORTANT NOTE

Throughout this book the *gallon* to which reference is made is the British Imperial gallon and not the U.S. gallon. The British gallon is equivalent to 1·2 U.S. gallons.

BIBLIOGRAPHY OF BLEACHING

THE HISTORY OF BLEACHING, by S. H. Higgins (*Longmans Green*, London, 1924).
BLEACHING AND RELATED PROCESSES, by J. M. Matthews (*Chemical Catalog Co.*, New York, 1921).
TEXTILE BLEACHING, by A. B. Steven (*Pitman*, London, 1921).
A PRACTICAL TREATISE ON THE BLEACHING OF LINEN AND COTTON YARN AND FABRIC, by L. Tailfer (*Scott Greenwood*, London, 1917).
THE PRINCIPLES OF BLEACHING AND FINISHING OF COTTON, by S. R. Trotman and E. L. Thorp (*Griffin*, London, 1927).
BLEACHING POWDER AND ITS ACTION IN BLEACHING, by R. L. Taylor (*Heywood*, London, 1922).
BLANCHIMENT, by A. Chaplet (*Gauthier-Villar*, Paris, 1929).
DAS BLEICHEN DER PFLANZENFASERN, by W. Kind (*Springer*, Berlin, 1931).

Some useful information relating to Bleaching may also be found in the following:

TEXTILE BLEACHING, DYEING, PRINTING AND FINISHING MACHINERY, by A. J. Hall (*Benn*, London, 1926).
PRINCIPLES AND PRACTICE OF TEXTILE PRINTING, by E. Knecht and J. B. Fothergill (*Griffin*, London, 1936).
BLEACHING, DYEING AND CHEMICAL TECHNOLOGY OF TEXTILE FIBRES, by S. R. and E. R. Trotman (*Griffin*, London, 1925).
CHEMICAL TECHNOLOGY OF TEXTILE FIBRES, by von Georgievics (*Benn*, London, 1920).
THE MANUAL OF DYEING, by E. Knecht, C. Rawson and R. Loewenthal (*Griffin*, London, 1925).
BLEACHING AND DYEING OF VEGETABLE FIBROUS MATERIAL, by J. Huebner (*Constable*, London, 1912).
DYEING OF TEXTILE FIBRES, by R. S. Horsfall and L. G. Lawrie (*Benn*, London, 1927).
DYEING WITH COAL TAR DYESTUFFS, by C. M. Whittaker and C. C. Wilcock (*Baillière, Tindall & Cox*, London, 1942).
BLEACHING, DYEING, PRINTING AND FINISHING FOR THE MANCHESTER TRADE, by J. W. McMyn and J. W. Bardsley (*Pitman*, London, 1932).
THE FINISHING OF WOOL GOODS, by J. and J. C. Schofield (Huddersfield, 1935).
A DICTIONARY OF APPLIED CHEMISTRY, by Thorpe and Whiteley (*Longmans, Green*, London, Fourth Edition, Vol. II, 1938).
BLANCHIMENT, TEINTURE, IMPRESSION, APPRÊTS, by Lederlin (*Baillière*, Paris, 1923).
ENZYKLOPAEDIE DER TEXTILCHEMISCHEN TECHNOLOGIE, by Heerman (*Springer*, Berlin, 1931).

BIBLIOGRAPHY

TEXTILE RECORDER YEAR-BOOK (*Harlequin Press*, Manchester).
TEXTILE MANUFACTURER YEAR-BOOK (*Emmott*, Manchester).
WOOL YEAR-BOOK (*Textile Mercury*, Manchester).
COTTON YEAR-BOOK (*Textile Mercury*, Manchester).
AMERICAN COTTON HANDBOOK, by Merrill, Macormac and Mauersberger (*American Cotton Handbook Co.*, New York).
AMERICAN WOOL HANDBOOK, by von Bergen and Mauersberger (*American Wool Handbook Co.*, New York).
RAYON AND STAPLE FIBRE HANDBOOK, by Mauersberger and Schwarz (*Rayon Publishing Co.*, New York).
DEUTSCHER FARBER KALENDER (Wittenberg).

The following journals and magazines are also of some interest:

The Journal of the Society of Dyers and Colourists.
The Journal of the Textile Institute.
The Journal of the Society of Chemical Industry.
The Transactions of the Faraday Society.
The Paper Trade Journal.
The Textile Recorder.
The Textile Manufacturer.
The Silk Journal and Rayon World.
The Dyer.
Silk and Rayon.
The Textile Mercury.
The Textile Weekly.
American Dyestuffs Reporter.
Textile World.
Rayon Textile Monthly.
The Textile Colorist.
Revue Générale des Matières Colorantes.
Teintex.
T.I.B.A. (Teinture, Impression, Blanchiment, Apprêts).
R.U.S.T.A. (Revue Universelle de la Soie et des Textiles Artificiels).
Cellulosechemie.
Textilberichte.
Kunstseide.
Seide.
Zellwolle.
Deutscher Farber Zeitung.
Leipziger Farber Zeitung.

NAME INDEX

ABERHALDEN & Fodor, 71
Adam, 119, 146, 149
Adams & Holmes, 108
Adolf, 255
Agardh, 10
Allen (see Knecht)
Allworden, 82
Ambronn, 27
Anderlini, 374
Anderson & Kerr, 6, 8
Arnold, 80, 410
Arrhenius, 115
Astbury, 4, 31, 65, 66, 72
Astbury, Bailey & Chibnall, 84
Astbury & Woods, 74, 75, 330
Atkins (see Cassie)
Atkinson, 272
Auerbach, 267
Austerlitz (see Whewell)

BAIER & Hundt, 258
Bailey (see Astbury)
Baker, 144
Balls, 5, 6, 9, 451
Barker, 315
Barker & Norris, 62
Barrett (see Kershaw)
Barritt, 64, 483
Barritt & King, 64, 482
Bartunek (see Heuser)
Bascom (see Rhodes)
Bath (see Ellis)
Beadle (see Cross)
Beadle & Stevens, 189
Becke, 479, 480 (see also Herzog)
Becker (see Schwalbe)
Bell (see Trotman)
Bemberg, 18
Bennett, 132
von Bergen, 76, 366, 477
Bergmann, 55, 376, 377
Bergmann & Niemann, 58, 72
Berl, 466
Berl & Klaye, 461
Berl & Rueff, 466
Berthollet, 152
Bevan (see Cross)
Birtwell, Clibbens & Geake, 43, 44, 46, 48, 469
Birtwell, Clibbens & Ridge, 44, 46, 212, 213, 459, 472
Blakeley, 214
Blanc, 378
Bloch (see Vickery)

Bohme, 289
Bondi, 374
Bone, 26, 425
Bonsma, 64
Boulton, Delph, Fothergill & Morton, 33
Boulton & Jackson, 489
Boulton & Morton, 34
Bradbury (see Speakman)
Bragg, 24
Braidy, 455
Brandt, 195, 250
Briggs (see Cross)
Bright, 452
Briscoe, 142, 194
British Celanese, 395
Britton, 115
Britton & Dodds, 218
Brock, 297
Brown 319 (see also Trotman)
Brownsett & Clibbens, 51, 467, 476
Brownsett, Clibbens & Davidson, 474
Buffalo Electro Chemical Co., 286
Buisine, 297
Bunn, Clark & Clifford, 201
Buntrock, 79
Burgess, 63
Burgess & Rimington, 481
Butterworth, 227, 268, 290
Butterworth & Elkin, 267

CADWALLADER, Howitt & Smith, 486
Campbell, 284, 378 (see also Wontner-Smith)
Carothers & Hill, 87
Caryl & Ericks, 135
Cassie, 76
Cassie, King & Atkins, 76
Cassie & Palmer, 147
Chamberlain (see Speakman)
Chibnall (see Astbury)
Chwala & Martina, 124, 141
Clark, 29, 191, 279 (see also Bunn)
Clarke, 360
Clayton, 264
Clibbens, 44, 485 (see Birtwell, also Brownsett)
Clibbens & Geake, 44, 212, 455, 458, 461
Clibbens & Little, 465, 467
Clibbens & Ridge, 47, 214, 228, 471
Clegg, 452
Clifford, 43
Clifford & Fargher, 460
Cocking, 115

NAME INDEX

Cohen (see Stamm)
Collins, 41
Cook, 358
Cooper (see Speakman)
Corner (see Ridge)
Courtalds, 16
Coward, 211
Cowen, 424
Cramer, 374
Crespi, 261
Cross & Bevan, 451
Cross, Bevan & Beadle, 12, 460
Cross, Bevan & Briggs, 267
Crowder & Harris, 308
Curtis, 450

DAHLENVORD, 289
Dalglish (see Laurie)
Damon, 287
Danzinger, 255
Davidson, 47, 48, 218, 466, 473 (see also Brownsett)
Delph (see Boulton)
Denham, 452
Denham & Dickinson, 59, 400
Denham & Lonsdale, 59
Denham & Woodhouse, 22
Deripasko & Drujan, 19
Derrett-Smith & Nodder, 228
Dewitz, 378
Dickinson (see Denham)
Dietz, 456
Dodds (see Britton)
Dore (see Sponsler)
Dorée, 261
Draves, 144, 377
Dreyfus, 21
Drujan (see Deripasko)
Dubeau, McMahon & Vincent, 234
Dubeau & Vincent, 231
Dubeau, Vincent & Synan, 235
Dubois, 368
Duerden, 61
Duhamel, 311
Dunbar, 142, 194
Du Pont, 256, 259, 282
Dyer, 343

ECKERSALL (see Urquhart)
Eckerson (see Farr)
Edwards, 82, 481
Ekenstam, 467
Ellis & Bath, 39
Elkin (see Butterworth)
Elöd, 82, 227
Elöd, Nowotny & Zahn, 369
Elöd & Schmid-Bielenberg, 41
Elsasser, 81
Erick (see Caryl)

Ermen, 456
Evans, 150, 162, 194
Everest & Hall, 461

FALKENHAUSEN, von (see Kalb)
Fall, 145
Fargher, 162, 169 (see also Clifford)
Fargher, Hart & Probert, 168
Fargher & Higginbotham, 185, 187
Fargher & Probert, 168
Farr & Eckerson, 10, 11
Farrow & Neale, 44, 461, 465, 467
Feibelmann, 233
Ferretti, 84
Fillipov & Voronkov, 283
Fischer, 57, 65
Fleming & Thaysen, 451
Flechig, 22
Foa, 55
Fodor (see Aberhalden)
Folgner & Schneider, 482
Fort, 263 (see also Lumsden)
Foster, Kaji & Venkataraman, 456
Fothergill (see Boulton)
Franchimont, 22
Freiberger, 281, 295, 460
Freney, 297
Freney & Lipson, 79
Freudenberg, 23
Frey-Wyssling, 31, 34, 36, 72

GABRIEL, 63
Gaede, 393
Garrett & Howitt, 485
Gavoret, 50
Geake (see Birtwell, also Clibbens)
Geiger (see Patterson)
Gibson, 461
Gilmore, 110
Glanztoff, 17, 18
Goodwin (see Maquenne)
Gotte, 196
Gotze, 461
Graham (see Tankard)
Greenwood, 211
Gross, von Roll & Schreiber, 481
Grunert, 181

HALL (see Everest)
Hall & Wood, 79
Haller, 460
Haller & Seidel, 197
Hand, 250
Hannay, 142, 150, 194
Harris, 58, 76, 319 (see also Crowder, Patterson, Rutherford, Smith, Sookne)
Harris & Smith, 361, 364, 477, 479, 481

Harrison, 457
Hart, 195, 385 (*see also* Fargher)
Hart & Searell, 373
Hartley, 122
Harvey, 451
Harwood, 108
Haworth, 22, 24
Haworth & Hirst, 22
Haworth, Long & Plant, 23
Haworth & Machemer, 25
Hayakawa (*see* Okada)
Haynn (*see* Munz)
Hengstenberg & Mark, 29
Henkel, 254
Herbig, 144
Hermans & Kratky, 33, 34
Herzog, 29, 35, 478
Herzog & Becke, 52
Herzog & Jancke, 28
Hess & Trogus, 39
Heuser & Bartunek, 50
Heuser & Stockigt, 458, 460
Heyes, 456
Higginbotham (*see* Fargher)
Higgins, 153, 186, 210, 224, 225, 263, 265, 266
Hill (*see* Carothers)
Hirose (*see* Nakahama, *also* Oku)
Hirst (*see* Howarth, *also* Irvine, *also* Speakman)
Holden & Vowler, 114
Holmes (*see* Adams)
Hooke, 2
Horner (*see* Trotman)
Howitt (*see* Cadwallader, *also* Garrett)
Hubner & Pope, 211
Hundt (*see* Baier)
Hurst, 410

I.G., 395
    Irvine & Hirst, 22

JACKSON (*see* Boulton)
    Jambuserwala, 195
Jancke (*see* Herzog)
Jandrier, 460
Jeanmaire, 471
Jecusco, 211
Johnson, 373 (*see also* Shelton)
Joubert, 258
Joyner, 461
Jucci & Manunta, 378

KAJI (*see* Foster)
    Kalk & von Falkenhausen, 46, 454
Katagiri & Nakahama, 397, 398
Katz, 66
Kauffmann, 215, 228, 254, 470
Kayser, 290

Kerr 9 (*see also* Anderson)
Kershaw, 254
Kershaw & Barrett, 181
Kertess, 142, 194
Kimball (*see* Snell)
King, 64, 358, 360 (*see* Barritt *also* Cassie)
King & Mukherjee, 144
Klaye (*see* Berl)
Kling, 194, 290
Knecht, 193, 402, 456, 460
Knecht & Allen, 169
Knecht, Rawson & Loewenthal, 225
Knecht & Thompson, 455
Kollman, 188, 189, 195, 254, 281
Kornreich, 195, 227, 267
Kraemer, 27
Krais, Market & Viertel, 478
Krais & Schleber, 482, 483
Kratky, 31, 33, 39

LABARRAQUE, 152
    Lagache, 225
Lange, 357
Langmuir, 124
Laporte, 276, 277
Laue, von, 28
Laurie, 420, 441
Laurie & Dalglish, 424
Lawrence, 124
Lefroy, 62
Lenk, 165
Lester, 484
Lever, Bros., 142
Lewis, 453
Liddiard, 112, 120
Lindner, 110
Linsenmeyer, 142
Lipson (*see* Freney)
Little (*see* Clibbens, *also* Ridge)
Liu (*see* Speakman)
Lloyd, 334
Loasby, 100
Loewenthal (*see* Knecht)
Lomax, 450
Long (*see* Haworth)
Lonsdale (*see* Denham)
Lord (*see* Peirce)
Lower, 299
Lumsden, Mackenzie, Robinson Fort, 263
Lunge, 225
Lynch & Nodder, 220

MACHEMER (*see* Haworth)
    Macilwaine, 225
Mackenzie (*see* Lumsden)
Malard, 310
Manunta (*see* Jucci)
Maquenne & Goodwin, 22

NAME INDEX

Mark 39, 65 (see also Meyer)
Market (see Krais)
Marsh, 51
Marsh & Wood, 318
Martina (see Chwala)
Mathieson Alkali Works, 233, 286
Matthews, 4, 59, 181, 392, 401
McBain, 122, 124, 131
McMahon (see Dubeau)
McMahon & Speakman, 64 (see also Speakman)
Mecheels, 257
Menkert (see Speakman)
Mercer, 51, 52, 82, 133
Meunier, 83
Meunier & Rey, 76
Meyen, 10
Meyer, 30, 39, 52, 195
Meyer & Mark, 24, 29, 58
Meyer & Misch, 29
Misch (see Meyer)
Miles, 19
Millison, 413
Minaev, 226, 294
Mizell (see Patterson)
Monier-Williams, 22
Morey, 35
Morgan, 146
Morgan & Seyferth, 386
Morton (see Boulton)
Mosher, 374, 375, 387, 398
Muir, 198
Mukherjee (see King)
Mulder, 376
Muller, 457
Munch, 195, 397
Munz & Haynn, 83

NABAR (see Scholefield)
Nageli, 10, 27
Nakahama & Hirose 400 (see also Katagiri)
Nakahama & Sakaguchi, 400
Nathusius, 63
Neale, 43, 44, 53 (see also Farrow)
Neale & Stringfellow, 458
Neville, 145
Nickerson, 39
Niemann (see Bergmann)
Nilssen (see Speakman)
Nodder, 293, 468 (see Derrett-Smith, also Lynth)
Norris & Rensburg 62 (see also Baker)
Nowotny (see Elöd)
Nusslein, 142, 194

OAKES (see Williamson)
O'Connor, 201
Odling, 201

Okada & Hayakawa, 476
Oku, 376, 377, 378
Oku & Hirose, 400
O'Neill, 211
Oparin 210
Ost, 461
Ost & Wilkening, 22

PALMER, 119, 146 (see also Cassie)
Parker, 105
Parsons (see Ridge)
Patat, 481
Patel (see Scholefield)
Patterson, 359
Patterson, Geiger, Mizell & Harris, 81
Pauly & Benz, 481
Peace (see Williams)
Peirce, 44
Peirce & Lord, 35
Phillips, 79, 308, 313, 360, 361
Plant (see Haworth)
Pokorny, 28
Polanyi, 28
Pope (see Hubner)
Powney, 146
Priestman, 306
Probert (see Fargher)
Proctor & Gamble, 382, 391
Przibram & Schmalfuss, 378
Purves, 39

RACE, Rowe, Speakman & Vicker staff, 360
Ramsden, 56
Raschig, 265
Rawson (see Knecht)
Raynes, 359
Reamur, 2
Reeves & Thompson, 39
Reich, von (see Weiss)
Rensburg (see Norris)
Reuss, 204
Rhodes & Bascom, 146
Rhodes & Winn, 147
Ridge, 167 (see Birtwell, also Clibbens)
Ridge & Little, 221.
Ridge, Little & Wharton, 268
Ridge, Parsons & Corner, 19, 44, 459, 467
Rimington, 481
Rinse, 457
Ristenpart, 288
Ritter, 10
Roard, 375
Roberts, 386, 395
Robinson (see Lumsden)
Rohm, 288
Roll, von (see Gross)
Rowe (see Race)
Rueff (see Berl)

Ruschmann, 262
Russell & Woodberry, 467
Rutherford & Harris, 374, 480
Ryberg, 343, 369, 479, 481

SAKAGUCHI (see Nakahama)
Sakurada 42 (see also Taniguchi)
Sand, 218
Sandoz, 140
Sauer, 482, 483
Saunderson (see Trotman)
Scheele, 152
Scheurer, 169, 186
Schilow, 210
Schleber (see Krais)
Schmalfus (see Przibram)
Schmid, 393
Schmid-Bielenberg (see Elöd)
Schmidt, 25, 227
Schneider (see Folgner)
Schofield, 306, 307, 309, 318, 324, 358, 360
Scholefield (see Turner)
Scholefield & Patel, 228
Scholefield & Turner, 228
Scholefield & Ward, 194, 282
Schramek, 257
Schreiber (see Gross)
Schunk, 168
Schutzenberger, 19
Schwalbe, 44, 455, 458
Schwalbe & Becker, 461
Schweizer, 52
Scott, 165, 410
Searell (see Hart)
Searle, 453
Seidel (see Haller)
Seyferth (see Morgan)
Shelton & Johnson, 374
Shinohara, 482
Shorter, 80, 186
Siefert, 83
Siemens & Halske, 261
Skinkle, 450, 483
Smolens, 253
Smith (see Cadwallader)
Smith & Harris, 81, 82 (see also Harris)
Snell & Kimball, 372
Soc. Chem. Ind. Basle, 139
Sookne & Harris, 50
Sorensen, 115
Speakman, 68, 69, 71, 72, 73, 74, 75, 80, 83, 330, 344, 360, 361 (see also McMahon, Race, Stott, Stoves)
Speakman & Chamberlain, 148, 334
Speakman & Cooper, 73
Speakman & Goodings, 82
Speakman & Hirst, 77, 78
Speakman & McMahon, 76
Speakman, Menkert & Liu, 344
Speakman & Nilssen, 71

Speakman & Stott, 69, 77, 308
Speakman & Stoves, 83, 330
Speakman, Stoves & Bradbury, 81
Speakman & Wang, 335
Speakman & Whewell, 78
Sponsler, 29
Sponsler & Dore, 24, 29
Spooner, 333, 443
Stamm, 467
Stamm & Cohen, 467
Staudinger, 27
Stevens (see Beadle)
Stockhausen, 146, 384, 385
Stockigt (see Heuser)
Strauss & Levy, 466
Stringfellow (see Neale)
Swan, 16
Synan (see Dubeau)

TAENZER, 62
Tankard & Graham, 465
Taniguchi & Sakurada, 19
Taylor, 210, 211
Tennant, 152, 201
Thaysen (see Fleming)
Thenard, 250
Thiele, 18
Thomas, 461
Thompson (see Knecht, also Reeves)
Thorp (see Trotman)
Todtenhaupt, 84
Tollens, 23
Townend, 315
Toyada, 385
Trogus (see Hess)
Trotman, 82, 193, 247, 485
Trotman & Bell, 485
Trotman, Bell & Saunderson, 478
Trotman & Horner, 335
Trotman & Thorp, 185, 239, 260
Trotman & Trotman, 370, 450
Trotman, Trotman & Brown, 83
Tschilikin, 195, 267
Tsunokaye, 384
Turner (see Scholefield)
Turner, Nabar & Scholefield, 214, 228
Tweedie, 485

ULLMANN, 197
Urquhart & Eckersall, 15, 44
Urquhart & Williams, 38, 40, 44

VANEY & Pelosse, 378
Van Natta, 87
Vauquelin, 296
Venkataraman, 129, 142, 148, 196 (see also Foster)
Vickerstaff (see Race)

NAME INDEX

Vickery & Bloch, 58
Victoroff, 195
Vincent (*see* Dubeau)
Voronkov, 283 (*see also* Fillipov)
Vowler (*see* Holden)

WAKSMAN, 164
  Wallerstein, 397
Wang (*see* Speakman)
Ward, 356 (*see also* Scholefield)
Wasser, 227
Webb, 128
Weber, 253, 363, 409
Weber, F., 407, 408
Weimarn, 52
Weiss & von Reich, 256
Wharton (*see* Ridge)
Whewell, 78 (*see also* Speakman)
Whewell & Austerlitz, 477, 484
Whewell & Woods, 479
Wickert (*see* Wilkes)
Wiesner, 10
Wig, 315

Wilkes & Wickert, 139
Wilkening (*see* Ost)
Williams, 52, 450
Williams (*see* Urquhart)
Williams & Peace, 349
Williamson & Oakes, 220
Willstatter & Zechmeister, 22
Wilsdon, 98
Winn (*see* Rhodes)
Witz, 212, 457, 460, 461, 471
Wood, 318
Wood (*see* Hall, *also* Marsh)
Woodberry (*see* Russell)
Woodhouse (*see* Denham)
Woods (*see* Astbury, *also* Whewell)

YORSTON, 218, 220

ZAHN (*see* Elöd)
  Zechmeister (*see* Willstatter)
Zsigmondy, 165

SUBJECT INDEX

ABSORBENCY, 151, 189
—— Acetate rayon, 19
—— —— bleaching, 276
—— —— scouring, 272
Acid, and cotton, 43, 51, 459
—— nylon, 91
—— silk, 59
—— wool, 77, 318
Acid steep, 161
Acidity, 115
Activated hypochlorite, 217, 224, 226
Activation by dyestuffs, 228
Activin, 166, 197, 232
Aerosol, 135
Ageing in bleaching, 362, 364, 405
Air in kiers, 198, 199, 237
Air-drying machines, 434
Air-lay drier, 436
Aktivin (see Activin)
Alkali, action on cotton, 48, 50, 111, 153
—— —— on fats, 125, 169, 336
—— —— on nylon, 91
—— —— on silk, 59, 380, 384
—— —— on wool, 78, 299, 305, 336
Alkali-boil test, 469
Alkalinity, 114
—— hypochlorite, 214
—— peroxide, 253, 256, 276, 277, 363, 366, 404, 405
Alkalis, 111
—— in degumming, 384
—— in kier-boiling, 169, 185, 190
—— in milling, 344, 352
—— in wool scouring, 336, 337, 339
—— in wool washing, 300, 305
Alkali cellulose, 12, 50
Alkali, cleansing action, 111, 169, 300, 305, 336, 339, 384
Alkali solubility tests, 468, 469, 481
Allworden test, 477
Aluminium chloride, 343
Aluminium vats for peroxide, 252, 253, 254
Amino-acids, 57, 65
Antiseptics in size, 158
Apron drier, 426
Aqua regia, 401
Aralac, 84
Arcy, 163
Ardil, 84
Ash of cotton, 167, 283, 460
Ash alkalinity, 460
Astol, 135, 338
Auxiliaries, 193, 385
Available chlorine, 201, 206, 295

Available chlorine, in bleaching, 206 207
Avirol, 134, 289
Avivan, 135

BACKWASHING, 322
Barlow kier, 175
Bast fibres, 5
Biancal, 197
Biolase, 163
Biuret test, 480
Blanket drying machine, 433
Bleaching acetate rayon, 276
Bleaching cotton, continuous, 256, 281
—— —— hosiery, 244
—— —— hypochlorite, 207
—— —— open-width, 179, 239
—— —— peroxide, 248
—— —— peroxide or hypochlorite, 252
—— —— pH control, 218, 294
—— —— principles, 200, 212, 228
—— —— rate of 207, 214, 253
—— —— temperature, 200, 208, 234, 253, 256, 287, 289, 295
—— —— testing, 212
Bleaching, historical, 201, 248, 355
Bleaching powder, 201
—— —— action of, 210, 212, 218
—— —— available Cl, 201, 202
—— —— composition, 201
—— —— historical, 201
—— —— solutions, 206
Bleaching rayon, 276
Bleaching silk, aqua regia, 401
—— —— hydrosulphite, 403, 404
—— —— permanganate, 410, 412
—— —— peroxide, 404
—— —— stoving, 403
Bleaching wool, bisulphite, 357
—— —— chamber, 355
—— —— hydrosulphite, 358
—— —— peroxide, 363
—— —— pH, 361, 363
—— —— stoving, 355
—— —— SO_2, removal, 355
—— —— testing, 364, 366
Blowing, 327
Boil-off (see degumming)
Boil-out (see scouring)
Borax, 113, 330
Bowking, 152
Bucking, 152
Buffers, 116, 206, 218, 236, 308
Buttermilk, 152

506

SUBJECT INDEX

CALCIUM hypochlorite, 203
— Calgon, 109, 194
Calsolene oil, 134
Carbonising, 317, 320, 321, 340, 355
Carboxyl group test, 458
Caro's acid, 197
Casein fibre, 84, 412
Castor oil soap, 196
Catalytic action of metals, 214, 249, 252
Caustic alkali and cotton, 48, 50, 111, 152, 155, 190
—— —— linen, 111, 262
—— —— rayon, 269, 270, 272
—— —— silk, 59, 379, 384
—— —— wool, 78, 111, 299
—— —— in kier, 187, 188, 190
—— —— in scouring, 185
Ce-Es bleach, 289
Cell drier, 433
Cellulose, chemical reactions, 41
—— — constitution, 22
—— — derivatives, 41
—— dispersed, 50
—— fluidity, 461
—— micelles, 29
—— modified, 42
—— moisture relations, 38
—— mol. wt. 26
—— nitrate, 16, 41, 42, 466
—— nitrate fluidity, 466
—— orientation, 35
—— pore space, 34
—— structure, 27
Centrifuge, 418
Chemic (see Hypochlorite)
Chemicking, 207, 213
—— by circulation, 207
—— machine, 207
Chloramides, 232
Chloramines, 267, 289
Chlorates, 203, 204, 208
Chlorine available, 201, 206, 295
—— quantity needed, 207
—— (see Hypochlorite)
Chlorite, 233
Cholesterol, 298
Circulation in kier, 173, 175, 178, 192
Circulation of chemic, 207, 213
Clark hardness, 102
Cold bleaching, 287
Coloured-woven goods, 156, 197, 240
Colouring matter in cotton, 210
Combined bleach, 288
Congo Red test, 452
Continuous bleach, 281
Copper Number, 43, 47, 211, 455
Cotton, 5
Cotton bleaching, 152, 214, 221, 227, 229, 242, 243
—— hosiery, 244
—— impurities, 151, 167

Cotton mercerised, 157
—— wax, 168
Crabbing, 327
Crofting, 152
Cuprammonium hydrate, 52, 462
—— rayon, 17
Cycloran, 135
Cylinder drying, 430
—— singeing, 159

DASH wheel, 182
Decalin, 136, 194
Degomma, 163
Degumming, 371, 379, 389
—— fabric, 395
—— hanks, 389
—— hosiery, 393
—— tussah, 396
—— waste, 396
—— assistants, 385
—— and pH, 387
—— with alkali, 385
—— —— enzymes, 397
—— —— foam, 383, 392
—— —— soap, 379, 387
Desizing, 160, 271
Desuinting, 310, 311, 314, 315
Detergency, 119, 144, 305, 383
—— and salts, 146
Diastase, 163
Diastofor, 163
Diffusion, 217, 423
Dispersed cellulose, 50
Dispersing agents, 144
Disulphide test, 482
Dolly machine, 245, 332, 349
Draves test, 144, 377
Drop number, 121
Drum drying machine, 437
Drying, 415, 420, 425
—— controlled 422, 424
—— cylinder, 421, 430
—— fabrics, 430
—— hanks, 428
—— knitwear, 446
—— loose fibres, 426
—— rate, 421, 424, 425

ECRU silk, 379, 401
—— and formaldehyde, 400
Edmeston Benz kier, 181, 239, 282
Elasticity of fibres, 66, 99
Electrolytic bleach, 204
Emulphor, 141
Emulsification, 149, 339
Emulsion scour, 300
Enzymes, 162, 397
Ethanolamine, 132
Evaporation, 26, 415, 420, 422, 425
Expanders, 417

FATTY acids, 119, 121, 125, 129, 132
—— —— in cotton wax, 169
—— —— in soap, 126
—— —— in wool, 298
Fehling's solution, 455
Fermasol, 163
Fibres, classification, 2
—— crystallinity, 98
—— denier, 1, 93
—— density, 93
—— dimensions, 92
—— extension, 97, 99
—— general requirements, 1
—— moisture, 95
—— strength, 93, 94, 96
—— swelling, 97
—— tenacity, 1, 93, 94
—— world production, 4
Fibroin, 57, 371
Fixanol, 140
Flax, 11
Flerhenol, 134
Floranit, 134, 289
Fluidity, cellulose, 213, 461
—— cellulose nitrate, 466
—— nylon, 489
—— silk, 485
Foam degumming, 383, 392
Formaldehyde and sericin, 400
Freezing wool fat, 315
Frosted wool process, 315
Fuller's earth, 336
Fulling, 327, 343, 351

GANTT piler, 182, 207, 282
Gardinol, 139, 194
Gas singeing, 159
Gebauer kier, 178
Gold number, 145
Grassing, 265
Grease recovery, 297
Grey sour, 152, 154, 186, 193, 230, 231, 238, 240
Guiders for fabric, 417

HARD soap, 126
Hard water, 102
Harrow machine, 301
Heating kier liquors, 173, 177
Herbig number, 129, 130, 144
Hexalin, 135, 194
Hexoran, 135
High pressure kiers, 173, 177
Historical, 152, 200, 250, 311, 320, 336, 343, 355
Homogenit, 290
Horizontal drying cylinders, 430
Hosiery, 244, 273, 349, 393
Humectol, 134

Hurricane drier, 436
Hydralin, 195
Hydranaphthal, 136
Hydrate cellulose, 53
Hydrocellulose, 43, 470
Hydrochloric acid, 154, 155, 230, 320
Hydro-extraction, 415, 418
Hydrogen ion, 115 (see pH)
Hydrogen peroxide, 251
—— —— and cotton, 252
—— —— and linen, 268
—— —— and rayon, 276
—— —— and silk, 404
—— —— and wool, 362
—— —— pH, 252, 253, 363, 405
—— —— temperature, 251, 253, 256, 362, 404
Hydrolysis of soap, 131, 305, 372, 379, 384
Hydrosulphites, 358, 403
Hypochlorites, 202, 210, 221, 264, 354, 357
—— concentration, 206, 207, 218, 222, 239, 276, 279, 281, 289, 293, 295
—— effect of acid, 212, 217, 219, 225
—— effect of heat, 208
—— effect of light, 208, 294
—— effect of pH, 212, 213, 214, 294
—— effect of salt, 224
—— estimation of chlorine, 207, 210
—— estimation of pH, 220
Hypochlorous acid, 211, 212

IGEPAL, 142
Igepon, 136, 137, 194
Impurities, cotton, 167
—— rayon, 270
—— silk, 371
—— wool, 296, 299
Injector kier, 173
Intrasol, 134
Invadine, 134
Italian silk, 371, 401

JACKSON kier, 180, 239
Japanese silk, 375, 376, 378
Javelle, Eau de, 152
J-box, 282, 284, 286
Jefferson-Walker, kier, 175
Jute, 268

KAUFFMANN test, 470
—— bleaching process, 254
Keratin, 65
Kier, 173
—— horizontal, 178
—— vertical, 170
Kier-boiling, assistants, 193
—— circulation, 192

SUBJECT INDEX

Kier-boiling, effect of air, 170, 175, 199
—— history, 153, 186
—— methods, 169
—— packing, 171
—— precautions, 171
—— with lime, 187
—— with resin soap, 193
—— with soda, 190
—— with soda ash, 197
Kier-liquor concentration, 191
Kier stains, 245
Knit goods drying, 446
Knit goods treatment, 199, 244, 349
Koloran, 135
Kontakt, 195
Korte process, 290

LABARRAQUE, Eau-de-, 152
Lamepon, 137
Lanital, 84
Lanolin, 299
Lant, 311
Laventin, 134
Lead acetate test, 479
Leonil, 134, 136, 194
Leviathan (see Harrow Machine)
Light, effect on bleaching, 208, 294
Lime boil, 153, 154, 155, 185, 186, 187, 189, 265, 266
Lime soaps, 102, 127, 128, 133, 134, 136, 137, 138, 142, 154, 186, 193, 299, 301, 309
Linen, 262
Linen bleaching, 263, 291
Linen scouring, 263, 291
Liquid chlorine, 201, 203, 206
Lissapol, 139, 194
Loop drying machine, 435
Lorol, 139, 194
Lousiness, 382
Ludigol, 197

MACHINE chemicking, 207, 208
Machines drying, 426
—— milling, 346
—— scouring, 331
—— washing, 182
—— wool washing, 301
Madder bleach, 154
Magnesium chloride, 342
Malard process, 310
Malt enzymes, 162
Malting, 162
Mangles, 415, 416
Market bleach, 151, 155
Marseilles soap, 193
Mather kier, 178, 196
Mather and Platt kiers, 176
Maxochlor, 202

Medialan, 137
Melioran, 137
Mercerised cotton, 40, 50, 53
—— —— bleaching, 157
Methylene Blue test, 458
Milling, 343, 351
Milling knitted goods, 349
Milling machines, 346
Mineral oil, 148, 334, 335, 336, 337
Mohr bleach, 289
Monopol soap, 133, 195
Motes, 161, 176

NAPHTHALENE sulphonic acids, 134
Nekal, 134, 136, 194, 195
Neopermin, 134, 136, 194
Neutral soaps, 132, 301, 304
Nitrate rayon, 16
Nitrogen content test, 483, 485
Nitrogen in cotton, 167
Nylon, 86, 413

OCENOL, 139, 194
Oils for soaps, 125, 126, 127, 128
Oil soaking, 371, 380
Oils for wool, 334
Old bleach liquors, 210, 213, 224
Oleic acid, 133
Oleines, 125, 126, 335
Olein oil, 126
Olive oil, 133, 334
Ondal, 290
Openers, 241
Open kiers, 173
Open-width scour, 180, 239
Oranit, 136
Oxidation in bleaching, 210, 228
—— acceleration by dyes, 228
Oxidation of cellulose, 46, 212, 470
Oxidation of wool, 366
Oxycellulose, 199, 212, 425, 470
Ozone, 261

PADDLE machine, 302
Palm oil, 126
Palmer machine, 433
Palmitic acid, 126, 129, 346
Pancreatin, 163
Pancreol, 163
Paraffin wax, 161, 166
Pauly test, 480
Pectin, 167, 198
Pendlebury kier, 174
Peracids, 258
Peractivin, 197
Perborates, 258, 277, 410
Perchloron, 202

510   TEXTILE BLEACHING

Peregal, 141
Permanent hardness, 102
Permanent set, 74, 330, 413
Permanganate, 260, 410
Perminal, 194
Peroxide, 248
Peroxide bleach, 252, 276, 362, 404
pH, 115, 146, 198, 213, 313, 330, 344, 361, 367, 380, 405
Physical properties of fibres, 92
Physical tests, 449
Pigments, 377
Pine oil, 195
Plate singe, 159
Polyzime, 162
Potash soaps, 125, 126, 131, 305
Potassium salts in wool, 296, 297, 311
Potassium permanganate, 260, 354, 410
Pre-chemicking, 279
Pressure boiling, 176, 177
Pressure kiers, 173
Prestabit oils, 133
Printers' bleach, 208
Proteins regenerated, 83
Protein fibres, testing, 477
Pump circulation in kiers, 170, 173, 175, 176, 179, 192

RAPIDASE, 163
    Raw cotton, 167, 168
Raw silk, 371
Raw wool, 296
Rayon, 12, 92, 269
Recovered grease, 297
Reducing action tests, 456
Reduction in bleaching, 355, 359, 360, 362
Reduction in kiering, 191, 197
Reeling linen, 263
Retting, 262
Revolving expander, 418
Roller washing machine, 182
Rosin soap, 127, 193
Rubbing boards, 265, 266
Rust stains, 231, 246

SALT-WATER soap, 126
    Salts and bleaching, 224
Salts and scouring, 196, 335
Salts and soap, 126, 146, 147
Sandofix, 140
Sapamines, 139
Saponification, 125
Saponification value, 127
Saturating machine, 169, 352
Savonade, 135
Scald, 264
Scotch stenter, 438
Scouring cotton, early methods, 154

Scouring cotton, effect of pressure, 176, 177
—— —— loss of wax, 187
—— —— loss of weight, 161
—— —— theory, 186
—— —— with lime, 155, 187
—— —— with soda, 155, 185
—— —— with soda or lime, 185
Scouring linen, 263
Scouring rayon, 270
Scouring silk (see degumming)
Scouring raw wool, 299
—— with fuller's earth, 336
—— with soap, 339
—— with soda, 337
—— woollens, 334
—— worsteds, 339
Scutcher, 241
Segmentation test, 453
Sericin, 374, 379
—— fixation, 400, 403
Sextol, 135, 338
Shives, 161, 176
Sichlor, 202
Silk, chemical props., 59
—— chemistry, 57
—— formation, 55
—— molecular weight, 58
—— pigments, 377
—— properties, 56
—— reeling, 56
—— structure, 58
—— wax, 375
Silk bleaching, aqua regia, 401
—— —— permanganate, 410
—— —— peroxide, 404
—— —— stoving, 403
Silk degumming, 379
Silver nitrate test, 457, 479
Silver number, 457
Singeing, 158
Sirial, 197
Size, 161
Skeins, degumming, 389
Skin wools, 299
Slack washing machine, 182
Slipe, 299, 309
Soap, 125, 193
Soap in degumming, 380, 386
—— in kier boil, 154, 193, 195, 196
—— in wool milling, 345
—— wool scouring, 335
—— in wool washing, 300, 304, 305
Soap Solubility Ratio, 128
Sodium bicarbonate, 231
—— bisulphite, 357, 403
—— carbonate, 112, 154, 186, 197, 306, 337, 345
—— chlorite, 233
—— hydrosulphite, 403
—— hydroxide, 112

SUBJECT INDEX

Sodium hypochlorite, 203
—— perborate, 258, 277, 410
—— peroxide, 249
—— phosphate, 113, 196
—— sesquicarbonate, 112, 221
—— silicate, 113, 196, 335, 345, 363, 386, 405
—— sulphide, 80, 299
—— sulphite, 403
Soft soap, 126
Soft water, 102
Softening water, 105
Solubility of modified cellulose, 47
Solubility number, 48, 213, 468
Solventol, 135
Solvents and soap, 135, 338
Solvent scouring, flax, 263
—— —— silk, 386
—— —— wool, 314
Soupled silk, 379, 401
Souring, 152, 154, 230
Sprits, 265
Squeezer, 183
Stability of hypochlorite, 203, 204, 208, 210
—— of peroxide, 251, 252
Staining tests, 452, 478
Stains, 245
Stannous chloride test, 479
Starch, 161
Steam in kiers, 177
Steam injector, 173
Steam pressure, 177
Stearic acid, 126, 127, 129, 297, 305, 346
Stearines, 125, 305
Steeping, 160, 362, 364, 405
Stenter, 423, 437
Stocks, 245, 346
Stoving, 355, 405
Suint, 296, 297, 311
Sulphated alcohols, 138
Sulphonated oils, 133, 194, 386
Sulphur bleach, 355, 403
Sulphur content test, 482
Sulphur dioxide, 355, 357, 359
Sulphuric acid, 152, 154, 230
Sulphurous acid, 358, 403
Supercontraction, 330
Surface tension, 120, 147, 305
Swelling tests, 451, 477
Swing rake machine, 301

TABLE drier, 426
Taka, 163
Tallow, 128, 345
Temperature and degumming, 382
—— and hypochlorite, 203, 206, 208
—— and milling, 346
—— and peroxide, 253, 256, 362, 365, 404
—— and scouring, 188, 192, 337

Temperature and wool washing, 307
Temporary hardness, 102
Tenacity, 94
Tendering, 157
Tensile strength, 93
Tenter (*see* stenter)
Tergitol, 139
Tetralin, 136, 194
Tetrapol, 135
Texapon, 135
Thiocyanate test, 480
Thiol test, 482
Time wheel, 182
Tinting, 324, 354
Tom-tom machine, 245
Trilon, 110, 194
Trioran, 135
Trypsin, 163
Turkey red oil, 126, 133, 135, 395
Turpurile, 135
Tussah, 396, 411

ULTRAVON, 137

VELAN, 141
Vinyon, 85
Viscose rayon, 12, 15
—— —— bleaching, 277
Viscosity (*see* Fluidity)

WALSH kier, 173
Warm chemic, 295
Wash wheel, 333, 350
Washing after bleaching, 238
Washing machines, 182
Washing wool, 299, 304
Water, 102
Water, hardness 103
Water softening, 105
Wax in cotton, 167, 168
—— in flax, 263
—— in silk, 375
—— in wool, 299
Weeting, 311
Weight losses in scouring cotton, 161, 185, 186
—— —— —— flax, 262, 263, 266
—— —— —— silk, 379, 389
—— —— —— wool, 297, 314, 315, 316, 336
Weighted silk, 407, 408
Wetting, 119
Wetting agents, 133, 335
White sour, 230
Wool chemical constitution, 64
—— chemical reactions, 76
Wool cloth scour, 327, 330

Wool, effect of acid, 77, 318
—— —— of alkali, 78, 307
—— —— of felting, 78
—— —— of halogens, 82
—— —— of metallic salts, 82
—— —— organic compounds, 83
—— —— of oxidising agents, 81
—— —— of pH, 68, 308
—— —— of reducing agents, 80
—— —— of water, 76
Wool fat, 307
Wool fibre, 61
Wool moisture relations, 73
—— molecular structure, 66
—— molecular weight, 71
—— occurrence, 61
—— pore space, 72

Wool scouring, 330
Wool scouring machines, 331
Woollen yarn scouring, 325
Worsted scouring, 339

YARN driers, 426
Yarn scouring, 243, 325, 389
Yellow gum silk, 377, 401, 402
Yellowing of cotton, 245
Yellowing of wool, 111, 308
Yorkshire grease, 297

ZEOLITES, 106
Zimmermann test, 484
Zincate test, 453

www.ingramcontent.com/pod-product-compliance
Lightning Source LLC
Chambersburg PA
CBHW030328240426
43661CB00052B/1568